Beyond Mechanism

Beyond Mechanism

Putting Life Back into Biology

Edited by Brian G. Henning and Adam C. Scarfe

With a foreword by Stuart A. Kauffman

LEXINGTON BOOKS
Lanham • Boulder • New York • Toronto • Plymouth, UK

Published by Lexington Books
A wholly owned subsidiary of The Rowman & Littlefield Publishing Group, Inc.
4501 Forbes Boulevard, Suite 200, Lanham, Maryland 20706
www.rowman.com

10 Thornbury Road, Plymouth PL6 7PP, United Kingdom

Copyright © 2013 by Lexington Books

All rights reserved. No part of this book may be reproduced in any form or by any electronic or mechanical means, including information storage and retrieval systems, without written permission from the publisher, except by a reviewer who may quote passages in a review.

British Library Cataloguing in Publication Information Available

Library of Congress Cataloging-in-Publication Data

Beyond mechanism : putting life back into biology / edited by Brian G. Henning and Adam C. Scarfe ; with a foreword by Stuart A. Kauffman.
   pages cm
Includes bibliographical references and index.
ISBN 978-0-7391-7436-4 (cloth : alk. paper)—ISBN 978-0-7391-7437-1 (electronic)
1. Life (Biology) 2. Life sciences—Research. I. Henning, Brian G., editor of compilation. II. Scarfe, Adam Christian, editor of compilation.
QH501.B49 2013
576.8—dc23   2012044647

∞™ The paper used in this publication meets the minimum requirements of American National Standard for Information Sciences—Permanence of Paper for Printed Library Materials, ANSI/NISO Z39.48-1992.

Printed in the United States of America

*To Lynn Margulis (1938-2011),
fearless scientist and beloved colleague.*

Contents

Foreword: Evolution beyond Newton, Darwin, and Entailing Law 1
Stuart A. Kauffman

Introduction: On a "Life-Blind Spot" in Neo-Darwinism's Mechanistic Metaphysical Lens 25
Adam C. Scarfe

Section 1: Complexity, Systems Theory, and Emergence

1. Complex Systems Dynamics in Evolution and Emergent Processes 67
   Bruce H. Weber

2. Why Emergence Matters 75
   Philip Clayton

3. On the Incompatibility of the Neo-Darwinian Hypothesis With Systems-Theoretical Explanations of Biological Development 93
   Gernot Falkner and Renate Falkner

4. Process-First Ontology 115
   Robert E. Ulanowicz

5. Ordinal Pluralism as Metaphysics for Biology 133
   Lawrence Cahoone

Section 2: Biosemiotics

6. Why Do We Need a Semiotic Understanding of Life? 147
   Jesper Hoffmeyer

7. The Irreducibility of Life to Mentality: Biosemiotics or Emergence? 169
   Lawrence Cahoone

Section 3: Homeostasis, Thermodynamics, and Symbiogenesis

8   Biology's Second Law: Homeostasis, Purpose and Desire   183
    J. Scott Turner

9   "Wind at Life's Back"—Toward a Naturalistic,
    Whiteheadian Teleology: Symbiogenesis and the Second
    Law   205
    Dorion Sagan and Lynn Margulis

10 Of Termites and Men: On the Ontology of Collective
    Individuals   233
    Brian G. Henning

Section 4: The Baldwin Effect, Behavior, and Evolution

11 The Baldwin Effect in an Extended Evolutionary Synthesis   251
    Bruce H. Weber

12 On the Ramifications of the Theory of Organic Selection
    for Environmental and Evolutionary Ethics   259
    Adam C. Scarfe

Section 5: Autogenesis, Teleology, and Teleodynamics

13 Teleology versus Mechanism in Biology: Beyond
    Self-Organization   287
    Terrence Deacon and Tyrone Cashman

14 Teleodynamics: A Neo-Naturalistic Conception of
    Organismic Teleology   309
    Spyridon Koutroufinis

Section 6: Epigenetics

15 Epigenesis, Epigenetics, and the Epigenotype: Toward An
    Inclusive Concept of Development and Evolution   345
    Brian K. Hall

16 Epigenetics, Soft Inheritance, Mechanistic Metaphysics, and
    Bioethics 369
    Adam C. Scarfe

Section 7: Organism and Mechanism

17 From Organicism to Mechanism—and Halfway Back? 409
    Michael Ruse

18 Machines and Organisms: The Rise and Fall of a Conflict 431
    Philip Clayton

Index 445

About the Contributors 469

Foreword
Evolution beyond Newton, Darwin, and Entailing Law

Stuart A. Kauffman

My large aim in this foreword is to take us from our deeply received scientific worldview and, derived from it, our view of the "real world" in which we live—namely, from the understanding of the world that was spawned by Newton and modern physics, to an entirely different, newly vibrant, surprising, partially unknowable world of becoming which the living, evolving world—biological, economic, cultural—co-creates, in an often unprestatable mystery, its own possibilities of becoming. If the latter perspective is right, we are beyond Newton, and even beyond Darwin, who, in all his brilliance, did not see that without natural selection "acting" at all, the evolving biosphere creates its own future possibilities. And we will see, at the foundations of all this, that no laws entail the evolution of the biosphere, economy, or culture. This is because, since Descartes and Newton, entailing laws are the heart of our worldview of *mechanism*. Thus, if entailing law *is* our modern understanding of the meaning of "mechanism" and if the evolving biosphere becomes beyond entailing law, then the evolving biosphere becomes *beyond mechanism.*

We will begin to see ourselves in the living, evolving world, in a world of inexplicable and unforeseeable opportunities that emerge without the "action" of natural selection in the evolving biosphere, or often without intent in the human world, that we partially co-create. It will follow that we live not only in a world of webs of cause and effect, but webs of opportunities that enable, but do not cause, often in unforeseeable ways, the possibilities of becoming of the biosphere, let alone human life. But most importantly, I seek in this new worldview a re-enchantment of humanity, of which this foreword may be a part. Our disenchantment following from Newton led to modernity. I believe we are partially lost in modernity, seeking half-articulated, a pathway forward. Re-enchantment may be an essential part of this transformation.

I have many points to make and ideas to explore, and hope they shall prove relevant and find resonance. If I am right, we are in the world in a way that we do not now clearly recognize. I begin with an amazing statement by early sociologist, Max Weber, who said, roughly, "with Newton we became disenchanted and entered modernity." Weber was right. Following Newton's triumph founding of classical physics, a mechanical worldview of celestial mechanics and billiard ball trajectories, the first clear fruition of Descartes' *res extensa* and his own mechanical worldview, came our Enlightenment, our Age of Reason, then with the rise and application of science, the Industrial Revolution, and modernity. Newton's amazing successes left no room for mystery in the proffered scientific account of the working of the world. Without some sense of mystery, we were, as Weber notes, disenchanted.

Newton

How the Western, and now, the modern world, three hundred and fifty years later, changed with the inventions of, largely, one mind, Newton. Not only did he invent the mathematics, the differential and integral calculus, that gives us our way of thinking as moderns, from physics upward, but he gave us his famous three laws of motion, and universal gravitation. Ask Newton: "I have six billiard balls rolling on a billiard table—what will happen to them?" and Newton might have rightly responded: "measure the positions, momenta, diameters of all the balls, and the boundary conditions of the table, write down my three laws of motion in differential equation form representing the forces between the balls and between the balls and the edges of the table, then integrate my equations to yield the deterministic future trajectories of the balls."

What had Newton done? He had mathematized Aristotle's "efficient cause" in his differential equations giving forces between the entities, the laws of motion. He had invented a conceptual framework to derive the deterministic trajectory consequences by integration. But integration is deduction is "entailment," so the laws of motion in differential form entail the deterministic trajectories. In this entailment, Newton mathematized, in a very general framework, Aristotle's argument that scientific explanation must be deduction: All men are mortal; Socrates is a man; hence, Socrates is mortal.

In the early 1800s, Simon Pierre Laplace further generalized Newton. Given a massive computing system, the Laplacian demon, informed of the instantaneous positions and momenta of all the particles in the universe, the entire future and (because Newton's laws are time-reversible) past of the universe is fully predictable and determined. This statement by Laplace is the birth of modern "reductionism," the long held view that there is some "final theory" down there, as in Stephen Weinberg's "Dream of a Final Theory," that will entail all that becomes in the universe.

We need two additional points. First, by the time of Poincaré, studying the orbits of three gravitating objects (a topic Newton knew was trouble), Poincaré was the first to show what is now known as deterministic chaos. Here tiny changes in initial conditions lead to trajectories that diverge from one another exponentially. Since we cannot measure positions and momenta to infinite accuracy, Poincaré showed that we cannot *predict* the behavior of a chaotic deterministic dynamical system. Determinism, contra Laplace, does not imply predictability. Second, quantum mechanics overthrew the ontological determinism of Newton, on most interpretations of quantum mechanics. Nevertheless, quantum systems obeying Schrödinger's equation *deterministically* evolve a *probability distribution* of the ontologically indeterminate probabilities of quantum measurements. That evolution of the probability distribution is again entirely entailed.

With general relativity and quantum mechanics, the twin pillars of twentieth century physics were and remain firmly in place. No attempt to unite general relativity and quantum mechanics has been successful after eighty-five years of trying. Success may or may not come. In modern physics, the conviction remains that all that arises in the universe is entailed.

Darwin

After Newton, and perhaps as profoundly, Darwin changed our thinking. We all know the central tenants of his theory: heritable variation among a population, competition for resources insufficient for all to survive, and hence, natural selection culling out those variants "fitter" in the current environment. Thus, we achieve adaptation, and critically, the *appearance of design* without a designer. The well-known story of the difficulties of Darwin's theory with "blending inheritance" and its unexpected rescue by Mendelian genetics, even the fact that a copy of Mendel's work lay unopened on Darwin's desk, is well known. Mendelian genetics prevents blending inheritance and paved the way for the mid-twentieth century neo-Darwinian—or "Modern"—Synthesis.

The entire panoply of life's evolution at last lay open to at least the start of understanding given Darwin. The history from Darwin and Mendel to the neo-Darwinian synthesis of the mid-twentieth century is well known. I would comment briefly here that the neo-Darwinian synthesis left out Waddington's "epigenesis"—the attempt, starting with Wolff, to link development and evolution. Just as the mathematical inventions of population genetics by Fisher, Haldane, and Wright served to unite Mendel and Darwin, the inventions of mathematical models of the genetic regulatory networks envisioned by Waddington—for example my own early study of Random Boolean Networks as models of genetic regulatory networks,[1] and now Systems Biology, including the modern sense of epigenetics as heritable changes in chromosomal "markings" by methylation or acetylation of histones, without changing the DNA sequence—are a further part of extending the neo-Darwinian synthesis to include development, environment, and ecology, into a broadened synthesis. Much of this is the subject of the current book. This foreword hopes to require an even broader re-examination of an evolution that is beyond entailing law. If entailing law *is* mechanism, evolution truly is *beyond mechanism*.

Monod and "Teleonomic"

The concepts of "function," "doing," "purpose," and "agency" in biology, and with it, a potential "meaning" for signs or symbols, totally absent in physics where only "happenings" occur, have been muted in standard biology by a concept voiced by Jacques Monod. Consider a bacterium swimming up a glucose gradient. It "seems" to be "acting to get food." But said Monod, this view of the organism seems to be entirely wrongheaded. The cell in its environment is just an evolved molecular machine. Thanks to natural selection, the swimming up the gradient gives the appearance of purpose, of teleology, but this is false. Instead, this behavior is a mere "as if" teleology that Monod called "teleonomy." (In a wider view, Monod struggled to reconcile teleonomy with his sense that organisms do act with purpose, an issue he thought was the central one of biology[2]). In short, for Monod, via teleonomy, and for legions of later biologists and philosophers, "doing" is unreal in the universe. Only the mechanical, selected appearance of "doing" is real.

Indeed, in so arguing, Monod is entirely consistent with physics. As noted, there are no "functions," "doings," "meanings," or "agency" in physics. Balls rolling down a hill are merely Newtonian "happenings." So too are the happenings in the evolved molecular machine that is the bacterium swimming up the glucose gradient. Yet we humans think functions and doings and agency are real in our world. If so, from whence functions, doings, agency, and meanings? Functions, meanings, and doings *are* real in the universe. I now give, as far as I know, an entirely new set of arguments that, I believe, fully legitimize functions, doings, agency, and even meanings as real in the universe, but beyond physics. The discussion has a number of steps.

The Non-Ergodic Universe above the Complexity of the Atom

Has the universe in 13.7 billion years of existence created all the possible fundamental particles and stable atoms? Yes. Now consider proteins. These are linear sequences of twenty kinds of amino acids that typically fold into some shape and catalyze a reaction or perform some structural or other function. A biological protein can range from perhaps fifty amino acids long to several thousands. A typical length is three hundred amino acids long. Then let us consider all possible proteins that are two hundred amino acids in length. How many are possible? Each position in the two hundred has twenty possible choices of amino acids, so there are 20 x 20 x 20 200 times, or 20 to the 200th power, which is roughly 10 to the 260th power possible proteins of 200 amino acids in length.

Now let us ask if the universe can have created all these proteins since its inception 13.7 billion years ago. There are roughly 10 to the 80th particles in the known universe. If these were doing nothing, ignoring space-like separation, but making proteins on the shortest time scale in the universe, the Planck time scale of 10 raised to the -43 seconds, it would take 10 raised to the 39th power times the lifetime of our universe to make all possible proteins length 200, just *once*. In short, in the lifetime of our universe, only a vastly tiny fraction of all possible proteins of length 200 can have been created. This means profound things. First, the universe is vastly non-ergodic in the physicists' sense of the ergodic hypothesis at the foundation of statistical mechanics. It is not like a gas at equilibrium in statistical mechanics. With this vast non-ergodicity, when the possibilities are vastly larger than what can actually happen, history enters. Not only will we not make all possible proteins length 200 or 2000, we will not make all possible organs, organisms, social systems. There is an *indefinite* hierarchy of non-ergodicity as the complexity of the objects we consider increases.

Kantian Wholes and the Reality of Functions, "Doings, and Agency"

The great philosopher, Immanuel Kant, wrote that "in an organized being, the parts exist for and by means of the whole, and the whole exists for and by means of the parts."[3] Kant was at least considering organisms, which I will call Kantian wholes. Functions are clearly definable in a Kantian whole. The function of a part is its causal role in sustaining the existence of the Kantian whole. Other causal consequences are side effects. Note that this definition of function rests powerfully on the fact that Kantian wholes, like a bacterial cell dividing, are complex entities that *only get to exist in the non-ergodic universe above the level of atoms because they are Kantian self-recreating wholes*. It is this combination of self-recreation of a Kantian whole, and therefore, its very existence in the non-ergodic universe above the level of atoms that, I claim, fully legitimizes the word, "function" of a part of a whole in an organism. Functions are real in the universe.

Now consider the bacterium swimming up the glucose gradient to "get food," Monod's merely teleonomic *as if* "doing." But we can rightly define a behavior that sustains a Kantian whole, say the bacterium existing in the non-ergodic universe, as a "doing." Thus, I claim, "doings" are real in the universe, not merely Monod's teleonomy. Furthermore, "agency" enters with "doing." In my third book, *Investigations*,[4] my own attempt to define agency stated that a molecular autonomous agent was a self-reproducing system that is able to do at least one thermodynamic work cycle. With Philip Clayton, we broadened this definition to involve the inclusion of the self-reproducing system in some boundary, say a liposome, and the capacity to make at least one discrimination, food or not food, and to "act" upon that discrimination. Bacteria clearly do this, and, without invoking consciousness, are therefore agents. Agency is real in the universe.

A rudimentary beginning of "emotion" emerges here.[5] The bacterium must sense its world and act to avoid toxins and to obtain food. The evaluation of "good" versus "bad," arguably,[6] the "first sense," enters here. Agency and the existence of the cell precedes this "semiotic" evaluation logically, for if there were no existence in the non-ergodic universe of the Kantian whole, agency and evaluation of food versus poison would not be selected, and so they would not exist in the universe. Thus, contra Hoffmeyer's fine contribution in this volume, I would argue that life is not sufficiently based on semiosis, for, as noted, if there were not a prior Kantian whole existing in the non-ergodic universe above the level of atoms, semiosis would not have evolved. I note also that Hume's famous "one cannot deduce an 'ought' from an 'is," namely, the famous Naturalistic Fallacy, rests on the critical fallacy that Hume, like Descartes, thought of a mind "knowing" its world. Hume did not think of an agent "acting" in its world. Given "action" and "doing," doing "it" well or poorly enters inevitably, and with it, "ought." With ought, the need for evaluation, the rudiments of emotion without positing consciousness, enter.[7]

Interestingly, Kant opined that there would never be a Newton of biology. Despite Darwin, a major point of this paper, which will take us beyond physics, is that here Kant was right. There never, indeed, will be a Newton of biology, for, as we will see below, unlike physics and its law-entailed trajectories, the evolution of the biosphere cannot be entailed by laws of motion and their integration. No *laws* entail the evolution of the biosphere, a first and major step beyond physics, beyond the entailed trajectories of physics, hence beyond mechanism, and toward "re-enchantment" at the "watershed of life."

Collectively Autocatalytic DNA Sets, RNA Sets, or Peptide Sets

Gonen Ashkenasy at the Ben Gurion University in Israel has created in the laboratory a set of nine small proteins, called peptides. Each peptide speeds up, or catalyzes, the formation of the *next* peptide by ligating two fragments of that next peptide into a second copy of itself. This catalysis proceeds around a *cycle* of the nine peptides.[8] It is essential that in Ashkenasy's real system, no peptide catalyzes its *own* formation. Rather the set as a whole *collectively catalyzes its own formation*. I shall call this a collectively autocatalytic set, "CAS."

These astonishing results prove a number of critical things. First, since the discovery of the famous double helix of DNA, and its Watson Crick template replication, many workers have been convinced that molecular reproduction *must* rest on something like a template replication of DNA, RNA, or related molecules. It happens to be true that all attempts to achieve such replication without an enzyme have failed for 50 years. Ashkenasy's results demonstrate that small proteins can collectively reproduce. Peptides and proteins have no axis of symmetry like the DNA double helix. These results suggest that molecular reproduction may be far easier than we have thought. I mentioned that briefly in 1971, 1986, and 1993,[9] inventing a theory for the statistically expected emergence of collectively autocatalytic sets in sufficiently diverse "chemical soups." This hypothesis, tested numerically, is now a theorem.[10] If so, routes to molecular reproduction in the universe may be abundant.

I raise a new question. We must ask: what kind of "law" does this theory of the spontaneous emergence of collectively autocatalytic sets involve? We will see that whatever form this "law" may be, it is *not* a Newton-like law with initial and boundary conditions, laws of motion in differential equation form, and the integration of those laws yielding entailed trajectories—i.e., our hallmark case since Newton of "mechanism." Newton mathematized "efficient cause," one of Aristotle's four causes: material, final, formal, and efficient. I suspect that the theory of the spontaneous emergence of collectively autocatalytic sets is a new kind of law: a Formal Cause law. If this is right, one, and perhaps the best, theory for the emergence of life as an expected property in ensembles of chemical reaction networks is *not* a mechanistic law at all, and the emergence of such sets is beyond mechanism in a new sense, even though each instantiation of such an emergence in the *ensemble* of all possible chemical reaction networks has a set of mechanisms. Collectively autocatalytic DNA sets and RNA sets have also been made.[11]

Collectively Autocatalytic Sets Are the Simplest Cases of Kantian Wholes and the Peptide Parts Have Functions

A collectively autocatalytic set is precisely a Kantian whole that "gets to exist" in the non-ergodic universe above the level of atoms precisely because it is a self-reproducing Kantian whole. Moreover, given that whole, the "function" of a given peptide part of the nine peptide set is exactly its role in catalyzing the ligation of two fragments of the next peptide into a second copy of that peptide. The fact that the first peptide may jiggle water in catalyzing this reaction is a causal side effect that is *not* the function of the peptide. Thus, functions are typically a subset of the causal consequences of a part of a Kantian whole.

Task Closure in a Dividing Bacterium

Collectively autocatalytic sets exhibit a terribly important property. If we consider catalyzing a reaction a "catalytic task," then the set as a *whole* achieves "task closure." All the reactions that must be catalyzed by at least one of Ashkenasy's nine peptides *are* catalyzed by at least one of those peptides. No peptide catalyzes its own formation. The set as a whole catalyzes its own reproduction via a clear task closure.

Consider a dividing bacterium. It too achieves some only partially known form of task closure in part in, and via, its environmental niche. But the tasks are far wider than mere catalysis. Among these tasks are DNA replication, membrane formation, the formation of chemiosmotic pumps and complex cell signaling mechanisms in which a chemically *arbitrary* molecule in the environment can bind to part of a transmembrane protein, and thereby alter the behavior of the intracellular part of that molecule, which, in turn, unleashes intracellular signaling. Thus this task closure is over a wide set of tasks.

Biosemiosis Enters at This Point

I thank Professor Kalevli Kull of the Tartu University Department of Semiotics for convincing me that, at just this point, biosemiotics enters.[12] As Kull points out, the set of environmental molecules that can bind the outside parts of transmembrane proteins are chemically arbitrary—a point Monod emphasized as well in considering allosteric enzymes. Thus, as Kull points out, the set of states of the different molecules outside the cell that can bind to the outside parts of these transmembrane proteins and unleash intracellular signaling and a coordinated cellular response, constitute a *semiotic code* by which the cell navigates its "known" world, "known"—without positing consciousness—via the code and, in general, probably evolved by selection encoding of the world as "seen" by the organism. Change the molecule species binding the outside of the transmembrane proteins, and the world the cell "knows" and evaluates changes. Biosemiosis is real in the universe.

Toward: No Entailing Laws, but Enablement in the Evolution of the Biosphere

I now shift attention to a new, and I believe, transformative topic. With my colleagues Giuseppe Longo and Maël Montevil, mathematicians at the École Polytechnique in Paris, I wish to argue that *no law* entails the evolution of the biosphere.[13] If we are right, entailing law, the centerpiece of physics since Newton, ends at the watershed of evolving life. If this claim is right, it is obviously deeply important. Furthermore, it raises the issue of how the biosphere, the most complex system we know in the universe, can have arisen beyond entailing law. I will discuss these issues as well. And, central to this book, *if* entailing law is the instantiation of Descartes' *res extensa* and mechanistic worldview as mathematized by Newton's "entailing" laws of motion, we will find a new sense in which evolution is not mechanism. Again, the foregoing discussion proceeds in several steps.

The Uses of a Screwdriver Cannot Be Listed Algorithmically

Here is the first "strange" step. Can you name all the uses of a screwdriver, alone, or with other objects or processes? Well, screw in a screw, open a paint can, wedge open a door, wedge closed a door, scrape putty off a window, stab an assailant, be an *objet d'art*, tied to a stick a fish spear, the spear rented to "natives" for a 5 percent fish catch return becomes a new business, and so on. I think we all are convinced that the following two statements are true: (1) the *number* of uses of a screw driver is *indefinite*; and (2) unlike the integers which can be ordered, there is no natural ordering of the uses of a screw driver. The uses are *unordered*. But these two claims entail that there is no "Turing Effective Procedure" to *list* all the uses of a screwdriver alone or with other objects or processes. In short, there is no algorithm to list all the uses of a screwdriver.

Now consider *one* use of the screwdriver, say to open a can of paint. Can you list all the other objects, alone or with other objects or processes, that may carry out the "function" of opening a can of paint? Again, the number of ways to achieve this function are indefinite in number and unorderable; so, again, no algorithm can list them all.

Adaptations in an Evolving Cell Cannot Be Prestated

Now consider an evolving bacterium or eukaryotic, say, single-celled organism. In order to adapt in some new environment, all that has to occur is that some one or many cellular or molecular "screwdrivers" happen to "find a use" that enhances the fitness of the evolving cell in that new environment. Then there must be heritable variation for those properties of the cellular screwdrivers, and then natural selection, acting at the level of the Kantian whole cell, will select, or cull out, the fitter variants with the new uses of the molecular screwdrivers which constitute adaptation.

I wish to make an additional point here. It is widely known that Darwin gave us natural selection "culling," and something like the "survival of the fittest," the appearance of design without a designer. But from whence came the *arrival* of the *fittest*, or at least the *fitter*? The above selection at the level of the Kantian whole cell of cellular or molecular screwdrivers for a new or modified use *is* the arrival of the "fitter." Adaptation consists in just this "finding of a use" that enhances fitness. More, by the above discussion, we cannot *prestate* what this use will be. But no algorithmic *list* of the possible uses of these cellular and molecular screwdrivers can be had, thus we cannot know, ahead of time, what natural selection *acting at the level of the Kantian whole organism*, will reveal as the new uses of the cellular screwdrivers acting in part via the niche of the organism, which *succeed better*, hence were selected. We cannot in general prestate the adaptive changes that will occur. This is the deep reason evolutionary theory is so weakly predictive.

We Cannot Prestate the Actual Niche of an Evolving Organism

The task closure of the evolving cell is achieved, in part, via causal or quantum consequences passing through the environment that constitutes the "actual niche" of the evolving organism. But the features of the environmental "niche" that participate with the molecular screwdrivers in the evolving cell which will allow a successful task closure are circularly defined with respect to the organism itself. We only know after the fact of natural selection what aspects of the evolving cell, its screwdrivers, as well as of the causal consequences of specific aspects of the actual niche, are successful when selection has acted at the level of the Kantian whole evolving cell population. Thus, we cannot prestate the actual niche of an evolving cell by which it achieves task closure in part via that niche.

But these facts have deep meaning. In physics, the phase space of the system is *fixed*, in Newton, Einstein, and Schrödinger. This allows for entailing laws. In evolution, each time an adaptation occurs and a molecular or other screwdriver finds a new use in a new actual niche, the very phase space of evolution has *changed* in an unprestatable way. But this means that we can *write no equations of motion for the evolving biosphere*. More, the actual niche can be considered as the boundary conditions on selection. But we cannot prestate the actual niche. In the case of the billiard balls, Newton gave us the laws of motion, and told us to establish initial and boundary conditions then integrate laws of motion stated in differential equation form to get the entailed trajectories. But, in biology, we cannot write down the laws of motion, so we cannot write them down in differential equation form. Even if we could, we could not know the niche boundary conditions, so we could not integrate those laws of motion that we do not have anyway. It would be like trying to solve the billiard ball problem on a billiard table whose shape changed forever in unknown ways. We would thereby have no mathematical model. Here, the profound implication is that *no laws entail the evolution of the biosphere*.

If this is correct, we are, as stated above, at the end of reductionism, and at the watershed of evolving life. Now the machine metaphor since Descartes, perfected by Newton, leads us to think of organisms, as Monod stated, as molecular *machines*. Let me distinguish diachronic from synchronic science. Diachronic studies the evolution of life, and its "becoming," over time. Synchronic studies the presumably fully reducible aspects of, for example, how a heart, once it has come to exist in the non-ergodic universe, "works." In these synchronic studies, reductionism may work, but some of the chapters in this book argue to the contrary. But in the diachronic becoming of the *biosphere, life is an ongoing, unprestatable, non-algorithmic, and non-mechanical process selected for fitness*.

There are, I think, deep implications to this. What is the "phenotype"? This is a much debated question. But few would doubt that the causal consequences of a part that sustain the Kantian whole are its *function*, and that therefore, *functions are either the phenotype or at least part of the phenotype*. This realization has an odd and new consequence. When some cellular or molecular screwdrivers alone or together "find some use" in the current, or new, environment that enhances the fitness of the Kantian whole organism, and so are selected if heritably variable, in a deep sense it is "immaterial" just what cellular or molecular screwdrivers, and which of their specific causal consequences, happen to fulfill the new adaptive use. Thus, in a deep sense, the Arrival of the Fitter does not depend in any specific way on any specific predefinable set of cellular or molecular screwdrivers. Although any specific instantiation of an adaptive step to a fitter function or new function will in fact utilize some specific cellular or molecular screwdrivers, which ones may happen to be the ones so selected cannot be algorithmically prestated. Thus, not only is there the familiar "multiple realizability" philosophers are used to, the situation is more radical—the multiple realizations for a given new use cannot be algorithmically listed, so are unprestatable. In this sense, the adaptive change of the organism is beyond entailing law and beyond statable mechanism.

Information Theory Is Useless in Evolution

Consider familiar information theory, for example, that of Shannon or Kolmogorov. It is central to these basic concepts of information theory that there be a finitely prestatable "alphabet of symbols." The familiar examples are bit strings over the alphabet [1,0], such as (1011010). Then Shannon computes the entropy of the information source given the number of copies of each bit string message in the source. Kolmogorov computes the complexity of the symbol string as the shortest program that can produce it. For both Shannon and Kolmogrov there must be a prestated alphabet of symbols.

But by the above discussion, *there is no prestated alphabet* of phenotypes and functions—themselves part of phenotypes—in biological evolution. The quantum behavior of a single electron, as in photosynthesis, or the relative placement of cilia on a Paramecium, can be a phenotype with functional consequences. We cannot prestate their becoming in the non-ergodic universe above the level of the atom, for no laws entail that becoming. Thus we have no prestatable alphabet upon which to build any known information theory. Therefore, there is no sense in which any current form of information theory is applicable to biological evolution.

Darwinian Preadaptations and Radical Emergence: The Evolving Biosphere

If we asked Darwin what the function of my heart is, he would respond, "to pump your blood." But my heart makes heart sounds and jiggles water in my pericardial sac. If I asked Darwin why these are not the function of my heart, he would answer that I have a heart because its pumping of blood was of selective advantage in my ancestors. In short, he would give a selection account of the causal consequence for virtue of which I have a heart. Note that he is also giving an account of why hearts exist at all as complex entities in the non-ergodic universe above the level of atoms. Hearts are functioning parts by pumping blood, of humans as reproducing Kantian wholes. Note again that the function of my heart is a subset of its causal consequences, pumping blood, not heart sounds or jiggling water in my pericardial sac.

Darwin had an additional deep idea: a causal consequence of a part of an organism of no selective significance in a given environment might come to be of selective significance in a different environment, so be selected, and typically, a new function would arise. These are called "Darwinian preadaptation" without meaning foresight on the part of evolution. Stephen Jay Gould renamed them "exaptations." I give but two examples of thousands of a Darwinian preadaptation. First, your middle ear bones that transmit sound from the eardrum to the cochlea of the inner ear evolved by Darwinian preadaptations from the jaw bones of an early fish. Second, some fish have a swim bladder, a sac partly filled with air and partly with water, whose ratio determines neutral buoyancy in the water column. Paleontologists believe that the swim bladder evolved from the lungs of lung fish. Water got into some lungs; now sacs partly filled with air, and partly with water, were poised to evolve into swim bladders. Let us assume the paleontologists are right.

I now ask three questions: first, did a new function come to exist in the biosphere? Yes, hearing and neutral buoyancy in the water column. Second, did the evolution of the middle ear or swim bladder alter the future evolution of the biosphere? Yes, new species of animals with hearing, new species of fish evolved with swim bladders. They evolved new mutant proteins. And critically, the middle ear, or the swim bladder, *once each came to exist*, constituted what I will call *a new adjacent possible empty niche*, for a worm, bacterium or both could evolve to live only in the middle ear or swim bladders. Thus the middle ear or swim bladder, once each exists, alters the possible future evolution of the biosphere. I return to this point in a moment for enchantment hides here. Third, now that you are an expert on Darwinian preadaptations, can you name all possible Darwinian preadaptations just for humans in the next three million years? Try it and feel your mind go blank. We all say no. A start to why we cannot is to ask the following questions: How would you name all possible selective environments? How would you know you listed them all? How would you list all the features of one or many organisms that might serve as "the preadaptation"? We cannot.

The underlying reasons why we cannot do this are given above in the discussion about screwdrivers, their non-algorithmically listable uses, either alone or with other objects and processes, as well as the non-algorithmically listable other objects and processes that can accomplish any specific task, opening a can of paint, that we can use a screwdriver to accomplish. In addition, the organism completing task closure in part via its actual niche is circularly defined and cannot be prestated until selection at the level of the Kantian whole organism reveals what has "worked," hence what the relevant variables, functions, and aspects of the environment actually now are.

The Adjacent Possible

Consider a flask of one thousand kinds of small organic molecules. Call these the Actual. Now let these react by a single reaction step. Perhaps new molecular species may be formed. Call these new species the molecular "adjacent possible." It is perfectly defined if we specify a minimal stable lifetime of a molecular species. Now, let me point at the adjacent possible of the evolving biosphere. Once lung fish existed, swim bladders were in the adjacent possible of the evolution of the biosphere. But two billion years ago, before there were multicelled organisms, swim bladders were not in the adjacent possible of the evolution of the biosphere.

I think we all agree to this. But now consider what we seem to have agreed to: with respect to the evolution of the biosphere by Darwinian preadaptions, *we do not know all the possibilities*. Now let me contrast our case for evolution with that of flipping a fair coin 10,000 times. Can we calculate the probability of getting 5,640 heads? Sure, use the binomial theorem. But note that here we know *ahead of time* all the possible outcomes, all heads, all tails, alternative heads and tails, all the 2 to the 10,000 power possible patterns of heads and tails. Given that we know all the possible outcomes, we thereby know the "sample space" of this process, so we can construct a probability measure. We do not know what *will* happen, but we know what *can* happen.

But in the case of the evolving biosphere, not only do we not know what *will* happen, we do not even know what *can* happen. There are at least two huge implications of this: first, we can construct no probability measure for this evolution by any known mathematical means. We do not know the sample space. Second, reason, the prime human virtue of our Enlightenment, cannot help us in the case of the evolving biosphere, for we do not even know what *can* happen, so we cannot reason about it. The same is true of the evolving econosphere, culture, and history. As I will try to show us, we often do not know ahead of time the new variables that will become relevant, so we cannot reason about them. Thus, real life is not an optimization problem, top down, over a known space of possibilities. It is far more mysterious. How do we navigate, not knowing what can happen? Yet we do. This has very large implications for how we govern ourselves and live wisely when we cannot know all that *can* happen.

Without Natural Selection, the Biosphere Enables and Creates Its Own Future Possibilities

Now I introduce Radical Emergence that I find enchanting. Consider the middle ear, or swim bladder, once either has evolved. We agreed above, I believe, that a bacterium or worm or both could evolve to live only in that middle ear or swim bladder, so the middle ear or swim bladder as *a new adjacent possible empty niche*, once it had evolved, alters the future possible evolution of the biosphere. Next, did natural selection act on an evolving population of hearing animals or fish to select a well-functioning middle ear system or swim bladder? Of course (I know I am here anthropomorphizing selection, but we all understand what is meant). But did natural selection "act" to create the middle ear or swim bladder as *a new adjacent possible empty niche*? No! Selection did not "struggle" to create the middle ear or swim bladder as a new empty adjacent possible niche. But that means something I find stunning. Without selection acting in any way to do so, evolution is creating its own future possibilities of becoming! And the worm or bacterium or both that evolves to live in the middle ear or swim bladder is a Radical Emergence unlike anything in physics.

It seems important to stress that the new realization that the biosphere, without natural selection, creates its own future possibilities, was not seen by Darwin, nor by contemporary evolutionary theory, including the neo-Darwinian Synthesis. We are, with "no laws entail the evolution of the biosphere," if true, beyond Newton, Einstein, and Schrödinger at the watershed of evolving life. And with the enchantment of the fact that the evolving biosphere creates, beyond selection, its own future possibilities, we are beyond Darwin. We have entered an entirely new worldview. Note that I am *not* talking here about familiar "niche construction," discussed in this book and elsewhere, where the behaviors of organisms modify their niches, nor am I talking about the Baldwin effect.

I am talking about "niche creation" itself.

Evolution Often Does Not Cause, But Enables Its Future Evolution

The bacterium or worm that evolves to live in the actual niche of the middle ear or swim bladder whereby it achieves a task closure selected at the level of the Kantian whole worm or bacterium, evolves by quantum indeterminate, and ontologically a-causal quantum events. Later, it is selected, if heritable and fitter, by natural selection acting at the level of the whole organism in its world. Thus, the swim bladder does not *cause*, but *enables*—that is, "makes possible"—the evolution of the bacterium or worm or both to live in the swim bladder. This means that evolving life is not only a web of cause and effect, but of empty niche opportunities that enable new evolutionary radical emergence as the evolving biosphere creates, beyond selection, its own future possibilities. The same is true in the evolving econosphere, cultural life, and history, as I discuss more fully below. We live in both a web of cause and effect and a web of enabling opportunities that enable new directions of becoming.

Moreover, the swim bladder or middle ear, as an "adjacent possible" empty niche opportunity for adaptation by a-causal quantum mutation events, then selected if heritable, is itself, as that new adjacent possible niche, an "enabling constraint." As that enabling constraint, the niche shapes evolution and enables the worm or bacterium in the adaptive solution it finds in a new adjacent possible created by the niche. The same claim is true in the entire evolution of the multi-specied biosphere. Each species or set of species, plus the abiotic environment, creates enabling constraints opportunities for yet new species (see the section on the evolution of collectively autocatalytic sets creating autocatalytic sets below). This raises the fascinating new question of whether we can find an account of how enabling constraints generate new adjacent possible ways of life and shape them, thereby generating yet newer enabling constraints yielding newer "adjacent possibilities" for evolution, perhaps maximizing the average rate of growth of the adjacent possible into which the biosphere evolves, and in so doing creating even more possibilities for itself.

Neither Quantum Mechanics Alone Nor Classical Physics Alone Account for Evolution

Mutations are often quantum a-causal and indeterminate, random events. Yet evolution is not random: the eye evolved some eleven times and the vertebrate and octopus camera eyes are nearly identical (except that the blood vessels in the octopus are behind the retina). These examples, widespread, of convergent evolution show that evolution is not random. Thus, neither quantum mechanics alone, nor classical physics alone suffices to account for evolution.

Toward a Positive Science for the Evolving Biosphere beyond Entailing Law

The arguments above support the radical claim that no laws entail the evolution of the biosphere. If right, then Kant was right. There will be no Newton of biology. Not even Darwin was that Newton yielding entailing laws. But the biosphere is the most complex system we know in the universe, and it has grown and flourished, even with small and large avalanches of extinction events, for 3.8 billion years. Indeed, there is a secular increase in species diversity over the Phanerozoic.

How are we to think of the biosphere building itself, probably beyond entailing laws? Organisms are Kantian wholes, and the building of the biosphere of these past 3.8 billion years seems almost certainly to be related to how Kantian wholes co-create their worlds with one another, including the creating, with no selection, of new empty adjacent possible niches that alter the future evolution of the biosphere. There may be a way to start studying this topic: a new quest. Collectively, autocatalytic sets are the simplest models of Kantian wholes. In very recent work with Wim Hordijk and Michael Steel, computer scientist and mathematician, respectively, we are studying what they call RAFs, which are collectively autocatalytic sets in which the chemical reactions without catalysis, occur spontaneously at some finite rate, and that rate is much speeded up by catalysis. Fine results by Hordijk and Steel show that RAFs emerge and require only that each catalyst catalyzes between one and two reactions. This is fully reasonable both chemically and biologically.[14]

Most recently, the three of us have examined the substructure of RAFs.[15] There are irreducible RAFs, which, given a food set of sustained small molecules, have the property of autocatalysis, but if any molecule is removed from the RAF, the total system collapses. It is irreducible. Then, given a maximum length of polymers allowed in the model as the chemicals, from monomers to longer polymers, there is a maximal RAF, which increases as the length of the longest allowed polymer, and hence, the total diversity of possible polymers allowed, increases. The most critical issue is this: there are *intermediate* RAFs called "submaximal RAFs," each composed either of two or more irreducible RAFs, or of one or more irreducible RAF and one or more larger "submaximal" RAF, or composed of two or more smaller submaximal RAFs. Thus we can think mathematically of the complete set of irreducible RAFs, all the diverse submaximal RAFs, and the Maximal RAF. For each, we can draw arrows from those smaller RAFs that jointly comprise it. This set of arrows is a partial ordering among all the diverse RAFs possible in the system.

The next important issue is this: if new food molecule species, or larger species, enter the environment, even *transiently*, the total system can grow to create *new* submaximal RAFs that did not exist in the system before. This is critical. It shows that existing Kantian wholes can create new empty adjacent possible niches and with a chemical fluctuation in which molecular species are transiently present in the environment, the total "ecosystem" can grow in diversity. A model biosphere is building itself!

In this system, the diverse RAFs can "help" one another, for example, a waste molecule of one can be a food molecule of another, or via inhibition of catalysis, or toxic products of one with respect to another, can hinder one another in complex ways. They form a complex ecology. Further, these RAFs, if housed in compartments that can divide, such as bilipid membrane vesicles called liposomes,[16] have been shown recently to be capable of open-ended evolution via natural selection, where each of the diverse RAFs act as a "replicator" to be selected and, in that selection, chemical reaction "arcs" that flower from the RAF core act as the phenotype with the core. Thus, to my delight, we have the start of a theory for the evolution of Kantian wholes. But there is a profound limitation to these models: they are in a deep sense algorithmic, and their possible phase spaces can be prestated. The reason is simple: the only functions that happen in these RAF systems are molecules undergoing reactions, which are catalyzed by molecules. But the set of possible molecules up to any maximum length polymer can be prestated. And the set of possible catalytic interactions can be prestated. Therefore, even in models where the actual assignment of which molecule catalyzes which reaction is made at random or via some "match rule" of catalyst and substrate(s), then assigned with a probability of catalysis per molecule, thus, we have already prestated all the possibilities for reaction systems and all possible patterns of catalysis. Therefore, the phase space is prestated.

By contrast, in the discussion above, we talked about the vast task closure achieved by an evolving bacterium or eukaryotic cell or organism. These tasks were not limited to catalysis. And as we saw with the discussion of the possible uses of a molecular screwdriver in a cell, those uses are both indefinite in number and not orderable, so no algorithm can list all those uses. Nor can we prestate how the evolving Kantian whole cell, where selection acts at the level of the Kantian whole and culls out altered screwdriver parts with heritable variations that achieve some often new functional task closure via the actual niche. Thus the real evolutionary process is non-algorithmic, non-machine, non-entailed. With respect to our initial evolving RAF ecosystems, we do not yet know how to make this evolution non-algorithmic and non-entailed. While we have a start, and a useful one, it is not enough.

Recent work, as noted, has shown that collectively autocatalytic sets in containers that divide with the set, such as liposomes, can support open-ended evolution. This is even more strikingly true if the set of "functions" involved is far wider than mere catalysis, like that in real evolving cells and not algorithmically statable. Thus, the neo-Darwinian Synthesis builds upon too narrow a basis of gene-centric contemporary organisms on Earth in its understanding of evolution. If we ever create or find new life-forms that are not based on DNA, RNA, and encoded proteins, we will have to vastly generalize our thinking about evolution.

The Economy Is an Evolving Autocatalytic Set

In the theory of autocatalytic sets, typically modeled are the "molecules" as binary strings (1000100) as substrates, products, and catalysts for speeding the reactions among the substrates and products. There is nothing at all that is fundamentally "chemical" about the binary symbol strings (e.g., 10001010). These symbol strings can stand for any objects that can "react" and transform. Thus consider the physical transformations in the real economy. Here two boards and a nail act as "substrate inputs" to a "production capacity" consisting of a hammer. The production capacity is the analogue of the reaction among chemical substrates to produce chemical products. Here, the two boards and nail as input goods are "acted upon, catalytically" by the hammer to produce the product: two boards nailed together. Now note that the hammer itself is a product of the economic system. Thus, the real economy is also a collectively autocatalytic set, with food interpreted instead as renewable resources that feed into the economy.

In turn, economic growth of ever-new goods and production capacities in the past fifty thousand years, from perhaps one thousand goods and production capacities then perhaps to ten billion now, has occurred. Just as in the use of collectively autocatalytic sets non-algorithmically creating ever-new niches for ever-new autocatalytic sets of autocatalytic sets in an ever-expanding "adjacent possible" in a biosphere of growing diversity, in the same way the growing diversity of goods and production capacities in the econosphere may well reflect the same collectively autocatalytic, non-algorithmic, niche formation. Both are likely to reflect the beginning theory of collectively autocatalytic sets, or RAFs as above, plus the non-algorithmic character by which many new goods and practical "uses" of economic "screwdrivers" emerge. Again, we are beyond entailment, for those new uses are indefinite in number and unorderable, so non-algorithmic, and not entailed. No law, it seems, entails the evolution of the economy. *A fortiori*, no law entails the evolution of culture, law, and history.

The Mathematical Theory of the Spontaneous Formation of Collectively Autocatalytic Sets May Constitute a New Form of Law: Formal Cause Law

The mathematical theory of the emergence of collectively autocatalytic sets, say with respect to the origin of life, and not to an economy, posits an ensemble of chemical reaction networks each of which has molecules as substrates and products, linked by chemical reactions in a "reaction graph." In addition, in the first version of this theory, each molecule has a fixed probability of catalyzing each reaction. Which molecule actually catalyzes any given reaction is assigned totally at random. Then numerical studies and theorems show that at a sufficient diversity of molecules and reactions, so many reactions are catalyzed according to the random catalysis probability rule, that collectively autocatalytic sets "emerge" spontaneously as a literal phase transition in the specific reaction graph. Also, this theory shows that over a vast ensemble of chemical reaction graphs the same emergence would hold. Furthermore, were the constants of nature in physics changed just a bit so that chemistry changed just a bit, the same theorems would hold, so they seem not even to depend upon the physics of our specific universe.

I now ask, as hinted above, is this theory and its "law of the expected emergence of collectively autocatalytic sets" (testable experimentally by, for example, using libraries of random peptides), anything like Newton's formulation? Recall, Newton mathematized Aristotle's efficient cause in his three laws of motion in differential equation form, giving the efficient cause forces between the particles and in his universal law of gravitation in differential equation form. Then by integration, given initial and boundary conditions, we yield the deduced, hence entailed, deterministic trajectories of the particles. This mathematization of efficient cause, which for Aristotle *is* causal mechanism, *is* our mathematization of mechanism, as promulgated by Descartes in his *res extensa*. But is the theory above of the phase transition in virtually any member of a vast ensemble of chemical reaction graphs, or more generally a set of objects, transformations among those objects and a generalized "catalysis" of the transformations among those objects, as in an economy, a Newton-like Law with efficient cause differential equation laws of motion? No it is not. So, we might ask: what form of law is the law of the expected emergence of collectively autocatalytic sets?

Aristotle's formal cause concerns what it is for a statue to *be* a statue. I suggest that the theory of the emergence of collectively autocatalytic sets is a Formal Cause Law of what it is to emerge as a collectively autocatalytic set. This requires much further discussion, for it proffers a new kind of law. Note that from the example of the economy as a collectively autocatalytic set, even the "materials" and "processes" in the emergence of the collectively autocatalytic set are "immaterial." They could be chemicals, reactions and catalysis, or input goods output goods production functions and speeding up of production functions by some goods in the system.

Wise Governance when We Not Only Do Not Know What Will Happen, but Often Do Not Even Know What Can Happen

My next comments are wider than required for the foreword to this fine book, but worth stating for their broader human implications. We do not live in the world we have thought in the Newtonian framework. We live in a world of unprestatable new, unintended possibilities, opportunities, biosphere, econosphere, history. We do not know all the variables that will become relevant.

As an economic example that we all know: the invention of the Turing machine enabled the invention of the main frame computer whose wide sale created the market opportunity for the personal computer whose wide sale created the market opportunity for word processing, whose wide use created the opportunity to share files, whose increasing use created the opportunity for the invention of the World Wide Web, whose coming into existence created the opportunity to sell things on the web, whose abundance was part of creating content on the Web which created the opportunity for Web browsers. Now we have Facebook and the Arab Spring. No one at the time of Turing or the early main frame foresaw, or intended, the invention of electronic commerce on the Web, or browsers or Facebook. These well-known facts are of profound importance practically. In 1950, had one tried to plan and optimize over a twenty-year time horizon, one would not have known the variables that would become relevant. Therefore, we live with the illusion of an adequacy of "scientific top down management and control." Instead, we live with the promise of discovery of that which we help co-create, but opportunities that often arise without intention, and the promise of wise enablement.

All of this presents us with an argument for rethinking governance. Centralized top down control fails for two reasons. First of all, in respect to that above, we cannot know all the relevant variables beforehand. But the second is equally astonishing. Complex problems with many factors that can interact positively or negatively create "multi-peaked" fitness or payoff landscapes. The local peaks often reflect local features of the landscape, for example, in policy formation, the specific variable values, behaviors, and other aspects of the problem in different geographic or cultural regions. Thus, even were the relevant variables all known, which they are not, "one size fits all" optimization misses these local features that engender multiple peaks, often of very similar fitness or payoff, requiring different modes of regulation and action that are locally appropriate.

Now consider governance: it strongly suggests that "one size fits all" rules, such as the United Nations and the IMF may mandate, may often be poor choices, missing the variegated and local structure of the landscape in question. Furthermore, if we often cannot know the relevant variables as well, this suggests an even more radical modification of governance. The general issue is obvious, but I give one example that is due to my colleague Asim Zia at the University of Vermont. An Amazon country was monitoring deforestation by satellite imaging of the forest canopy. Local businesspeople entered the forest and cut down smaller trees and sold them for profit, leaving the canopy intact. Who would have thought ahead of time of that novel, presumably non-algorithmic, solution of the locals to making a living and subverting the efforts of the government monitoring? We really do not know the variables that will become relevant.

Wise enablement via laws and regulations, which have cascading consequences we cannot foresee,[17] as is often seen in the *sequelae* to legal rulings, requires a new examination of what we enable when we enable, and wise revision of what we enable when we see it go awry, even though we still will not know the further new unprestatable variables that we then enable and that will become relevant. Living is more mysterious and open-ended than we have thought in our Newtonian worldview, where we pretend that we can know and master all. In truth, we often know not what we do.

Re-Enchantment and Creating a New World

I return to Max Weber's astonishing statement: "with Newton we became disenchanted and entered Modernity." Was Weber right? I think so. With Newton, we left the late Middle Ages and entered a worldview of the deterministic dynamics of celestial mechanics, the Theistic God acting in the universe retreated in the Enlightenment to a Deistic God who set up the universe with Newton's laws and let them unfold. The war between theistic religion and science, let alone science and the arts, was underway. Next came our beloved Enlightenment, with its mantra, "down with the Clerics, up with science for the perpetual betterment of Man." The Enlightenment was the "Age of Reason." Next came the Industrial Revolution based on science derived from physics and chemistry. Thence we entered Modernity.

We know the goods and ills of our fully lived Enlightenment dreams. We have democracy, a higher standard of living, are better educated, have better health and longer lives. Yet our democracies are often corrupted by power elites. We are, as Gordon Brown said as Prime Minister of the United Kingdom, "reduced to price tags" in our increasingly global economy, where we often make, sell, and buy plastic, purple penguins for the poolside. If we ask why we do this, part of the answer is that we do not know what else to do. Furthermore, we *are* disenchanted. We are, a billion of us, secular realists in a meaningless universe, to quote Stephen Weinberg's famous dictum. We have lost our spirituality. But our physics-based worldview, if right for the abiotic universe, seems badly wrong for the living, evolving world, past the watershed of evolving life. We do live in a world of cause and effect, but also unprestatable opportunities that emerge in an unprestatable, ever-growing and changing adjacent possible that we partially co-create, with and without intent.

It really is true that, with no selection acting to do so, the newly evolved middle ear or swim bladder are new adjacent possible empty niches that alter the future possibilities of biological evolution. The worm or bacterium that is enabled to evolve really is radically emergent. It really is true that the Turing machine enabled the mainframe computer whose wide sale created the market opportunity for the personal computer whose wide sale created the market opportunity for word processing and file sharing, whose wide use create a niche for the World Wide Web, whose creation generated an opportunity to sell things on the Web which created content on the Web, which created a market opportunity for browsers such as Google and Yahoo. Then Facebook came and the Arab Spring. None could have foreseen this. None intended this radically emergent becoming, so similar to the radical emergence in the evolving biosphere. In both cases, with neither selection, nor intent, the evolving system creates, typically unprestatably, its own future possibilities. How much enchanting mystery do we want to be re-enchanted? Moreover, the Age of Reason assumed that we could come to *know*, that the world was solvable by reason. But if we often do not know what *can* happen, we cannot reason about it. Reason, the highest virtue of our Enlightenment, is an inadequate guide for living our lives. And top down decision making, as if we knew ahead of time the variables that would become relevant, then "optimize," is often an illusion. We need to rethink how we make and live in our worlds. Then what if we ask whether current first world civilization best serves our humanity, or do we largely serve it, price tags and all? I think we are lost in modernity, without a clear vision of what our real life is.

Ralph Waldo Emerson is famous for "Emersonian Perfectionism," by which is meant that we are born with a set of virtues, or strengths, and should devote our lives to perfecting them. But this Perfectionism seems static, like a European hotel breakfast room with all the food choices laid out. We have only to choose among our preset virtues and to perfect them. Emerson saw a life of ongoing self-affirmation. But this rather fixed world is not how real life is: we live a life of ever-unfolding, often unprestatable opportunities that we partially create and co-create, with and without intent. I am thus falling in love with "Living the Well Discovered Life." Then my own dream for "Beyond Modernity" starts to become our thirty civilizations around the globe, woven gently together to protect the roots of each, yet firmly enough to generate ever new cultural forms by which we can be human in increasingly diverse, creative ways, each helping himself or herself and the other to live a well-discovered life, and ameliorating our deep shadow side. We need an enlarged vision of ourselves and of what we can become.

Notes

1. Stuart A. Kauffman, "Metabolic Stability and Epigenesis in Randomly Constructed Genetic Nets," *Journal of Theoretical Biology* 22 (1969): 437-467; Stuart A. Kauffman, *The Origins of Order: Self Organization and Selection in Evolution* (New York: Oxford University Press, 1993).
2. Jacques Monod, *Chance and Necessity: An Essay on the Natural Philosophy of Modern Biology* (New York: Alfred A Knopf, 1995).
3. Immanuel Kant, *The Critique of Judgment*, trans. J. H. Bernard (New York: Prometheus Books, 2000), §65.
4. Stuart A. Kauffman, *Investigations* (New York: Oxford University Press, 2000).
5. K. T. Piel, "Emotion: A Self-Regulatory Sense," *Biophysical Psychological Review*, 2012 (in press).
6. Piel, "Emotion."
7. Piel, "Emotion."
8. Nathaniel Wagner and Gonen Ashkenasy, "Symmetry and Order in Systems Chemistry," *The Journal of Chemical Physics* 130 (2009): 164907-164911.
9. Stuart A. Kauffman, "Cellular Homeostasis, Epigenesis and Replication in Randomly Aggregated Macromolecular Systems," *Journal of Cybernetics* 1 (1971): 71-96; Stuart A. Kauffman, "Autocatalytic Sets of Proteins," *Journal of Theoretical Biology* 119 (1986): 1-24; Kauffman, *The Origins of Order*.
10. Elchrann Mossel and Mike Steel, "Random Biochemical Networks: The Probability of Self-Sustaining Autocatalysis," *Journal of Theoretical Biology* 233, no. 3 (2005): 327-336.
11. Bianca J. Lam and Gerald F. Joyce, "Autocatalytic Aptazymes: Ligand-Dependent Exponential Amplification of RNA," *Nature Biotechnology*, 2009, http://www.nature.com/nbt/journal/vaop/ncurrent/full/nbt.1528.html; G. von Kiedrowski, "A Self-Replicating Hexadesoxynucleotide," *Agnewandte Chemie International Edition in English* 25 (1986): 982.
12. From personal communication with Kalevi Kull, Tartu, Estonia, April 25th, 2012; also see chapters 6 and 7 in this volume.
13. Giuseppe Longo, Maël Montevil, and Stuart A. Kauffman, "No Entailing Laws, but Enablement in the Evolution of the Biosphere," *Physics ArXiv*, 1201.2069v1 (January 10, 2012): 1-19, http://arxiv.org/pdf/1201.2069.pdf; also see Giuseppe Longo, Maël Montevil, and Stuart A. Kauffman, GECCO, in press, 2012.
14. Wim Hordijk and Mike Steel, "Detecting Autocatalytic, Self-Sustaining Sets in Chemical Reaction Systems," *Journal of Theoretical Biology* 227, no. 4 (2004): 451-461.
15. Wim Hordijk, Mike Steel, and Stuart Kauffman, "The Structure of Autocatalytic Sets: Evolvability, Enablement and Emergence," *Physics ArXiv* 1205.0584v2 (May 4, 2012): 1-13, http://arxiv.org/pdf/1205.0584v2.pdf.
16. Pier Luigi Luisi, Pasquale Stano, Silvia Rasi, and Fabio Mavelli, "A Possible Route to Prebiotic Vesicle Reproduction," *Artifical Life* 10, no. 3 (2004): 297-308.
17. C. Devins, "Beyond Law: Towards a Unifying Theory of Legal and Biological Systems Through Emergence, US The Case Study of Federal Crack Cocaine Sentencing Policy," *Duke Law Review*, 2012 (in press).

Bibliography

Devins, C. "Beyond Law: Towards a Unifying Theory of Legal and Biological Systems Through Emergence, US The Case Study of Federal Crack Cocaine Sentencing Policy." *Duke Law Review*, 2012 (in press).

Hordijk, Wim, and Mike Steel. "Detecting Autocatalytic, Self-Sustaining Sets in Chemical Reaction Systems." *Journal of Theoretical Biology* 227, no. 4 (2004): 451-461.

Hordijk, Wim, Mike Steel, and Stuart A. Kauffman. "The Structure of Autocatalytic Sets: Evolvability, Enablement and Emergence." *Physics ArXiv* 1205.0584v2 (May 4, 2012): 1-13, http://arxiv.org/pdf/1205.0584v2.pdf.

Kant, Immanuel. *The Critique of Judgment*. Translated by J. H. Bernard. New York: Prometheus Books, 2000.

Kauffman, Stuart A. "Metabolic Stability and Epigenesis in Randomly Constructed Genetic Nets." *Journal of Theoretical Biology* 22 (1969): 437-467.

———. "Cellular Homeostasis, Epigenesis and Replication in Randomly Aggregated Macromolecular Systems." *Journal of Cybernetics* 1 (1971): 71-96.

———. "Autocatalytic Sets of Proteins." *Journal of Theoretical Biology* 119 (1986): 1-24.

———. *The Origins of Order: Self Organization and Selection in Evolution*. New York: Oxford University Press, 1993.

———. *Investigations*. New York: Oxford University Press, 2000.

Kiedrowski, G. von. "A Self-Replicating Hexadesoxynucleotide," *Agnewandte Chemie International Edition in English* 25 (1986): 982.

Lam, Bianca J., and Gerald F. Joyce. "Autocatalytic Aptazymes: Ligand-Dependent Exponential Amplification of RNA." *Nature Biotechnology*, 2009, http://www.nature.com/nbt/journal/vaop/ncurrent/full/nbt.1528.html.

Longo, Giuseppe, Maël Montevil, and Stuart A. Kauffman. "No Entailing Laws, but Enablement in the Evolution of the Biosphere." *Physics ArXiv*, 1201.2069v1 (January 10, 2012): 1-19, http://arxiv.org/pdf/1201.2069.pdf.

———. "No Entailing Laws, but Enablement in the Evolution of the Biosphere." *Proceedings of the Fourteenth International Conference on Genetic and Evolutionary Computation Conference Companion* (2012): 1379-1392.

Luisi, Pier Luigi, Pasquale Stano, Silvia Rasi, and Fabio Mavelli. "A Possible Route to Prebiotic Vesicle Reproduction." *Artifical Life* 10, no. 3 (2004): 297-308.

Monod, Jacques. *Chance and Necessity: An Essay on the Natural Philosophy of Modern Biology*. New York: Alfred A Knopf, 1995.

Mossel, Elchrann, and Mike Steel. "Random Biochemical Networks: The Probability of Self-Sustaining Autocatalysis." *Journal of Theoretical Biology* 233, no. 3 (2005): 327-336.

Piel, K. T. "Emotion: A Self-Regulatory Sense." *Biophysical Psychological Review*, 2012 (in press).

Wagner, Nathaniel, and Gonen Ashkenasy. "Symmetry and Order in Systems Chemistry." *The Journal of Chemical Physics* 130 (2009): 164907-164911.

Introduction
On a "Life-Blind Spot" in Neo-Darwinism's Mechanistic Metaphysical Lens

Adam C. Scarfe

The question "What is Life?" is not often asked in biology, precisely because the machine metaphor already answers it: "Life is a machine." Indeed, to suggest otherwise is regarded as unscientific and viewed with the greatest hostility as an attempt to take biology back to metaphysics.[1] —Robert Rosen

An organized being is thus not a mere machine, . . . it has a self-propagating formative power, which cannot be explained through the capacity for movement alone (that is, mechanism).[2] —Immanuel Kant

The task confronting biology as a science is to develop an entirely new projection of the objects of its inquiry. (Expressed from another point of view, which is not necessarily identical with what we have just said, the task today is to liberate ourselves from the mechanistic conception of life).[3] —Martin Heidegger

Mechanism and the Modern Synthesis

Any sampling of recent studies carried out in the biological sciences would reveal that a predominant aim of inquiries therein involves uncovering the *mechanisms* that underpin the functioning of natural systems, that determine the physiological, morphological, and behavioral functioning of organisms, as well as that govern their evolution.[4] In addition, there is also a heavy demand for this sort of knowledge, perhaps, in part, due to the prospect of deriving technological applications to exploit these mechanisms, putatively in order to heighten human flourishing. Although there is a fair amount of literature focusing on the subject of the meaning of the concept of mechanism as it pertains to the biological field,[5] in a good number of studies employing the term there is often little examination into what exactly is meant by it, nor any discussion of why this concept is paramount.

Darwin's success was that he discovered the chief *efficient cause* of evolutionary change: natural selection, as well as several of its major subordinate vehicles, namely, sexual selection, community selection, and artificial selection, although he was open to the notion that many other factors played a role in evolutionary change. The work of Gregor Mendel (1822-1884) and of later population geneticists (e.g., J. B. S. Haldane, Ronald A. Fisher, and Sewall Wright) served to shed light on the operations of genes and on the fundamentals concerning heredity, cellular differentiation and specialization, the growth of organisms, and of speciation in our contemporary evolutionary epoch. Initially conceived by many to be incommensurable, these two discoveries were merged together in the 1930s and 1940s, constituting the two pillars of the "Modern Synthesis" that characterizes modern biology. Julian Huxley (1887-1975), Theodosius Dobzhansky (1900-1975), George Gaylord Simpson (1902-1984), and Ernst Mayr (1904-2005) were some of the main galvanizing figures behind the consensus that was reached. As Mayr describes it,

> The term "evolutionary synthesis" was introduced by Julian Huxley in *Evolution: The Modern Synthesis* (1942) to designate the general acceptance of two conclusions: gradual evolution can be explained in terms of small genetic changes ("mutations") and recombination, and the ordering of the variation by natural selection; and the observed evolutionary phenomena, particular macroevolutionary processes and speciation, can be explained in a manner that is consistent with the known genetic *mechanisms*.[6]

The modern synthesis firmly established the foundations of modern biology as a fully independent science, even though it inherited many of its core principles from the "older sciences" like physics and chemistry. In the process of merging Darwin and Mendel, there was a general rejection of views "running counter" to these outlooks, such as those found in neo-Lamarckian, orthogenetic, vitalistic, and saltational theories, especially most theories or concepts that involved "metaphysical" principles, for example, internal drives, vital forces, purposes, ends, teleology, and final causality for which it was deemed that there was no empirical evidence. In so doing, among prominent biologists, the modern synthesis led to a "hardening"[7] of perspective in terms of the importance of natural selection and of the laws of genetics as causal factors in explaining evolutionary change, such that they were considered its chief *mechanisms*, to the elimination of (e.g., Darwin's) limited openness to other explanatory factors. As some scholars suggest, the reconciliation and unification of Darwinian natural selection and Mendelian and population genetics also brought with it a "constrictive"[8] and exclusionary reduction of evolutionary phenomena to explanation by way of these "mechanisms." While the term "neo-Darwinism" existed long before the synthesis took place, and its meaning has been the subject of much debate and change, evolutionary psychologists, historians, and philosophers of biology have more recently employed it to designate strict adherence to the core principles of the modern synthesis, in this "hardened" respect.

It is clear that since the modern synthesis, biology has been powerfully served by the employment of the mechanistic lens with which to interpret nature. Admittedly, a great number of its major developments have been initiated via reductionistic, mechanistically oriented research. At the same time, the emphasis on such a lens has not been without detractors. Criticism of the fact that the science that studies life has mechanistic underpinnings, as the title of this volume suggests and which provide the thematic "site" for most of the chapters in it, is nothing new.[9] However, recent trends in what might loosely be called the "New Frontiers" of biology: Systems Biology (see chapters 1 and 3), Emergence Theory (see chapters 1, 2, and 7), Biosemiotics (see chapters 6 and 7), Niche Construction (see chapters 6, 8, and 11), Organic Selection (see chapters 11 and 12), Epigenetics (see chapters 3, 15, and 16), as well as some of the other avenues of research mentioned in this volume, may be said to call for further questioning of the mechanistic outlook that is generally assumed in the mainstream of biology. Some have claimed that the modern synthesis is "unfinished"[10] or "incomplete,"[11] while others have called for an "extended synthesis,"[12] in which the traditional principles and assumptions of the modern synthesis are "revised," "supplemented," "softened,"[13] and/or "relaxed,"[14] so as to accommodate inquiry in novel directions and to assist the progress of biology.

One of the chief criticisms of the employment of the mechanistic lens in biology is that it seemingly offers an inadequate analogy for what it studies: *life*. On the one hand, living organisms have been described, stipulatively, as finite, experiencing, suffering, dynamically enduring, auto-catalytic dissipative-[15], self-organizing-critical-[16], self-creating (*autopoietic*)-[17] systems, with permeable or semi-permeable boundary membranes.[18] Living organisms valuate and selectively appropriate elements and/or data[19] from their surroundings (upon which they depend for existence), which are metabolized[20] (physically, proto-mentally, and/or mentally), promoting internal processes of self-repair, self-renewal, differentiation, reproduction, self-development, and/or growth. According to systems biologist Evelyn Keller (2007), a living organism has been defined as

> a bounded body capable not only of self-regulation, self-steering, but also, and perhaps most important, of self-formation and self-generation. An organism is a body which, by virtue of its peculiar and particular organization, is made into a "self" that, even though not hermetically sealed (or perhaps because it is not hermetically sealed), achieves autonomy and the capacity for self-generation.[21]

But she asks us to further consider that "demarcating organisms in the real world [is] not always easy. Should we think of ant or termite colonies as organisms? Beehives? Coral communities? Humans too are social organisms—should the societies we form be regarded as organisms in and of themselves? Are they also not 'self-organizing'?"[22] Some of these latter questions about what constitutes a singular organism are explored in chapter 10 of this volume.

On the other hand, the concept of mechanism conjures up images of vehicle motors, factory machinery, grandfather clocks, and computers, with nuts, bolts, springs, hinges, gears, and circuits, as well as levers, knobs, dials, switches, and/or programs that determine and control the "sequentially ordered"[23] functioning of the component parts. As defined by William Bechtel and Adele Abrahamsen (2005), a mechanism "is a structure performing a function in virtue of its component parts, component operations, and their organization. The orchestrated functioning of the mechanism is responsible for one or more phenomena."[24] A machine is, in general, a system that carries out prescribed functions on the basis of the interaction of preset internal mechanisms, which are the *efficient causes* involved in its operation. Unlike organisms, a great majority of machines that we know of have been designed, engineered, programmed, orchestrated, organized, and/or assembled from without—via the instigation of a human agent—for the sake of carrying out a closed loop of prescribed output functions (given specific inputs), usually that assist with the satisfaction or the perceived satisfaction of human ends. While machines can potentially be built by other machines and can potentially build new internal mechanisms for themselves, their origin can ultimately be traced back to design by an intelligent organism: one or more human agents (which, in the vast majority of known cases, is representative of both their final and efficient causes). However, in contrast, an organism autonomously produces its own efficient causes inside of itself.[25] As Keller suggests, "What separates organisms from machines is the fact that the organized complexity of the former, unlike the latter, arose and evolved spontaenously. Machines may be self-steering, but even after all these years of mechanical ingenuity, it remains the case that only organisms can be said to be self-organizing in [the Kantian] sense of the term."[26]

Of course, separating the organic from the mechanistic is not all so simple, and given the great advances made in cybernetics, robotics, artificial intelligence, as well as those made in biology which have depended on the mechanistic metaphor, our characterization of machines above will undoubtedly be labeled by some as involving a "straw machine" argument. The metaphor of "the mechanical" admittedly overlaps in many respects with "the organismic," and may be said to provide useful knowledge about life,[27] and the attempt at distinguishing life from machinery pushes thinking to the limit. Nevertheless, given the key differences that have been alluded to above, the organismic cannot be said to be fully *reducible*, without remainder, to the mechanical, and therefore, it ought to be concluded that the study of organisms, and of their functioning, cannot fully depend on an analogy to the mechanical. After all, an analogy depends for its viability on the generalized meaning of a term, and not on special cases that technological advance might one day afford to us. In a parallel manner, a mechanistic method of analysis would seem to involve a selective focus on identifying the efficient causal powers acting on, and within, the phenomena being studied. If the object of investigation is an organism, then to some extent, the mechanistic focus stands in conflict with what it studies. As such, we must ask why, today, the analogy of the mechanical seemingly dominates modern biological discourse in relation to the meaning of life, and has been widely defended as the bulwark of rigorous, objective biology and/or as the very condition for the possibility of biological research in general.[28] Neo-Darwinists suggest that the use of the mechanistic framework is necessary in that it curtails vitalistic and teleological views of life from entering into objective science, in so doing offering a "non-metaphysical" account of nature. However, this claim depends on what we mean by "metaphysics."

Evolutionary Neo-Kantianism and the Nature of Metaphysics

Conrad Hal Waddington (1905-1975) once wrote that when biologists are confronted by metaphysical speculation, they "usually react as though they feel obscurely uneasy," although in his view, a biologist should recognize that his or her metaphysical assumptions "are not mere epiphenomena." On the contrary, Waddington writes that they "have a definite and ascertainable influence on the work he [or she] produces" and that "metaphysical presuppositions may have a definite influence on the way in which scientific research proceeds."[29] Metaphysics is not so simply to be deemed the science that is concerned with "realities beyond nature," namely, a field associated with the type of explanatory "skyhooks"[30] of the sort that Daniel Dennett is so critical of, as is commonly conceived. And it need not be construed as intrinsically onto-theological in character.[31] Rather, the notion of metaphysics can be taken in an "evolutionary neo-Kantian"[32] sense that is fully consistent with evolutionary biology. From this perspective, metaphysics can be considered as a field that has deep roots in evolutionary epistemology, comprising a critical exploration of the guiding ideas by which human beings render the world intelligible, such as substance, self, causality, teleology, and mechanism. From an evolutionary neo-Kantian perspective, these *a priori* concepts of the understanding are to be considered categories of thought that have been selected for over eons of evolutionary time, much like any other advantageous phenotypic trait. Here we must recall that for Immanuel Kant (1724-1804), the *a priori* concepts of the understanding that rational beings bring to the table of experience always concern objects of experience, rather than being divorceable from them, as is commonly conceived.

According to Konrad Lorenz (1903-1989), the Kantian *a priori* categories can be construed as biologically "'inherited working hypotheses' [belonging to the very makeups of rational beings] which have shown their mettle in dealing with the physical world."[33] Such concepts are to be considered "indispensable"[34] habits of thought, i.e., "good [mental] tricks"[35] that provide an advantage in the struggle for existence. To be sure, David Hume (1711-1776) pointed to the notion that the concept of causality *qua* necessary connection had no empirical warrant, due to the fact that the necessary connection that is implied between putative causes and putative effects, is not observable. At best, for Hume, the notion of causality designated an inductively-realized regularity of accompaniment between like As and Bs. That said, Hume agreed that causality *qua* necessary connection is a metaphysical concept that is highly indispensable to humanity. For example, a keen sense of causality *qua* necessary connection provides clear advantages for problem solving, for setting and accomplishing goals, for knowing how to satisfy needs and desires, for predicting consequences, for warning others and heeding warnings about environmental dangers, and for living a functional, successful life. On the contrary, if the concept of causality *qua* necessary connection were somehow jettisoned from our rational natures, there would be no reason not to throw oneself off of a balcony, or to pick up and drop a rock on one's head, or to do things that would subject ourselves to harm or death. As Hume suggests, "All human life must perish, were [the skeptic of causality's] principles universally and steadily to prevail."[36] With it, we consider the potential consequences of our actions, and we refrain from carrying out those that will put us at risk of harm or death, even in relation to new experiences, events, and environments, without having first seen a model. At the same time, an overdeveloped sense of causality might make us so risk averse such that we are unable to act in the world. In any event, the concept of causality, upon which the sciences as well as the category of mechanism depends, provides to us great survival value, even though, according to Hume, the "necessary connection" that is assumed between putative causes and putative effects is not present in perceptual experience. It is an extrapolation (of the highest degree) from what is observable. In *Science and the Modern World* (1925), process philosopher Alfred North Whitehead (1861-1947) noted the irony here by stating that while "some variant of Hume's philosophy has generally prevailed among men of science," science "has remained blandly indifferent to its refutation by Hume."[37]

The lacunae concerning causality *qua* necessary connection and induction, which had been pointed out by Hume, also applied to the notion of teleology. Kant attempted to address them in *Critique of the Power of Judgment* (1790). In a nutshell, we might ask ourselves: how do we know that tadpoles will grow into frogs, or that seeds have the potential to develop into plants? From an "average everyday" way of thinking, teleology in nature would seem to be a given. Children are told that caterpillars will eventually turn into butterflies, that babies will grow into kids, and that if they just are persistent in striving for their life-goals, they will accomplish them. However, upon further investigation, there is no empirical warrant for the necessary connection that teleological thinking posits between the potentialities and actualities involved, and no inductively realized regularity of accompaniment can guarantee the universality of the movement that is assumed. Nevertheless, the concept of teleology has been indispensable for human beings, as can be seen in the human ability to anticipate what organisms will be, or look like, in the future. And it is highly important in relation to pursuing life-sustaining activities like hunting and agriculture. One would not plant seeds if one did not believe that this act would yield a harvest, given favorable environmental conditions. Teleology has been a concept that has been selected for due to its indispensability. It may be a simplification of our experience that cannot be easily jettisoned, notwithstanding the fact that the teleological view of evolution, importing final causality, ends, purposes, progress, unseen vitalistic forces, and deities into the explanatory framework, has, since Darwin's time, been heavily criticized.[38] From these considerations, given that the metaphysical concept of mechanism seems indispensable for the natural sciences, but is dependent on the notion of efficient causality, and hence, is also subject to many of the same problems that Hume pointed out, we might ask whether mainstream evolutionary biology has been selective in admitting it into its explanatory framework while dismissing "common-sense" teleology.[39] Is it appropriate to employ the term "teleology" in describing an organism's process of self-creation? This question is touched upon in section 5 of this book.

At any rate, from an evolutionary neo-Kantian standpoint, such *a priori* concepts can be considered provisional products of human adaptation to the world and of evolution. One might speculate that such concepts have, in part, been "genetically assimilated" so to speak, namely, "fixed [in the character of the rational being] prior to individual experience" and "adapted to the external world for exactly the same reasons as the hoof of the horse is already adapted to the ground of the steppe before the horse is born and the fin of the fish is adapted to the water before the fin hatches."[40] Yet, at the same time, they may be considered to require "releasing events"[41] for them to be realized by the organism, rather than being Platonically innate in us. From an evolutionary point of view, we could speculate that such ideas are not assimilated equally in the natures of all human beings, that they can emerge as a result of processes of development and of intellectual and/or psychosocial selection, and that they continue to be subject to such processes, being refined, critiqued, and thrown creatively into fresh combinations. Such a position holds the door open for rational beings to be plastic and versatile in relation to their conceptual frameworks, so that they can adapt to the exigencies of changing environmental conditions and of world events, rather than putting all our eggs in fixed habits of thought uncritically deemed to be "rigorous."

So, based on the discussion of the evolutionary neo-Kantian perspective outlined above, metaphysics is not somehow divorceable from biology, since metaphysical concepts and frameworks are adaptive products of human evolution relating to survivability. Although it may be "irksome"[42] for many biologists to "do metaphysics," metaphysics simply cannot be escaped. We cannot somehow step outside of ourselves to experience the world in some purer way, or to render it intelligible without the employment of some set of concepts, if, of course, we are to be functional and/or to survive in it. Nor is it possible to embrace metaphysical nihilism. Even if it were possible to embrace a purely descriptive phenomenological approach in biological inquiry (e.g., of a "to the organisms themselves!"[43] variety), having basic metaphysical presuppositions will be unavoidable, even without reference to the fact that all language systems (which render scientific explanation possible) have metaphysical foundations. Since metaphysics cannot be avoided, it is important to take up the task of making logically coherent and adequate metaphysical frameworks explicit, and/or at least (since metaphysical construction is a fallible endeavor) to adopt a multi-perspectival outlook that offers a balance between several competing ones, so as not to be given over to a single inadequate set of ideas, held dogmatically.

Due to its indispensability as a ratiocinative tool, the mechanistic metaphysical framework, that has been in vogue in biology at least since the modern synthesis, has obviously been selected for in the past. But there is no reason why we cannot attempt to challenge its adequacy now, and to begin to think about nature and life in a different way, if, for example, it is impeding a fuller understanding of nature and/or if it is contributing to global problems that may present challenges to human survival and well-being because of the abstractions it creates. After all, from the point of view of evolutionary biology, is that not what most academic, scholarly, and scientific criticism and debate is all about—the attempt to grasp ever more adequate concepts and frameworks that can serve as lenses through which the human organism may interpret the world, so as to heighten its understanding of, and further its adaptation to, it? It would seem that, from an evolutionary point of view, the chief reason for learning and education (and the purpose of schools and universities) is to facilitate our engagement in processes of "psycho-social" or "intellectual" selection in which ideas are appropriated, criticized, judged, modified, defended, and synthesized, with the hope of achieving "breakthroughs to new dominant patterns of organization,"[44] as Julian Huxley articulated.

Alfred North Whitehead, who defined metaphysics as "the science which seeks to discover the general ideas which are *indispensably* relevant to the analysis of everything that happens"[45] once wrote in *Adventures of Ideas* (1933) that "no science can be more secure than the unconscious metaphysics which tacitly it presupposes,"[46] a primary target of his criticism being the mechanistic assumptions embedded in modern biology. Whitehead warned against the "canaliz[ation of] thought and observation within predetermined limits, based upon inadequate metaphysical assumptions dogmatically assumed."[47] As such, through processes of intellectual selection, the concepts belonging to the mechanistic framework are indeed de-centerable, alterable, refinable, and/or replaceable, of course, if there is good reason to do so—in the same way as the concepts belonging to the teleological account of nature have been scrutinized. Moreover, systems biologists have recently called on philosophy to provide foundational concepts in order to assist with the task of studying "living organisms as wholes" and of "what life is."[48] At any rate, in order to better situate ourselves in relation to the origin of the mechanistic philosophy, we turn now to an elucidation of some of its main tenets, according to one of its chief founders, René Descartes (1596-1650).

Descartes' Mechanistic Metaphysics

René Descartes, a father of the scientific method, was, along with Galileo (1594-1642), Robert Boyle (1627-1691), and others, one of the original advocates of the mechanistic metaphysic that was inherited by biology from physics in the late nineteenth and the early twentieth centuries. Inspired by the invention of many mechanical devices by human industry during his time, the basis of his method of inquiry was to consider the entire universe "as if it were a machine."[49] In expressing the sheer explanatory power of the mechanistic analogy he employed, Descartes wrote "there is nothing in the whole of nature [other than thought and mind] . . . which is incapable of being deductively explained on the basis of [the] principles" pertaining to "the shape, size, position and motion of particles of matter."[50] With the use of this metaphor, all change in the universe could be reduced to rearrangements of particles of matter subject to a set of mechanical rules. Descartes thought that if human beings could discover these mechanical rules (which amounted to the laws of nature),[51] then they could "render [them]selves the lords and possessors of nature . . . enabled to enjoy without any trouble the fruits of the earth, and all its comforts, but also and especially [the ability to] preserv[e] health, which is without doubt, of all the blessings of this life, the first and fundamental one."[52]

In *Treatise on Man* (1633), *Discourse on the Method of Rightly Conducting One's Reason and on Seeking Truth in the Sciences* (1637), and elsewhere, Descartes further demonstrated the explanatory power of the mechanistic framework by providing a reductionistic account of a variety of human and non-human physiological functions and activities. The functioning of an organism's autonomic nervous system, such as heartbeat and respiration, was deemed explainable as "'push and pull' operations—operations not in principle any different from the simple workings of cogs and levers and pumps and whirlpools."[53] Furthermore, he thought that a great number of human psychological functions like memory, perception, dreaming, and many internal sensations could be similarly understood. However, ultimately, because of the indefinite range of responses to the exigencies of life that the mind affords to human beings through its capacity for reasoning, Descartes differentiated the human mind from, and privileged it over, material things.[54] By way of their reasoning powers, human beings could discover the chief efficient causes responsible for the operations of nature, and then manipulate these mechanical levers, hinges, and switches, for their own benefit.

From these considerations, Descartes divided the world into two basic types of things: (1) extended, physical, material substances (*res extensa*, i.e., indefinitely divisible things, having length, breadth, depth, thereby not only being spatial but constituting space); and (2) thinking, cogitating substances (*res cogitans*, i.e., indivisible things not having length, breath, and depth). Unlike many other strong advocates of materialism, like Isaac Newton (1642-1727), who believed that atoms represented irreducible, eternal, "solid, hard, massy, impenetrable, moveable particles" that were impervious to "wear" or to "break[ing] in pieces"[55] of which substances were compounded, Descartes remained consistent with his dualism, and did not believe that such fundamental units existed. Rather, he thought that any minute particle would also be divisible.[56] Working with, but obscuring Aristotelian renderings in which substances were seen as dynamic entities, Descartes defined the notion of a substance generally as "a thing which exists in such a way as to depend on no other thing for its existence."[57] However, what Descartes really meant, at least in his formal writings, was that substances had a degree of independent existence, God (the Prime Mover), as he indicated outwardly, being the only truly independent Entity.

Interpreting organisms through the lens of his metaphysical distinction between extended substances and thinking substances, Descartes concluded that animals were devoid of the thinking substance. For him, animals were "nothing other than mechanized, and wholly non-conscious, extended substances."[58] While in some of his later letters, Descartes wrote of animals expressing feelings such as joy, fear, and hope, seemingly implying that they had a degree of mentality, in his formal writings, Descartes interpreted the behaviors of animals basically as stimulus-responses stemming from the external movements of matter acting on them. He wrote that animals "have no intelligence at all, and . . . it is [mechanical] nature which acts in them according to the disposition of their organs."[59] Non-human organisms were, for Descartes, to be considered mindless machines or automatons, entirely conditioned in their operation by external material forces acting upon them.

In relation to human beings, for Descartes, the human body was to be considered a machine that was "inhabited" by a mind.[60] Yet, Descartes did not resolve how mind and body were interconnected, although he did state that they were "closely conjoined."[61] Only later in his life did he entertain an interactionist standpoint in relation to the mind-body problem. Interestingly enough, he deployed the machine analogy in describing the difference between the body of a living person and a dead one, stating,

> and let us recognize that the difference between the body of a living man and that of a dead man is just like the difference between, on the one hand, a watch or other automaton (that is, a self-moving machine) when it is wound up and contains in itself the corporeal principle of the movements for which it is designed, together with everything else required for its operation; and, on the other hand, the same watch or machine when it is broken and the principle of its movement ceases to be active.[62]

The healthy body-machine was one in which the organs-parts successfully carried out their designed function, whereas the unhealthy body-machine was one that was corrupted, such that it was unable to carry out its functions. On this view, a doctor could be construed as a "mechanic of the body," fixing the body's internal dysfunctions, ensuring its proper operation (without reference to external nature). Indeed, in 1646, after reading Descartes' *Discourse*, Dutch physician Cornelis van Hogelande (1590-1676) proclaimed that "all bodies, however they act, are to be viewed as machines, and their actions and effects . . . are to be explained *only* according to mechanical laws."[63] In his formal philosophical works, Descartes maintained that minds and bodies were designed and controlled by God in their motion (being their final cause), the universe being a giant machine created by the Divine Mind. Much later, in *Natural Theology* (1802), William Paley (1743-1805) offered up the design argument for the existence of God, based on the nature-as-machine metaphor, which has more recently been critiqued by Dawkins in *The Blind Watchmaker* (1986).[64]

Cartesian Reductionism versus Holism and Emergentism

The apparent opposition between reductionism and holism is one of the chief issues in the search for novel research methods in contemporary systems biological research. On the one hand, while there are many different forms of reductionism, reductionism is exemplified by Descartes' strategy of interpreting the world as *nothing but* a machine (although one could equally carry out an explanatory "reduction" to something else).[65] Undoubtedly, reductionist methods have provided the biological sciences with an indispensable and "powerful research methodology, a clear epistemological guiding principle for acquiring knowledge, an effective strategy for judging the quality of manuscripts and grant applications and a consistent view of the world."[66] However, on the other hand, holism is a philosophical standpoint that seeks to recognize the limits of reductionism. It involves a focus on the notion that nature is a complex interdependent system, which, to some extent, resists being fully explained by reductionist methods of the "x is nothing but y" sort. As Julian Huxley states, "We must beware of reductionism. It is hardly ever true that something is 'nothing but' something else."[67]

To illustrate the contrast between reductionism and holism, one of the chief principles Descartes identified "for the direction of our native intelligence" (i.e., "for the direction of our minds scientifically") is stated as follows: "Rule Thirteen: if we [are to] perfectly understand a problem we must abstract it from every superfluous conception, reduce it to its simplest terms and, by means of an enumeration, divide it up into the smallest possible parts."[68] In other words, we ought, first, to narrow our focus and to delimit the phenomena to be explained in terms of local region in space and time. We should partition it off from its total environment, and exclude a consideration of "extraneous factors," namely, those factors outside of its boundaries. Quite crudely, on this reductionist view, if we want to understand the functioning of an organism, a "precise analysis" will involve taking it out of its ecological context, rendering it into a corpse, slicing it open, and dissecting it, revealing the workings of its internal mechanical structure.

On the contrary, in confronting this first aspect of the reductionist method, a holist would suggest that one creates abstractions in understanding when one omits a consideration of the ecological context within which the organism lives. Accordingly, contemporary systems biologist Jan-Hendrik Hofmeyr states that "nothing in an organism makes sense except in the light of context . . . always taking context into account amounts to using a 'macroscope'," that prevents the "interacting elements that form an integrated whole"[69] from being excluded in the explanation. Furthermore, in examining a dead body, a holist would point out that one is not examining a *living* organism. Hans Jonas (1903-1993) pointed to the notion that such reductionistic thinking is "under the ontological dominance of death."[70] And, from the holistic perspective, as opposed to mechanistic reductionism, the fact that organisms are finite, physical creatures that are subject to the conditioning influences of their environment, does not mean that they are simply reducible to machines.

The second task Descartes is prescribing in Rule Thirteen is that phenomena and/or complex problems are to be broken up, or decomposed, into smaller, more manageable, more basic, parts for the sake of analysis. The assumption here is that "all apparent 'wholes' are in principle reducible to their parts,"[71] and that in "finding the parts that construct the whole, we [can] explain everything about the whole, including how the whole functions."[72] On this view, for example, everything about a multicellular organism can be explained through an examination of the various specialized cells, and perhaps an exploration of their various relations. However, from a holistic perspective, one problem here is that multicellular organisms are not merely conglomerations of specialized cells working together, but they also have what some holists call "emergent properties"—they have meta-cellular functions. For example, the heart's pumping of blood, which circulates oxygen and nutrients to the whole body, is not simply explainable by examining the cells making up the heart, for organs are integral wholes that carry out their functions only on the basis of this integrity. The function of pumping blood is not entailed merely by having heart cells, or by having half of a heart. Furthermore, at the "higher level" the organism engages in activities, behaviors, and ways of being that are meta-cellular, such as hunting, using tools, reading a book, attending to another organism, being in love, constructing a 747 airplane, or consciousness in general. While the functioning of the cells and organs enables the organism to carry out such activities and there are definite biological repercussions for the cells and organs as a result of the organism carrying them out, these meta-cellular functions cannot simply be explained by way of a focus on the operation of the cells. Accordingly, Julian Huxley states that "an organization is always more than the mere sum of its elements, and must be studied as a unitary whole as well as analysed into its component parts."[73]

This form of holistic thinking is highlighted by *emergentism*, a mode of thought championed by figures such as Conway Lloyd Morgan (1852-1936) and Samuel Alexander (1859-1938). (See chapters 1, 2, and 7 of this volume for a discussion of Emergence). In *Emergent Evolution* (1927), Morgan characterized the emergence perspective as a "protest against . . . an uncritical acceptance of what is sometimes spoken of as 'the mechanistic dogma'," in which life is interpreted merely "in terms of physics and chemistry."[74] Emergentism involves the notion that complex "higher order" systems arise out of the confluence of lower-order conditions, each layer of the natural world emerging out of the set of relations belonging to the previous, "lower" one. While the confluence of the "lower order" developments provides the conditions that make it possible for the "higher order" complexities to emerge, the "higher order" systems are distinctively novel in that they contain "supervenient" or "emergent properties" that make them *something more* than just the sum of their constituent "lower order" parts. While Morgan employed the terms "higher" and "lower" in a figurative way (i.e., they do not mean "better" and "worse" respectively), he charged that the "mechanistic interpretation" of the reductionist variety generally "ignores the something more that must be accepted as emergent."[75] Alexander described the idea in *Space, Time, and Deity* (1920), "The higher quality emerges from the lower level of existence and has its roots therein, but it emerges therefrom, and it does not belong to that level, but constitutes its possessor a new order of existent with its special laws of behavior."[76] Emergentists also focus on the possibility of supervenient properties, in turn, interacting with the "lower level" parts (i.e., "downward causation"), as is hypothesized, for example, in the theory of Organic Selection (i.e., the "Baldwin Effect"), which maintains that changes in terms of an organism's behavior, or dominant structures of activity, can instigate morphological evolution (see chapters 11 and 12 of this volume).

Some holists see evolution as a gradual emergence in which preceding forms of life of "lower complexity" provide the conditions for the possibility of succeeding ones of "higher complexity" (with reference to extinction events, of course) as can be seen in the progression and great fan out of life from prokaryotes to eukaryotes to multicellular organisms to plants, animals, to insects, reptiles, birds, and mammals over the last 3.8 billion years. Accordingly, for example, as Stephen Jay Gould (1941-2002) speculated, the survival of *Pikaia gracilens*, a primitive Pre-Cambrian chordate, through the Burgess Decimation, may have provided an evolutionary opportunity for a "chordate future,"[77] establishing the conditions for the possibility of vertebrates, including human beings.

A third aspect of reductionism involves the notion that one domain of inquiry or science "reduces" to a more "fundamental" one. For example, the humanities and the social sciences are said to be reducible to biology, since the phenomena that are studied in economics, ethics, psychology, and education are deemed ultimately to be a function of biological processes, such as human organisms finding ecological niches to exploit as well as heightening their inclusive fitness and the prospects of reproductive success. At the same time, biology is said to fully reduce to chemistry in that organisms are biochemically composed (for instance, that psychological stress is explainable as a cortisol imbalance in the brain). In turn, chemistry is said to fully reduce to physics, since the chemical is a function of the material. In this way, explanatory reductionism holds that the phenomena that are studied by the disciplines at the top of the hierarchy are ultimately explainable by the ones at the "base" of the order of nature. As Daniel Dennett suggests, "After all, societies are composed of human beings, who, as mammals, must fall under the principles of biology that cover all mammals. Mammals, in turn, are composed of molecules, which must obey the laws of chemistry, which in turn must answer to the regularities of the underlying physics."[78]

Holists employ a similar reasoning to that mentioned above regarding emergent properties in offsetting this form of explanatory reductionism—while emergent properties on the "higher rungs" are enabled by what has transpired on the "lower rungs" of the totem pole of knowledge (i.e., the "lower" in general providing the conditions for the possibility of the "higher"), they are not fully explainable, without remainder, through them. To be sure, as several systems biologists point out, "although molecules constitute living systems, they are by themselves not alive. No matter how thin the dividing line is, there is a qualitative jump between a living and a nonliving system."[79] Considering that biology, in general, occupies an intermediary level in this hierarchy, it is obviously of great explanatory advantage to be skillful at identifying connections upward and connections downward. And one might also hold open the possibility that physics could be "revived by the infusion of life" in biology.[80] In the next sections of this chapter, I work toward emphasizing that a phase of holistic reflection ought to accompany biological research.

Mechanism, Technology, and a "Life-Blind Spot"

The full thrust of mechanistic thinking of the reductionist variety (i.e., that which reduces nature and living organisms to machines) does not stop at the scientific explanation of natural phenomena. Research that is based in the mechanistic framework is, in many cases, given over to technical-rational thinking and motivated by the prospects of the technological and biotechnological applications it seemingly secures. By uncovering the set of mechanisms that are deemed to underlie natural phenomena, it is entailed that such mechanisms are there to be manipulated, i.e., set up in advance in order to bring about a predetermined result.[81] In other words, the mechanistic lens presents nature as a series of levers, switches, and dials that are at-the-ready for instrumental manipulation through technological and/or biotechnological applications. In that mechanistic explanations are a necessary condition for the heightening of human power and control over the natural world by way of the technological and biotechnological applications they enable, the mechanistic metaphysical framework is deemed to be indispensable to human existence. However, today, the mechanistic framework is marred by its contribution to ecological dysfunctions that pose threats to the well-being, the survival, and the evolutionary futures of contemporaneous life-forms on this planet.

As bio- and environmental- ethicists who are concerned for the integrity of ecosystems as well as for the humane treatment of non-human organisms, the editors of this volume believe that there are drastic consequences for most life-forms on this planet, including ourselves, if we do not de-emphasize the mechanistic framework, and at the very least, cultivate the value of an engagement in a phase of holistic reflection in scientific research that lays bare the abstractions which are created by mechanistically oriented research. There is a "blind spot" in our understanding of the natural world when looked at solely through the mechanistic lens. And it is largely because of this "blind spot" that human beings continue to develop and employ technologies that may seem to heighten their power, render their lives easier, and improve their well-being in the short term, but which unwittingly do substantial harm to the biosphere's life-support systems and/or to the diverse life-forms which help to compose it. To be sure, on the basis of mechanistically oriented research of the reductionist variety, which views living creatures as machines (without remainder) and hence, there to be manipulated, human beings, often driven by commercial interests, have sought to maximize their control over the natural world by way of the genetic, physiological, and behavioral manipulation of organisms.

In many cases involving experimentation on non-human animals, such manipulations (1) fail to respect *living* creatures as integral wholes with intrinsic worth, (2) inflict suffering and death on them, (3) cause ill health, and (4) decide the evolutionary destinies of the organisms involved. The selective cloning of domesticated animals is deemed to provide great benefits for commercial agriculture. However, it has a low success rate, and produces many organisms with abnormalities and increased susceptibility to pathogens and diseases, phenomena which are likely to be, in part, due to the fact that the chromatin modifications that occur through epigenetic reprogramming in normal embryonic development do not take place in Somatic Cell Nuclear Transfer.[82] The biotechnologist who is peering through the lens of the microscope, and using a pipette to slurp out nuclear material of an embryo in order to replace it with that derived from a somatic cell, is, most probably, at that moment, indifferent to the long-term consequences for the organisms involved and for the long-term hereditary effects of the procedure.

Given the fact that, as holists claim, nature is a complex interdependent system, many technological and biotechnological manipulations also risk scattering "unintended," "unexpected," "unpredicted," or "non-targeted" effects. For example, the development, deployment, and commercialization of genetically modified crops for the purpose of rendering them more resistant to herbicides and pesticides, have had unintended effects on the host organism, and have unexpectedly cross-pollinated with non-GM strains.[83] Systemic pesticides and GMOs are also a chief suspect in relation to the unintended collapse of insect populations, such as honeybee colonies, on which we depend for pollination.[84] And there are well-reasoned concerns about the unintended health hazards that GM foods pose.[85] Moreover, farmed salmon and GM salmon (which are crosses of ocean pout and salmon), shedding viral particles into wild habitats, have recently been linked to the contraction of infectious salmon anemia and sea lice in wild populations.[86] Another case is presented at length in chapter 16 of this book. It explores some of the ethical issues and the potential human health dangers in terms of the potential "non-targeted effects" that may be produced as a result of the use of epigenetic drugs, in view of the mechanistic understanding of the epigenome that is being emphasized in biological research today. In this light, *the ecological degradation being perpetuated by the continual, indefinite scattering of "non-targeted" effects throughout the biosphere, as a result of the employment of technologies and biotechnologies that originate directly from mechanistically-oriented biological research of the reductionist variety, provides proof of the fact that the mechanistic lens has a great blind spot in terms of its ability to provide a comprehensive understanding of organic nature.*

Mechanistically oriented researchers in whom this metaphysic is very deeply entrenched may respond to the above claim by suggesting that all we need to do in these instances is to continue to inquire and to discover the mechanisms behind such unintended effects in order to "fix" them, without having to reflect on the adequacy of the mechanistic metaphysic to provide a comprehensive understanding of life. They may claim that the creation of these problems only represents the current limitations of the technology in question and that we can expect that, with time and progress, such problems will inevitably be resolved and/or overcome. They may think that if we somehow create an "error" in manipulating the mechanisms controlling the operations of the natural world, then through human technological-rational ingenuity, it will be assumed that we can just "pull a different lever" or "flip a different switch" in order to "fix" that error. To be sure, the current emphasis on, and heavy financial investment in "geo-engineering projects"[87] in the hopes of mitigating global warming and climate change (e.g., pumping sulfur dioxide into the atmosphere to reflect more of the sun's radiation back into space; or "iron fertilization" involving dumping iron into the oceans so as to stimulate the growth of phytoplankton and other organisms that remove carbon dioxide from the atmosphere), instead of placing emphasis on education to promote lifestyle changes that will reduce greenhouse gas emissions at the outset, are a stark case in point.

The lesson, according to holists, is that we need to realize the profound implications of the fact that the complex interdependent system that is nature is not just comprised by a network of efficient causal mechanisms that can be manipulated without ecological consequences. The message of the holists is that, rather than merely amassing "biological knowledge" which will lead to ever-increasing "technological control of nature," we ought to cultivate what Conrad Hal Waddington called "biological wisdom."[88] Accordingly, Craig Holdrege of the *Nature Institute* says,

> To ask, "Do we understand what we are doing?" is not merely to ask whether, in some narrow sense, we understand the genetic "mechanisms" involved. Rather: Do we truly fathom the consequences of our actions? We are producing hereditary alterations in organisms, with all their consequences for the life of the organism itself and its environment, and the array of unintended effects can alert us to the far-reaching impact we are having on *life*....[89]

There is a *life-blind spot* in the mechanistic lens of the reductionist sort. It is not just that there are "many more mechanical moving parts" to nature that need to be taken into account. Rather, nature itself is not divided up into the substances—the putative efficient causes and their effects—that we tend to assume it is both in our thinking and in our language. In that nature is a complex interdependent system in which biological processes permeate all physical barriers erected and conceptually posited boundaries, the non-targeted "effects" of our technological manipulations are indefinitely scattered throughout nature, in the interstices between our conceptually posited "substantial objects."

All of this is not to place blame on biology, biologists, biotechnologists, scientists, or engineers, for creating or contributing to ecological problems. Rather, the intent of this analysis is to shed light on the relation between our metaphysical and epistemological commitments—comprising the ways in which we interpret the natural world and see our place in it—and the perpetuation of anthropogenic ecological problems, which are everyone's fault and everyone's problem to deal with. Furthermore, in so far as metaphysics is the domain of philosophy, it is largely a failure of philosophy that it has not done more to provide the natural sciences with more adequate conceptual frameworks with which to study life. Philosophy of biology as a coherent discipline has only existed for about forty years, and both of the major traditions of philosophy in the twentieth century—Analytic and Continental—largely repudiated the construction of metaphysical schemes. Two exceptions were Alfred North Whitehead's process-relational philosophy of organism, which provided much of the conceptual ground for Conrad Hal Waddington's researches[90] (as are alluded to in chapters 15 and 16 of this volume) and Justus Buchler's (1914-1991) metaphysics of natural complexes, both of which are referred to in the various chapters in this volume. (For a substantive treatment of the latter, see chapter 5). The editors of this volume do not think that these metaphysical systems provide, in and of themselves, complete solutions to the problems posed by the mechanistic understanding, but they perhaps provide good starts, and models for what may currently be needed. That said, one of the chief factors that "led to the diminution of Waddington's stature" alongside his neo-Darwinist colleagues, such as Ernst Mayr, was his support for Whiteheadian process-relational metaphysics, since they simply "denied the relevance of their own implicit metaphysical commitments to their scientific theories."[91] Waddington had also maintained that "the depletion of natural resources, hazardous pollution, and other threats to human flourishing could be found . . . in the pervasive worldview—extrinsic to evolution, but intrinsic to neo-Darwinism—that 'everything is molecules and nothing but molecules'."[92]

The heuristic analogy of mechanism and the language of mechanism will probably be central in biology for the foreseeable future, and it is indispensable in that it perhaps enables biology to do its work most "easily." At the same time, alone, it creates abstractions which frame a worldview that is geared toward technical-rationality, without considering consequences for the biosphere as a whole. There is a "blind spot" in mechanistic analyses of the reductionist sort, when a phase of holistic reflection in scientific research is not engaged (something which has very little cost, if any, to undertake). Given the potentially drastic consequences of environmental problems today, it is in our interest to be plastic in respect to our metaphysical frameworks through which we interpret the world, rather than employing the mechanistic one as a fixed habit. The whole point of this introduction is to argue that we ought to include a phase of holistic reflection in biological research, whereby the abstractions in understanding created as a result of our heuristic metaphors are laid bare. In order to cultivate a holistic, organicist perspective more fully, I turn now to an elucidation of some aspects of Alfred North Whitehead's process-relational philosophy.

Whitehead's Process-Relational Metaphysics

Process-relational philosopher, Alfred North Whitehead, was inspired by the work of emergence theorists, Alexander and Lloyd Morgan,[93] which provided a model for his central notion of "concrescence," or the "growing together of actual occasions," by which "the many become one, and are increased by one"[94] (representing their "supervenient properties"). Whitehead also sought to provide a criticism of the dominance of the mechanistic framework in the biological sciences. He wrote that while "final causality" had been "wildly overstress[ed]" in the Middle Ages, "efficient causality" was being "correlatively overstress[ed]" in the modern scientific period, as exemplified by the dominance of the mechanistic framework.[95] While Whitehead thought that it was a very "baffling"[96] task to fully grasp the nature of *life*, in essence, it was a "category mistake"[97] for biologists to employ the analogy of mechanism in a reductionist or "nothing-but" way in order to explain natural phenomena. He also wrote that the mechanistic lens of the reductionist variety confined "nature under an abstraction in which all reference to life [is] suppressed."[98] Rather, for him, biology should seek to guide its researches and to express itself with reference to categories more "proper to the organism"[99] itself.

In his philosophical writings, Whitehead attempted to articulate what the mechanistic metaphysic is lacking in its description of organic nature: it wrongly postulates that the fundamental entities of the world are "static," "simply located," and defined by their external relations, rather than being considered "processual" and internally related[100] to one another. For him,

> A thoroughgoing evolutionary philosophy is inconsistent with materialism. The aboriginal stuff, or material, from which a materialistic philosophy starts, is incapable of evolution. This material is in itself the ultimate substance. Evolution, on the materialistic theory, is reduced to the rôle of being another word for the description of the changes of the external relations between portions of matter.[101]

According to Whitehead, the mechanistic model wrongly presupposes that the natural world is composed of material substances, which are deemed to be the only "real" things. These material substances are considered to be static, self-sufficient, self-enclosed, namely, dependent on themselves for their existence *à la* Descartes. In addition, it is assumed in the mechanistic framework that the spatio-temporal position of any of these material substances can be determined without reference to their relations to other entities occupying other regions of space and to other durations of time. On this view, the entities can be adequately "described without reference to the goings on in any other region of space,"[102] which may provide some insight into the reasons why "non-targeted effects" are scattered by technological applications stemming from the mechanistically oriented research, as mentioned in the previous section. For Whitehead, the assumptions that the entities composing the natural world are "simply located," that is, able to be adequately defined *at an instant* (i.e., devoid of reference to their temporal character), to the elimination of their relations with other entities, and in a disembodied way (i.e., without reference to a percipient), point to what is chiefly problematic with the mechanistic model of the reductionist variety.

To better illustrate the contrast between external and internal relations, in respect to ordinary conceptions of causality it is assumed that one material substance *acts on* another in producing its effect. Here, the one substance is considered to be external to, or separate from, the other, leaving the percipient, David Hume, with no impression of a necessary connection between the putative cause and the putative effect. However, Whitehead's criticism involves pointing out that causal interactions are reciprocal (the putative cause, in turn, is affected by the substance acted), and that, in truth, the two substances are *primitively interrelated*, since, as a continuum of extension, there are no real boundaries in the natural world (not that Whitehead would be satisfied with the metaphysical concept of substances to begin with). That is to say, there is equally no impression of a necessary separation or division between them. They are internally related in the context of the causal event. Kant makes a very similar point in the *Prolegomena to Any Future Metaphysics* in attempting to deal with Hume's skepticism of causality *qua* necessary connection.[103] And while the data that we perceive do, to a certain extent, lend themselves to be conceptually divided in the manner that is assumed in the mechanistic way of thinking, i.e., into externally related material substances, Whitehead warns that science must recognize the abstractions that are implicit in this mode of objectification.

Consider the nature of power. Power is vacuous unless there is something or someone that the entity that is deemed "powerful" has sway over. The concept of power requires a reference to relationality. As such, interpreting power merely in terms of how much an entity exhibits itself as a self-sufficient substance, as we may typically do under the mechanistic rubric, is logically incoherent.

Whitehead not only offered a critique of the mechanistic philosophy, but he constructed an alternative metaphysical scheme in which "the order of nature is bound up with the concept of nature as the locus of organisms in process of development."[104] Defending a position of "organic realism,"[105] Whitehead dispensed with Descartes' substance-ontological view of the natural world, and redefined the basic units out of which it was composed. He employed the concept of "events" or "actual occasions" in order to better reflect the organic, relational, temporal, processual, and emergent characteristics of the entities therein. Actual occasions were, for him, evolving "drops of experience, complex and interdependent"[106] that are reciprocally interactive and, in part, constituted by their relations with other entities. In addition, not only are actual occasions, for Whitehead, co-dependent with their percipients (or more accurately, the subjects "prehending" them), but the experiencing organism is itself an actual entity (or, more precisely, can be described as a "society"[107] of actual entities). Furthermore, in an event, the relationships between component entities "are internal" and "constitutive of what the event is in itself."[108] Accordingly, Whitehead writes that "the philosophy of organism is mainly devoted to the task of making clear the notion of 'being present in another entity',"[109] which is something that he notes is rejected in the traditional Aristotelian and Cartesian substance ontologies, as well as in the subsequent Newtonian mechanistic metaphysical framework.

Whitehead's process-relational cosmology is based on the notion that the natural world "is not made up of independent things, each completely determinate in abstraction from all the rest,"[110] as is assumed in the mechanistic metaphysical framework. Rather, he espouses a worldview of organic interdependence, in which the entities of the natural world are considered to be complexly interrelated. To illustrate this holism, in 1925, Whitehead wrote of the rich organic interdependency of, and the reciprocal interactions between, the diverse species, trees, elements, and weather patterns found in the Brazilian rainforest. He stated,

> The trees in a Brazilian forest depend upon the association of various species of organisms, each of which is mutually dependent on the other species. A single tree by itself is dependent upon all the adverse chances of shifting circumstances. The wind stunts it: variations in temperature check its foliage: the rains denude its soil: its leaves are blown away and are lost for the purpose of fertilization. You may obtain individual specimens of fine trees either in exceptional circumstances, or where human cultivation has intervened. But in nature the normal way in which trees flourish is by their association in a forest. Each tree may lose something of its individual perfection of growth, but they mutually assist each other in preserving the conditions for survival. The soil is preserved and shaded; and the microbes necessary for its fertility are neither scorched, nor frozen, nor washed away. A forest is the triumph of the organization of mutually dependent species. Further a species of microbes which kills the forest, also exterminates itself. . . . Every organism requires an environment of 'friends', partly to shield it from violent changes, and partly to supply it with its wants.[111]

In sharp contradistinction to overt reductionism, Whitehead writes that in scientific explanation, "the essential connectedness of things can never be safely omitted."[112] Every entity, or phenomenon studied, requires its environment in order to exist, and every environment is comprised by an indefinite multitude of interdependent factors, each conspiring together to bring about what is. He writes that "nature is divisible and thus extensive" but warns that "any division, including some activities and excluding others, also severs the patterns of process which extend beyond all boundaries."[113] That is to say, our carving up of reality through our sharp conceptual negations, selections, judgments, exclusions, eliminations, divisions, discriminations, bifurcations, taxonomies, classifications, and expressing them in language for the sake of precise analysis abstracts from the total picture of the natural world. In a similar sense, we tend to lose sight of the whole when we rely solely on mechanistic language and explanations of the reductionist variety.

Neo-Darwinian biology, interpreting the natural world on the basis of a mechanistic lens of the reductionist variety, can definitely be said to have emphasized external relations and to have neglected internal relations, for example, in having postulated a strict separation of the germ line from the somatic line, in suggesting that genes code for proteins with no reciprocal action from phenotypes to genotypes (see chapter 16), and in considering phenotypes as "mere machines" for the replication of genes. As Evan Thompson (2007) states, "according to genocentrism, organisms evolve as elaborate contraptions—'robots' or 'survival machines,' as Dawkins calls them—constructed and controlled by genes."[114] Neo-Darwinism has also emphasized how the individual organism competes against another in the struggle for existence, to the rejection of more ecologically oriented and "societal" conceptions of the organism (see chapter 10), which involve emphases on organic interdependence and on cooperative behavior (e.g., involving "community" or "social" selection as well as "kin" selection which contribute to the organism's inclusive fitness). Neo-Darwinism has promoted a view of organisms as objects upon which natural selection acts, to the neglect of the notion that they are also subjects or agents of selection (see chapter 12). Furthermore, neo-Darwinism has emphasized that the meaning of evolution involves changes in gene frequency and morphological change to the neglect of other factors, such as changes in respect to dominant behavioral habits and structures of activity.[115]

While being highly critical of the overemphasis on the mechanistic conceptual framework in describing organic nature, Whitehead did not simply seek to purge it outright from his system. While he preferred the term "philosophy of organism,"[116] he did use the term "philosophy of organic mechanism"[117] in several places in *Science and the Modern World* as a descriptor of his metaphysical framework. This would seem to indicate that, at the very least, he was seeking to challenge our occupation with it as a singular and narrow explanatory focus. For Whitehead, the mechanistic framework

> is not wrong, if properly construed. If we confine ourselves to certain types of facts, abstracted from the complete circumstances in which they occur, the materialistic assumption expresses these facts to perfection. But when we pass beyond the abstraction either by more subtle employment of our senses, or by the request for meanings and for coherence of thoughts, the scheme breaks down at once.[118]

The interpretation of the natural world through the heuristic machine analogy provides useful knowledge, just as systems biologists emphasize that their explanations are "often mechanistic explanations."[119] However, Whitehead sought to interpret nature via organismic categories, rather than merely mechanistic ones. Additionally, Whitehead realized that scientific explanation in general required reductionism as part of its method, and that apart from it, we would fail to penetrate into the dominant characteristics of things. That said, he did not think that reductionism should occupy all phases of scientific research.

Admittedly, the proposal to include a phase of holistic reflection into the scientific method is not in line with today's dominant reductionist research agendas. Holisms have, in the interests of "rigorous, objective research," traditionally been dismissed, and have typically, in the process, been characterized as "pseudo-scientific" in nature. Nevertheless, Whitehead emphasized the importance of holistic reflection and speculation in the research process. Rather than merely observe nature through the narrow "arrow slit" of reductionism, he affirmed holistic reflection and speculation. Speculation, with its etymological root, *specula*, meaning a watchtower, enables us to have a wider, panoramic vantage point on reality, and, for example, to anticipate problems that might issue from technological applications stemming from mechanistic science of the reductionist sort. Whitehead postulated that the research process, like other forms of learning, takes place best in the context of a rhythm of cyclical phases.[120] In a nutshell, in addition to phases of (1) curiosity (or "romance" as he called it, whereby a problem or set of phenomena that needs explanation is identified, hypotheses are generated, and a plan for investigation is initiated) and (2) reductionist analysis (or "precision" as he termed it, in which there is an objectification of research objects, observation, experimentation, data-gathering, as well as a comparison and an evaluation of data), Whitehead also suggested that a phase of (3) holistic and speculative reflection (or "generalization" as he described it) was important, in order to synthesize competing claims, to place findings in relational context, and to lay bare the abstractions created as a result of the phase of precision. The phase of generalization, for Whitehead, ignited the creative spark that could lead a researcher back to the phase of curiosity or "romance," which would reinitiate the whole process of inquiry. According to Whitehead, the rhythm of research was not to be seen as an implicitly linear sequence of stages. Rather, each of these phases, for him, was to be emphasized equally, namely, given its due season in research, although there could be minor eddies of any of the phases occurring in the others.

In relation to the importance of a phase of holistic reflection in scientific research, Whitehead warns that we risk committing what he calls *the fallacy of misplaced concreteness* when we confuse our constructed, conceptual schemas, models, and analogies with the way things are, and when we neglect "the degree of abstraction involved when an actual entity is considered merely insofar as it exemplifies certain categories of thought." He continues, "there are aspects of actualities which are simply ignored [or neglected, when] . . . we restrict [our] thought to these categories."[121] A phase of holistic reflection provides checks and balances in relation to the misplaced concretenesses that are created by the mechanistic framework of the reductionist sort.

While, as has been discussed above, we cannot go without metaphysical concepts, from a Whiteheadian perspective, researchers ought to cultivate thinking of the holistic sort and to recognize the abstractions created, for instance, by the assumption that the world can be adequately described with reference to selected metaphysical concepts, such as substance, accident, causality, essence, teleology, matter, and mechanism. Elsewhere, he attempted to articulate the nature of the blind spot that can be created as a result of overemphasizing a narrow, reductionist focus in science, by writing:

> The topic of every science is an abstraction from the full concrete happenings of nature. But every abstraction neglects the influx of the factors omitted into the factors retained. Thus a single pattern discerned by vision limited to the abstractions within a special science differentiates itself into a subordinate factor in an indefinite number of wider patterns when we consider its possibilities of relatedness to the omitted universe.[122]

In sum, Whitehead thought that the reduction of phenomena down to their basic building blocks so as to analyze and to understand them is an essential phase of scientific research. Reductionism is necessary if we are to explain any phenomenon adequately; it belongs to the established methodology of the natural sciences. But, his point was that reductionism is a phase in scientific analysis, not the entire operation. After performing a rigorous analysis of some phenomenon, we should reconnect our findings with the complex interdependent system that is nature, and should attempt to realize the abstractions that have been created as a result of the methods of inquiry, as well as the limitations of the concepts that have been employed in the process. We should ask questions with an orientation toward the relational whole, such as:

- Because nature as a whole is a complex interdependent system rather than a set of self-contained, mechanical, material substances, what elements have been left out (in terms of the integral whole) in the analysis, and what abstractions have accrued as a result of the reductionist methodologies employed?
- What conceptual frameworks and/or metaphysical assumptions obscuring the relational interdependence of nature have been present in the analysis?
- How do the findings of such research connect to the findings of other disciplines and to the whole picture of reality that our patchwork of knowledge seeks to paint?
- How will the application of this scientific research via technology and/or biotechnology impact on living organisms, and on other regions, elements, and levels of the complex interdependent system that is nature, now and in the short- and long-term evolutionary future?
- Given that nature is a complex interdependent system, what ethical concerns come to light as a result of the recognition of the abstractions of reductionism when viewed in the light of the more holistic perspective?

The Intent and Contents of This Volume

While much of this introduction has focused on the relationships between biology, metaphysics, ethics, and process philosophy, attempting to prime the reader for what follows, admittedly, most of the chapters in this volume do not touch explicitly on ethics, metaphysics, or process philosophy. Rather, they explore what might be called "the New Frontiers" of biology, that is, contemporary areas of research and recent developments in biology which appear to call for an updating, a supplementation, or a relaxation of some of the main tenets of the modern synthesis. Examples of these "New Frontiers," which provide the general frame in terms of the organization of this volume, include: (1) Emergence and Systems Biology, (2) Biosemiotics, (3) Homeostasis and Symbiogenesis, (4) the Theory of Organic Selection (i.e., the "Baldwin Effect"), (5) Self-Organization and Teleology, and (6) Epigenetics. To facilitate the reader's acquisition of a well-rounded grasp of these "New Frontiers," including their ramifications for biology as a whole in its quest to understand the nature of life, as well as for the public perception of (e.g., evolutionary) biology, an attempt has been made in each section to pair the elaborations of scientists and specialists with those of philosophers.

While no overarching consensus has been reached among the authors of the chapters that are contained in this volume (in some cases there are conflicts between them), and while there is no single proposal that binds each of the sections and chapters together, two common themes among them are: critical reflection in relation to the neo-Darwinist paradigm in biology, and the cultivation of novel avenues of investigation in relation to the nature of life, which deviate, to some extent, from this paradigm. It has been said that new discoveries and developments in the human, social, and natural sciences hang "in the air"[123] prior to their consummation. It is hoped that the chapters of this volume will contribute to a movement to update, to supplement, and/or to relax some of the main tenets of the modern synthesis, in order to promote a greater degree of inclusivity in the biological sciences.

Acknowledgments

The idea of putting this book together was generated at the "Beyond Mechanism" Consultation, which was held July 19-21, 2010, at Claremont Graduate University in Claremont, California, USA. The Consultation brought together Bruce Weber, Philip Clayton, Jesper Hoffmeyer, Robert Ulanowicz, Scott Turner, Lawrence Cahoone, John Cobb, Spyridon Koutroufinis, William Grassie, Brian Henning, and Adam Scarfe. The Consultation was assisted by John Quiring, John Sweeney, and Steve Hulbert of the Center for Process Studies. The editors of this volume wish to express their thanks to everyone mentioned above, as well as to Brian Hall, Michael Ruse, Terrence Deacon, Tyrone Cashman, Stuart Kauffman, Gernot Falkner, Renate Falkner, Dorion Sagan, and the late Lynn Margulis, for their valuable contributions to it. In addition, the editors would like to thank Mitchell Palmquist for his typographical assistance in the early stages of putting together this volume.

Notes

1. Robert Rosen, *Life Itself: A Comprehensive Inquiry into the Nature, Origin, and Fabrication of Life* (New York: Columbia University Press, 1991), 23.
2. Immanuel Kant, *Critique of the Power of Judgment*, trans. Paul Guyer and Eric Mathews, ed. Paul Guyer (New York: Cambridge University Press, 2000), 5:374, 246.
3. Martin Heidegger, *The Fundamental Concepts of Metaphysics: World, Finitude, Solitude*, trans. William McNeill and Nicholas Walker (Indianapolis, IN: Indiana University Press, 1995), 188-189.

4. Alternatively stated, the activity of biological researchers today seems to be dominated by what Peter J. Bowler, *Evolution: The History of an Idea* (Berkeley, CA: University of California Press, 1989), 31, calls a "search for the mechanical origins" underlying natural phenomena.

In "Mechanism and Biological Explanation," *Philosophy of Science* 78, no. 4 (October 2011): 533, William Bechtel states "within biology (and the life sciences more generally), there is a long tradition of explaining a phenomenon by describing the mechanism responsible for it. A perusal of biology journals and textbooks yields many mentions of mechanisms but few of laws."

In "Explanation: A Mechanist Alternative," *Studies in the History and Philosophy of Biological and Biomedical Sciences* 36 (2005): 421-422, William Bechtel and Adele Abrahamsen state "explanations in the life sciences frequently involve presenting a model of the mechanism taken to be responsible for a given phenomenon." They continue,

perusing the biological literature, it quickly becomes clear that the term biologists most frequently invoke in explanatory contexts is *mechanism*. That is, biologists explain *why* by explaining *how*. In cell biology, for example, there are numerous proposed mechanisms to explain various phenomena of fermentation, cellular respiration, protein synthesis, secretion, action potential generation, and so forth.

They then proceed to list "a sampling of titles from the 1950s and 1960s that offered mechanistic explanations of phenomena of protein synthesis."

5. See, for example, the following sources: Francisco J. Varela and Humberto Maturana, "Mechanism and Biological Explanation," *Philosophy of Science* 39, no. 3 (September 1972): 378-382; Bechtel and Abrahamsen, "Explanation: A Mechanist Alternative," 421-441; Bechtel, "Mechanism and Biological Explanation," 533-557.

6. Ernst Mayr, "Prologue: Some Thoughts on the History of the Evolutionary Synthesis," in *The Evolutionary Synthesis: Perspectives on the Unification of Biology*, ed. Ernst Mayr and William B. Provine (Cambridge, MA: Harvard University Press, 1980), 1, emphasis added.

7. Bowler, *Evolution*, 317. Bowler states, "The synthesis now 'hardened' to eliminate even the limited role for non-Darwinian mechanisms conceded by some of its founders." See also Stephen J. Gould, "The Hardening of the Modern Synthesis," in *Dimensions of Darwinism: Themes and Counterthemes in Twentieth Century Evolutionary Theory*, ed. Marjorie Grene (Cambridge, MA: Cambridge University Press, 1983), 71.

8. Anya Plutynski, "The Modern Synthesis," 6, 7, 10, unpublished manuscript, http://hum.utah.edu/~plutynsk/ModernSynthesisfinal.pdf, 6, 7, 10.

9. For example, see Vassiliki Betty Smocovitis, *Unifying Biology: The Evolutionary Synthesis and Evolutionary Biology* (Princeton, NJ: Princeton University Press, 1996), 34-35.

10. See Robert G. Reid, *Evolutionary Theory: The Unfinished Synthesis* (Ithaca, NY: Cornell University Press, 1985); also see Niles Eldredge, *Unfinished Synthesis: Biological Hierarchies and Modern Evolutionary Thought* (New York; Oxford University Press, 1985).

11. Plutynski, "The Modern Synthesis," 9.

12. See Massimo Pigliucci and Gerd B. Müller, eds. *Evolution: The Extended Synthesis* (Cambridge, MA: MIT Press, 2010).

13. Joeri Witteveen, "The Softening of the Modern Synthesis," *Acta Biotheoretica* 59, no. 3-4 (2011): 333.

14. Pigliucci and Müller, *Evolution: The Extended Synthesis*, 3-4.

15. See Stuart Kauffman, *At Home in the Universe: The Search for Laws of Self-Organization and Complexity* (New York: Oxford University Press, 1995), 69; Bruce Weber, "Emergence of Life and Biological Selection from the Perspective of Complex Systems Dynamics," in *Evolutionary Systems: Biological and Epistemological Perspectives on Selection and Self-Organization*, ed. Gertrudis Van de Vijver, Stanley N. Salthe, and Manuela Delpos (Dordrecht: Kluwer, 1998), 63.
16. See Per Bak, Chao Tang, and Kurt Wiesenfeld, "Self-Organized Criticality: An Explanation of the $1/f$ Noise," *Physical Review Letters* 39, no. 4 (1987): 381-384.
17. See Humberto Maturana and Francisco J. Varela, *Autopoeisis and Cognition: The Realization of the Living* (Dordrecht, Holland: D. Reidel Publishing Company, 1980), xvii; see the foreword and chapters 13 and 14 of this volume.
18. See Francisco Varela, "Autonomie und Autopoiesie," in *Der Diskurs des Radikalen Konstruktivismus*, ed. Siegfried J. Schmidt (Frankfurt am Main: Suhrkamp Verlag, 1987), 122.
19. See Alfred North Whitehead, *Process and Reality: Corrected Edition*, ed. David Ray Griffin and Donald W. Sherburne (1929; corr. ed. New York: The Free Press, 1978,) 233. Citations refer to corrected edition. Alfred North Whitehead, *Science and the Modern World* (1925; repr. New York: The Free Press, 1967), 69; Also see Adam Scarfe, "Negative Prehensions and the Creative Process," *Process Studies* 32, no. 1 (2003): 94-105.
20. See Hans Jonas, *The Phenomenon of Life: Toward a Philosophical Biology* (New York: Harper & Row, 1966; repr. Evanston, IL: Northwestern University Press, 2001), 97-106.
21. Evelyn Keller, "The Disappearance of Function from 'Self-Organizing' Systems," in *Systems Biology: Philosophical Foundations*, ed. Fred C. Boogerd, Frank J. Bruggeman, Jan-Hendrik S. Hofmeyr, and Hans V. Westerhoff (New York: Elsevier, 2007), 305.
22. Keller, "The Disappearance of Function," 305.
23. Bechtel, *Mechanism and Biological Explanation*, 536.
24. Bechtel and Abrahamsen, "Explanation: A Mechanist Alternative," 423-424. In *Discovering Complexity: Decomposition and Localization as Strategies in Scientific Research* (Princeton, NJ: Princeton University Press, 1993), 17, William Bechtel and Robert Richardson characterize a machine as "a composite of interrelated parts, each performing its own functions, that are combined in such a way that each contributes to producing a behavior of the system. A mechanistic explanation identifies these parts and their organization, showing how the behavior of the machine is a consequence of the parts and their organization."
25. As Robert Rosen states in *Life Itself*, 244, organisms are "closed to efficient causation." Also see Evan Thompson, "Life and Mind: From Autopoeisis to Neurophenomenology—A Tribute to Francisco Varela," *Phenomenology and the Cognitive Sciences* 3 (2004), 390-391.

In reflecting on the difference between a living organism and a mechanical device or machine, given that today, mechanical parts, such as pace-makers are increasingly being used in biomedicine, an interesting (though well-worn) thought experiment in would be to ask how many mechanical parts replacing a person's organs and tissues would it take to transform them into a machine.
26. Keller, "The Disappearance of Function," 316.

To provide some context here vis-à-vis the Kantian notion of "self-organization," in *Critique of the Power of Judgment*, 5:423-424, 291-293, Kant writes that an organized being is thus not a mere machine, for that has only a motive power, while the organized being possesses in itself a formative power, and indeed one that it communicates to the matter, which does not have it (it organizes the latter): thus it has a self-propagating formative power, which cannot be explained through the capacity for movement alone (that is, mechanism).

In providing a brief history of inquiry into the meaning of life since Kant, Keller further states that the terms

> organism and self-organization—remained tightly linked and clearly set apart from the realm of inanimate objects, especially from those objects, machines, that were designed and built to serve human goals. Machines were designed, and they were designed from without. Of course, organisms too were still seen as designed, but in contrast to machines, biological design—or organization—was internally generated. The burden of the concept of self-organization thus fell on the term self, for it was the self as source of the organization that prevents an organism from ever being confused with a machine. (305)

We must be very careful about the meaning of the word "self," when we list "self-organizing" or "self-creating" systems, so as not to construe it as a static, thinking substance, or core *cogito*, of the type that Descartes postulated, namely, as a phenomenon that is disembodied and rigidly set apart from the natural world. Rather, the self is here to be considered an abstract term, designating the indefinite set of internal and external relations that constitute the organism as a life as such. Stipulatively, the term "self" describes an emergent phenomenon, not divorceable from bodily structures, enduring, yet comprised by a continuous process of creation, and thoroughly dependent on elements and/or data appropriated from the natural world for existence. The term "self" designates the finite valuative and selective agency of the organism in this life-process.

Tibor Gánti, in *The Principles of Life* (New York: Oxford University Press, 2003), 120-121, contrasts the living organism with the machine as follows:

> First, living beings are soft systems, in contrast with the artificial hard dynamic systems. Furthermore, machines must always be constructed and manufactured, while living beings construct and prepare themselves. Living beings are growing systems, in contrast with technical devices which never grow after their completion; rather, they wear away. Living beings are multiplying systems and automata (at least at present) are not capable of multiplication. Finally, evolution—the adaptive improvement of living organisms—is a spontaneous process occurring of its own accord through innumerable generations, whereas machines, which in some sense may also go through a process of evolution, can only evolve with the aid of active human contribution.

27. Critics might also try to suggest that nature itself is a machine (that was certainly Descartes' idea as will be alluded to below) that produced all organisms, including the human agents we are speaking of, and that hence they are machines too. Other opponents will also point to the fact that Humberto Maturana and Francisco Varela developed the notion of *autopoiesis*, which was deemed "necessary and sufficient to characterize the *organization* of living systems" (*Autopoeisis and Cognition*, 82) within the context of maintaining a mechanistic program. Furthermore, we cannot be said to have fully removed the ambiguity surrounding the meaning of life. The nature of life has been pondered for millennia and will continue to be long after this book has been forgotten. However, our considerations have promoted learning.

The great German Idealists, J. G. Fichte (1762-1814) and G. W. F. Hegel (1770-1831), emphasized that one aspect of knowing what something is, ontologically speaking, is to posit what it is not, and to negate what it is not. On the basis of the comparison between living organisms and machines, we have undoubtedly learned more about what living beings are by knowing what they are not.

28. For example, August Weismann in *Studies in the Theory of Descent, Volumes 1-2*, trans. Raphael Meldola (London, UK: Sampson Low, Marston, Searle & Rivington, 1882), states that the admission of

> an innate metaphysical developmental force . . . [or] *power is directly opposed to the laws of natural science*, which forbid the assumption of unknown forces as long as it is not demonstrated that known forces are insufficient for the explanation of the phenomena. . . . Thus even without the foregoing special investigations we should deny a phyletic vital force; the more so as its admission is fraught with the greatest consequences, since it involves a renunciation of the possibility of comprehending the natural world. We should, on this assumption [i.e., that of belief in a vital principle], at once cut ourselves off from all possible mechanical explanation of organic nature, *i.e.* from all explanation conformable to law. But this signifies no less than the renunciation of all further inquiry; for what is investigation in natural science but the attempt to indicate the mechanism through which the phenomena of the world are brought about? Where this mechanism ceases science is no longer possible, and transcendental philosophy alone has a voice.

> This conception represents very precisely the well-known decision of Kant:—"Since we cannot in any case know *a priori* to what extent the mechanism of Nature serves as a means to every final purpose in the latter, or how far the mechanical explanation possible to us reaches," natural science must everywhere press the attempt at mechanical explanation as far as possible.

29. Conrad Hal Waddington, *The Evolution of an Evolutionist* (Ithaca, NY: Cornell University Press, 1975), 11, 1, 10.

30. Daniel Dennett, *Darwin's Dangerous Idea: Evolution and the Meanings of Life* (New York: Simon & Schuster, 1995), 77-80.

31. See Martin Heidegger's seminal critique of metaphysics as onto-theology in "The Ontotheological Constitution of Metaphysics" in *Identity and Difference*, trans. Joan Stambaugh (Chicago: University of Chicago Press, 1969), 42-76, 107-146.

32. See Konrad Lorenz, "Kant's Doctrine of the *A Priori* in the Light of Contemporary Biology," in *Philosophy After Darwin: Classic and Contemporary Readings*, ed. Michael Ruse (Princeton, NJ: Princeton University Press, 2009), 231-247.

33. Michael Ruse, *Philosophy after Darwin*, 224.

34. In *Critique of Pure Reason*, trans. and ed. Paul Guyer and Allen W. Wood (New York: Cambridge University Press, 1998), Introduction 2nd Edition, B5, 138, Kant says that the *a priori* principles are "indispensable for the possibility of experience itself."

While David Hume was skeptical of causality from his empiricist perspective, since the necessary causal connection that is assumed between any putative cause A and effect B has no empirical warrant, he did indicate that it was indispensable for human existence. We can interpret this notion of indispensability in the biological sense. In *Prolegomena to Any Future Metaphysics*, trans. and ed. Gary Hatfield (New York: Cambridge, 1997), 4:258-259, 9, Kant states that Hume's skepticism was not about "whether the concept of cause is right, useful, and, with respect to all cognition of nature, indispensable, for this Hume had never put into doubt; it was rather whether it is thought through reason *a priori*, and in this way has an inner truth independent of all experience. . . . The discussion was only about the origin of this concept, not about its indispensability in use."

Also see Adam Scarfe, "Skepticism Concerning Causality: An Evolutionary Epistemological Perspective," *Cosmos and History* 8, no. 1 (2012), 227-288.

35. Dennett, *Darwin's Dangerous Idea*, 77.

36. David Hume, *Enquiries Concerning Human Understanding and Concerning Principles of Morals*, ed. Lewis Amherst Selby-Bigge and Peter Harold Nidditch (New York: Oxford University Press, 1975), H128, 159-160.

37. Whitehead, *Science and the Modern World*, 4, 16.

38. For example, see Weismann, *Studies in the Theory of Descent*, 718, who holds that the mechanistic conception of nature is the only viable option for biological explanation, "so long as the interference of teleological forces in the course of the process of organic development has not been demonstrated to him." Also see Dennett, *Darwin's Dangerous Idea*, 34-60, who asserts that evolution is a "mindless, mechanical—algorithmic process" (60), not a purposeful or teleological one.

39. Evan Thompson in "Life and Mind: From Autopoeisis to Neurophenomenology—A Tribute to Francisco Varela," *Phenomenology and the Cognitive Sciences* 3 (2004), articulates a position that is open to teleology in nature. He states,

Teleology is not an intrinsic organizational property, but an emergent one that belongs to a concrete autopoietic system interacting with its environment. . . . If living beings are not reducible to algorithmic mechanism, and if teleology is an emergent relational property, not an intrinsic organizational one, then we are faced with the prospect of a new kind of biological naturalism beyond the classical opposition of mechanism and teleology. (392)

40. Lorenz, "Kant's Doctrine of the *A Priori*," 233.

41. According to Lorenz, such *a priori* concepts as well as instinctual behaviors may require a particular type of event, experience, stimulus, or an appropriate environmental condition or situation that triggers their realization. It was unfortunate that Lorenz called these conditions "releasing mechanisms," in *Evolution and Modification of Behavior* (Chicago: University of Chicago Press, 1965), 48, although this fact points to the dominance of mechanistic thinking in biology. In a similar fashion, in *Critique of Pure Reason*, A66 / B91, 203, Kant himself was open to the notion that the *a priori* concepts of the understanding had a certain mixed status, rather than simply being innate. As he describes, the "first seeds and dispositions" of the transcendental ideas "lie ready" in the understanding "until with the opportunity of experience they are finally developed and exhibited in their clarity by the very same understanding."

42. Whitehead, *Science and the Modern World*, 157.

43. This is a play on Edmund Husserl's slogan "to the things themselves!," expressing the phenomenological attitude. See also, Evan Thompson, *Mind in Life: Biology, Phenomenology, and the Sciences of Mind* (Cambridge, MA: Belknap Press of Harvard University Press, 2007).

44. Julian Huxley, *Evolutionary Humanism* (Amherst, NY: Prometheus Books, 1964), 76.

Intellectual selection is the psychosocial process by which ideas are selected for and against on the basis of their practical utility and of their biological indispensability, but also in terms of their truth, via speculation, research, critical thinking, skepticism, learning, and education (see Adam Scarfe, "Skepticism Concerning Causality: An Evolutionary Epistemological Perspective," *Cosmos and History* 8, no. 1 (2012), especially 279-283). A synthesis of correspondence and pragmatist theories of truth might be assumed here. As articulated by William James in "Pragmatism's Conception of Truth," in *The Writings of William James: A Comprehensive Edition*, ed. John McDermott (Chicago:

University of Chicago P..ss, 1977), 435, "Any idea that helps us to deal, whether practically or intellectually, with reality, that doesn't entangle our progress in frustrations, that fits, in fact, and adapts our life to the reality's whole setting, will agree sufficiently to meet the requirement [of truth]."

In describing processes of "intellectual selection," as I call it, we may further look to Robert J. Richards' *Darwin and the Emergence of Evolutionary Theories of Mind and Behavior* (Chicago: University of Chicago Press, 1987), who states,

> rational appraisal by the scientific community tests the mettle of new ideas. The survivors are incorporated into the advancing discipline. The social and professional conditions of the discipline also work to cull ideas, sanctioning some and eliminating others. Both of these selection processes—selection against intellectual standards and against social demands—may act either in complementary fashion or in opposition. But both must be heeded . . . if one is to understand the actual history of a science. (577)

He continues,

> Ideas and ultimately well articulated theories are originally generated and selected within the conceptual domain of the individual scientist. Only after an idea system has been introduced to the scientific community . . . does public scrutiny result. To the extent, however, that the problem environments of the individual and the community coincide, individually selected ideas or theories will be fit for life in the community. . . . In construing the acquisition of knowledge in science . . . one must suppose that selection components operate in accord with certain essential criteria: logical consistency, semantic coherence, standards of verifiability and falsifiability, and observational relevance. These criteria may function only implicitly, but they form a necessary subset of criteria governing the development of scientific thought throughout its history. Without such norms, we would not be dealing with the selection of *scientific* ideas. . . . It should be stressed, however, that these selection criteria are themselves the result of previous idea generation and continuous selection, processes by means of which science has descended from protoscience. . . . The complete set of selection criteria define what in a given historical period constitutes the standard of scientific acceptability. (582)

45. Alfred North Whitehead, *Religion in the Making* (New York: Macmillan, 1926; repr., New York: Fordham University Press, 1996), 84, emphasis added. Citations refer to the Fordham University Press edition.

46. Alfred North Whitehead, *Adventures of Ideas* (1933; repr., New York: The Free Press, 1967), 154. He continues, "The individual thing is necessarily a modification of its environment, and cannot be understood in disjunction. All reasoning, apart from some metaphysical reference, is vicious." Elsewhere, Whitehead defines metaphysics as "the science which seeks to discover the general ideas which are indispensably relevant to the analysis of everything that happens."

47. Whitehead, *Adventures of Ideas*, 118.

48. See Fred C. Boogerd, Frank J. Bruggeman, Jan-Hendrik S. Hofmeyr, and Hans V. Westerhoff, eds. *Systems Biology: Philosophical Foundations* (New York: Elsevier, 2007), 9, 324, and 335.

49. René Descartes, "Principles of Philosophy," in *Selected Philosophical Writings*, trans. John Cottingham, Robert Stoothoff, and Dugald Murdoch (New York: Cambridge University Press, 1988), Part IV, §188, 200.

50. Descartes, "Principles of Philosophy," Part IV, §187, 200, my addition.

51. See Sophie Roux, "Cartesian Mechanics," unpublished manuscript, http://www.gobiernodecanarias.org/educacion/3/usrn/fundoro/archivos%20adjuntos/publicaciones/otros_idiomas/ingles/Seminario11-12/Roux_MechanicalPhilosophy.pdf, 12.

52. René Descartes, *A Discourse on Method: Meditations and Principles*, trans. John Veitch (Rutland, VT: Orion Publishing Group, 1912, repr. 1992), Part VI, 46.

53. John Cottingham, "Cartesian Dualism: Theology, Metaphysics, and Science," in *The Cambridge Companion to Descartes*, ed. John Cottingham. (Cambridge, MA: Cambridge University Press, 1992), 250.

54. See René Descartes, "A Discourse on the Method," in *Selected Philosophical Writings*, trans. John Cottingham, Robert Stoothoff, and Dugald Murdoch (New York: Cambridge University Press, 1988), Part V, §57, 45.

55. Isaac Newton, *Opticks*, Query 31 (London: William Innys, 1730), Project Gutenberg, http://www.gutenberg.org/files/33504/33504-h/33504-h.htm, 402; also see 390.

56. Descartes, "Principles of Philosophy," Part II, §20-21, 197-198.

57. See Descartes, "Principles of Philosophy," Part I, §51, 177.

58. David Sztybel, "Did Descartes Believe That Nonhuman Animals Cannot Feel?" unpublished manuscript, http://sztybel.tripod.com/animal_feelings.html, 1.

59. Descartes, "A Discourse on the Method," in *Selected Philosophical Writings* §59, 45, my addition.

60. David Goodman and Colin A. Russell, eds. *The Rise of Scientific Europe: 1500-1800* (Kent, UK: Hodder & Stoughton, 1991), 182.

61. Descartes, "Principles of Philosophy," Part II, §2, 190.

62. René Descartes, "The Passions of the Soul," in *Selected Philosophical Writings*, trans. John Cottingham, Robert Stoothoff, and Dugald Murdoch (New York: Cambridge University Press, 1988), §6, 219-220.

63. Gideon Manning, "Descartes' Healthy Machines and the Human Exception," in *The Mechanization of Natural Philosophy*, ed. Sophie Roux and Dan Garber (New York: Springer, in press, 2012), http://www.hss.caltech.edu/~gmax/publications/Manning_Descartes_Healthy_Machines_and_the_Human_Exception.pdf, 2, emphasis added.

64. In *The Blind Watchmaker: Why the Evidence of Evolution Reveals a Universe Without Design.* (New York: W. W. Norton & Company, Inc., 1986), 5, my emphasis, Richard Dawkins writes,

> Paley's argument is made with passionate sincerity and is informed by the best biological scholarship of his day, but it is wrong, gloriously and utterly wrong. *The analogy between telescope and eye, between watch and living organism, is false.* All appearances to the contrary, the only watchmaker in nature is the blind forces of physics, albeit deployed in a very special way. A true watchmaker has foresight: he designs his cogs and springs, and plans their interconnections, with a future purpose in his mind's eye. Natural selection, the blind, unconscious, automatic process which Darwin discovered, and which we now know is the explanation for the existence and apparently purposeful form of all life, has no purpose in mind. It has no mind and no mind's eye. It does not plan for the future. It has no vision, no foresight, no sight at all. If it can be said to play the role of watchmaker in nature, it is the blind watchmaker.

Of course, Dawkins' dismissal of the analogy between watch (and hence, machine) and living organism in the context of his argument against Paley would seem to contradict the "organism-as-replication-machine-for-genes" metaphor that he advances elsewhere.

65. On this note, James Griesemer in "Heuristic Reductionism and the Relative Significance of Epigenetic Inheritance in Evolution," in *Epigenetics: Linking Genotype and Phenotype in Development and Evolution*, ed. Benedikt Hallgrímsson and Brian K. Hall (Berkeley, CA: University of California Press, 2011), suggests, "It is helpful to view reductionism itself as a heuristic research strategy for theory construction . . . rather than, as philosophers usually do, as an account of explanation by derivation of less from more general theories or, as scientists usually do, as an account of higher-level phenomena explained in terms of lower-level mechanisms" (25); "Reductionism in science is better viewed as a heuristic strategy than a logic of explanation . . ., and it should not be confused with mechanistic research per se" (36).

66. Boogerd, Bruggeman, Hofmeyr, and Westerhoff, eds. *Systems Biology: Philosophical Foundations*, 7.

67. Huxley, *Evolutionary Humanism*, 101.

68. René Descartes, "Rules for the Direction of Our Native Intelligence," in *Selected Philosophical Writings*, trans. John Cottingham, Robert Stoothoff, and Dugald Murdoch (New York: Cambridge University Press, 1988), Rule 13, 18.

69. Jan-Hendrik S. Hofmeyr, "The Biochemical Factory That Autonomously Fabricates Itself: A Systems Biological View of the Living Cell," in *Systems Biology: Philosophical Foundations*, ed. Fred C. Boogerd, Frank J. Bruggeman, Jan-Hendrik S. Hofmeyr, and Hans V. Westerhoff (New York: Elsevier, 2007), 219, citing J. de Rosnay, *The Macroscope: A New World Scientific System*, trans. Robert Edwards (New York: Harper and Row, 1979).

70. Hans Jonas, *The Phenomenon of Life*, 12.

71. David Ray Griffin, "Of Minds and Molecules: Postmodern Medicine in a Psychosomatic Universe," in *The Re-Enchantment of Science*, ed. David Ray Griffin (Albany, NY: State University of New York Press, 1988), 144.

72. Scott F. Gilbert and Sahotra Sarkar, "Embracing Complexity: Organicism for the 21st Century," *Developmental Dynamics* 219 (2000): 1.

73. Julian Huxley, *Evolutionary Humanism*, 101.

74. Conway Lloyd Morgan, *Emergent Evolution* (London, UK: Williams and Norgate, 1927), 7-8.

In *At Home in the Universe: The Search for Laws of Self-Organization and Complexity* (New York: Oxford University Press, 1995), vii, Stuart Kauffman writes,
The past three centuries of science have been predominantly reductionist, attempting to break complex systems into simple parts, and those parts, in turn, into simpler parts. The reductionist program has been spectacularly successful, and will continue to be so. But it has often left a vacuum: How do we use the information gleaned about the parts to build up a theory of the whole? The deep difficulty here lies in the fact that the complex whole may exhibit properties that are not readily explained by understanding the parts. The complex whole, in a completely nonmystical sense, can often exhibit collective properties, "emergent" features that are lawful in their own right.

75. Lloyd Morgan, *Emergent Evolution*, 8.

76. Samuel Alexander, *Space, Time, and Deity* (New York: Macmillan, 1920), 46.

77. Stephen Jay Gould, *Wonderful Life: The Burgess Shale and the Nature of History* (New York: W. W. Norton and Company Ltd., 1989), 322.

78. Dennett, *Darwin's Dangerous Idea*, 80-81. Dennett makes a distinction between "reductionism, which is in general a good thing, from *greedy reductionism*, which is not. The difference, in the context of Darwin's theory, is simple: greedy reductionists

think that everything can be explained without cranes; good reductionists think that everything can be explained without skyhooks" (81-82).

79. Boogerd, Bruggeman, Hofmeyr, and Westerhoff, eds. *Systems Biology: Philosophical Foundations*, 15-16, quoting Martin Mahner and Mario Bunge, *Foundations of Biophilosophy* (Berlin: Springer-Verlag, 1997).

80. Keller, "The Disappearance of Function," 307.

81. According to Martin Heidegger in "The Question Concerning Technology" in *Basic Writings*, trans. David Farrell Krell (New York: Harper San Francisco, 1977), 303, mechanistic science "sets nature up to exhibit itself as a coherence of forces calculable in advance," and "orders its experiments precisely for the purpose of asking whether and how nature reports itself when set up in this way," being in service to technology.

82. See Nessa Carey, *The Epigenetics Revolution: How Modern Biology Is Rewriting our Understanding of Genetics, Disease, and Inheritance* (London, UK: Icon Books, 2011), 122-124.

83. For example, see Martin Hoyle and James E. Cresswell, "The Effect of Wind Direction on Cross-Pollination in Wind-Pollinated GM Crops," *Ecological Applications* 17 (2007): 1234-1243.

84. For example, see Chensheng Lu, Kenneth M. Warchol, and Richard A. Callahan, "*In Situ* Replication of Honey Bee Colony Collapse Disorder," *Bulletin of Insectology* 65, no. 1 (2012): 99-106.

85. For example, see Artemis Dona and Ioannis S. Arvanitoyannis, "Health Risks of Genetically Modified Foods," *Critical Reviews in Food Science and Nutrition* 49 (2009): 164-175.

86. For example, see Bruce Cohen, *The Cohen Commission of Inquiry into the Decline of Sockeye Salmon in the Fraser River*, Public Hearings transcript, December 15, 2011: http://alexandramorton.typepad.com/Cohen%20Dec%2015.pdf.

87. For example, see The Royal Society's (UK) Policy Document, "Geo-Engineering the Climate: Science, Governance, and Uncertainty," September 2009, 1-98.

88. Conrad Hal Waddington, *The Ethical Animal* (New York: Atheneum, 1960), 23, 30.

89. Craig Holdrege, Nature Institute 2008, "Understanding the Unintended Effects of Genetic Manipulation," http://natureinstitute.org/txt/ch/nontarget.php, 8, italics added.

90. In *The Evolution of an Evolutionist* (Ithaca, NY: Cornell University Press, 1975), Conrad Hal Waddington confesses having spent "much more attention during the last two years of [his] undergraduate career [to Whitehead's process-relational metaphysics] than [he] did to the textbooks in the subjects on which [he] was going to take [his] exams" (3). He continues, "thus my particular slant on evolution—a most unfashionable emphasis on the importance of the developing phenotype—is a fairly direct derivative from Whiteheadian-type metaphysics. . . . [In my early career] my approach to experimental epigenetics was again strongly influenced by Whiteheadian metaphysics" (8). Whiteheadian metaphysical concepts were even at the root of some of Waddington's key concepts. Waddington states, "What I have been calling by the Whiteheadian term 'concrescence' is what I have later called a *chreod*" (10). Waddington also indicates that he "tried to put the Whiteheadian outlook to actual use in particular experimental situations" (10).

91. Erik L. Peterson, "The Excluded Philosophy of Evo-Devo? Revisiting C. H. Waddington's Failed Attempt to Embed Alfred North Whitehead's 'Organicism' in Evolutionary Biology," *History and Philosophy of the Life Sciences* 33 (2011): 303.

92. Peterson, "The Excluded Philosophy," 317.

93. See Whitehead, *Science and the Modern World*, viii.
94. See Whitehead, *Process and Reality*, 21.
The theory of prehensions, which is the centerpiece of Whitehead's cosmological scheme and the chief exemplification of his doctrine of internal relations, is an epistemological description of the *emergence* of various interrelated modes of organismic experience from simple physical feelings to consciousness. It depicts organisms engaged in unique processes of self-creation, largely depending on activities of valuation and selective appropriation of the elements and data from their environments in so doing.
95. Whitehead, *Process and Reality*, 84.
96. Alfred North Whitehead, "Nature Alive" in *Modes of Thought* (1938; repr., New York: The Free Press, 1966), 148.
97. See Gilbert Ryle, *The Concept of Mind* (New York: Barnes & Noble, Inc., 1949), 15-24, 76-82.
98. Whitehead, *Modes of Thought*, 144.
99. Whitehead, *Process and Reality*, 84.
100. In the *Science of Logic*, trans. Arnold V. Miller (Atlantic Highlands, NJ: George Allen & Unwin, 1969), G. W. F. Hegel (1770-1831) described the concept of mechanism in relation to chemism and teleology, deployed by the objective Concept (*Begriff*). Here, Hegel was heavily reliant on the writings of Kant and Fichte, who described how causality and substance mutually entailed one another. However, mechanistic thinking, for him, ultimately involved a contradiction between self-subsistency (i.e., the notion that substances are dependent on nothing for their existence) and self-determination. Causality, for him, was the negative dialectical moment of substances in the process of their finite self-determination. Hegel's analysis revealed that the category of mechanism is caught up with "the externality of causality" (see 569, also 710-726), whereby a static, self-subsistent material object (i.e., a substance) acts causally on another substance, bringing about a certain effect. For Hegel, the category of mechanism involves the omission of consideration of the possibility of "reciprocal interaction" (see 569-571) between them and/or of "internal relations."
101. Whitehead, *Science and the Modern World*, 107.
102. Whitehead, *Modes of Thought*, 139.
103. In the *Prolegomena* (§27-28, 4:30-311, 64-65), in attempting to overcome the Humean skepticism of causality *qua* necessary connection, Kant states, writing in an ironic tone, that we have "no insight . . . into how substances, each of which has its own separate existence, should depend on one another, and should indeed do so necessarily . . . I do not have the least concept of such a connection of things in themselves, how they could exist as substances or work as causes or be in community with others . . . for I have no concept of the possibility of such a connection of existence."
104. Whitehead, *Science and the Modern World*, 73.
105. Whitehead, *Process and Reality*, 309.
106. Whitehead, *Process and Reality*, 18.
107. Whitehead, *Process and Reality*, 34. See also chapter 10 in this volume.
108. Whitehead, *Science and the Modern World*, 104.
109. Whitehead, *Process and Reality*, 50.
110. Alfred North Whitehead, *Essays in Science and Philosophy* (New York: Philosophical Library, 1941), 157.
111. Whitehead, *Science and the Modern World*, 206.
112. Whitehead, *Adventures of Ideas*, 154.
113. Whitehead, *Modes of Thought*, 140.

114. Thompson, *Mind in Life*, 173.
115. If Mary Jane West-Eberhard in *Developmental Plasticity and Evolution* (New York: Oxford University Press, 2003), 157, 24, is correct to say that genes can be "followers in evolution" and that "behavioral change often precedes and directs morphological change," then behavioral change, or *ethotypic* change, should factor in to the root meaning of the term "evolution," in abstraction from genotypic and phenotypic change.
116. Whitehead, *Process and Reality*, xi.
117. Whitehead, *Science and the Modern World*, 80, 107.
118. Whitehead, *Science and the Modern World*, 17.
119. Boogerd, Bruggeman, Hofmeyr, and Westerhoff, eds. *Systems Biology: Philosophical Foundations*, 326.
120. See Alfred North Whitehead, *The Aims of Education and Other Essays* (1929; repr., New York: The Free Press, 1957), 15-28. See also Adam Scarfe, *The Adventure of Education* (Amsterdam: Rodopi Press, 2009), 12-15, 169-211.
121. Whitehead, *Process and Reality*, 7-8.
122. Whitehead, *Modes of Thought*, 144.
123. Peter J. Bowler, "Darwin's Originality," *Science* 323 (2009), 223.

Bibliography

Alexander, Samuel. *Space, Time, and Deity*. New York: Macmillan, 1920.
Bak, Per, Chao Tang, and Kurt Wiesenfeld. "Self-Organized Criticality: An Explanation of the 1/*f* Noise." *Physical Review Letters* 39, no. 4 (1987): 381-384.
Bechtel, William. "Mechanism and Biological Explanation." *Philosophy of Science* 78, no. 4 (October 2011): 533-557.
Bechtel, William, and Adele Abrahamsen. "Explanation: A Mechanist Alternative." *Studies in the History and Philosophy of Biological and Biomedical Sciences* 36 (2005): 421-441.
Bechtel, William, and Robert C. Richardson. *Discovering Complexity: Decomposition and Localization as Strategies in Scientific Research*. Princeton, NJ: Princeton University Press, 1993.
Boogerd, Fred C., Frank J. Bruggeman, Jan-Hendrik S. Hofmeyr, and Hans V. Westerhoff, eds. *Systems Biology: Philosophical Foundations*. New York: Elsevier, 2007.
Bowler, Peter J. *Evolution: The History of an Idea (Revised Edition)*. Berkeley, CA: University of California Press, 1989.
———. "Darwin's Originality." *Science* 323 (2009): 223-226.
Buchler, Justus. *Metaphysics of Natural Complexes: Second Expanded Edition*. Edited by Kathleen Wallace, Armen Marsoobian, and Robert S. Corrington. Albany, NY: State University Press of New York, 1990.
Burian, Richard M. "Challenges to the Evolutionary Synthesis." *Evolutionary Biology* 23 (1988): 247-269.
Carey, Nessa. *The Epigenetics Revolution: How Modern Biology Is Rewriting Our Understanding of Genetics, Disease, and Inheritance*. London, UK: Icon Books, 2011.
Cellini, F., A. Chesson, I. Colquhoun, A. Constable, H. V. Davies, K. H. Engel, A. M. R. Gatehouse, S. Kärenlampi, E. J. Kok, J.-J. Leguay, S Lehesranta, H. P. J. M. Noteborn, J. Pedersen, and M. Smith. "Unintended Effects and Their Detection in

Genetically Modified Crops." *Food and Chemical Toxicology* 42 (2004): 1089-1125.
Cohen, Bruce. *The Cohen Commission of Inquiry into the Decline of Sockeye Salmon in the Fraser River.* Public Hearings transcript, December 15, 2011, http://alexandramorton.typepad.com/Cohen%20Dec%2015.pdf.
Cottingham, John. "Cartesian Dualism: Theology, Metaphysics, and Science." In *The Cambridge Companion to Descartes*, edited by John Cottingham, 236-257. Cambridge, MA: Cambridge University Press, 1992.
Dawkins, Richard. *The Blind Watchmaker: Why the Evidence of Evolution Reveals a Universe Without Design.* New York: W. W. Norton & Company, Inc., 1986.
Dennett, Daniel. *Darwin's Dangerous Idea: Evolution and the Meanings of Life.* New York: Simon & Schuster, 1995.
de Rosnay, J., *The Macroscope: A New World Scientific System.* Translated by Robert Edwards. New York: Harper and Row, 1979.
Descartes, René. *A Discourse on Method: Meditations and Principles.* Translated by John Veitch. Rutland, VT: Orion Publishing Group, 1912, repr. 1992.
———. *Selected Philosophical Writings*, translated by John Cottingham, Robert Stoothoff, and Dugald Murdoch. New York: Cambridge University Press, 1988.
Di Paolo, Ezequiel A. "Autopoesis, Adaptivity, Teleology, Agency." *Phenomenology and the Cognitive Sciences* 4, no. 4, 2005.
Dona, Artemis, and Ioannis S. Arvanitoyannis. "Health Risks of Genetically Modified Foods." *Critical Reviews in Food Science and Nutrition* 49 (2009): 164-175.
Eldredge, Niles. *Unfinished Synthesis: Biological Hierarchies and Modern Evolutionary Thought.* New York; Oxford University Press, 1985.
Gánti, Tibor. *The Principles of Life.* New York: Oxford University Press, 2003.
Gilbert, Scott F., and Sahotra Sarkar. "Embracing Complexity: Organicism for the 21st Century." *Developmental Dynamics* 219 (2000): 1-9.
Goodman, David, and Colin A. Russell, eds. *The Rise of Scientific Europe: 1500-1800.* Kent, UK: Hodder & Stoughton, 1991.
Gould, Stephen J. "The Hardening of the Modern Synthesis." In *Dimensions of Darwinism: Themes and Counterthemes in Twentieth Century Evolutionary Theory.* Edited by Marjorie Grene. Cambridge, MA: Cambridge University Press, 1983.
———. *Wonderful Life: The Burgess Shale and the Nature of History.* New York: W. W. Norton and Company Ltd., 1989.
Griesemer, James. "Heuristic Reductionism and the Relative Significance of Epigenetic Inheritance in Evolution." In *Epigenetics: Linking Genotype and Phenotype in Development and Evolution*, edited by Benedikt Hallgrímsson and Brian K. Hall, 14-40. Berkeley, CA: University of California Press, 2011.
Griffin, David Ray, ed. *The Re-Enchantment of Science.* Albany, NY: State University of New York Press, 1988.
Hegel, Georg Wilhelm Friedrich. *The Science of Logic.* Translated by Arnold V. Miller. Atlantic Highlands, NJ: George Allen & Unwin, 1969.
Heidegger, Martin. "The Ontotheological Constitution of Metaphysics." In *Identity and Difference*, translated by Joan Stambaugh, 42-76, 107-146. Chicago: University of Chicago Press, 1969.
———. "The Question Concerning Technology." In *Basic Writings*, translated by David Farrell Krell, 283-317. New York: Harper San Francisco, 1977.

———. *The Fundamental Concepts of Metaphysics: World, Finitude, Solitude*. Translated by William McNeill and Nicholas Walker. Indianapolis, IN: Indiana University Press, 1995.

Hofmeyr, Jan-Hendrik S. "The Biochemical Factory That Autonomously Fabricates Itself: A Systems Biological View of the Living Cell." In *Systems Biology: Philosophical Foundations*, edited by Fred C. Boogerd, Frank J. Bruggeman, Jan-Hendrik S. Hofmeyr, and Hans V. Westerhoff, 217-242. New York: Elsevier, 2007.

Ho, Mae-Wan, "Organism and Psyche in a Participatory Universe." In *The Evolutionary Outrider: The Impact of the Human Agent on Evolution—Essays Honouring Ervin Laszlo*, edited by David Loye, 49-66. Westport, CA: Adamantine Press Limited, 1998.

Holdrege, Craig. "Understanding the Unintended Effects of Genetic Manipulation," *The Nature Institute*, 2008: http://natureinstitute.org/txt/ch/nontarget.php.

Hoyle, Martin, and James E. Cresswell. "The Effect of Wind Direction on Cross-Pollination in Wind-Pollinated GM Crops." *Ecological Applications* 17 (2007): 1234-1243.

Hume, David. *Enquiries Concerning Human Understanding and Concerning Principles of Morals*, edited by Lewis Amherst Selby-Bigge and Peter Harold Nidditch. New York: Oxford University Press, 1975.

Huxley, Julian. *Evolutionary Humanism*. Amherst, NY: Prometheus Books, 1964.

Jonas, Hans. *The Phenomenon of Life: Toward a Philosophical Biology*. Evanston, IL: Northwestern University Press, 2001. First published in 1966 by Harper & Row.

Kant, Immanuel. *Prolegomena to Any Future Metaphysics*, translated and edited by Gary Hatfield. New York: Cambridge, 1997.

———. *Critique of Pure Reason*. Translated and edited by Paul Guyer and Allen W. Wood. New York: Cambridge University Press, 1998.

———. *Critique of the Power of Judgment*. Translated by Paul Guyer and Eric Mathews, edited by Paul Guyer. New York: Cambridge University Press, 2000.

Kauffman, Stuart. *At Home in the Universe: The Search for Laws of Self-Organization and Complexity*. New York: Oxford University Press, 1995.

Keller, Evelyn. *Making Sense of Life: Explaining Biological Development with Models, Metaphors, and Machines*. Cambridge, MA: Harvard University Press, 2002.

———. "The Disappearance of Function from 'Self-Organizing' Systems." In *Systems Biology: Philosophical Foundations*, edited by Fred C. Boogerd, Frank J. Bruggeman, Jan-Hendrik S. Hofmeyr, and Hans V. Westerhoff, 303-317. New York: Elsevier, 2007.

Lorenz, Konrad. "Kant's Doctrine of the *A Priori* in the Light of Contemporary Biology." In *Philosophy after Darwin: Classic and Contemporary Readings*, edited by Michael Ruse, 231-247. Princeton, NJ: Princeton University Press, 2009.

———. *Evolution and Modification of Behavior*. Chicago: University of Chicago Press, 1965.

Lloyd Morgan, Conway. *Emergent Evolution*. London: Williams and Norgate, 1927.

Lu, Chensheng, Kenneth M. Warchol, and Richard A. Callahan. "*In Situ* Replication of Honey Bee Colony Collapse Disorder." *Bulletin of Insectology* 65, no. 1 (2012): 99-106.

Manning, Gideon. "Descartes' Healthy Machines and the Human Exception." In *The Mechanization of Natural Philosophy*, edited by Sophie Roux and Dan Garber. New York: Springer (in press, 2012),
http://www.hss.caltech.edu/~gmax/publications/Manning_Descartes_Healthy_Mach

ines_and_the_Human_Exception.pdf.
Maturana, Humberto, and Francisco J. Varela. *Autopoeisis and Cognition: The Realization of the Living.* Dordrecht, Holland: D. Reidel Publishing Company, 1980.
Mayr, Ernst, and William B. Provine, eds. *The Evolutionary Synthesis: Perspectives on the Unification of Biology.* Cambridge, MA: Harvard University Press, 1980.
Newton, Isaac. *Opticks*, Query 31. London: William Innys, 1730. Project Gutenberg, http://www.gutenberg.org/files/33504/33504-h/33504-h.htm.
Peterson, Erik L. "The Excluded Philosophy of Evo-Devo? Revisiting C. H. Waddington's Failed Attempt to Embed Alfred North Whitehead's 'Organicism' in Evolutionary Biology." *History and Philosophy of the Life Sciences* 33 (2011): 303-320.
Pigliucci, Massimo, and Gerd B. Müller, eds. *Evolution: The Extended Synthesis.* Cambridge, MA: MIT Press, 2010.
Plutynski, Anya. "The Modern Synthesis," unpublished manuscript, http://hum.utah.edu/~plutynsk/ModernSynthesisfinal.pdf.
Reid, Robert G. *Evolutionary Theory: The Unfinished Synthesis.* Ithaca, NY: Cornell University Press, 1985.
Richards, Robert J. *Darwin and the Emergence of Evolutionary Theories of Mind and Behavior.* Chicago: University of Chicago Press, 1987.
Rischer, Heiko, and Kirsi-Marja Oksman-Caldentey. "Unintended Effects in Genetically-Modified Crops Revealed by Metabolomics?" *Trends in Biotechnology* 24, no. 3, 102-104.
Rosen, Robert. *Life Itself: A Comprehensive Inquiry into the Nature, Origin, and Fabrication of Life.* New York: Columbia University Press, 1991.
Roux, Sophie. "Cartesian Mechanics," unpublished manuscript, http://www.gobiernodecanarias.org/educacion/3/usrn/fundoro/archivos%20adjuntos/publicaciones/otros_idiomas/ingles/Seminario11-12/Roux_MechanicalPhilosophy.pdf.
Royal Society. "Geo-Engineering the Climate: Science, Governance, and Uncertainty." UK Policy Document, September 2009, 1-98.
Ruse, Michael, ed. *Philosophy After Darwin: Classic and Contemporary Readings.* Princeton, NJ: Princeton University Press, 2009.
Ryle, Gilbert. *The Concept of Mind.* New York: Barnes & Noble, Inc., 1949.
Scarfe, Adam. "Negative Prehensions and the Creative Process." *Process Studies* 32, no. 1 (2003): 94-105.
———. *The Adventure of Education.* Amsterdam: Rodopi Press, 2009.
———. "Skepticism Concerning Causality: An Evolutionary Epistemological Perspective," *Cosmos and History* 8, no. 1 (2012), 227-288.
Smocovitis, Vassiliki Betty. *Unifying Biology: The Evolutionary Synthesis and Evolutionary Biology.* Princeton, NJ: Princeton University Press, 1996.
Sztybel, David. "Did Descartes Believe That Nonhuman Animals Cannot Feel?" Unpublished manuscript, http://sztybel.tripod.com/animal_feelings.html.
Thompson, Evan. *Mind in Life: Biology, Phenomenology, and the Sciences of Mind.* Cambridge, MA: Belknap Press of Harvard University Press, 2007.
———. "Life and Mind: From Autopoeisis to Neurophenomenology—A Tribute to Francisco Varela." *Phenomenology and the Cognitive Sciences* 3 (2004): 381-398.
Varela, Francisco J. "Autonomie und Autopoiesie." In *Der Diskurs des Radikalen Konstruktivismus.* Edited by Siegfried J. Schmidt, 119-132. Frankfurt am Main: Suhrkamp Verlag, 1987.

Varela, Francisco J., and Humberto Maturana, "Mechanism and Biological Explanation." *Philosophy of Science* 39, no. 3 (September 1972): 378-382.

Waddington, Conrad Hal. *The Ethical Animal*. New York: Athenaum, 1961.

———. *The Evolution of an Evolutionist*. Ithaca, NY: Cornell University Press, 1975.

Weber, Andreas, and Francisco J. Varela. "Life after Kant: Natural Purposes and the Autopoietic Foundations of Biological Individuality." *Phenomenology and the Cognitive Sciences* 1 (2002): 97-125.

Weber, Bruce, "Emergence of Life and Biological Selection from the Perspective of Complex Systems Dynamics." In *Evolutionary Systems: Biological and Epistemological Perspectives on Selection and Self-Organization*, edited by Gertrudis Van de Vijver, Stanley N. Salthe, and Manuela Delpos, 59-66. Dordrecht: Kluwer, 1998.

Weismann, August. *Studies in the Theory of Descent, Volumes 1-2*. Translated by Raphael Meldola. London, UK: Sampson Low, Marston, Searle & Rivington, 1882.

West-Eberhard, Mary Jane. *Developmental Plasticity and Evolution*. New York: Oxford University Press, 2003.

Whitehead, Alfred North. *Science and the Modern World*. 1925. Reprinted, New York: The Free Press, 1967.

———. *Religion in the Making*. New York: Macmillan, 1926. Reprinted with new Introduction by Judith A. Jones. New York: Fordham University Press, 1996. Citations refer to Fordham University Press edition.

———. *The Aims of Education and Other Essays*. 1929. Reprinted, New York: The Free Press, 1957.

———. *Process and Reality: Corrected Edition*. Edited by David Ray Griffin and Donald Sherburne. 1929. Corrected with new introduction. New York: The Free Press, 1978. Citations refer to corrected edition.

———. *Adventures of Ideas*. 1933. Reprinted, New York: The Free Press, 1967.

———. *Modes of Thought*. 1938. Reprinted, New York: The Free Press, 1966.

———. *Essays in Science and Philosophy*. New York: Philosophical Library, 1941.

Witteveen, Joeri, "The Softening of the Modern Synthesis." *Acta Biotheoretica* 59, no. 3-4 (2011): 333-345.

Wright, Cory, and William Bechtel. "Mechanisms and Psychological Explanation." *Handbook of the Philosophy of Science, Volume 12: Philosophy of Psychology*, edited by Dov M. Gabbay, Paul Thagard and John Woods, 31-79. New York: Elsevier, 2006.

Section 1:
Complexity, Systems Theory, and Emergence

Chapter One
Complex Systems Dynamics in Evolution and Emergent Processes

Bruce H. Weber

For the past quarter of a century or so there has been ferment in evolutionary theory as new insights from molecular biology, developmental biology, systems ecology, and origin-of-life research have informed and can potentially reform our basic ideas about evolutionary phenomena. Discussions now abound about whether the received evolutionary theory, usually denoted as neo-Darwinism but more appropriately called the Modern Evolutionary Synthesis, is adequate for dealing with the information, phenomena, and implications of current biological knowledge, or needs some sort of revision, expansion, and/or extension, or more fundamentally needs replacement with a new paradigm.[1]

In *Darwinism Evolving*, and elsewhere, David Depew and I have argued that Darwinism is more perspicuously conceived as a Research Tradition—a Lakatosian research program extended over time with a core theoretical commitment, as suggested by Larry Laudan—rather than as a specific theoretical instantiation within it, such as the Modern Evolutionary Synthesis.[2] The Darwinian Research Tradition is characterized by a core commitment to natural selection as a major, though not sole, principle producing adaptation, biological order, and novelty. We further argued that specific phases or programs within the Darwinian Research Tradition relied upon background and auxiliary assumptions about systems dynamics, namely, what counts as a system and by what dynamics such a system can change. We made the case that Darwin's Darwinism was informed by Newtonian systems dynamics, which were characterized by four assumptions: causal closure, deterministic change due to closure, decomposable organization due to an atomistic ontology, and reversible processes.[3] As Robert Ulanowicz has pointed out, these assumptions imply a reductionism in which the explanatory arrows point downward and the causal arrows point upward only, as well as a view of nature as governed by a set of mechanistic laws. Ulanowicz goes on to argue that Darwin opened a "second window" on nature, in addition to the one that Newton had opened, bringing process, interaction, chance, and history into view.[4] Depew and I described, drawing upon recent Darwin scholarship, the reason Darwin sought to couch his insights and theory in Newtonian dynamical terms was that Newtonian methodology was the only model available in the early efforts of philosophy of science as presented by David Brewster, John Herschel, and William Whewell, among others. However, the fit between Darwin and Newton was makeshift at best because Darwin's window brought into view phenomena that could not be fully addressed by Newton's.[5]

The first quarter or so of the twentieth century was a period of crisis for Darwinism, as the findings of the rediscovered Mendelian genetics seemed to be inconsistent with the gradualism of Darwianian evolutionary change, as was argued by Vernon Kellogg in 1907.[6] Through the efforts of J. B. S. Haldane, Sergi Chetverikov, Ronald Fisher, and Sewell Wright in the 1920s and 1930s, the Modern Evolutionary Synthesis of genetics and natural selection was forged through changing the assumption of determinism to one of probabilities and statistics as applied to gene frequencies in populations.[7] This provided the basis for others, such as Theodosius Dobzhansky, Julian Huxley, George Gaylord Simspon, Ledyard Stebbins, and Ernst Mayr, to complete the synthesis by extending it to a wider range of biological phenomena, which accomplishment was commemorated by the 1959 symposium published in three volumes by the University of Chicago Press.[8] The shift in the background assumption from determinism to probability can be seen as part of the larger probability revolution in science during this time. The other Newtonian assumptions remained in place in the Modern Evolutionary Synthesis.[9] As Stephen Gould pointed out, there was a period in the earlier phase of the formation of the synthesis that allowed for more causal pluralism, but by the 1950s the synthesis had "hardened" to use his term for selection acting on and measured by gene frequencies in populations.[10] The expectation then was that the recent discoveries in molecular biology would just fill in the details and that developmental processes were no more than a simple "read-out" of the DNA code. Not only was developmental biology marginalized, despite the protests of Conrad Waddington and concerns of Julian Huxley, but ecology was also marginalized, except as it could be formulated as population ecology. Although origin-of-life research gained considerable momentum, during the 1950s and 1960s, it was still considered as irrelevant to evolutionary theory.[11]

The primary problem addressed by Darwin was that of adaptation and apparent design so as to counter the arguments of William Paley that such phenomena were not understandable by natural causes and that there would never be a "Newton of a blade of grass." Fisher, the "Boltzmann of a blade of grass" if we may so characterize him, in his effort to reconcile Darwin and Mendel, sought to refine the understanding of the process of adaptation; Dobzhansky, Mayr, and Simpson respectively shifted the focal problems to those of speciation, adaptive radiation, and macroevolution, although aspects of these problems remained at issue. Depew and I noted that by the 1990s the advances in molecular biology, developmental genetics, and systems ecology were once again putting pressure on the Darwinian Research Tradition, or more precisely on the background dynamical assumptions of the Modern Evolutionary Synthesis.[12] We speculated that shifting to assumptions of complex systems dynamics, such as that systems are open to extra-systemic inputs (particularly matter-energy flows), partial non-decomposability (i.e., relationships among components affect the properties of the components), and irreversible processes due to systems operating away from thermodynamic equilibrium, in addition to the shift already made to probabilities, constituted a more robust context for understanding evolutionary phenomena. Ulanowicz concurs and argues further that probabilities need to be replaced by Karl Popper's notion of propensities when dealing with the complexity of living systems.[13] This complexity is due to the enormous variational potential of the combinatorics of polymers, such that both he and Stuart Kauffman view nature at the level of living entities, doing so as radically non-ergodic.[14] This shifts us further away from universal, deterministic laws to principles of propensities and historical contingency. Our understanding of complex systems and their dynamics has been enhanced by progress in computers and our ability to simulate phenomena due to nonlinearity and self- (or more aptly systems-) organization. By combining computation and laboratory experiments, a plausible research program on the origin, or more properly termed the emergence, of life has arisen.[15] As Jeffrey Wicken proposed and Depew and I among others have argued, such a shift of emphasis can help bring the results and insights of such research to bear on our understanding of emergent processes generally as well as enrich our understanding of evolutionary ones. This brings us to considering how such an expanded or extended evolutionary synthesis might be formulated, or how a new synthesis that goes beyond the Darwinian Research Tradition might be proposed, that would integrate development, systems ecology, and origin-of-life research, with the rest of our understanding of evolution.[16] Such a research program would have as its focal problem the process of emergence: the emergence of living systems, of evolutionary novelties such as the eukaryotic cell structure, multicellularity, and even the emergence of mind.[17]

The sense that the Modern Evolutionary Synthesis was incomplete or unfinished came to the fore in 1972 with the proposal of punctuated equilibrium by paleontologists Stephen Jay Gould and Niles Eldredge that posited the possibility of natural selection acting at the species rather than just organismic level.[18] At about the same time the work of Thomas Jukes and Motoo Kimura presented a picture of evolution at the level of proteins in which not only could natural selection act at the molecular level but that, because of hierarchical depth, neutral molecular evolution was possible, which was the basis of the molecular evolutionary clock.[19] Together this led to a period of controversy over levels of selection that forms the basis of an expanded synthesis. As Patricia Princehouse has pointed out, this effort by Gould helped revive interest in a putative alternative synthesis in the 1920s and 1930s, the "German Synthesis," which incorporated developmental biology with genetics and natural selection, and which was characterized by causal plurality acting at multiple levels and for which higher taxa arose not just by accumulated micro-mutations, but by chromosome rearrangements and/or changes in the regulation and timing of developmental processes.[20]

Princehouse goes on to argue that current efforts to accomplish the hierarchical expansion of the evolutionary synthesis and its extension to incorporate developmental phenomena are converging toward what might have arisen earlier, had the German Synthesis not been overwhelmed by the Modern Evolutionary Synthesis. In addition to the attempt to conceive of evolutionary phenomena in terms of complex systems dynamics, as mentioned above, Massimo Pigliucci has recently called explicitly for an Extended Evolutionary Synthesis to encompass evolutionarily important developmental processes as well as such phenomena as evolvability, phenotypic plasticity, and epigenesis.[21] For Pigliucci, such a move will take the synthesis from a theory of genes to, what Popper had called for, a theory of forms. For Pigliucci quantitative genetics is detached from populational genetic "forces" and is reattached to development allowing comparisons of various species in various ecological conditions.[22] Both Pigliucci and Scott Gilbert see emerging a field of "eco-evo-devo" in which eco-system dynamics play an integral role.[23] Complexity theory provides both new sources of variability as well as another organizing principle, along with the high-dimensional adaptive landscapes of complex systems dynamics. Such high-dimensional fitness landscapes facilitate the transition from one adaptive peak to another and major transitions affect the evolvability of all descendent clades. Such an expanded or extended evolutionary synthesis can still include the population genetic phenomena of the traditional synthesis as a special case, and can be considered as being within the Darwinian Research Tradition despite the changes in background dynamical assumptions.

However, there can be a question as to how far one can stretch the integument before allowing that a new research tradition is emerging that will replace the Darwinian. Some, such as Stanley Salthe, and possibly Gilbert, are open to this move in which selection is a trimming force at best, rather than having a creative and directional role.[24] It is simply too early in the process to tell, but as Ulanowicz points out, we can begin to see the contours of a new metaphysics facilitated by the opening of such a third window.[25] This window emphasizes the nonlinearity and dissipative nature of autocatalytic systems in which selection acts upon each component as well as on the processes and propensities of the system as a whole. The emergence of novelty in such complex systems is due to an interplay of chance, self-organization, and selection as well as an interplay of upward and downward influence.[26] In this view, agency resides in the processes rather than in the objects, which are taken to be configurations of processes. Further, it seems to me that the emergence and evolvability of semiotic processes becomes crucial for understanding living systems and their evolutionary dynamics, as has been championed by Jesper Hoffmeyer.[27]

Opening Ulanowicz's third window allows us to view nature and its phenomena in fuller richness rather than ignoring those aspects that do not conform with the assumptions of the first two windows. As David Depew and I once wrote, we need not to trim our biology to fit our physics but to enrich our physics to fully engage biological phenomena.[28] Process ecology, which informs the third window, can help guide us toward such an enriched science, one that holds the promise of more fully understanding evolution as well as the emergent processes and phenomena of natural systems more generally.

Notes

1. See David J. Depew and Bruce H. Weber, *Darwinism Evolving: Systems Dynamics and the Genealogy of Natural Selection* (Cambridge, MA: MIT Press, 1995); David J. Depew and Bruce H. Weber, "Challenging Darwinism: Expanding, Extending, or Replacing the Modern Evolutionary Synthesis," in *The Cambridge Encyclopedia of Darwin and Evolutionary Thought* (in press), ed. Michael Ruse (Cambridge: Cambridge University Press, 2012); Scott F. Gilbert and David Epel, *Ecological Developmental Biology: Integrating Epigenesis, Medicine, and Evolution* (Sunderland, MA: Sinauer, 2009); Stephen J. Gould, *The Structure of Evolutionary Theory* (Cambridge, MA: Harvard University Press, 2002); Massimo Pigliucci and Gerd B. Müller, eds., *Evolution: The Extended Synthesis* (Cambridge, MA: MIT Press, 2010); Bruce H. Weber, "Extending and Expanding the Darwinian Synthesis: The Role of Complex Systems Dynamics," *Studies in History and Philosophy of Biological and Biomedical Sciences* 42 (2011): 75-81.

2. See Larry Laudan, *Progress and Its Problems: Towards a Theory of Scientific Growth* (Berkeley, CA: University of California Press, 1977); Depew and Weber, *Darwinism Evolving*; Bruce H. Weber and David J. Depew, "Natural Selection and Self-organization: Dynamical Models as Clues to a New Evolutionary Synthesis," *Biology and Philosophy* 11 (1996): 33-65; Bruce H. Weber and David J. Depew, "Developmental Systems, Darwinian Evolution, and the Unity of Science," in *Cycles of Contingency: Developmental Systems and Evolution*, ed. S. Oyama, P. E. Griffiths, and R. D. Gray, 573-639 (Cambridge, MA: MIT Press, 2001).

3. See Depew and Weber, *Darwinism Evolving*.

4. See Robert E. Ulanowicz, *A Third Window: Natural Life beyond Newton and Darwin* (West Conshohocken, PA: Templeton Foundation Press, 2009).

5. See Depew and Weber, *Darwinism Evolving*; Ulanowicz, *A Third Window*.

6. See Vernon Kellogg, *Darwinism Today: A Discussion of Present-Day Scientific Criticism of the Darwinian Selection Theories* (New York: Holt, 1907).

7. See Depew and Weber, *Darwinism Evolving*.

8. Sol Tax, ed. *Evolution after Darwin, Volume 1: The Evolution of Life, Its Origin, History and Future* (Chicago: University of Chicago Press, 1960a); Sol Tax, ed. *Evolution After Darwin, Volume 2: The Evolution of Man, Mind, Culture and Society*

(Chicago: University of Chicago Press, 1960b); Sol Tax, ed. *Evolution after Darwin: Volume 3: Issues in Evolution* (Chicago: University of Chicago Press, 1960c).

9. See Depew and Weber, *Darwinism Evolving*.

10. See Stephen J. Gould, "The Hardening of the Modern Synthesis," in *Dimensions of Darwinism: Themes and Counterthemes in Twentieth Century Evolutionary Theory*, ed. M. Grene (Cambridge: Cambridge University Press, 1983), 71-93.

11. See Depew and Weber, *Darwinism Evolving*; Bruce H. Weber, "On the Emergence of Living Systems," *Biosemiotics* 2 (2009): 343-359.

12. See Depew and Weber, *Darwinism Evolving*; Weber and Depew, "Natural Selection and Self-organization," 33-65.

13. See Ulanowicz, *A Third Window*.

14. See Stuart A. Kauffman, *The Origins of Order: Self-organization and Selection in Evolution* (New York: Oxford University Press, 1993); Stuart A. Kauffman, *At Home in the Universe: The Search for the Laws of Self-organization and Complexity* (New York: Oxford University Press, 1995); Stuart A. Kauffman, *Investigations* (New York: Oxford University Press, 2000).

15. See Weber, "Emergence of Living Systems."

16. See Jeffery S. Wicken, *Evolution, Information and Thermodynamics* (New York: Oxford University Press, 1987); Depew and Weber, *Darwinism Evolving*; Depew and Weber, "Challenging Darwinism"; Weber, "Expanding the Darwinian Synthesis," 75-81.

17. See Philip Clayton, *Mind and Emergence: From Quantum to Consciousness* (Oxford: Oxford University Press, 2004); Weber, "Emergence of Living Systems"; Weber, "Expanding the Darwinian Synthesis," 75-81; Cynthia Macdonald and Graham Macdonald, *Emergence in Mind* (New York: Oxford University Press, 2010).

18. See Niles Eldredge and Stephen J. Gould, "Punctuated Equilibria: An Alternative to Phyletic Gradualism," in *Models in Paleobiology,* ed. Thomas J. M. Schopf (San Francisco: Freeman Cooper, 1972), 82-115; Stephen J. Gould and Niles Eldredge, "Punctuated Equilibria: The Tempo and Mode of Evolution Reconsidered," *Paleobiology* 3, no. 2 (1977): 115-151; Stephen J. Gould and Niles Eldredge, "Punctuated Equilibrium Comes of Age," *Nature* 366 (1993): 223-227.

19. See Motoo Kimura, "Evolutionary Rate at the Molecular Level," *Nature* 217 (1968): 624-626; Motoo Kimura, *The Neutral Theory of Molecular Evolution* (Cambridge: Cambridge University Press, 1983); Motoo Kimura, "Molecular Evolutionary Clock and the Neutral Theory," *Journal of Molecular Evolution* 26 (1987): 24-33; Jack L. King and Thomas H. Jukes, "Non-Darwinian Evolution," *Science* 164 (1969): 788-798; Thomas H. Jukes, "Transitions, Transversions and the Molecular Evolutionary Clock," *Journal of Molecular Evolution* 26 (1987): 87-98.

20. See Patricia Princehouse, "Punctuated Equilibria and Speciation. What Does It Mean to Be a Darwinian?" in *The Paleobiological Revolution: Essays on the Growth of Modern Paleontology*, ed. David Sepkowski and Michael Ruse, 149-175 (Chicago: University of Chicago Press, 2009).

21. See Massimo Pigliucci, "Do We Need an Extended Evolutionary Synthesis?" *Evolution* 61 (2007): 2743-2749.

22. See Massimo Pigliucci, "The Proper Role of Population Genetics in Modern Evolutionary Theory," *Biological Theory* 3 (2008): 316-324.

23. See Gilbert and Epel, *Ecological Developmental Biology*; Pigliucci and Müller, *Evolution*.

24. See Stanley N. Salthe, *Development in Evolution: Complexity and Change in Biology* (Cambridge, MA: MIT Press, 1993).
25. See Ulanowicz, *A Third Window*.
26. See Weber and Depew, "Natural Selection and Self-organization," 33-65; Ulanowicz, *A Third Window*.
27. See Jesper Hoffmeyer, *Signs of Meaning in the Universe* (Bloomington: Indiana University Press, 1996); Jesper Hoffmeyer, *Biosemiotics: An Examination into the Signs of Life and the Life of Signs* (Scranton: University of Scranton Press, 2009).
28. Depew and Weber, *Darwinism Evolving*, 495.

Bibliography

Clayton, Philip. *Mind and Emergence: From Quantum to Consciousness*. Oxford: Oxford University Press, 2004.

Depew, David J., and Bruce H. Weber. "Challenging Darwinism: Expanding, Extending, or Replacing the Modern Evolutionary Synthesis." In *The Cambridge Encyclopedia of Darwin and Evolutionary Thought* (in press), edited by M. Ruse. Cambridge: Cambridge University Press, 2012.

———. *Darwinism Evolving: Systems Dynamics and the Genealogy of Natural Selection*. Cambridge, MA: MIT Press, 1995.

Eldredge, Niles, and Stephen J. Gould. "Punctuated Equilibria: An Alternative to Phyletic Gradualism." In *Models in Paleobiology*, edited by Thomas J. M. Schopf, 82-115. San Francisco: Freeman Cooper, 1972.

Gilbert, Scott F., and David Epel. *Ecological Developmental Biology: Integrating Epigenesis, Medicine, and Evolution*. Sunderland, MA: Sinauer, 2009.

Gould, Stephen J. "The Hardening of the Modern Synthesis." In *Dimensions of Darwinism: Themes and Counterthemes in Twentieth Century Evolutionary Theory*, edited by M. Grene, 71-93. Cambridge: Cambridge University Press, 1983.

———. *The Structure of Evolutionary Theory*. Cambridge, MA: Harvard University Press, 2002.

Gould, Stephen J., and Niles Eldredge. "Punctuated Equilibria: The Tempo and Mode of Evolution Reconsidered." *Paleobiology* 3, no. 2 (1977): 115-151.

———. "Punctuated Equilibrium Comes of Age." *Nature* 366 (1993): 223-227.

Hoffmeyer, Jesper. *Biosemiotics: An Examination into the Signs of Life and the Life of Signs*. Scranton: University of Scranton Press, 2009.

———. *Signs of Meaning in the Universe*. Bloomington: Indiana University Press, 1996.

Jukes, Thomas H. "Transitions, Transversions and the Molecular Evolutionary Clock." *Journal of Molecular Evolution* 26 (1987): 87-98.

Kauffman, Stuart A. *At Home in the Universe: The Search for the Laws of Self-organization and Complexity*. New York: Oxford University Press, 1995.

———. *Investigations*. New York: Oxford University Press, 2000.

———. *The Origins of Order: Self-organization and Selection in Evolution*. New York: Oxford University Press, 1993.

Kellogg, Vernon. *Darwinism Today: A Discussion of Present-Day Scientific Criticism of the Darwinian Selection Theories*. New York: Holt, 1907.

Kimura, Motoo. "Evolutionary Rate at the Molecular Level." *Nature* 217 (1968): 624-626.
———. "Molecular Evolutionary Clock and the Neutral Theory." *Journal of Molecular Evolution* 26 (1987): 24-33.
———. *The Neutral Theory of Molecular Evolution*. Cambridge: Cambridge University Press, 1983.
King, Jack L., and Thomas H. Jukes. "Non-Darwinian Evolution." *Science* 164 (1969): 788-798.
Laudan, Larry. *Progress and Its Problems: Towards a Theory of Scientific Growth*. Berkeley, CA: University of California Press, 1977.
Macdonald, Cynthia, and Graham Macdonald. *Emergence in Mind*. New York: Oxford University Press, 2010.
Pigliucci, Massimo. "Do We Need an Extended Evolutionary Synthesis?" *Evolution* 61 (2007): 2743-2749.
———. "The Proper Role of Population Genetics in Modern Evolutionary Theory." *Biological Theory* 3 (2008): 316-324.
Pigliucci, Massimo, and Gerd B. Müller, eds. *Evolution: The Extended Synthesis*. Cambridge, MA: MIT Press, 2010.
Princehouse, Patricia. "Punctuated Equilibria and Speciation. What Does It Mean to Be a Darwinian?" In *The Paleobiological Revolution: Essays on the Growth of Modern Paleontology*, edited by David Sepkowski and Michael Ruse, 149-175. Chicago: University of Chicago Press, 2009.
Salthe, Stanley N. *Development in Evolution: Complexity and Change in Biology*. Cambridge, MA: MIT Press, 1993.
Tax, Sol, ed. *Evolution after Darwin, Volume 1: The Evolution of Life, Its Origin, History and Future*. Chicago: University of Chicago Press, 1960a.
———. *Evolution after Darwin, Volume 2: The Evolution of Man, Mind, Culture and Society*. Chicago: University of Chicago Press, 1960b.
———. *Evolution after Darwin: Volume 3: Issues in Evolution*. Chicago: University of Chicago Press, 1960c.
Ulanowicz, Robert E. *Ecology: The Ascendant Perspective*. New York: Columbia University Press, 1997.
———. *Growth and Development: Ecosystems Phenomenology*. New York: Springer-Verlag, 1986.
———. *A Third Window: Natural Life beyond Newton and Darwin*. West Conshohocken, PA: Templeton Foundation Press, 2009.
Weber, Bruce H. "Extending and Expanding the Darwinian Synthesis: The Role of Complex Systems Dynamics." *Studies in History and Philosophy of Biological and Biomedical Sciences* 42 (2011): 75-81.
———. "On the Emergence of Living Systems." *Biosemiotics* 2 (2009): 343-359.
Weber, Bruce H., and David D. Depew. "Developmental Systems, Darwinian Evolution, and the Unity of Science." In *Cycles of Contingency: Developmental Systems and Evolution*, edited by S. Oyama, P. E. Griffiths, and R. D. Gray, 573-639. Cambridge, MA: MIT Press, 2001.
———. "Natural Selection and Self-organization: Dynamical Models as Clues to a New Evolutionary Synthesis." *Biology and Philosophy* 11 (1996): 33-65.
Wicken, Jeffery S. *Evolution, Information and Thermodynamics*. New York: Oxford University Press, 1987.

Chapter Two
Why Emergence Matters

Philip Clayton

The explosion of studies of emergent complexity represents one of the most significant developments in recent biology. A staggering range of phenomena reflect the dynamics of complex systems, and the research being produced in these fields is revolutionizing the way we think about biological systems. The various chapters in this volume present some of these new results and begin to explore their implications. At the center of our common inquiry lies a philosophical question: how should one conceive the world in light of the dynamics of pervasive emergence? That perplexing question forms the focus of the present chapter.

Early modern science sought to reduce all natural phenomena to matter and the laws of physics. But a shift of emphasis has taken place in the last decades. Scientists now recognize nature's tendency to produce more and more complex forms of organization, not all reducible to fundamental laws. This new "non-reductionist" picture of the world gives rise to some rather different assessments of the goals and methods of the biological sciences. Minimally, fields such as systems biology open up significant new areas of study, such as proteomics and metabolomics, and provide important new data—for example, in the prediction of breast cancer. Less cautious commentators are speaking of fundamentally new paradigms for conceiving and studying the biosphere.

The chapters in this volume spell out in more detail some of the concepts, methods, and empirical results of the relevant new subfields in biology. It may be helpful, however, to begin first with a rough summary statement of what the new developments in biology share in common. Of course, any attempt at summary will be controversial; there is no single formulation or meta-theory that all the authors in this volume would endorse. I have chosen the term "emergence" because of its long history: for about 150 years it has been used as an umbrella term to describe non-mechanistic and non-reductionist understandings of the biological sciences. If this brief overview helps to orient the reader for the fascinating array of more specific theories contained in the other chapters of this volume, it will have served its function.

The Concept of Emergence: A First Approximation

The concept of emergence is often presented by contrasting it with two alternative (and still widely held) views. According to *reductionist theories*, the phenomena studied by a given discipline are considered to be scientifically understood only when they can be expressed using the laws of a lower-level discipline. When scientific reduction is successful, the higher-order phenomena become a special case of the more general explanatory framework represented by the lower-level laws. If one seeks to reduce each level to the level beneath it, one must eventually come down to the fundamental laws of physics, which then serve as the bedrock for all of nature.

According to *dualist theories*, by contrast, there are gaps in the relations between the various disciplines, such that the reductionist ideal is impossible. Not only can phenomena pertaining to *mind* not be explained in terms of any lower-level laws, dualists believe; they also challenge the claim that mind depends essentially on any of its physical or material substrates. Thus dualists have classically held that minds are essentially different from bodies and can continue to exist without them. Minds do not rely on the physical energies that sustain bodies and allow them to move; instead they belong to a different ontological order altogether.

Emergence theories attempt to split the difference between these opposing positions. They grant the downward dependence of the reductionists, but they challenge whether downward explanatory reduction can actually be achieved, even in principle. Instead, they maintain that it is a contingent fact of natural history that new forms of organization emerge that are not fully predictable or explainable in terms of lower-level laws, forces, or particles. Emergence in this sense breaks with the dualist belief that two separate kinds of "stuff" exist; in this sense, emergentists are monists. But we do argue that complex forms of organization produce new forms of causation in the natural world, such as the multiple forms of agency that we study across the biosphere.

What I have just written is the standard way to define emergence theories. But it is helpful to supplement this description with a second way of defining the concept. On this view, emergence is a theory about evolution. Specifically, it is a theory about how the various scientific disciplines that study evolution in the cosmos are related to one another. Suppose we order the various disciplines that study the natural world according to the order in cosmic history in which the phenomena that each studies first arose: quantum physics, classical or macrophysics, physical chemistry, biochemistry, genetics, cell biology, anatomy, etc... Let us label the resulting list using the letters A, B, C, etc... to stand for the specific disciplines. We can then number the particular relations between any two neighboring disciplines using the integers 1, 2, 3, etc…, yielding a three-leveled picture of emergence (see Figure 2.1.).

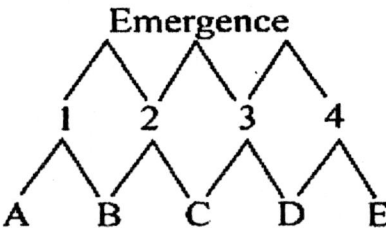

Figure 2.1.: Emergence is not itself a scientific theory; it identifies features shared in common in the transitions between scientific theories at various levels of reality.

As the diagram shows, emergence theory thus becomes a sort of third-order theory: a theory about the relations between the relations between scientific disciplines.

The Re-Emergence of Emergence

A special issue of the *Proceedings of the National Academy of Sciences* explores the principles of self-organization and the formation of complex matter, asking "What are the steps and the processes that lead from the elementary particle to the thinking organism, the (present!) entity of highest complexity?"[1] Complexity, the authors argue, is inherently a systemic function; it involves the interaction between multiple components of multiple kinds and the principles that affect their correlation, coupling, and feedback relationships. The goal of the sciences of complexity is "to progressively discover, understand, and implement the rules that govern [matter's] evolution from inanimate to animate and beyond, to ultimately acquire the ability to create new forms of complex matter."[2] Emergent complexity spans the entire spectrum of cosmic history, "from divided to condensed matter then to organized and adaptive matter, on to living matter and thinking matter, up the ladder of complexity."[3]

This vision clearly forms the core of research programs currently running in most, if not all, the major natural sciences, from the emergence of classical physical systems out of quantum physical systems, through the emergence of chemical properties and the origins of life, and on up to the higher cognitive behaviors of the Great Apes. Indeed, Gerald Edelman and Giulio Tononi are not doing anything different when, in their widely cited *Science* article, "Consciousness and Complexity," they offer a theory of consciousness as an emergent property of the brain.[4]

Examples of emergence do not start at the level of life or mind, however; they are already present in scientists' attempts to explain the very earliest stages of cosmic evolution. Classical physics, for example, has been described as emergent out of quantum physics.[5] And more recently Stephen Adler argued that quantum theory itself is an emergent phenomenon.[6] The physics Nobel laureate Robert Laughlin drew significant attention to the emergence debate within the sciences by arguing that scientific reduction is a dogma and that his field, condensed matter physics, could be grasped only by using the paradigm of emergence.[7] Among his examples of emergent phenomena are superconductivity, the quantum Hall effect, phase transitions, crystallization, collective instabilities, and hydrodynamics. Other studies of complex matter in which nonlinearities dominate, such as soft materials, come to similar conclusions. Elbio Dagotto, for example, reports the "spontaneous emergence of electronic nanometer-scale structures" in transition metal oxides:

> In complex systems the properties of a few particles are not sufficient to understand large aggregates when these particles strongly interact. Rather, in such systems, which are not merely complicated, one expects emergence, namely the generation of properties that do not preexist in a system's constituents. This concept is contrary to the philosophy of reductionism, the traditional physics hallmark. Complex systems spontaneously tend to form structures (self-organization) and these structures vary widely in size and scales.[8]

Chemists have long held up their discipline as a model of emergence. Pier Luigi Luisi, for example, maintains that chemistry is "the embodiment of emergence" because it studies properties that, although rooted in physical structures, cannot be explained without the help of a new conceptual framework. Chemical properties emerge only in sufficiently complex natural systems, and one requires a level of analysis distinct from physics in order to understand them.[9] Luisi also endorses downward causation: "Chemical examples show that emergence must go hand in hand with downward causation—one is the consequence of the other, and the two phenomena take place simultaneously."[10] He adds finally that "life can be seen as a particular kind of emergent property" and "life itself is indeed the most dramatic outcome of emergence."[11]

This new perspective goes under multiple names within specific sciences; "emergence" is, as mentioned above, only an overarching rubric to describe many different research programs in many different sciences. Molecular biologists speak, for instance, of the emergence of a "network perspective," which is necessary for describing how particular types of chemical reactions are catalyzed by evolutionarily related enzymes.[12] The analogue of metabolic networks in the study of cells is *systems biology*, one of the largest growth areas in contemporary biology. Hiroaki Kitano describes its core assumption: "While an understanding of genes and proteins continues to be important, the focus [of systems biology] is on understanding a system's structure and dynamics. Because a system is not just an assembly of genes and proteins, its properties cannot be fully understood merely by drawing diagrams of their interconnections."[13]

In cell biology, what was at first merely a way of expressing reservations about purely gene-driven analyses ("epigenetics" or "epigenetic factors") has become a rigorous study of "system-level insights" in its own right.[14] As Kitano writes, "A transition is occurring in biology from the molecular level to the system level that promises to revolutionize our understanding of complex biological regulatory systems and to provide major new opportunities for practical application of such knowledge."[15] And in an interesting review in *Science*, Kevin Laland baptizes the new approach as "the new interactionism," which describes "how genes are triggered into action by environmental events; how they switch other genes on and off; how they guide neurons to build brains; and how learning operates through gene expression."[16]

The new interactionism goes between the horns of the classical dilemmas, "genes *versus* environment" and "nature *versus* nurture." These new perspectives may defuse one further classical dilemma: the dilemma between "upward" and "downward" causation. Reductionists have generally held that all causal influences occur at the level of microphysics; these effects, taking place within highly complex systems, are said to account for the deceptive appearance that distinctively biological or psychological causes exist. In opposition to them, idealists and Cartesian dualists protested that there is a distinct type of cause, mental causes, which are different in kind from physical causes, are not dependent on them, and which exercise their own form of causal agency in the world. But from the systems or interactionist perspective, the entire dichotomy appears to be mistaken. Causality is "circular"; it involves interacting effects between different levels of natural organization (e.g., between the microscopic and the macroscopic). This new framework, as Moreno and Umerez note,

> makes it possible to talk properly about the appearance of new kinds of causal relationships.... Thus, what enables us to speak in terms of a double causal action—upward and downward—is precisely the conjunction of a circular causality with [at least] two different levels of organization, one of which is constituted by informational components.... [This] special, downward kind of causation appears just when very complex (interwoven) meta-networks of recursive reaction networks arise in Nature.[17]

Harold Morowitz likewise denies that the notion of downward causation is speculative, describing it instead as a "key feature" that is characteristic of all biological emergence.[18] One can trace examples of emergence along the path of natural history from the first cells to the great apes.[19] In fact, Morowitz identifies as many as twenty-eight distinct levels of emergence in his monograph, *The Emergence of Everything*.[20] In an article in *Physics Today*, the astrophysicist

George Ellis describes the emergent features of quantum measurement, DNA coding, social creations, and economics, and Ellis et al. have edited a recent issue of the *Journal of the Royal Society Interface,* which is devoted entirely to scientific articles on emergence and top-down causation.[21] Finally, in this presentation, I have not even begun to trace the construal of consciousness as an emergent phenomenon from sufficiently complex brains and central nervous systems.[22] It is basic to the cognitive sciences, to evolutionary psychology, and to traditional social science (psychology, sociology, cultural anthropology) to treat consciousness as depending on its neural substratum but not as identical to it. Consciousness may also represent a clear example of an emergent phenomenon that is not fully accessible to scientific study.

What Makes Emergence Scientific?

It is one thing to speculate about emergence as part of a metaphysical theory, and something else to claim scientific support for the framework of emergence. Although I believe such claims are justified, they bring with them unique challenges. As long as one continues to do science, one attempts to draw the closest possible connections between the set of phenomena, which one is studying, and the lower-level laws that are available. The scientific study of chemistry is impossible, for example, without its connections to physics; to study the origins of life *means* to try to explain the transition from non-reproducing biochemical molecules to reproducing life-forms (and if one could understand life in terms of biochemistry, so much the better); and to understand an animal's behavior scientifically *just is* to explain as much of it as possible in terms of the body's morphology, hormone releases, selective pressures, and the like. The quest to explain phenomena in terms of reconstructible, testable causal systems is so basic to the project of science that we could almost use it as *the* defining characteristic of science.

The phenomenon of emergence makes this project more difficult, but it does not eliminate it. If it had turned out to be possible to explain higher-order phenomena in terms of lower-order laws across the scientific disciplines, then science would be in the position fully to achieve the goal in terms of which it is often defined: reducing all empirical phenomena to a single set of fundamental laws and initial conditions. Even if, as emergence theorists believe, the natural world is such that this downward reduction is often impossible in principle, the goal does not simply disappear, to be replaced by a happy-go-lucky holism. One can still determine the scientific or non-scientific status of a theory about the world by the continued presence of this goal—namely, to connect the phenomena of one level as closely and precisely as possible with the phenomena at the next lower level.

Consider two concrete examples. Walter Elsasser, in his classic *Reflections on a Theory of Organisms*,[23] makes the case for the autonomy of biological explanations. Elsasser believes that the "information stability" of living things cannot be reduced to the physico-chemical stability of molecules. Alongside this emergentist manifesto, however, Elsasser is careful to show how the scientific study of life-forms is still possible. He does not argue for the untestability of biological theories, for example, but rather for the necessity of utilizing different kinds of tests. It is just that biological causality "cannot be fully verified by the standard procedure of the physicist or chemist," that is, by "measurement followed by mathematical extrapolation, technically called integration, of the equations of quantum mechanics that govern molecular motion."[24]

Moreover, Elsasser gives very precise, empirical reasons for a certain autonomy of biology, which he defines in a precise and rigorous fashion. Thus his fourth chapter demonstrates how an immense number of molecular configurations are compatible with a given set of physico-chemical constraints. Even if the structure and dynamics of all molecules can be understood by applying quantum mechanical principles, "a cell is much too complex to admit of meaningful analysis in such terms."[25] Because of this "combinatorial explosion" (Harold Morowitz), it is demonstrably impossible to compute cell behavior in quantum-physical terms. Likewise, in a later chapter Elsasser describes how physicists study stable systems (those in which each mode of motion is stable) and then come to understand what happens when the system is perturbated. All biological systems, by contrast, are massively unstable; they are, in Stuart Kauffman's beautiful expression, always existing "at the edge of chaos."[26] Hence it is not possible to understand them by extrapolation from stable systems and through computation of each perturbation—which is to say: it is impossible to understand them physically. Elsasser states, "Owing to the amplificatory effect, the ultimate changes are no longer predictable."[27] Elsasser concludes that "the morphological future of such [biological] systems [is] unpredictable on the basis of physics and chemistry."[28]

Finally, Elsasser remains committed to the scientific study of biological phenomena, even in light of this unpredictability. He emphasizes that biological results "cannot differ from any known rule of physics and chemistry."[29] He continues to emphasize the importance of structure (morphology) and function as basic to testable biological theories. And he does not advance biological holism as a way of avoiding tests and experiments, but rather as representing a call for a new type of testability: "If the holistic properties are to be verified experimentally, a different type of experiment from that conventionally used by physicists and chemists is required."[30] Elsasser thus represents a paradigmatic example of scientific emergence. Under the influence of this example, Morowitz comments, "emergence requires pruning rules to reduce the transcomputable to the computable. . . . [I]n both Elsasser's approach and [John] Holland's view, biology requires its own laws that are not necessarily derivable from physics, but do not contradict the physical foundations."[31]

Contrast this careful demonstration of how emergence remains a part of the scientific project with the approach to emergence taken by Brian C. Goodwin and Rupert Sheldrake. Both thinkers wish to appeal to what they call "morphogenetic fields" and "morphic resonance." Goodwin explicitly refuses to interpret the morphogenetic field in terms of any known forces: "electrical forces can affect it . . . but I would not wish to suggest that [it] is essentially electrical. Chemical substances" can affect it, yet it is not "essentially chemical or biochemical in nature." He is nonetheless certain that morphogenetic fields "play a primary role in the developmental process."[32] Similarly, Sheldrake postulates "patterns of oscillatory activity" throughout the world, which he calls morphic resonance.[33] Resonances are strongest, he is sure, with one's own past, somewhat weaker with genetically similar animals, and weaker still with animals from other races. Apparently, though, some resonance still exists between all living things. A genuinely scientific theory of morphic resonances looks unlikely, however, since, like Goodwin, he is loath to correlate them with any known forces.

Sheldrake does claim that his theory has "testable predictions," although he admits that his predictions "may seem so improbable as to be absurd." For example,

> If thousands of rats were trained to perform a new task in a laboratory in London, similar rats should learn to carry out the same task more quickly in laboratories everywhere else. If the speed of learning of rats in another laboratory, say in New York, were to be measured before and after the rats in London were trained, the rats tested on the second occasion should learn more quickly than those tested on the first. The effect should take place in the absence of any known type of physical connection or communication between the two laboratories.[34]

Unfortunately, this kind of testability is not sufficient to make a theory scientific. It is equally crucial that the theories in question specify their connections with the existing body of scientific knowledge. In addition to meeting this condition, a theory of scientific emergence must provide details, given as much as possible in terms of lower-level theories, that show why a given set of phenomena would be irreducible to those theories. When somebody suggests, as Sheldrake does, that there are both energetic and non-energetic fields, this means that they must seek to show how the one arises out of the other. The same principle holds for relations between (say) biochemistry and the actions of organisms as agents.

The Logic of Scientific Emergence

Terrence Deacon offers the clearest expression of the logic of scientific emergence available today; it is therefore valuable to consider his proposal in some detail. Deacon begins with the empirical evidence that "complex dynamical ensembles can spontaneously assume ordered patterns of behavior that are not prefigured in the properties of their component elements or in their interaction patterns."[35] On his view, only emergence theories can adequately interpret and explain this type of self-organization.

Not all emergences are identical, however. Deacon is the first to have identified three distinct "orders" of emergence. First-order emergence involves the appearance of new properties in the aggregate that are not present in the individual particles. Deacon draws his primary examples from quantum theory and statistical thermodynamics, though he admits that even simpler examples can be adduced, such as how the properties of water molecules produce liquid properties. He writes, "Although the nature of the wave and its detailed underlying dynamical realization in each [particular wave] may differ depending on whether the fluid is water, air, or an electromagnetic field, the ability to propagate a wave is a first-order emergent feature they all share in common."[36] As with the other two "orders" of emergence, Deacon is careful to specify exactly what are the conditions under which this kind of emergence occurs. Thus he argues, for example, that "it is only when certain of the regularities of molecular interaction relationships add up rather than cancel one another that certain *between*-molecule relationships can produce aggregate behaviors with ascent in scale."[37]

In second-order emergence, specific perturbations of a system are amplified, resulting in types of causal effects not seen in the first order. In the formation of snow crystals, for example, the specific temperature and humidity present at each stage of the crystal's descent through the air are "recorded" in the emerging structure of the crystal as it evolves; these features then influence its subsequent structural formation. The structural features emerging at a given point are amplified, in other words, such that they affect all subsequent crystal growth. These "feed-forward circles of cause and effect" are distinctive of this new type of emergent property. Deacon offers detailed examples, drawing from "self-undermining (divergent) chaotic systems, as in turbulent flow, and self-organizing (partially convergent) chaotic systems."[38] Second-order emergence is also found in the so-called autopoietic systems. This type of emergence works not just by aggregating individual components (say, molecules); here systematic features play a causal role. Put differently, forms or structures, and not merely particles, become the operative links in the feed-forward cycle.

Third-order emergence shares this feature from the previous order. Yet now what is passed on is *information or memory*, not merely forms and structures. As a result, Deacon argues,

> Third-order emergence inevitably exhibits a developmental and/or evolutionary character. . . . It occurs where there is both amplification of global influences on parts, but also redundant "sampling" of these influences and reintroduction of them redundantly across time and into different realizations of the same type of second-order system.[39]

The classic example of third-order emergence is the self-reproducing cell.[40] Cells exhibit features not present in pre-biological physical systems; they contain information—specifically, information sufficient for building other cells like themselves. The information is stored inside a boundary (the cell wall), which allows the cell as a whole to function as an entity in its own right on which environmental forces act. Because cells can make copies of themselves in ways that pre-biotic structures cannot, the forces of natural selection can begin to operate for the first time. A cell thus becomes a sort of hypothesis about what informational structure will survive and reproduce most effectively in a given environment. If the cell exists in an environment congenial to its reproductive success, it will make more successful copies of itself than its rivals and come to prosper in its ecosystem. Deacon describes this process as "a sort of self-referential self-organization, an autopoiesis of autopoieses."[41] As a result, cells can only be understood through "a combination of multi-scale, historical, and semiotic analyses. . . . This is why living and cognitive processes require us to introduce concepts such as representation, adaptation, information, and function in order to capture the logic of the most salient emergent phenomena."[42]

By introducing the framework of semiotics, derived from the work of C. S. Peirce, Deacon implicitly claims that the cell is an interpreter of the world: it stands in an informationally mediated relationship to its environment. Emergent entities at this level refer to or represent their world; they are themselves hypotheses about how best to survive and thrive in a particular environment. Despite his occasional reticence about anti-reductionist language, Deacon's position clearly stands as a sharp alternative to reductionist analyses of the natural world:

> Life and mind cannot be adequately described in terms that treat them as merely supervenient because this collapses innumerable convoluted levels of emergent relationships. Life is not mere chemical mechanism. Nor is cognition mere molecular computation.[43]

If Deacon is right—and I think recent origins-of-life research, systems biology, and ecosystems theory all offer empirical support for his analysis—then to understand life scientifically means to understand it according to different principles than those that pertain to purely physical or chemical systems. Only third-order emergent processes in evolution have the capacity "to progressively embed [other] evolutionary processes within one another via representations that amplify their information-handling power."[44] Indeed, the process of emergence does not stop with the first self-reproducing cell. Under evolutionary selection pressures, natural systems continue to increase in complexity, discovering ever new ways of "making a living" (as Stuart Kauffman likes to say) in the world. In Deacon's masterful study, *The Symbolic Species*,[45] for example, he traces the co-evolution of brains and language or culture. Language use, of course, remains dependent on a complex brain and central nervous system, but language can never be reduced to an instinct,[46] or a mere byproduct of brain processes; the two evolving phenomena mutually influence one another. The result is a continual growth in complexity of both of them. For example, Deacon argues, language moves from the "iconic" mode of representation to a more complex form of reference involving indexicals, and finally (in *Homo sapiens*) to the rich symbolic modes of representation that are the bread and butter of human cultural existence.

In Defense of Strong Emergence

One finds in the literature an ongoing battle between weaker and stronger versions of emergence theory. The more robust version, which I have labeled "strong emergence" and defended in *Mind and Emergence,* makes a clear claim about causation and agency. According to strong emergentists, new *kinds of agents* emerge in natural history, which means that a full explanation of the behaviors of an organism cannot be given in biochemical or chemical or microphysical terms. Biochemistry, genetics, and (in more complex organisms) neurology are part of the causal story. But a crucial part of the explanatory story must also be given in terms of the particular biological systems with and in which the organism interacts. Biological agents are agents as this particular organism in interaction with other organisms in their ecosystem. Explanations that omit the features of the interacting systems, *described at the highest level of emergent complexity that they manifest*, are necessarily incomplete. If the relevant systems include culture (socially transmitted and learned behaviors), mental representations, or intentions, then a complete explanation must be given, at least in part, in these terms.

Of course, interpretations of evolution are fraught with controversy. If evolution is really all about the genes, as Richard Dawkins' work implies, then all evolved structures are nothing more than expressions of this same fundamental dynamic. However rich and staggeringly diverse these manifestations are, they can and should still be understood from a gene-centric perspective. If, on the other hand, the dualists are right, then at some point in the evolutionary process one encounters a radical, ontological break: mind arises. In this sense dualists remain, at heart, Cartesians: one can study the entire physical world from atoms to chimpanzees with the same set of mechanistic explanatory tools; but as soon as one turns to man or woman, who alone possesses Mind, a new explanatory tool box is required—one that relies now on the nature of souls, the eternal Laws of Thought, and perhaps the intentions of God.

In contrast to both views, emergentists claim that the story of evolution is one of continuity *and* discontinuity. Continuity first: everything in the natural world is composed of the same "stuff" of matter and energy, and no new substances are added along the way. This means no souls and no personal substances—a difficult entailment for more traditionally minded Western philosophers. When one pursues the scientific project, one seeks to develop a continuity of understanding to the greatest possible extent. But sharing the scientist's "natural piety" for the world *as it actually expresses itself empirically* also means that one works with whatever explanatory framework best explains the data at present—as long as it is testable and can demonstrate its explanatory superiority over its rivals.

Practicing this natural piety—this commitment to study the world in whatever ways it presents itself to us—means that we are not merely biologists *simpliciter*, much less geneticists only. We are cytologists, systems biologists, botanists, zoologists, primatologists. We are not only molecular biologists and geneticists, but also population biologists and ecosystem theorists. We are interested in the large and complex as much as in the small, in emergent phenomena as well as reducible phenomena. This, I suggest, was the mind-set that lay behind Charles Darwin's great breakthroughs. Recall his famous description of the river bank in the *Origin of Species*:

> It is interesting to contemplate an entangled bank, clothed with many plants of many kinds, with birds singing on the bushes, with various insects flitting about, and with worms crawling through the damp earth, and to reflect that these elaborately constructed forms, so different from each other in so complex a manner, have all been produced by laws acting around us.... There is grandeur in this view of life, with its several powers, having originally breathed into a few forms or into one; and that, whilst this planet has gone on cycling on according to the fixed law of gravity, from so simple a beginning endless forms most beautiful and most wonderful have been, and are being, evolved.[47]

What Darwin could only envision in 1858 had become scientific reality by the end of the twentieth century. In one sense, discovering complex systems is just "science as usual"; the new systems biology provides causal explanations of natural systems, which are then empirically tested and verified. In another sense, however, emergence represents a new way of doing and understanding the biological sciences. It is biology without reductionism. Now systems of emergent phenomena are understood not only in terms of lower-level laws and processes, but equally in terms of the higher-level systems that constrain their functioning and their outcomes. The science is complicated; there are many more subtleties in the study of emergent phenomena than we have covered here. Other chapters in this volume give much fuller accounts of the details and the actual complexities of organism-centered biology today. But enough is already on the table for one to see that the results of this approach are of revolutionary significance for our emerging understanding of living systems.

Notes

1. See the introduction to the collection in Jean-Marie Lehn, "Toward Complex Matter: Supramolecular Chemistry and Self-organization," *Proceedings of the National Academy of Sciences* 99, no. 8 (April 16, 2002): 4763-4768.
2. Lehn, "Toward Complex Matter," 4768.
3. Lehn, "Toward Complex Matter," 4768.
4. See Giulio Tononi and Gerald M. Edelman, "Consciousness and Complexity," *Science* 282 (4 December 1998): 1846-1851.
5. Wojciech Zurek, "Decoherence and the Transition from Quantum to Classical—Revisited," *Los Alamos Science* 27 (2002): 86-109; see Zurek, "Decoherence and the Transition from Quantum to Classical," *Physics Today* (1991), 44.
6. See Stephen Adler, *Quantum Theory as an Emergent Phenomenon: The Statistical Dynamics of Global Unitary Invariant Matrix Models as the Precursor of Quantum Field Theory* (Cambridge: Cambridge University Press, 2004). If string theory is correct, quantum mechanics would be an emergent property of strings.
7. Numerous examples are offered by Robert B. Laughlin in his recent book, *A Different Universe: Reinventing Physics from the Bottom Down* (New York: Basic Books, 2005). See also the review by Philip Anderson, "Emerging Physics: A Fresh Approach to Viewing the Complexity of the Universe," *Nature* 434 (7 April 2005): 701-702.
8. See Elbio Daggoto, "Complexity in Strongly Correlated Electronic Systems," *Science* 309 (8 July 2005): 257-262, citing 257.
9. See Pier Luigi Luisi, "Emergence in Chemistry: Chemistry as the Embodiment of Emergence," *Foundations of Chemistry* 4 (2002): 183-200. See also his forthcoming book: *The Emergence of Life: From Chemical Origins to Synthetic Biology* (Cambridge, MA: Cambridge University Press, 2006).

An earlier case for the importance of emergence in chemistry was made by the process thinker, Joseph Earley in "Far-From-Equilibrium Thermodynamics and Process Thought," in *Physics and the Ultimate Significance of Time*, ed. David R. Griffin (Albany: State University of New York Press, 1985): 251-255; "The Nature of Chemical Existence," in *Metaphysics as Foundation*, ed. Paul Bogaard and Gordon Treash (Albany: State University of New York Press, 1992); "Self-Organization and Agency: In Chemistry and in Process Philosophy," *Process Studies* 11 (1981): 242-258; "Towards a Reapprehension of Causal Efficacy," *Process Studies* 24 (1995): 34-38; and "Collingwood's Third Transition: Replacement of Renaissance Cosmology by an Ontology of Evolutionary Self-Organization," in *With Darwin beyond Descartes—The Historical Concept of Nature and Overcoming "The Two Cultures,"* ed. Luigi Zanzi (Pavia, Italy: forthcoming).

10. See Luisi, "Emergence in Chemistry," 195.
11. Luisi, "Emergence in Chemistry," 197.
12. Rui Alves, Raphael A. G. Chaleil, and Michael J. E. Sternberg, "Evolution of Enzymes in Metabolism: A Network Perspective," *Journal of Molecular Biology* 320 (2002): 751-770.

13. See Hiroaki Kitano, "Systems Biology: A Brief Overview," *Science* 295 (1 March 2002): 1662-1664.
14. Kitano, "Systems Biology," 1664. See also chapters 15 and 16 in this volume.
15. Kitano, "Systems Biology," 1664.
16. See Kevin Laland, "The New Interactionism," *Science* 300 (20 June 2003): 1879-1880, drawing on Matt Ridley, *Nature via Nurture: Genes, Experience, and What Makes Us Human* (New York: HarperCollins, 2003).
17. See Alvaro Moreno and Jon Umerez, "Downward Causation at the Core of Living Organization," in *Downward Causation: Minds, Bodies and Matter*, ed. Peter Bøgh Andersen, Claus Emmeche, Niels Ole Finnemann, and Peder Voetmann Christiansen (Aarhus, Denmark: Aarhus University Press, 2000), 112, 115.
18. See Harold Morowitz, "The Emergence of Intermediary Metabolism," unpublished paper, 8.
19. Barbara Smuts, "Emergence in Social Evolution: A Great Ape Example," in *The Re-emergence of Emergence: The Emergentist Hypothesis from Science to Religion*, ed. Philip Clayton and Paul Davies (Oxford: Oxford University Press, 2006).
20. Harold Morowitz, *The Emergence of Everything: How the World Became Complex* (New York: Oxford University Press, 2002).
21. See George Ellis, "Physics, Reductionism, and the Real World," in *Physics Today* 59, no. 3 (March 2006): 12-14; George Ellis "Physics and the Real World," in *Physics Today* 58, no. 7 (July 2005): 49-54; George Ellis, Denis Noble, and Timothy O'Connor, "Top-Down Causation: An Integrated Theme within and across the Sciences," *Journal of the Royal Society Interface* 2, no. 1 (February 6, 2012): 1-3, http://rsfs.royalsocietypublishing.org/content/2/1. Accessed July 27, 2012.
22. Philip Clayton, *Mind and Emergence: From Quantum to Consciousness* (Oxford: Oxford University Press, 2004).
23. Walter Elsasser, *Reflections on a Theory of Organisms* (Quebec: Editions ORBIS, 1987).
24. Elsasser, *Reflections*, 142.
25. Elsasser, *Reflections*, 52.
26. Stuart A. Kauffman develops these ideas in *At Home in the Universe: The Search for the Laws of Self-organization and Complexity* (New York: Oxford University Press, 1995) and *Investigations* (New York: Oxford University Press, 2000).
27. Elsasser, *Reflections*, 105.
28. Elsasser, *Reflections*, 142.
29. Elsasser, *Reflections*, 148.
30. Elsasser, *Reflections*, 148.
31. Morowitz, "The Emergence of Intermediary Metabolism," 4.
32. Brian C. Goodwin, "On Morphogenetic Fields," *Theoria to Theory* 13 (1979): 109-114, cited in Rupert Sheldrake, *A New Science of Life: The Hypothesis of Morphic Resonance* (Rochester, VT: Park Street Press, 1995).
33. Sheldrake, *A New Science of Life*, 170.
34. Sheldrake, *A New Science of Life*, 14.
35. Terrence Deacon, "The Hierarchic Logic of Emergence: Untangling the Interdependence of Evolution and Self Organization," in *Evolution and Learning: The Baldwin Effect Reconsidered*, ed. Bruce H. Weber and David J. Depew (Cambridge, MA: MIT Press, 2003), 273-308.
36. Deacon, "The Hierarchic Logic of Emergence," 290.
37. Deacon, "The Hierarchic Logic of Emergence," 288.

38. Deacon, "The Hierarchic Logic of Emergence," 295.
39. Deacon, "The Hierarchic Logic of Emergence," 299.
40. Deacon has developed the notion of the "autocell" in Terrence W. Deacon, "Reciprocal Linkage between Self-organizing Processes Is Sufficient for Self-reproduction and Evolvability," *Biological Theory* 1, no. 2 (Spring 2006): 136-149.
41. Deacon, "The Hierarchic Logic of Emergence," 299.
42. Deacon, "The Hierarchic Logic of Emergence," 300.
43. Deacon, "The Hierarchic Logic of Emergence," 304.
44. Deacon, "The Hierarchic Logic of Emergence," 305.
45. Deacon, *The Symbolic Species: The Co-Evolution of Language and the Brain* (New York: W. W. Norton and Company, 1997). See also Zoltán N. Oltvai and Albert-László Barabási, "Life's Complexity Pyramid," *Science* 298 (25 October 2002): 763-764. Barabási is best known for his popular presentation, *Linked: The New Science of Networks* (Cambridge, MA: Perseus Books, 2002); for a more technical presentation see Barabási and Reka Albert, "Emergence of Scaling in Random Networks," *Science* 286 (15 October 1999): 509-512. Note that embracing the core principles of network theory does not mean that it will lead to "an accurate mathematical theory of human behavior," as Barabási claims in "Network Theory—The Emergence of the Creative Enterprise," *Science* 308 (29 April 2005): 639-641.
46. Steven Pinker, *The Language Instinct: How the Mind Creates Language* (New York: HarperPerennial ModernClassics, 2007).
47. Charles Darwin, *The Origin of Species by Means of Natural Selection* (New York: D. Appleton and Company, 1889); see the concluding paragraph.

Bibliography

Adler, Stephen. *Quantum Theory as an Emergent Phenomenon: The Statistical Dynamics of Global Unitary Invariant Matrix Models as the Precursor of Quantum Field Theory.* Cambridge, MA: Cambridge University Press, 2004.

Alves, Rui, Raphael A. G. Chaleil, and Michael J. E. Sternberg. "Evolution of Enzymes in Metabolism: A Network Perspective." *Journal of Molecular Biology* 320 (2002): 751-770.

Anderson, Philip. "Emerging Physics: A Fresh Approach to Viewing the Complexity of the Universe." *Nature* 434 (7 April 2005): 701-702.

Barabasi, Albert-László. *Linked: The New Science of Networks.* Cambridge, MA: Perseus Books, 2002.

———. "Network Theory—The Emergence of the Creative Enterprise." *Science* 308 (29 April 2005): 639-641.

Barabási Albert-László, and Reka Albert, "Emergence of Scaling in Random Networks." *Science* 286 (15 October 1999): 509-512.

Clayton, Philip. *Mind and Emergence: From Quantum to Consciousness.* Oxford: Oxford University Press, 2004.

Daggoto, Elbio. "Complexity in Strongly Correlated Electronic Systems." *Science* 309 (8 July 2005): 257-262.

Darwin, Charles. *The Origin of Species By Means of Natural Selection.* New York: D. Appleton and Company, 1889.

Deacon, Terrence. *The Symbolic Species: The Co-Evolution of Language and the Brain.* New York: W. W. Norton and Company, 1997.

———. "The Hierarchic Logic of Emergence: Untangling the Interdependence of Evolution and Self Organization." In *Evolution and Learning: The Baldwin Effect Reconsidered*, edited by Bruce H. Weber and David J. Depew. Cambridge, MA: MIT Press, 2003, 273-308.

———. "Self-Reproduction and Natural Selection from Reciprocally Linked Self-Organizing Processes." *Nature*, forthcoming.

Earley, Joseph. "Self-Organization and Agency: In Chemistry and in Process Philosophy." *Process Studies* 11 (1981): 242-258.

———. "Far-From-Equilibrium Thermodynamics and Process Thought." In *Physics and the Ultimate Significance of Time*, edited by David R. Griffin. Albany: State University of New York Press, 1985, 251-255.

———. "The Nature of Chemical Existence." In *Metaphysics as Foundation*, edited by Paul Bogaard and Gordon Treash. Albany: State University of New York Press, 1992.

———. "Towards a Reapprehension of Causal Efficacy." *Process Studies* 24 (1995): 34-38.

———. "Collingwood's Third Transition: Replacement of Renaissance Cosmology by an Ontology of Evolutionary Self-Organization." In *With Darwin beyond Descartes—The Historical Concept of Nature and Overcoming "The Two Cultures,"* edited by Luigi Zanzi. Pavia, Italy, forthcoming.

Ellis, George. "Physics, Reductionism, and the Real World." *Physics Today* 59, no. 3 (March 2006): 12-14

———. "Physics and the Real World." *Physics Today* 58, no. 7 (July 2005): 49-54.

Ellis, George, Denis Noble, and Timothy O'Connor. "Top-Down Causation: An Integrated Theme within and across the Sciences." *Journal of the Royal Society Interface* 2, no. 1 (February 6, 2012): 1-3, http://rsfs.royalsocietypublishing.org/content/2/1. Accessed July 27, 2012.

Elsasser, Walter. *Reflections on a Theory of Organisms.* Quebec: Editions ORBIS, 1987.

Goodwin, Brian C. "On Morphogenetic Fields." *Theoria to Theory* 13 (1979): 109-114.

Kauffman, Stuart A. *At Home in the Universe: The Search for the Laws of Self-organization and Complexity.* New York: Oxford University Press, 1995.

———. *Investigations.* New York: Oxford University Press, 2000.

Kitano, Hiroaki. "Systems Biology: A Brief Overview." *Science* 295 (1 March 2002): 1662-1664.

Laland, Kevin. "The New Interactionism." *Science* 300 (20 June 2003): 1879-1880.

Laughlin, Robert B. *A Different Universe: Reinventing Physics from the Bottom Down.* New York: Basic Books, 2005.

Lehn, Jean-Marie. "Toward Complex Matter: Supramolecular Chemistry and Self-organization." *Proceedings of the National Academy of Sciences* 99, no. 8 (April 16, 2002): 4763-4768.

Luisi, Pier Luigi. "Emergence in Chemistry: Chemistry as the Embodiment of Emergence." *Foundations of Chemistry* 4 (2002): 183-200.

———. *The Emergence of Life: From Chemical Origins to Synthetic Biology.* Cambridge: Cambridge University Press, 2006.

Moreno, Alvaro, and Jon Umerez, "Downward Causation at the Core of Living Organization." In *Downward Causation: Minds, Bodies and Matter*, edited by Peter Bøgh

Andersen, Claus Emmeche, Niels Ole Finnemann, and Peder Voetmann Christiansen. Aarhus, Denmark: Aarhus University Press, 2000.

Morowitz, Harold. *The Emergence of Everything: How the World Became Complex.* New York: Oxford University Press, 2002.

———. "The Emergence of Intermediary Metabolism," unpublished paper.

Oltvai, Zoltán N., and Albert-László Barabási. "Life's Complexity Pyramid." *Science* 298 (25 October 2002): 763-764.

Pinker, Steven. *The Language Instinct: How the Mind Creates Language.* New York: HarperPerennial ModernClassics, 2007.

Ridley, Matt. *Nature via Nurture: Genes, Experience, and What Makes Us Human.* New York: HarperCollins, 2003.

Sheldrake, Rupert. *A New Science of Life: The Hypothesis of Morphic Resonance.* Rochester, VT: Park Street Press, 1995.

Smuts, Barbara. "Emergence in Social Evolution: A Great Ape Example." In *The Re-emergence of Emergence: The Emergentist Hypothesis from Science to Religion*, edited by Philip Clayton and Paul Davies. Oxford: Oxford University Press, 2006.

Tononi, Giulio, and Gerald M. Adelman. "Consciousness and Complexity." *Science* 282 (4 December 1998): 1846-1851.

Zurek, Wojciech. "Decoherence and the Transition from Quantum to Classical—Revisited." *Los Alamos Science* 27 (2002): 86-109.

———. "Decoherence and the Transition from Quantum to Classical." *Physics Today* (1991): 36-47.

Chapter Three
On the Incompatibility of the Neo-Darwinian Hypothesis with Systems-Theoretical Explanations of Biological Development

Gernot Falkner and Renate Falkner

The Implications of the Mechanistic Worldview for Biology

The neo-Darwinian hypothesis that seeks to explain the evolution of species exclusively through the concepts of mutation and selection suffers from several inconsistencies and limitations. The main reason for these problems can be traced back to its grounding in the mechanistic worldview, in which basic propositions of the physics of the nineteenth century have been generalized into an ideology. In order to show how the shortcomings of the neo-Darwinian hypothesis can be avoided using a process philosophical interpretation of the recently established systems-environment theory, it is necessary to recapitulate briefly the implications of the mechanistic worldview for biology.

Classical physics owes its success to two fundamental metaphysical propositions. The first refers to Newton's conception of absolute space and absolute time forming a continuum that is separable into infinitesimal quantities that do not penetrate each other. The second proposition deals with the Cartesian separation of reality into mental substances (*res cogitans*) and material substances (*res extensa*). The two substances differ in that the mental substance is supposed to be the basis of conscious experience, whereas the material substance is composed of independently existing things "which require nothing but [themselves] in order to exist."[1] This conceptual scheme allowed the state of independently existing bodies to be localized at any instant of time and predicted their spatiotemporal alteration under the condition that the forces acting upon them are known.

The successful technical application of these fundamental propositions led to the idea that every process in nature can be understood by superimposing dynamical laws or equations of motion upon defined states of material configurations. This provided an "objective" description of natural changes, but required a notion of substance that is devoid of everything that can be related to subjectivity, feelings, and organismic inwardness. A generalization of this reductionist interpretation of natural processes resulted in a cosmology in which Descartes' mental substance was no longer needed. The thus obtained mechanistic worldview is therefore essentially unhistorical, because in historical processes subjective interpretations of antecedent events have a potential impact on the course of subsequent events. In contrast, mechanistic operations are supposed to proceed independently of possible acts of observation and their interpretations.

An application of this reductionist conception to biological processes, performed vigorously by biologists since the eighteenth century, had far-reaching consequences for an understanding of living systems. An interpretation of natural processes, solely based on efficient causes, is forced to deny the existence of *final causes* that direct two or more *simultaneous* changes within a system toward a particular finite state. In this case a not yet existing state would exert a regulatory function on antecedent processes and give biological processes an inherent orientation opposed to mechanistic explanations in which a cause always has to *precede* an effect. In biology, this raises a problem since almost all important biological processes, such as the growth of a cell, the emergence of highly differentiated organisms from a fertilized egg, or the evolution of more complex organisms from simpler living forms, exhibit directionality toward more organized structures and behaviors. In order to establish the general validity of mechanistic interpretations in biology, those apparently directed processes had to be explained on the basis of efficient causes. To achieve this, the analytical and cognitive strategies of classical physics were adopted in biological investigations, assuming that living systems consist of localizable *states* whose alterations could then be ascribed to *external influences*.

The Mechanistic Foundation of Neo-Darwinism

Darwin's hypothesis offered the possibility of establishing a comprehensive scheme of efficient causes for an evolution of species. Darwinian evolution can be explained by mutations leading to variations within a species and subsequent selection, favoring reproduction of those descendants that are better adapted to the prevailing environment. When phenotypic *variations* of a species are interpreted as a *state* and the *selective pressure* as a *force*, exerted by external conditions, a seductive analogy to mechanical operations is obtained. This explains why many scientists could not resist this analogy in the nineteenth and twentieth centuries. Ludwig Boltzmann, for instance, an advocate of mechanistic explanations, even went so far as to make the remarkable statement, "In my opinion all salvation for philosophy has to come from Darwin's doctrine."[2]

In order to give Darwin's hypothesis such a mechanistic direction, great effort was made toward finding proof of the impossibility of inheritance of acquired characteristics. This ultimately led to the neo-Darwinian paradigm that essentially represents a combination of three concepts: first, Mendel's distinction between phenotype and genotype, the latter consisting of discrete units that are inherited unmodified, and trigger, in one way or another, the emergence of different phenotypes.[3] The second concept was Weismann's hypothesis of the "continuity of germ-plasm," which strictly separated from the "soma" and not affected by its alterations. In order to establish a mechanistic explanation of evolution, Weismann postulated that there is no inheritance of acquired characters, because this would reduce the significance of selection for an explanation of evolutionary changes, and it would allow some sort of final causation.[4] The "germ-plasm" consequentially was related to chromosomes. The third notion was the postulation of DNA as the exclusive carrier of the genes, which remain intact from generation to generation, at least as long as no mutation takes place in the DNA sequence by a random event. In this way, the "central dogma" of molecular biology: the stipulation that inherited information is always transferred from DNA via messenger RNA to proteins, but never the other way around, provided a mechanistic justification of the neo-Darwinian ban on transmission of acquired characteristics.[5]

In the meantime, the neo-Darwinian paradigm has lost credibility, primarily because of an erosion of the DNA-based notion of genes. Comparative analyses of DNA-sequences of different species reveal a remarkable invariance in genetic content, indicating that evolution of species is caused rather by a progressively more complex regulation of gene expression than by the invention of new genes. The "modern synthetic theory of evolution," attributing to DNA sequences a central coordinating role in all biological processes, is therefore about to lose its molecular biological foundation, and, as such, it has to be replaced by a theory of organism that comprises recent results of epigenetic investigations and self-organization models.

Epistemological Objections to a Mechanistic Explanation of Biological Developments

In order to elaborate differences between neo-Darwinian and systems-theoretical explanations of evolution, it is not necessary to go into details of diverse contemporary modifications of the neo-Darwinian hypothesis because none of these modifications departs from an explanation of evolution by operations of the Cartesian "extended substance." It is sufficient to note that Alfred North Whitehead already has warned against an evolutionary theory that relies on a notion of substance in which portions of matter exist independently of each other:

> The aboriginal stuff, or material, from which a materialistic philosophy starts is incapable of evolution. . . . Evolution, on the materialistic theory, is reduced to the role of being another word for the description of the changes of the external relations between portions of matter. There is nothing to evolve, because one set of external relations is as good as any other set of external relations. There can merely be change, purposeless and unprogressive. But the whole point of the modern doctrine is the evolution of the complex organisms from antecedent states of less complex organisms. The doctrine thus cries aloud for a conception of organism as fundamental for nature.[6]

Neo-Darwinian explanations cannot say anything about an increase in complexity during evolution, because mechanistic operations can only be complicated but never complex.[7]

In the twentieth century, developments in physics have shaken the foundation of the mechanistic worldview, which is criticized by Whitehead in the above-mentioned passage. In quantum mechanics, an observer is part of the observed phenomena; the materialistic notion of substance and a strict separation between a "thinking substance" (represented by an observing subject) and an "extended substance" (i.e., the observed object), can no longer be sustained. In searching for an alternative ontology, Whitehead developed a "philosophy of organism," in which the two Cartesian substances are transformed into the physical and the mental aspect of a process of self-creation. Accordingly, not only biological self-constitution, but also quantum phenomena, can be considered as manifestations of a complex system that constitutes itself in distinct experiences of *its* environment, to which it then conforms ("in a transformation of incoherence into coherence, and in each particular instance ceases with this attainment").[8] Naturally, such an understanding of reality eviscerates the allegation of vitalism, formerly raised against teleological explanations of biological processes, because according to Whitehead, no sharp distinction can be drawn between the animate and the inanimate. Biology is now entitled to refer organismic self-constitution to the subjective aspect of experience, possibly by integrating an interpretation of quantum mechanics derived from process philosophy into models of biological self-organization. For this purpose, an appropriate relationship between final and efficient causes has to be arrived at, one that is aimed at providing an adequate treatment of the organization of living systems, such that the gap between the two Cartesian substances is bridged.

The American philosopher John Dewey made a major contribution to a physiological treatment of the mind-body problem. He suggested searching in human experience for analogies to the process of physiological adaptation of an organism to environmental changes and to use these analogies for an understanding of natural processes. To obtain an empirical approach to the relation between experience and physiological adaptation, Dewey focused on the difference between organisms and non-living things, in that "the activities of the former are characterized by needs, by efforts which are active demands to satisfy needs, and by satisfactions."[9] The physiological basis for these subjective forms of experience was found by Dewey in an accommodation process, initiated when an external influence disturbs an energetically favorable stationary state of the organism. Accordingly, "need" was attributed to a "tensional distribution of energies such that the body is in a condition of uneasy or unstable equilibrium," resulting from that disturbance. "Demands" and "efforts" were then considered as a response to the unstable equilibrium, leading to modifications of the environment in ways that react upon the organismic body, such that its characteristic pattern is restored by a new but different stationary state. "Satisfaction" was interpreted in physiological terms as "recovery of equilibrium pattern, consequent upon changes of environment due to interactions with the active demand of the organism."[10]

Dewey's analysis can be given a sound physiological meaning when an "equilibrium pattern" is interpreted as a stationary state in which the substrate and energy flows are conformed to the corresponding phenotype of the organism. It is notable that the restored stationary pattern represents an entirely new organismic structure that is independent of alterations of the genome. The example from ecology, given below, shows that during the recovery of a stationary state a multitude of physiological partial processes have to be aligned with a harmonized whole that is potentially useful under the prevailing environmental conditions. This requires some sort of "final cause," referring—in the sense of a local teleology—to the psycho-physical aspect of the animate:

> Whenever the activities of the constituent parts of an organized pattern of activity are of such a nature as to conduce to the perpetuation of the patterned activity, there exists the basis of sensitivity. Each "part" of an organism is itself organized, and so are the "parts" of the part. Hence its selective bias in interaction with environing things is exercised so as to maintain *itself*, while also maintaining the whole of w' ich it is a member.... This pervasive operative presence of the whole in the pa. and of the part in the whole constitutes susceptibility—the capacity of feeling—whether or not this potentiality be actualized in plant-life.[11]

By integrating the goal-oriented progression of need, effort, and satisfaction into the directed transition from a non-stationary to a stationary manifestation of the energy flow of which the organism *and* its environment are comprised, Dewey solves three problems that go beyond the framework of a mechanistic biology: firstly, a characterization of the internal relatedness of the constituents of an organism, based on a pervasive operative presence of the whole in the parts and the parts in the whole; secondly, a consideration of the psycho-physical dimension of the animate, using a relation between efficient and final causes; and thirdly, the interdependence between the self-constitution of organisms and the generation of their environment as a single process.

It is obvious that this approach cannot be reconciled with biological models inspired by classical physics. It is possibly for this reason that Dewey has been ignored by contemporary mainstream biology that favors mechanistic explanations of biological self-constitution. However, these explanations require specific boundary conditions under which deterministic reactions occur. Unfortunately, such models refer to self-organization "without a self,"[12] because they do not display the fact that a modeller has implemented the necessary boundary conditions. For this reason, this approach ignores the special achievement of organisms—to create appropriate boundary conditions on their own under which they give themselves a new structure by deterministic reactions.

Treatment of Biological Processes by Systems-Environment Theory

The traditional problem of the relation between the parts and the whole has been worked out in a recently established systems-environment theory that restates this problem by adopting ideas of Dewey for a theory of system differentiation, determined by self-referential features.[13] The philosophical basis for this theory can be traced back to Whitehead, who has broadened and deepened Dewey's ideas into a coherent process philosophy. Systems are self-referential when they are able to relate their experience of environmental alterations to processes of self-constitution, in which they maintain structural identity, even if in this process their constituents are completely replaced. This combination of "self-identity with self-diversity," as it has been termed by Whitehead,[14] is achieved by an organismic network of process-related *adaptive events*, in which energy-converting subsystems of an organism conform to the prevailing environment and to each other in order to attain stationary *adapted states* of optimal efficiency. Energy converting subsystems catalyze the energy flow through a biological system. They consist in the simplest case of two coupled biochemical reactions in which the free energy of one reaction is utilized to a variable degree by the other reaction. In this way, an "input flow," frequently associated with the degradation of a high-energy compound, drives an "output flow" that either establishes a concentration gradient across a membrane or leads to the formation of new cellular constituents from simpler metabolic substrates (two examples are shown in Figure 3.3). Other examples for energy-converting subsystems are the respiratory or photosynthetic electron transport chain in mitochondria, thylakoids, and bacteria, in which the driving electron flow is coupled to a variable degree with the biosynthesis of the energy-rich compound adenosine triphosphate (ATP). Since ATP supplies the energy for all kinds of energy-dependent biochemical processes, each of them also represents an energy-converting subsystem. Furthermore, the so-called "two-component regulatory systems," monitoring external signals, participate *via* phosphorylation- and dephosphorylation-reactions in the energy flow through a metabolic system. By "cross-regulation" among different signal transduction pathways, an information-processing network can be established, coordinating the biosynthesis of different proteins with the subsequent growth requirement of a cell in a great many ways. Moreover, cytoskeleton biosynthesis in growing and moving eukaryotic cells depends on ordered interplay of many energy-converting subsystems potentially involved in information processing. Last, but not least, a whole microbial cell can be considered as a single energy-converting unit, when all of its intracellular energy converting subsystems are conformed to each other, such that available substrates and energy are utilized with optimal efficiency.

When an alteration in the ambient milieu impairs the efficiency of one of these energy-converting subsystems, it is reconstructed to operate with less energy dissipation under the altered conditions, which affects the metabolic activity of the subsystem in question. But since all energy-converting subsystems of a cell are linked by the common energy flow (this provides an internal relatedness of all cellular processes), the adaptive response of one subsystem to an environmental change initiates an adaptive process in all other subsystems, implying that the overall energy flow of the cell is also organizing itself toward a state of minimum entropy production. In such a case, all subsystems conform to each other, until the whole cell has become a single adapted energy converter that dissipates a minimum of energy under the new external conditions. In this way, the physiological activity of an act of experience approaches its fulfilment in which "the many become one, and are increased by one."[15] The whole process re-commences, when an efficiently operating ensemble of subsystems is again perturbed by an environmental change. Due to the vast number of possible stable tertiary structures that a protein can take, the components of energy-converting subsystems can be conformed to each other in practically infinite different ways, thereby forming functioning-dependent structures.[16] Since in this process every adaptive event influences subsequent adaptive processes in a particular manner, there is no repetition of the structural manifestations of the different adapted states: an attribution of a persistent structure to an organism or a species is therefore based on arbitrary abstractions.

Dewey correlates "need" with a "tensional distribution of energies such that the body is in a condition of uneasy or unstable equilibrium."[17] This idea can be interpreted in physiological terms by postulating that the energy flow through a living system serves as a medium for a field of tension, involved in anticipatory information processing about a potentially useful reconstruction of energy converting subsystems. Functioning as an "attractor" that is generated when the energy flow is remote from a stationary state, the field provides connectivity among adaptive events in an "adaptive representational network"[18] of interrelated processes; thereby deviations from stationary states, caused by a new environmental influence on the outcome of former adaptive events, are corrected in subsequent adaptive events, such that the identity of the organism is maintained.

Thus, by means of a field of tension, as defined above, a cellular system is able of producing a "description of itself."[19] In an analogical manner, Rosen has attributed to anticipatory systems the capacity to construct a "model of itself and/or its environment, which it can utilize to modify its own present activities"[20] in a self-referential manner. Dewey's idea that a tensional distribution of energies regulates the mutual adjustment of energy converting subsystems can be found again in Whitehead's philosophy of organism. Also this philosophy requires the postulate that in its self-creation an organism is guided by its "ideal of itself."[21] It is an ideal adjustment factor, because after the concretion of an adaptive event, its guiding pre-conditions are no longer the same; a tension-free state is therefore never really attained but can be approached. The result of this activity is a self-sustaining *unitas multiplex* in which each adaptive event reflects the difference between system and environment, albeit from a different perspective. In this way, a system distinguishes itself from its environment by self-differentiation (a physiological example is given below). Self-reference, in these operations, is responsible for all kinds of manifestations of living systems, such as maintenance of a certain form or a distinct behavior. The justification of systems-environment theory by process philosophy gives a rational explanation of the developmental and evolutionary interdependence of organism and environment and the relation between cognitive and developmental construction.[22]

The Ontological Difference between Two Manifestations of Adaptive Events

How can experience-related "*adaptive events*" be distinguished from the constant flux of structural alterations associated with the overall energy conversion in a cell? This can be achieved by considering adaptive events as "*acts of becoming*"[23] that share essential features with Whitehead's "actual entities," also termed "actual occasions" (of experience). Each "actual entity" has a bipolar nature: in an initial phase it "interprets" and integrates the diverse facts of its environment for its own self-constitution in an individual aesthetic achievement (in Whitehead's terminology this was termed "prehension," adopted from the Latin "prehendere"[24]). The result of that information processing that occurs in a self-referential manner then leads in a final phase to a new fact for subsequent occasions of experience. There are positive and negative prehensions; the former are related to environmental factors that are considered relevant, the latter are excluded from this self-creative process. In physiological terms, the initial phase corresponds to an *adaptive operation mode* in which the organism "*subjectively interprets*" deviations from a stationary state in respect to an appropriate self-constitution. In this mode, an organism determines to what extent an external influence "has significance for itself."[25] According to this anticipatory interpretation, the energy-converting subsystems then attain new *adapted states* that are potentially useful under the new environmental conditions. By recurrent repetitions of such elementary processes, an organismic system constantly reproduces itself, concomitantly with a degradation of previously produced constituents. The two manifestations differ in that only adapted states of energy converting subsystems can be localized in the spatial arrangement of its molecular constituents, whose interaction can be explained by efficient causes. In contrast, adaptive operation modes are guided by "final causation," aimed at diminution of an inner-organismic tensional distribution of energies. The ontological difference between subjective interpretations, occurring in adaptive operation modes, and the objective features of adapted states, is a consequence of the psychophysical dimension of every act of experience.

An experimental analysis of adaptive processes on the basis of a mechanistic model confronts an experimenter with a primary problem—namely, that the organisms adapt to the experimental conditions during the course of an investigation, by changing their energetic and kinetic parameters. In such a case, the investigator becomes part of the investigated system and of the organismic response to the experimentally imposed conditions, making futile any analysis in mechanistic terms. In this regard, physiology is, to some extent, analogous to quantum theory, where the observer is part of the quantum phenomenon (naturally in physiology this complex behavior has another reason: it is a consequence of information processing during adaptive events). Since the experimental conditions also constitute information that is processed and stored during the experimental procedure, the outcome of an experiment depends on the way it has been performed. This impairs an objective description, which is independent of the experimental procedure employed. However, due to the dependence of cellular information processing on the prior experiences of an organism, the response of a cellular system to experimentally imposed conditions provides an insight into the prehistory of the organism.

The energetic constraint to operate with optimal efficiency has an important ecological implication. At each substrate concentration of the driven "output flow" an energy converter can only adopt one state of optimal efficiency,[26] each of them characterized by a defined degree of coupling between the involved driving and driven biochemical reactions. During fluctuations of the external substrate concentrations, as they normally occur under natural conditions, available energy is utilized with optimal efficiency only when the cells are able to conform to the average concentration level of a given pattern of substrate fluctuations. This presupposes that cells have the capacity to process information about the pattern of experienced substrate fluctuations and to "decide" on possible alterations of that pattern in regard to a necessary reconstruction. Two preconditions must be fulfilled for this kind of "decision-making": first, some sort of "memory" of previously performed pattern interpretations; and second, a self-referential valuation of external factors, according to which the organism "functions in respect to its own determination."[27] In this regard, the self-referential aspect of environmental experience, reflecting Whitehead's "ideal of itself" (see above) provides identity to the organism along a historic succession of adapted states and adaptive events.

In the following section, we give an example of pattern recognition by a bacterial population, using an example from aquatic ecology. In this example, information processing and memory phenomena are studied by observing the effect of antecedent on subsequent adaptive events. The empirical investigation of this phenomenon departs from an objectivistic research strategy that is inspired by the ideals of classical experimental physics, consisting of a catalogue of protocols by which *identical* experiences are made by an observer. Generalizing the implications of this concrete biological example will then allow for a systems-theoretical explanation of evolutionary developments.

Systems-Theoretical Treatment of a Particular Adaptive Process

In nutrient-poor (oligotrophic) lakes the concentration of available phosphate decreases to such low values that incorporation into cells of cyanobacteria and algae is impaired for energetic reasons.[28] In such a situation, these organisms can only grow when the external concentration exceeds, at least occasionally, a characteristic *threshold value* above which available energy suffices for the transport into the cell. When this occurs (for example, after excretion by other organisms, such as zooplankton or fish), algae incorporate phosphate *via* an activated uptake system and store it as polyphosphate granules that serve as the proper phosphorus source for the proliferating cells. Concomitantly, and because of the uptake activity of the whole phytoplankton population, the external concentration decreases more or less rapidly to the threshold value again, which, under phosphate-deficient conditions, is in the nanomolar range.[29] As a result of this energetic constraint, occasionally occurring increases in the external phosphate concentration are interrupted by periods without phosphate supply, so that fluctuations of the ambient phosphate concentration are "experienced" by the cells as a *pattern of pulses*.

The energetic and kinetic alteration of the phosphate uptake system during a sequence of pulses has to obey an important regulatory constraint: on the one hand the polyphosphate granules must not become too big, as this would lead to disruption of cellular structures. On the other hand the granules must be sufficiently large to sustain a continuous growth process at a rate that, in turn, depends on the amount of stored phosphate (in continuous cultures growth rate becomes greater with an increase in the amount of stored phosphate[30]). A direct regulatory effect of the size of the granules on the phosphate transport system is not conceivable, since the granules are osmotically inert and localized inside the cells of the microorganisms. But the activity of the carrier (as well as the growth rate) can be *indirectly* conformed to the polyphosphate content, if some sort of information processing about the pattern of antecedent pulses guides the adaptive uptake behavior during subsequent pulses. This would enable the cells to memorize how much phosphate they have stored in previous pulses and to adjust the growth rates accordingly. It is obvious that the existence of such a memory supersedes competition for the available nutrient. Naturally, this kind of information storage cannot be detected by experiments, in which microorganisms are exposed to constant external concentrations. However, information processing can be studied if two reference samples of one and the same population are exposed first to two different patterns of phosphate pulses in which they receive the same amount of phosphate. The discriminatory effect of this distinct pretreatment on subsequent adaptive events is then revealed in a final test pulse by analyzing in the two samples the time course of decrease in the external concentration from an identical initial value. In this case, uneven uptake kinetics in the two test pulses shows to what extent information about the two dissimilar antecedent patterns of phosphate pulses has been stored and influences the uptake behavior in subsequent test pulses in a distinct manner. In this context, information is defined as a difference (in the environment) that makes a difference (in the physiological behavior).[31]

The following examples demonstrate such pattern recognition by cyanobacteria. In the first example, two samples of the same population of *Anabaena variabilis* were exposed to five small pulses of 1 μM each and one greater pulse with 5 μM, but in reverse order: one sample experienced first the five smaller pulses and then the great pulse, the other sample *vice versa*. The uptake kinetics in a final test pulse, in which both samples received an identical amount of phosphate, revealed significant differences (Figure 3.1): the population that had been exposed to the smaller pulses *after* the great pulse had a lower uptake activity than the reference population. It appears that under this condition a difference in the pattern of phosphate fluctuations led to a corresponding difference in the subsequent adaptive behavior, indicating that already a simple prokaryotic organism is capable of some sort of pattern recognition. This phenomenon can be explained by a concatenation of adaptive events, in which each pulse generated a distinct adapted state that influenced the adaptive operation mode in the subsequent pulse in a characteristic way. Due to that interrelationship between adaptive events, different patterns of previously experienced phosphate fluctuations finally resulted in unequal adapted states, which then affected adaptation to the same test pulse in a dissimilar manner. Naturally this process evades an analysis in mechanistic terms, since the sequence of pulses is determined by a dialectics between the system and its environment: thereby each transiently occupied state of the organism, provoked by changes in the environment, affects conversely the environment, so that cause and effect no longer can be differentiated.

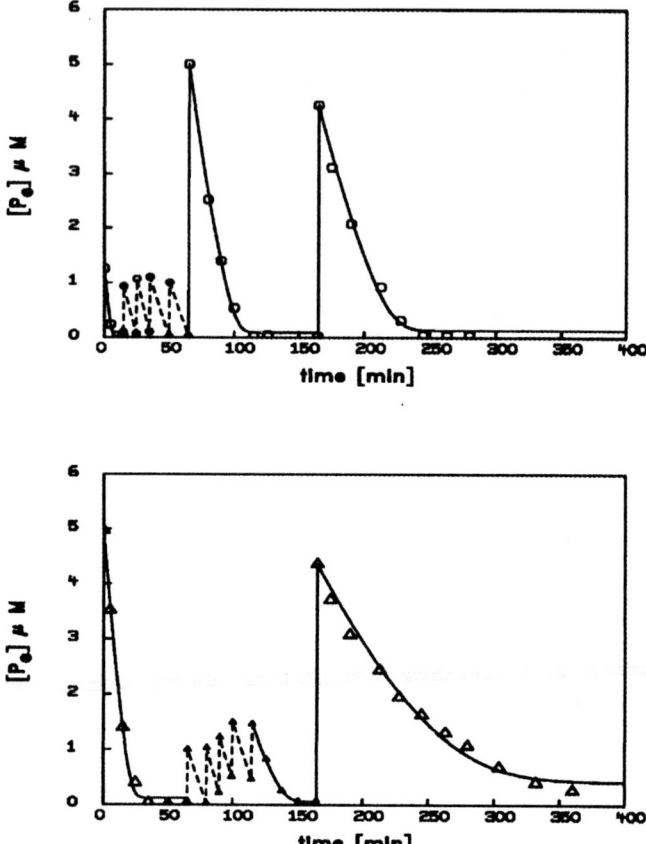

Figure 3.1: Time course of the decrease of the external phosphate concentration in two identical suspensions of *Anabaena variabilis* which had been exposed to two different pulse patterns. (Source: Renate Falkner, Martin Priewasser, and Gernot Falkner, "Information Processing by Cyanobacteria during Adaptation to Environmental Phosphate Fluctuations." *Plant Signaling and Behavior* 1 (2006): 212-220. Copyright Landes Bioscience.)

Differences in the pulse patterns influenced the rate of subsequent growth even if the amounts of phosphate incorporated during these pulses were identical.[32] Obviously cellular changes provoked by pulses—outgoing from the phosphate uptake system as epicentre—spread all over the cell such that also the subsequent growth was affected.

When the two patterns of phosphate pulses—to which two different samples of the same population were exposed—differ significantly, a distinct adaptive behavior could even be observed after subsequent cell divisions. In an experiment, in which information about former nutrient fluctuations was transferred to daughter generations, two identical suspensions of *Anabaena v.* were confronted with phosphate additions in different supply modes: one suspension experienced a pulse of 10 µM, the other ten pulses of 1 µM. After this treatment, both suspensions were cultivated for 24 hours at the expense of the stored phosphorus, which was the same in both cultures. During that time, the chlorophyll content of the two cultures increased from 0.13 to 0.33 and 0.31 mgL^{-1} resp., which corresponds to about a threefold increase in the amount of cells. The two suspensions were then harvested and exposed to three identical test pulses for a comparison of their adaptive behavior. Figure 3.2 shows that the culture that had been submitted to one great pulse (high-pulse culture) during its previous growth deactivated the uptake system more rapidly than the reference culture (low-pulse culture) in which the external concentration never attained high values. It appears as if the cells of the "high-pulse culture" "remember" the former challenge by elevated phosphate concentrations, several cell divisions ago and appear to anticipate during subsequent adaptations to pulses a continuation of excessive phosphate supply. It is notable that this anticipatory behavior is passed on—as an *acquired property*—to the great-grandchildren.

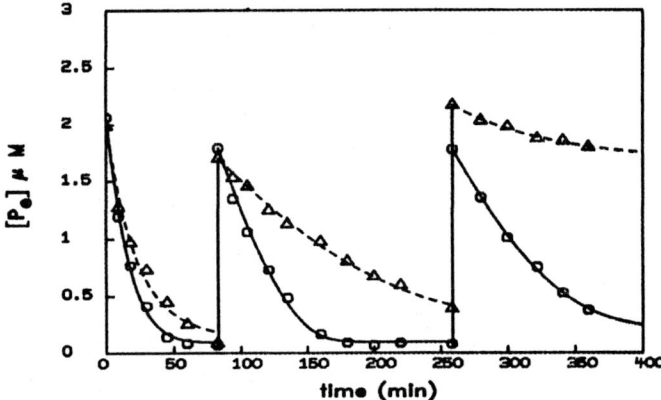

Figure 3.2: Time course of ^{32}P-phosphate removal from the external medium by two populations of *A. variabilis*, produced by different pre-treatments of the same mother culture. Circles: low-pulse culture, triangles: high-pulse culture. (Source: Renate Falkner, Martin Priewasser, and Gernot Falkner, "Information Processing by Cyanobacteria during Adaptation to Environmental Phosphate Fluctuations." *Plant Signaling and Behavior* 1 (2006): 212-220. Copyright Landes Bioscience.)

The Intracellular Communication of Energy Converting Subsystems

Information processing about the experienced pattern of phosphate fluctuations is based on interplay of the two energy-converting subsystems involved in the conversion of external phosphate into polyphosphates. This process proceeds in three steps:[33]

(1) Transport of external phosphate P_e from the external medium into the cell, catalyzed by a "carrier protein" (reaction 1 in Figure 3.3). At low external concentrations, energy for the translocation against the existing electrochemical gradient at the cell membrane is provided in the cyanobacterium *Anacystis nidulans* by an ATPase (reaction 2 in Figure 3.3);

(2) Conversion of the incorporated internal phosphate P_i to ATP by an ATP-synthase (reaction 3). This process is driven by the flow of protons from the thylakoid space into the cytoplasmic space (in the light by photophosphorylation; in the dark, by oxidative phosphorylation);

(3) Formation of polyphosphates from ATP (reaction 4 in Figure 3.3).

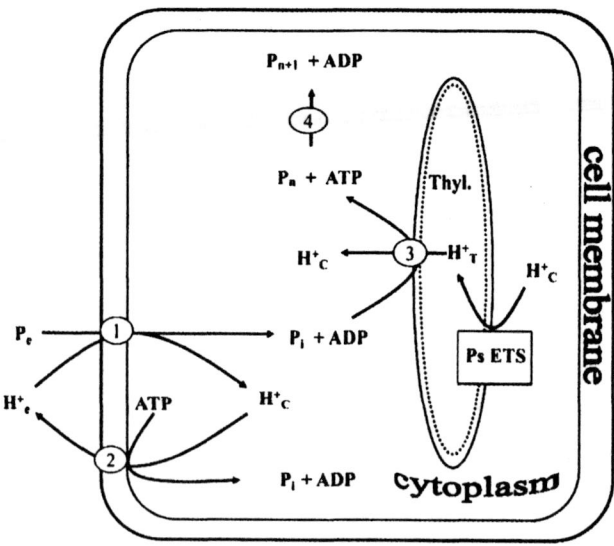

Figure 3.3: Schematic presentation of phosphate utilization by cyanobacteria. Reactions 1, 2, 3, and 4 show the conversion of external phosphate to polyphosphates. The correct stoichiometries are not indicated in the figure. P_e: external phosphate; P_i: internal phosphate; P_n, P_{n+1}: polyphosphates; H^+_C and H^+_T are the proton concentrations in the cytoplasmic and thylakoid space; Ps ETS: photosynthetic electron transport system.

During a pulse the changing external and internal phosphate concentrations provoke adaptive processes in both the phosphate transport system and the ATP-synthase, aimed at attainment of states of least energy dissipation. This changes the metabolic activity and alters the energy conversion of the two subsystems (*via* readjustment of the degrees of coupling between phosphate carrier and ATP-ase and between ATP-synthase and proton flow). The mutual adjustments of the two systems during a sequence of pulses can be considered as a form of communication in which the adaptive operation modes of each of the two subsystems alternately interpret the adapted states of the other. This could be confirmed by a computer simulation of the adaptive phosphate uptake behavior using a model based on non-equilibrium thermodynamics. In this case, the simulated time courses of phosphate uptake were in accordance with the observed data, when the model altered its parameters in response to its own simulation.[34] Communication starts at the beginning of a new pulse, when after a sudden rise in the external phosphate concentration the incorporation does not proceed with minimum entropy production at the heightened phosphate concentration. In a primary reconstruction process aimed at attainment of an energetically favorable state for an anticipated growth rate, the degree of coupling between the activity of the phosphate carrier and the ATPase (reaction 1 and 2 in Figure 3.3) is altered in a way that reduces energy dissipation. In a subsequent step, the ATP-synthase then conforms the degree of coupling between its activity and the proton flow at the thylakoid membrane to the elevated cytoplasmic phosphate concentration during a pulse, again directed toward an efficient operation in this situation. Concomitantly, and as a result of the uptake activity by the altered properties of the two subsystems, the external concentration decreases further and the transport system has to readapt again, and so forth, until a constant external phosphate concentration is finally attained at the threshold value where the adaptive communication of the two systems has come to an end. The two systems are then self-referentially conformed to each other and to the external concentration which, in turn, results from that conformed state. Due to that self-referential interdependence, the conformed ensemble is stable. Their adapted properties, which represent a kind of primitive "memory" of antecedent exposures to phosphate fluctuations, are maintained for a certain period of time during a subsequent period without phosphate incorporation. When such a self-referential state is disturbed by a new pulse, intracellular communication recommences and is now resumed with the properties originating from the antecedent pulse. This establishes a communication tradition, in which information about adaptation to an antecedent pulse is transferred in a specific manner to the subsequent pulse. Decisive for a concatenation of adaptive events is how fast each of the two subsystems alternately adapts to the manifestations of the other. In this respect, reaction time, reflecting inner-organismic tension, is a measure of the sensitivity of the system for external concentration changes.

In order to operate at the anticipated growth rate with optimal efficiency, other cellular energy-consuming subsystems also have to accommodate to the energetic and kinetic properties of the ATP-synthase with respect to least energy dissipation.[35] This implies that other energy-consuming subsystems of the cell display adaptive operation modes in which they "subjectively interpret" the objective manifestations of the ATP-synthase in its stable self-referential state. Since a new energetic situation influences all energy-dependent subsystems *simultaneously*, an adjustment of energy-consuming subsystems to the energetic manifestations of the ATP-synthase can be expected to occur in a coherent manner until a new adapted state emerges. Hence, the connectivity of adaptive events establishes a network of interacting subsystems, to which the phosphate uptake system belongs. Since, in this process, every cell coherently adapts to environmental changes that are produced by the whole community, a community of cells potentially merges to a conformed unity. This leads to interdependence between individual acts of experience of the organisms of a biocoenosis and establishes historicity of cooperative experiences in which all organisms partake in direct or indirect interactions. In this regard, adaptive events have a reference beyond themselves. An irreversible advancement of organisms therefore proceeds in various intertwined evolutionary strands that cannot be considered independently of each other.

The Systems-Theoretical Perspective of an Evolution of Species

A generalization of the above interpretation of the irreversible feature of biological information processing allows delineating evolution as a progressive heightening of the intensity of experience, paralleled by an increasing differentiation of energy-converting subsystems. The lowest position in this hierarchy is taken by unicellular microorganisms that react to restricted experiences of one single growth factor by a reconstruction of their cellular structure. However, already at this level differentiations into various cell types have occurred by the above described process, in which every microorganism constitutes itself via an experience of its particular environment in a specific way. This results in dissimilar functions of energy-converting systems and favors the emergence of differently specialized cells. Furthermore, an exchange of DNA elements between cells broadens their metabolic capacity much faster than this could have been achieved by mutation and selection. The resulting increase in the complexity of responses to environmental changes has generated a plethora of new modes of self-constitution by which available energy is exploited in many different ways.

Whitehead's "Categoreal Obligations" allow for an understanding in relation to how a heightening of the intensity of experience arises from functional integration of the metabolism of different prokaryotic cells, leading to the emergence of a eukaryotic cell. The Eighth Category of Subjective Intensity specifies that "the subjective aim, whereby there is origination of conceptual feeling, is at intensity of feeling (α) in the immediate subject, and (β) in the *relevant* future."[36] A heightening of the intensity of experience then arises from a particular order such "that the multiplicity of components in the nexus can enter explicit feeling as *contrasts*, and are not dismissed into negative prehensions as *incompatibilities*."[37] Accordingly, in a first step along that line, the scope of environmental experience is extended in evolution by formation of symbiotic associations that behave like multicellular organisms.[38] In such associations, environmental experience comprises a wider range of external factors than the experience of single organisms, for the following reason: when isolated, organisms compete for a single nutrient, they experience environmental changes only by metabolic operations that are restricted to the availability of this nutrient. In contrast, symbiotic associations arrive at "second order experiences" in which every organism observes the operations of its partner, which in turn observes a different environment. This leads to "interpenetrations" and "structural couplings"[39] of symbiotic partners. For example, an association of a motile heterotrophic microorganism with an immotile autotrophic photosynthetic bacterium provides the possibility of coordinating the experience of changes in the light intensity with an additional experience of the availability of useful organic compounds.

Thus, the formation of a symbiotic association represents a "conceptual reversion"[40] of the range of environmental experiences of non-symbiotic organisms: in this case the experience of constituents of a symbiotic association is only partially identical with, and partially diverse from, the experience of isolated organisms; the diversity is directed at the new "subjective aim" of each constituent to conform its various activities to the symbiotic partners.[41] According to Whitehead's Sixth Category of Transmutation[42] this situation now offers a possibility to intensify experience. Thereby, an extracellular symbiotic consortium is transmuted *via* an intracellular symbiosis into a eukaryotic cell in which the "simple physical feelings"[43] of the former symbiotic partners are subsequently replaced by a single, but more complex feeling. In this process, a symbiotic association in which the experience of the host cell is still separated from the experience of its intracellular inmates becomes *one* entity that constitutes itself in a concerted action when it experiences its external world. This establishes a new relation between organism and environment, determined by the "subjective aim"[44] to attain a new stationary state in which the form of the new type of cell is maintained in an individual aesthetic achievement. This necessitates a functional integration of all metabolic processes by a sole coordinating principle that also regulates the major part of the protein synthesis of the whole cell, using the common DNA-pool of the nucleus. It is notable that during that "concrescence" of symbiotic partners to a single cell, the structures of the previously independently existing organisms have not disappeared. They persist as perceiving organelles such as mitochondria and chloroplasts, which is a precondition for integrating the "prehensions" of organelles as *contrasts* for a heightening of the intensity of experience[45] (in accordance with the Category of Objective Diversity[46]). Also in this way "the many become one, and are increased by one."[47]

This has cleared the way for an evolution of eukaryotic organisms for which the need to coordinate a greater amount of external factors results in a multitude of modes of self-constitution and allows an increasing differentiation into multicellular structures. According to process philosophical categories, this evolutionary process is accompanied by further intensifications of experience that finally lead to higher cognitive achievements, establishing symbolic relationships between elements of the experienced environment.

Thus, the neo-Darwinian paradigm and systems-environment theory differ essentially in the answer to the question how organismic identity is maintained and altered. For neo-Darwinism, the identity of organisms is provided by inheritance of invariant material configurations that are ultimately ascribable to mechanistic operations, such as a "genetic program," determined by the DNA sequence. The development of plentiful organismic forms from simpler ancestors is therefore attributed to accidental changes of this DNA sequence. Accordingly, the evolution of species proceeds in branches of a phylogenetic tree in which the organismic deployments in individual branches do not influence each other by communicative interdependence. This explanation of evolution necessarily sticks to a consideration of objectifying compounds, which impedes an explanation for a preservation of the morphological features of a system. In a scenario in which all metabolic constituents are in flux, it is difficult to explain the maintenance of a certain form by a stable arrangement of constituents. However, an approach that is even unable to explain a preservation of forms is hopelessly overstrained when, beyond that, it is confronted with the task to explain the well-regulated transformation of this form during evolution.

In contrast, systems-environment theory no longer attributes maintenance of organismic identity to a persistent material configuration. According to this theory, inheritance is not explained by transmission of a certain DNA sequence from one generation to the next, but by a selective appropriation of the outcome of antecedent experience-related adaptations by subsequent adaptations. While in neo-Darwinism, better adapted organisms prevail over less adapted ones due to their more numerous offspring, systems-environment theory does not discern different degrees of adaptations. According to Luhmann, "Any system is necessarily exposed to an environment and has insofar always been . . . adapted to its environment—or it does not exist."[48] For systems theory, an organism consists of an advancing multitude of intra-cellular adaptive events that are single, but coordinated entities as within the totality of the whole organism which, in turn, is adapted to its prevailing environment. In this process, organismic identity is based on an internal representation, provided by a field of tension that emerges self-referentially from and acts upon its adaptive processes (in this regard, systems are "operationally closed"). Identity of an organismic system is maintained as long as the anticipatory properties of its adaptive events can cope with subsequent environmental changes in a way that preserves a connectivity of adaptive events. This establishes an organismic memory, reflecting the temporal vector character of adaptive events that connect the past with the future and are responsible for the irreversible character of biological processes. A certain phenotype is preserved in evolution if the defining characteristics of the energy-converting subsystems responsible for this phenotype are passed on from antecedent to subsequent experience-related adaptations. Simultaneously, a development to higher organismic experiences is possible when an intuitive anticipation of higher modes of experience guides the integration of more simple organismic systems to more complex organisms. Evolution is therefore primarily an evolution of intensity of experience, comprehended as a creative endeavour in which "in their nature entities are disjunctively 'many' in process of passage into conjunctive unity."[49] When the advancing integration of an ever-increasing diversity of organismic manifestations invents a new coherently operating structure, the self-constitution of other organismic ensembles is affected because permanently occurring new creations influence each other on the basis of interrelated experiences and self-constitutions. This explains the realization of diverse eukaryote types of cells within a short period of time or the explosive emergence of different organismic structures in the Cambrian period of the Paleozoic Era, which poses a great problem for the postulate of a gradual evolution of species in neo-Darwinism. Naturally, a model of such a creative outburst cannot be established unless the bifurcation of nature by two Cartesian substances has been abandoned.

Acknowledgments

The authors are grateful to Professor Dr. Derek G. Smyth for reading and improving the manuscript. The work was supported by the Austrian Science Fund.

Notes

1. René Descartes, *Principia Philosophiae* (Amsterdam: Elzevier, 1644), 15.
2. Ludwig Boltzmann, *Populäre Schriften* (Leipzig: Johann Ambrosius Barth, 1905), 396, trans. Renate Falkner.
3. The term "genes" for these units was later introduced by Wilhelm L. Johannsen, *Elemente der exakten Erblichkeitslehre* (Jena: Verlag Gustav Fischer, 1913), 143.
4. August Weismann, *Die Continuität des Keimplasma's als Grundlage einer Theorie der Vererbung* (Jena: Verlag Gustav Fischer, 1885), 10, 59.
5. For a more detailed presentation of this development, see John Stewart, *La vie existe-t-elle?* (Paris: Vuibert, 2004), 38.
6. Alfred North Whitehead, *Science and the Modern World* (1925; repr., London: Free Association Books, 1985), 134f.

7. Donald C. Mikulecky, "Complexity, Communication between Cells, and Identifying the Functional Components of Living Systems: Some Observations," *Acta Biotheoretica* 44 (1996): 189.

8. Alfred North Whitehead, *Process and Reality*, ed. David Ray Griffin and Donald W. Sherburne (1929; corr. ed., New York: Macmillan, 1978), 25. Page references refer to the 1978 corrected edition.

9. John Dewey, *Experience and Nature* (New York: Dover Publications, 1958), 252.

10. Dewey, *Experience and Nature*, 253.

11. Dewey, *Experience and Nature*, 256.

12. Spyridon Koutroufinis, *Selbstorganisation ohne Selbst* (Berlin: Pharus Verlag, 1996) 47, 126; for a more detailed description of the deficiencies of physical self-organization models, see chapter 14 in this volume.

13. See, for example, Niklas Luhmann, *Social Systems* (Stanford, CA.: Stanford University Press, 1995), 9.

14. Whitehead, *Process and Reality*, 25.

15. Whitehead, *Process and Reality*, 21.

16. Michel Thellier, Guillaume Legent, Vic Norris, Christophe Baron, and Camille Ripoll, "Introduction to the Concept of Functioning-Dependent Structures in Living Cells," *CR Academy of Science Paris, Sciences de la vie / Life sciences* 327 (2004): 1017-1024.

17. Dewey, *Experience and Nature*, 253.

18. Anthony Trewavas, "Green Plants as Intelligent Organisms," *Trends in Plant Science* 10 (2005): 415.

19. Luhmann, *Social Systems*, 9.

20. Robert Rosen, "Organisms as Causal Systems Which Are Not Mechanisms: An Essay into the Nature of Complexity," in *Theoretical Biology and Complexity*, ed. Robert Rosen (New York: Academic Press, Inc., 1985), 195.

21. Whitehead, *Process and Reality*, 85.

22. Susan Oyama, *Evolution's Eye* (Durham, NC: Duke University Press, 2000), 3, 168.

23. Whitehead, *Process and Reality*, 68f.

24. Whitehead, *Process and Reality*, 41.

25. Whitehead, *Process and Reality*, 25.

26. O. Kedem and S. Roy Caplan, "Degree of Coupling and Its Relation to Efficiency of Energy Conversion," *Transactions of the Faraday Society* 61 (1965): 1906; Gernot Falkner, Renate Falkner, and Andreas J. Schwab, "Bioenergetic Characterization of Transient State Phosphate Uptake by the Cyanobacterium *Anacystis nidulans*. Theoretical and Experimental Basis for a Sensory Mechanism Adapting to Varying Environmental Phosphate Levels." *Archives of Microbiology* 152 (1989): 356.

27. Whitehead, *Process and Reality*, 25.

28. Jeff J. Hudson, William D. Taylor, and David W. Schindler, "Phosphate Concentrations in Lakes," *Nature* 406 (2000): 54-56.

29. Gernot Falkner and Renate Falkner, "The Complex Regulation of the Phosphate Uptake System of Cyanobacteria," in *Bioenergetic Processes of Cyanobacteria*, ed. Günter Peschek et al. (London: Springer, 2011), 114.

30. M. R. Droop, "Some Thoughts on Nutrient Limitation of Algae," *Journal of Phycology* 9 (1973): 264-272.

31. Gregory Bateson, *Steps to an Ecology of Mind* (Chicago: The University of Chicago Press, 1972), 315.
32. Ferdinand Wagner, Emel Sahan, and Gernot Falkner, "The Establishment of Coherent Phosphate Uptake Behavior by the Cyanobacterium *Anacystis nidulans*," *European Journal of Phycology* 35 (2000), 250.
33. Falkner and Falkner, "Regulation," 113.
34. Kristjan Plaetzer, Randall S. Thomas, Renate Falkner, and Gernot Falkner, "The Microbial Experience of Environmental Phosphate Fluctuations. An Essay on the Possibility of Putting Intentions into Cell Biochemistry," *Journal of Theoretical Biology* 235 (2005): 540-554.
35. Jörg W. Stucki, "The Optimal Efficiency and the Economic Degrees of Coupling of Oxidative Phosphorylation," *European Journal of Biochemistry* 109 (1980): 275.
36. Whitehead, *Process and Reality*, 27.
37. Whitehead, *Process and Reality*, 83. In physiological terms, the expression 'nexus' refers to an organismic unity of interdependent adaptive events.
38. James A. Shapiro, "Thinking about Bacterial Populations as Multicellular Organisms." *Annual Review of Microbiology* 52 (1998): 81-104. For a treatment of the energetic implications of symbiotic relationships see chapter 9 in this volume.
39. Luhmann, *Social Systems*, 213, 221.
40. See Whitehead's "Fifth Categoreal Obligation," *Process and Reality*, 26.
41. Whitehead, *Process and Reality*, 26.
42. Whitehead, *Process and Reality*, 27.
43. Whitehead, *Process and Reality*, 236.
44. Whitehead, *Process and Reality*, 27.
45. Whitehead, *Process and Reality*, 83.
46. Whitehead, *Process and Reality*, 26.
47. See the "Category of the Ultimate," Whitehead, *Process and Reality*, 21.
48. Niklas Luhmann, *Die Wissenschaft der Gesellschaft* (Frankfurt: Suhrkamp Verlag, 1992), 563, trans. Renate Falkner.
49. Whitehead, *Process and Reality*, 21.

Bibliography

Bateson, Gregory. *Steps to an Ecology of Mind*. Chicago: The University of Chicago Press, 1972.
Boltzmann, Ludwig. *Populäre Schriften*. Leipzig: Johann Ambrosius Barth, 1905.
Descartes, René. *Principia Philosophiae*. Amsterdam: Elzevier, 1644.
Dewey, John. *Experience and Nature*. New York: Dover Publications, 1958.
Droop, M. R. "Some Thoughts on Nutrient Limitation of Algae." *Journal of Phycology* 9 (1973): 264-272.
Falkner, Gernot, and Renate Falkner. "The Complex Regulation of the Phosphate Uptake System of Cyanobacteria." In *Bioenergetic Processes of Cyanobacteria*, edited by Günter Peschek et al., 109-130. London: Springer, 2011.
Falkner, Gernot, Renate Falkner, and Andreas J. Schwab. "Bioenergetic Characterization of Transient State Phosphate Uptake by the Cyanobacterium *Anacystis nidulans*. Theoretical and Experimental Basis for a Sensory Mechanism Adapting to Varying Environmental Phosphate Levels." *Archives of Microbiology* 152 (1989): 353-361.

Falkner, Renate, Martin Priewasser, and Gernot Falkner. "Information Processing by Cyanobacteria During Adaptation to Environmental Phosphate Fluctuations." *Plant Signaling and Behavior* 1 (2006): 212-220.
Hudson, Jeff J., William D. Taylor, and David W. Schindler. "Phosphate Concentrations in Lakes." *Nature* 406 (2000): 54-56.
Johannsen, Wilhelm L. *Elemente der exakten Erblichkeitslehre.* Jena: Verlag Gustav Fischer, 1913, 143-146.
Kedem, O., and S. Roy Caplan. "Degree of Coupling and Its Relation to Efficiency of Energy Conversion." *Transactions of the Faraday Society* 61 (1965): 1897-1911.
Koutroufinis, Spyridon. *Selbstorganisation ohne Selbst.* Berlin: Pharus Verlag, 1996.
Luhmann, Niklas. *Social Systems.* Stanford, CA.: Stanford University Press, 1995.
———. *Die Wissenschaft der Gesellschaft.* Frankfurt: Suhrkamp Verlag, 1992.
Mikulecky, Donald C. "Complexity, Communication between Cells, and Identifying the Functional Components of Living Systems: Some Observations." *Acta Biotheoretica* 44 (1996): 179-208.
Oyama, Susan. *Evolution's Eye.* Durham, NC: Duke University Press, 2000.
Plaetzer, Kristjan, Randall S. Thomas, Renate Falkner, and Gernot Falkner. "The Microbial Experience of Environmental Phosphate Fluctuations. An Essay on the Possibility of Putting Intentions into Cell Biochemistry." *Journal of Theoretical Biology* 235 (2005): 540-554.
Rosen, Robert. "Organisms as Causal Systems Which Are Not Mechanisms: An Essay into the Nature of Complexity." In *Theoretical Biology and Complexity*, edited by Robert Rosen, 165-203. New York: Academic Press, Inc., 1985.
Shapiro, James A. "Thinking about Bacterial Populations as Multicellular Organisms." *Annual Review of Microbiology* 52 (1998): 81-104.
Stewart, John. *La vie existe-t-elle?* Paris: Vuibert, 2004.
Stucki, Jörg W. "The Optimal Efficiency and the Economic Degrees of Coupling of Oxidative Phosphorylation." *European Journal of Biochemistry* 109 (1980): 269-283.
Thellier, Michel, Guillaume Legent, Vic Norris, Christophe Baron, and Camille Ripoll. "Introduction to the Concept of Functioning-Dependent Structures in Living Cells." *CR Academy of Science Paris, Sciences de la vie / Life Sciences* 327 (2004): 1017-1024.
Trewavas, Anthony. "Green Plants as Intelligent Organisms." *Trends in Plant Science* 10 (2005): 413-419.
Wagner, Ferdinand, Emel Sahan, and Gernot Falkner. "The Establishment of Coherent Phosphate Uptake Behavior by the Cyanobacterium *Anacystis nidulans.*" *European Journal of Phycology* 35 (2000): 243-253.
Weismann, August. *Die Continuität des Keimplasma's als Grundlage einer Theorie der Vererbung.* Jena: Verlag Gustav Fischer, 1885.
Whitehead, Alfred North. *Process and Reality: Corrected Edition*, edited by D. R. Griffin and D. W. Sherburne. 1929. Corrected ed., New York: The Free Press, 1978. Page references to corrected edition.
———. *Science and the Modern World.* London: Free Association Books, 1925. Repr., 1985.

Chapter Four
Process-First Ontology

Robert E. Ulanowicz

Whence Biotic Organization?

The argument could be made that the philosophy of biology is preoccupied with ontogeny. We exist, after all, most identifiably as organisms, and we are naturally predisposed toward narcissism. Furthermore, ontogeny invites a focus upon mechanism in biology, because our own bodies exhibit a myriad of mechanical-like phenomena as part of our constitutive dynamics. So it is not unreasonable in light of a scenario that appears so strongly scripted and regulated to focus on the material and the mechanical as the crux of our scientific narrative of organisms. It is only relatively rare exceptions to mechanism that give one pause.

Such cogent reasoning notwithstanding, a mechanical narrative of organisms and their development eventually encounters difficulties. In particular, the neo-Darwinian script whereby the molecular genome directs the construction of the organism via a sequence of molecular mechanisms leads to a particularly less-than-satisfying end. Such was the conclusion by Sidney Brenner and his associates after years of trying to map out the domains of influence exerted by each gene upon the 959 cells comprising the simple roundworm, *Caenorhabditis elegans*:

> At the beginning it was said that the answer to the understanding of development was going to come from a knowledge of the molecular mechanisms of gene control. . . . [But] the molecular mechanisms look boringly simple, and they do not tell us what we want to know. We have to try to discover the principles of organization, how lots of things are put together in the same place.[1]

To paraphrase Brenner, we may have to pay less attention to objects moving according to universal laws in order to emphasize better the relationships among parts and processes. Here our fixation upon organisms tends to divert us, because organization there is quite rigid and not at all unmechanical. Organization elsewhere, however, exists alongside far greater flexibility. As Gunther Stent noted,

> Consider the establishment of ecological communities upon colonization of islands or the growth of secondary forests. Both of these examples are regular phenomena in the sense that a more or less predictable ecological structure arises via a stereotypic pattern of intermediate steps, in which the relative abundances of various types of flora and fauna follow a well-defined sequence. The regularity of these phenomena is obviously not the consequence of an ecological program encoded in the genome of the participating taxa.[2]

Stent was independently supporting Brenner's suggestion that the focus upon the mechanisms of transcription from genome to phenome should not overshadow more relevant organizational influences expressed at the level of the entire system. It would be a mistake, however, to follow our narcissistic inclinations immediately into the human sciences, such as economics, sociology, anthropology, because amidst such higher-level ensembles intentionality can cloud our search for the rudimentary non-mechanical organizing principles. Fortunately, ecology seems to occupy a propitious middle ground in that ecosystems appear to exhibit considerable flexibility in abstraction of human volition. As Stent hinted, it may be the preferred theater in which to describe non-mechanical agencies.

The focus here, then, is upon ecology, as we probe beyond the mechanical to gain deeper insights into biological reality. For the time being, however, our attention remains with mechanism, specifically as it is manifested via universal physical laws. In particular, the Enlightenment assumptions of causal closure, atomism, and universality taken together imply that "law determines all"—namely, nothing at all happens except that it be *elicited* by the workings of the universal laws. In particular, our argument here will be that biology resembles more a theater of repeated particulars than a playbook of universal laws. In Peircean terms, nature tends to take on habits.

The Logic of It All

The notion that universal laws can somehow elicit actions is misguided, because all actual events involve at least some particulars. It is almost tautological to note that universal laws can be described only in terms of universal variables. In that vein, Gregory Bateson observed how the "stuff" of conventional sciences consists of only generic and homogeneous categories (which he collectively called "pleroma"), such as mass or energy.[3] Solving actual problems involves more than generality, however—a fact that is implicit, even in classical physics. Real problems consist of a field or domain over which the general law in question determines behavior and a required boundary or initial point at which conditions *necessarily remain contingent*. Heretofore, the contingency of the boundary problem has almost always been overshadowed by our fixation on the deterministic behavior within domain. In classical physics the boundary *conditions* are usually assumed to be *specified* (made particular) by the individual who states the problem. Whence our Newtonian bias in favor of the detached observer tempts us to ignore the boundaries as part of the physical problem, but to do so is a palpable error. It is impossible to state any real problem in full without involving both determinacy and *contingency*.

While the foregoing might seem like semantic quibbling to some, the necessity of contingency has made for historical footnotes at times (as with the rise of Deism during the seventeenth and eighteenth centuries). Although contingency at the boundaries causes no headaches in posing simple classical problems, the importance of obligatory contingency grows as problems become ever more complex. Aside from the intentional action of specifying boundary contingencies, there are at least two other ways by which initial and peripheral constraints can arise *naturally*: (1) They can be fully contingent, i.e., they can appear from outside the considered domain by pure chance. This is pretty much the scenario for "natural selection" in Darwinian discourse. (2) The conditions can arise via the regular interference of some subsystems on others, behavior which usually is referred to as "self-organization" (as discussed further below).

Regardless of how contingency enters the problem, it must always be present, simply because it is impossible to determine the particular and/or the heterogeneous entirely in terms of the general and the homogeneous. Universal laws can never *determine* all—a conclusion commensurate with Gödel's proposition that any formal, self-consistent, recursive axiomatic system cannot encompass some true propositions. The ability of universal laws to determine actions erodes in combinatory fashion as one encounters ever more particulars, and it effectively vanishes with sufficient heterogeneity (such as always exists in living systems). Bateson, for example, pointed out how in biology one is forced to deal with heterogeneous tokens, each of which can be distinguished from the others. It was Walter Elsasser who suggested that such transition from the homogeneous to the heterogeneous requires that the investigator employ a qualitatively different logic:

> When in the early years of this century Whitehead and Russell succeeded in combining logic and mathematics into one edifice, modern mathematics 'took off.' It is almost entirely based on sets whose elements are assumed to have no internal structure. This makes modern mathematics ideally suited for dealing with the constituents of matter discovered by the physicist. It is an experimentally well-established fact that those constituents, electrons, protons, and so forth are indistinguishable; their properties are such as to make these particles rigorously identical with each other. The more refined experiments allow one to specify this identity quantitatively to many decimal places. It is a well-established principle of physics that when one forms a class of, say, electrons, all elements of that class are strictly indistinguishable; it is as a matter of principle impossible to "label" the members of such a class so as to distinguish

them individually. We shall speak of classes with this property as perfectly homogeneous classes.[4]

Further on, Elsasser purports that the logic of homogeneous classes is distinct from that of heterogeneous collections and that one cannot pass unaffected from the homogeneous into the heterogeneous: "and while homogeneous classes are the equivalent of a mathematical treatment, heterogeneous classes do not lend themselves easily to mathematical treatment."[5] He concluded that any laws as one might discover in biology cannot be of the universal form that appear in physics. In particular, this frustrates our efforts to predict with the same confidence as is possible in physics.

A caricature of the prediction that is possible in working with homogeneous classes is provided in Figure 4.1, where a class of five identical integers 2 operates in some way (say multiplication of arbitrarily paired tokens) upon a similar homogeneous class of integers 4. The result is yet another *homogeneous* class consisting entirely of the integer 8.

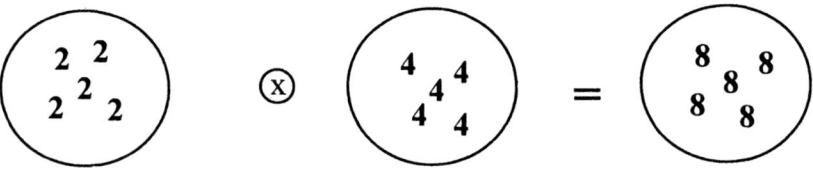

Figure 4.1: Operations between homogeneous classes are determinate. (Source: Robert E. Ulanowicz, *A Third Window: Natural Life beyond Newton and Darwin*, copyright 2009, Templeton Foundation Press. Reprinted with permission of the publisher.)

By contrast, the same operations between heterogeneous groupings do not in general yield tokens of any single heterogeneous grouping. To see this, we consider heterogeneous collections of the integers from 1 to 5, 6 through 10, 11 to 15, etc... When the same operation that was applied to the homogeneous integers is repeated for the first heterogeneous category acting on itself, the results scatter among the classes (Figure 4.2), culminating in a diffuse indeterminacy.

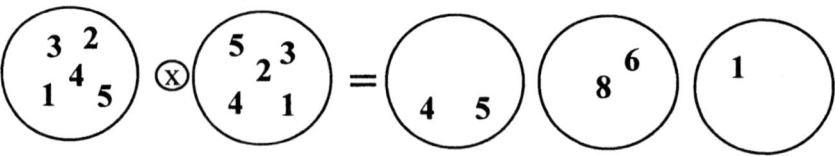

Figure 4.2: Operations between inhomogeneous groupings are indeterminate. (Source: Robert E. Ulanowicz, *A Third Window: Natural Life beyond Newton and Darwin*, copyright 2009, Templeton Foundation Press. Reprinted with permission of the publisher.)

When dealing with heterogeneous classes, the unexpected or indeterminate is always possible (see Kauffman's foreword in this volume). Kauffman also remarked on the combinatoric unmanageability of what he called the "adjacent possible."[6] Yet again, universal laws are seen to be intrinsically incapable of dealing with particularities.

The Limits of Probability

Indeterminacy is hardly new, of course, even though its existence is sometimes questioned. It is the ubiquity of variation that keeps statisticians employed in biology, and they have been very successful in quantifying statistical regularities throughout the living realm. Does that infer, however, that we thereby retain the advantage of prediction, albeit in a statistical sense? Elsasser answers with a definitive "No!" to this question by demonstrating that probability theory cannot be invoked for all chance phenomena.

In conventional probability theory, tacit assumptions are made that all chance events are simple, generic, and repeatable. Elsasser demonstrated, however, that the overwhelming majority of stochastic events in biology are totally unique, never again to be repeated.[7] This statement sounds absurd at first, given the enormity and age of our universe, but his assertion happens to be surprisingly easy to defend. Elsasser noted that there are fewer than 10^{85} elementary particles[8] in the whole known universe, which itself is about 10^{25} nanoseconds old.[9] This means that, at the very most, 10^{110} simple events could have occurred over all physical time. It thereby follows that if any event has considerably less than 10^{-110} probability of re-occurring, it will never do so in any physically realistic time. Of course, 10^{110} is a genuinely enormous number. It does not, however, require Avagadro's Number (10^{23}) of distinguishable entities to create a number of combinations exceeding Elsasser's limit on physical events. Nor does it require billions, millions, or even thousands. A system with merely 75 or so distinguishable components will suffice. It can be said with overwhelming confidence that any event randomly comprised of more than 75 distinct elements has never occurred earlier in the history of the physical universe. Ecosystems, which conservatively are comprised of hundreds or thousands of distinguishable organisms, must give rise, not just to an occasional unique event, but to legions of them. In ecology, unique, singular events are occurring all the time, everywhere!

A prerequisite for applying probability theory to chance phenomena is that the events in question re-occur at least several times, so that a legitimate frequency can be estimated. Singular events, however, occur only once, never to be repeated. Any probabilities assigned to them transcend physical reality. Furthermore, such particular singular events elude the abilities of universal laws to predict. Akin to Heisenberg uncertainties or the Pauli Exclusion Principle, such singularities are a necessary part of nature, not some epistemological lacuna awaiting theoretical elaboration. Yet again, determinism is judged not to be a ubiquitous characteristic of nature. Gradually, a larger picture is beginning to emerge: in very simple problems, the action of universal law provides most of the explanation needed for a particular behavior. Boundary considerations remain quite simple and constitute but a small part of the explanatory narrative. As one considers ever more complex, heterogeneous problems, the burden of explanation shifts away from the constraining universal laws and involves more the complicated boundary statements. Furthermore, such contingent constraints come to interact with one another, and it is those interactions, not the laws themselves, that actually *elicit* new behaviors. For in a world where radically contingent complex constraints can appear, entirely new behaviors can emerge quite naturally. It is not that the physical laws are necessarily violated. They continue to be part of the overall configuration of constraints. It is just that the most cogent explanation is to be found among the boundary phenomena.

We should have seen this coming. If we consider, for example, the number of conceivable combinations of the four force laws of physics and the two laws of thermodynamics, we are faced with a considerable count of possible juxtapositions (6! = 720). That magnitude pales, however, in comparison to the tally of all changes possible amongst a complex system having, say, 35 loci for incremental change (approximately 10^{40}). Any particular juxtaposition of laws likely will be satisfied precisely by a *very large* multiplicity of possibilities—conceivably billions or more. We conclude, therefore, that laws continue to constrain complex biological phenomena, but they are woefully insufficient to *determine* particular results. The contingencies that do specify outcomes must lie elsewhere. But where?

Enter Process

Although universal physical laws cannot deal fully with the individualities of a heterogeneous ecology (or of biology in general), many particulars, as Stent observed, do recur in ecosystems with evident regularity. What then, if not universal laws, determines particular outcomes or fosters their recurrence? In certain respects the answer is quite conventional—it lies with Darwinian process. What remains unmentioned, however, is that most evolutionary theorists either fail to apprehend or intentionally ignore the full implications of Darwin's paradigm, equating the Darwinian scenario instead with Ayala's conception of matter moving in accordance with universal laws. While evolution does not violate any universal laws, an unhealthy preoccupation with those laws can blind one to the larger nature of evolution as process.

In order to make clear how process differs from universal law, it behooves us to define the former more precisely. Accordingly, we suggest that: "a process is the interaction of random events upon a configuration of constraints that results in a non-random, but indeterminate outcome."[10] This definition is likely to strike some readers as foreign, and the juxtaposition of "non-random" with "indeterminate" is possibly confusing. It should prove helpful, therefore, to consider a simple example of an artificial process called Polya's Urn.[11] This exercise begins with a collection of red and blue balls and an urn containing one red ball and one blue ball. The urn is shaken and a ball is blindly drawn from it. If that ball is the blue one, a blue ball from the collection is paired with it and both are returned to the urn. The urn is shaken and another draw is made. If a ball drawn is red, it and another red ball are placed into the urn, etc... The first question arising is whether a long sequence of such draws and additions would culminate in a virtually constant ratio of red to blue balls? It is rather easy to demonstrate that after some one thousand or so draws, the ratio indeed converges to the close neighborhood of some constant, say 0.54591. That is, the ratio becomes progressively non-random as the number of draws progresses.

That the system does not converge closely to 0.5000 prompts a second question—what would happen if the urn was emptied and the starting configuration recreated? Would the subsequent series of draws converge to the same limit as the first? It can readily be demonstrated that it almost certainly will not. After a second thousand draws, the ratio might approach a limit in the vicinity of 0.19561. That is, the Polya process is clearly *indeterminate*. Repetition of the Polya process many times reveals that the ratio of balls is evenly distributed over the interval from zero to one. It can be any real number in that range. Furthermore, the ratio is progressively constrained by the particular series of draws (history) that have already occurred. We note further that some histories converge to behaviors that are difficult to distinguish from mechanical, law-like dynamics interrupted by occasional noise. The possibility thus arises that law-like behavior might constitute limiting forms of more general, less constraining processes.[12] For later reference, we emphasize three features of the artificial Polya process: (1) it involves chance; (2) it involves self-reference; (3) the history of draws is crucial to any particular series.

Natural Origins of Constraint

Polya's Urn, unfortunately, is not a natural process. Gregory Bateson, however, hinted how natural processes might create constraints that impart order to noisy affairs. He noted that the outcome of random noise acting upon a feedback circuit is generally non-random.[13] Following this lead, we now focus upon a particular form of feedback—autocatalysis.[14] By "autocatalysis" is meant any manifestation of a positive feedback loop wherein the direct effect of every link on its downstream neighbor is positive (Figure 4.3).

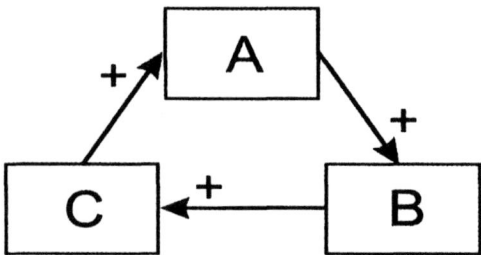

Figure 4.3: A three-component autocatalytic configuration of processes. (Source: Robert E. Ulanowicz, *Ecology, the Ascendent Perspective*, copyright 1997 Columbia University Press. Reprinted with permission of the publisher.)

An illustration of autocatalysis in ecology is found in the community that forms around the aquatic macrophyte, *Utricularia*.[15] All members of the genus *Utricularia* are carnivorous plants. Scattered along its feather-like stems and leaves are small bladders, called utricles (Figure 4.4a) Each utricle has a few hair-like triggers at its terminal end, which, when touched by a feeding zooplankter, opens the end of the bladder, and the animal is sucked into the utricle by a negative osmotic pressure maintained inside the bladder. In nature, the surface of *Utricularia* plants is always host to a film of algal growth known as periphyton. This periphyton serves in turn as food for any number of species of small zooplankton. The autocatalytic cycle is closed when the *Utricularia* captures and absorbs many of the zooplankton (Figure 4.4b).

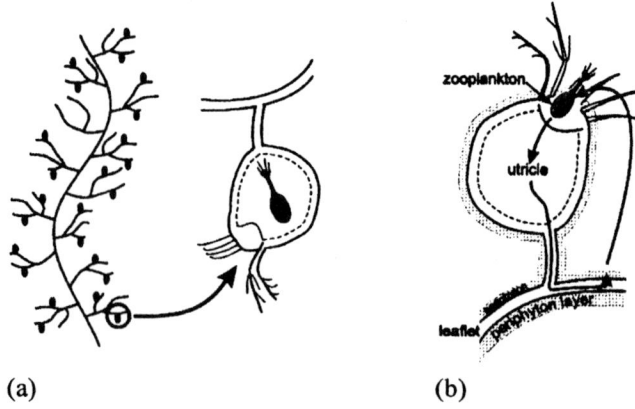

(a)　　　　　　　　　(b)

Figure 4.4: (a) Stem of Utricularia with closeup of utricle. (b) The autocatalytic processes inherent in the Utricularia system. (Source: Robert E. Ulanowicz, *Ecology, the Ascendent Perspective*, copyright 1997, Columbia University Press. Reprinted with permission of the publisher.)

Perhaps the most important feature of autocatalysis is that it exerts selection pressure upon all of its components and *any of their attendant mechanisms*. Any change in a characteristic of a component that either makes it more sensitive to catalysis by the upstream member, or a better catalyst of the element that it catalyzes, will be rewarded. Other changes will at best be neutral, but more likely will be decremented by the feedback. An immediate and cardinal effect of such internal selection is that it re-enforces those changes that bring more material or energy into a participating element, resulting in what can be called (in Newton's terminology) "centripetality" (Figure 4.5).

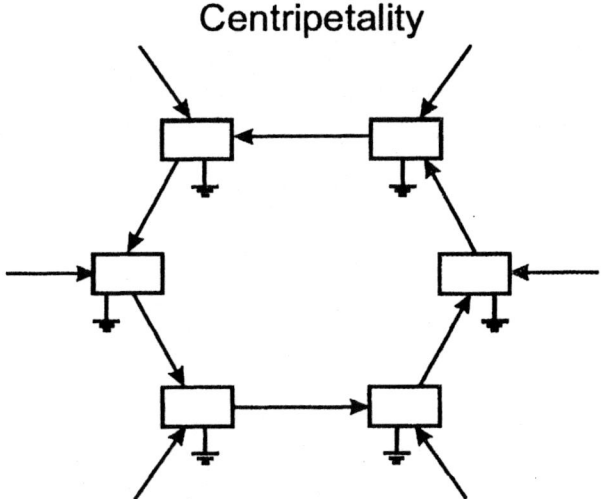

Figure 4.5: Autocatalysis induces centripetality. (Source: Robert E. Ulanowicz, *Ecology, the Ascendent Perspective*, copyright 1997, Columbia University Press. Reprinted with permission of the publisher.)

It is almost impossible to overstate the importance of centripetality for the nature of life. Conventional Darwinism conveniently ignores the role of "striving" in evolution.[16] Because the various organisms are competing with one another in epic struggle, one is moved to ask what accounts for their drive? Although striving is considered epiphenomenal and absent from most Darwinian accounts, here's how Bertrand Russell regarded the phenomenon: "Every living thing is a sort of imperialist, seeking to transform as much as possible of its environment into itself and its seed. . . . We may regard *the whole of evolution* as flowing from this 'chemical imperialism' of living matter."[17] Obviously, by "chemical imperialism" Russell is writing about centripetality; and, as we may infer from systems ecology, he correctly identifies it and not blind chance as the drive behind *all* of evolution.

Equally important is that centripetality is a prerequisite for competition. Without the generation of centripetality at one level, competition simply cannot arise at the next. We note that mutuality behind centripetality is essential, whereas competition is an accidental consequence. To see how centripetality induces competition, we regard the sequence in Figure 4.6. In the second graph element D appears spontaneously in conjunction with A and C. If D is more sensitive to A and/or a better catalyst of C, then the ensuing dynamics of centripetality will so favor D over B, that B will either fade into the background or disappear altogether. That is, selection pressure and centripetality can guide the replacement of elements.

Returning to Figure 4.3, we can envision how C might be replaced by E and A by F, so that it is likely that the lifetime of the autocatalytic configuration will exceed that of any of its components *or their attendant mechanisms*. Such is an example of supervenience by the whole over its parts, and it explicitly contradicts the Newtonian dictum of closure.[18] In fact, all the other Enlightenment postulates describing a mechanical world fare no better.[19] As already noted, determinism is rare in complex systems.

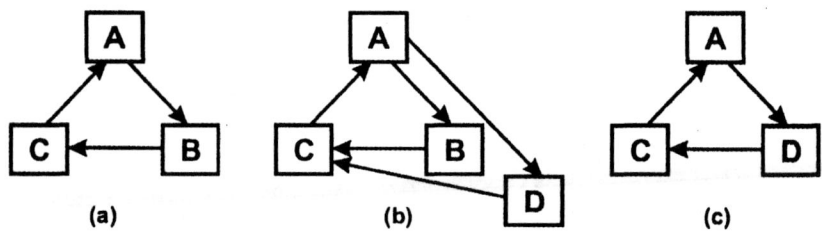

Figure 4.6: Centripetality induces competition. (Source: Robert E. Ulanowicz, *Ecology, the Ascendent Perspective*, copyright 1997, Columbia University Press. Reprinted with permission of the publisher.)

The asymmetric nature of autocatalysis contravenes reversibility. Because each component develops within the context of its co-participants, they will all become progressively co-dependent over time, so that an organic complex will no longer be amenable to atomistic decomposition. Finally, the domain of any individual process is hardly universal, being circumscribed in time and space and subject to mitigation by processes at other levels.

New Fundamental Assumptions

At least within the realm of ecology, all five Enlightenment postulates—closure, atomism, reversibility, determinism, and universality (see chapter 1 of this volume)—fail in some way or another to describe living dynamics. What is needed is an entirely new, but wholly naturalistic, metaphysic—an ecological metaphysic. In particular, the new framework requires that we shift our focus away from laws and objects toward configurations of processes. Furthermore, we want the new postulates to reflect the primacy of process, and so we return to the three features of the Polya process earmarked earlier—namely, that process requires chance, self-influence, and history.

Our first postulate establishes the ontological reality of chance:

(1) *Radical Contingency*: Nature in its complexity is rife with singular events.

Organic systems are constantly encountering unique contingencies, but the self-stabilizing properties of autocatalysis keep most of these events from upsetting the prevailing dynamics. A miniscule few, however, may divert a system into a wholly different mode of *emergent* behavior, so that emergence appears as an entirely *natural* phenomenon under the new assumptions.[20]

The Newtonian constraints of closure and atomism did not allow systems to maintain their integrities or grow.[21] By contrast, autocatalytic action, a particular form of self-influence, can impart form, constraint, and pattern to nature. Thus, we replace both closure and atomism by allowing for

(2) *Self-Influence*: A process in nature, via its interaction with other natural processes, can influence itself.

Thirdly, instead of reversibility we recognize, as did Darwin, that a system must retain some record of its past configurations—namely, it must possess a:

(3) *History:* The effects of self-influence are usually constrained by the culmination of past such changes as recorded in the configurations of living matter.

In this context the reader will likely think immediately of RNA, DNA, or similar molecules, but it is more likely that, well before material genomes came onto the scene, the first records of organic history were written into the topologies of stable, long-lived configurations of *processes.*

The three postulates thus constitute a natural platform from which to project a process-oriented view of ecology. In this framework, agency exerted by configurations of processes takes precedence over universal laws acting on objects. Furthermore, life itself can be closely identified with configurations of processes. This was made clear by the example of the dead deer provided by the late Enzo Tiezzi, who was a thermodynamicist and part-time hunter. Tiezzi asked what was different about a deer that he had just shot from the one that had been alive three minutes earlier? Its mass, form, bound energy, genomes—even its molecular configurations—all remained virtually unchanged immediately after death. What had ceased with death and was no longer present was the configuration of processes that had been coextensive with the animated deer—the actual agency by which the deer had been identified as being alive.[22]

Secondly, one discerns a definite opposition between the first two attributes of process. While autocatalysis imparts animation for systems to grow and maintain themselves, this action is opposed by radical chance, which serves to degrade and dissipate existing structures. This observation is hardly original. Diogenes reported how Heraclitus taught that nature was the outcome between agonistic tendencies that build-up as opposed to those that teardown (see also chapter 5 of this volume for a "pluralism that refuses the dualistic option"). The conflict between these drives is not absolute, however. At higher levels novel structures could never emerge without the action of radical contingency. Conversely, larger, complex and more constrained structures persist only by dissipating more resources.[23] Together the three fundamental postulates, along with their two corollary observations, constitute the framework of what has been called "process ecology."[24]

Moving Away from Objects

As one passes from the homogeneous world of physics into the highly particularized realm of ecology, it is becoming clearer that we must cease looking to universal laws and fixed mechanisms for full explanation and turn rather to the study of the organizing constraints exerted by process. That matter is moving according to universal laws simply does not tell us much, and our preoccupation with law diverts our attention away from the more complex nature of reality. In ecology, reality is scripted by process.

With the rise of computer technology has dawned the feasibility of creating "autonomous agent models" of ecosystems.[25] The focus in such models on object-object encounters has fostered the adoption of an "object-oriented-ontology" (O-O-O).[26] Although this emerging philosophical thrust does de-emphasize dependence on universal laws, it also plays down the relational nature of feedbacks that may arise in the actual systems. Focus in O-O-O is on the trees (objects), while the forest (processes) remains ignored.

O-O-O represents the natural culmination of a widespread trust in materialism, which, as Richard Lewontin wryly remarked, remains the *sine qua non* for most scientists:

> We take the side of science in spite of the patent absurdity of some of its constructs, in spite of its failure to fulfill many of its extravagant promises of health and life, in spite of the tolerance of the scientific community for unsubstantiated just-so stories, because we have a prior commitment, a commitment to materialism. It is not that the methods and institutions of science somehow compel us to accept a material explanation of the phenomenal world, but, on

the contrary, that we are forced by our a priori adherence to material causes to create an apparatus of investigation and a set of concepts that produce material explanations, no matter how counter-intuitive, no matter how mystifying to the uninitiated. . . .[27]

As we have seen, however, the priority given to objects over processes is ill-considered for several other reasons: For one, matter, as we usually conceive it, is exceedingly rare in a cosmos where more than ninety-nine percent of matter consists of hydrogen and helium radicals. Matter in solid form is the rare exception in the universe. Furthermore, stable atoms themselves, according to the Big Bang scenario, did not appear until (logarithmically speaking) well along in the development of the cosmos. Finally, what gave rise to matter appears, for the entire world, to be process.[28]

Presumably, the universe began as a chaotic, incredibly dense mass of extremely high-energy photons—pure flux.[29] As this continuum began to expand, some of the photons came together (collided) to form pairs of closed-looped circulations of energy called hadrons—the initial matter and anti-matter. For a while, these hadrons were destroyed by collisions with photons about as fast as they appeared. Continued expansion put space between the elementary particles so that matter and anti-matter pairs annihilated each other with decreasing frequency, and the diminishing energy of the photons made their collisions with extant material less destructive. Matter was beginning to appear, but was also disappearing at much the same rate.

Meanwhile, a very subtle (one in a billion) asymmetrical chance event produced slightly more matter than anti-matter, so that the mutual destruction of anti-matter by matter resulted in a growing residual of matter (feedback). Further expansion gave rise to yet larger configurations of emerging matter and the appearance of weaker forces. Eventually, matter coalesced under gravity (into stars) to a density that ignited chain fusion reactions (more feedback), producing larger, more complex aggregations—the heavier elements. From these it became possible to construct solid matter. And so the history of the physical universe reads, "process first—material later."

Life, according to the materialist scenario arose out of dead matter. Missing from this proposition is the "how." It likely will remain a mystery until our obsession with matter is eschewed in favor of process. That process is antecedent to life (as it presumably was to matter) was intuited by ecologist Howard T. Odum, who proposed that proto-ecological systems must already have been in existence before proto-organisms could have arisen.[30] In his scenario, at least two opposing (agonistic) reactions (like oxidation-reduction[31]) had to transpire in separate spatial regions. One volume or area had to contain a source of energy and another had to serve as a sink to convey created entropy out of the system. Physical circulation between the two domains was necessary. Such a "proto ecosystem" or circular configuration of processes provides the initial animation notably lacking in substance-based scenarios. We have seen that circular configurations of processes are capable of engendering selection, and they are capable of giving rise naturally to more complicated but smaller cyclical configurations (proto-organisms).

The spawning of proto-organisms poses no enigma. In irreversible thermodynamics, processes are assumed to engender (and couple with) other processes all the time (n.b. that one form of change begets another is dimensionally consistent, unlike the spontaneous appearance of a rate emerging from a substance). Large cyclical motions spawn smaller ones as the normal matter of course—as, for example, when large-scale turbulent eddies shed smaller ones. Corliss has suggested that a scenario like the one described by Odum might have played out around Archean thermal springs[32]—an idea that recently has found new enthusiasts in Harold Morowitz and Robert Hazen.[33] Yet again, origins reside in process, which mediates and gives form to material. We thus reckon that material and life share a common origin—process. No longer are we forced to accept the scenario of dead material mysteriously jumping up and coming alive.

What is common to all these reconsiderations is that process does not force us to view the cosmos going backwards. Conventional materialist models begin by considering systems that are homogeneous, rarified, and weakly interacting. We give priority to these assumptions, because they fit with our simplistic mental preconceptions. True, using this approach we have been extremely clever at projecting back into time to construct a history for our universe. But that is definitely not the way the world came at us. The cosmos apparently began as an incredibly dense, strongly interacting system, from which systems that are homogeneous, rarified, and weakly interacting could evolve only *after a very long time*. It is nigh time to put the horse before the cart and recognize, both historically and conceptually, that material and mechanism are secondary, and to a point accidental, in comparison to primal and generative process.

Notes

1. Sidney Brenner, quoted in Roger Lewin, "Why Is Development So Illogical?" *Science* 224 (1984): 1327.
2. Gunther Stent, quoted in Lewin, "Why Is Development So Illogical?" 1328.
3. Gregory Bateson, *Steps to an Ecology of Mind* (New York: Ballantine Books, 1972), 489.
4. Walter Elsasser, "A Form of Logic Suited for Biology?" in *Progress in Theoretical Biology, Vol. 6*, ed. Robert Rosen (New York: Academic Press, 1981), 23.
5. Elsasser, "A Form of Logic," 30.
6. Stuart Kauffman, *Reinventing the Sacred: A New View of Science, Reason and Religion* (New York: Basic Books, 2008), 127.
7. Walter Elsasser, "Acausal Phenomena in Physics and Biology: A Case for Reconstruction," *American Scientist* 57 (1969): 508.

8. Today the figure is put at closer to 10^{81}.
9. A nanosecond is one-billionth of a second—the timescale of atomic reactions.
10. Robert E. Ulanowicz, *A Third Window: Natural Life beyond Newton and Darwin* (West Conshohocken, PA: Templeton, 2009), 29.
11. Joel Cohen, "Irreproducible Results and the Breeding of Pigs," *Bioscience* 26 (1976), 391.
12. Paul C. W. Davies, *The Mind of God: The Scientific Basis for a Rational World* (New York: Simon and Schuster, 1993), 73.
13. Bateson, *Steps to an Ecology of Mind*, 410.
14. Stuart Kauffman, *At Home in the Universe: The Search for the Laws of Self-Organization and Complexity* (New York: Oxford University Press, 1995), 114; Robert E. Ulanowicz, *Ecology, the Ascendant Perspective* (New York: Columbia University Press, 1997), 42.
15. Robert E. Ulanowicz, "Utricularia's Secret: The Advantage of Positive Feedback in Oligotrophic Environments," *Ecological Modelling* 79 (1995): 50.
16. John Haught, *Is Nature Enough?: Meaning and Truth in the Age of Science* (New York: Cambridge University Press, 2006), 178.
17. Bertrand Russell, *An Outline of Philosophy* (Oxford: Routledge, 2009), 31, emphasis by author.
18. Philip Clayton, *God and Contemporary Science* (Grand Rapids, MI: Eerdmans, 2004), 254.
19. See David Depew and Bruce Weber, *Darwinism Evolving: Systems Dynamics and the Genealogy of Natural Selection* (Cambridge, MA: MIT Press, 1995), 92.
20. Philip Clayton and Paul Davies, eds. *The Re-emergence of Emergence: The Emergentist Hypothesis from Science to Religion* (Oxford: Oxford University Press, 2006), 1; Robert E. Ulanowicz, "Emergence, Naturally!," *Zygon* 42 (2007): 955.
21. Ulanowicz, *A Third Window*, 68.
22. Enzo Tiezzi, *Steps towards an Evolutionary Physics* (Southampton, UK: WIT Press, 2006), 47.
23. Eric Chaisson, *Cosmic Evolution: The Rise of Complexity in Nature* (Cambridge, MA: Harvard University Press, 2001), 26.
24. Ulanowicz, *A Third Window*, 11.
25. Marco Janssen, and Elinor Ostrom, "Empirically Based, Agent-Based Models," *Ecology and Society* 11 (2006), 37.
26. Ian Bogost, "What Is Object-Oriented Ontology?: A Definition for Ordinary Folk," December 8, 2009, http://www.bogost.com/blog/what_is_objectoriented_ontolog.shtml.
27. Richard Lewontin, "Billions and Billions of Demons," *The New York Review* (January 9, 1997), 31.
28. Ulanowicz, *A Third Window*, 60.
29. Chaisson, *Cosmic Evolution*, 108.
30. Howard Thomas Odum, *Environment, Power and Society* (New York: John Wiley and Sons, 1971), 157.
31. Daniel Fiscus, "The Ecosystemic Life Hypothesis II: Four Connected Concepts," *Bulletin of the Ecological Society of America* 83 (2002), 94.
32. John B. Corliss, John A. Baross, and Sarah E. Hoffman. "A Hypothesis Concerning the Relationship between Submarine Hot Springs and the Origin of Life on Earth," *Oceanologica Acta* 4 (1981, Suppl.), 59.

33. George Cody, Nabil Z. Boctor, Robert Hazen, Jay Brandes, Harold Morowitz, and Hatten Yoder, Jr. "The Geochemical Roots of Autotrophic Carbon Fixation: Hydrothermal Experiments in the System Citric Acid, H_2O-(+/-FeS)-(+/-NiS)," *Geochimica et Cosmochimica Acta* 65 (2001), 3557.

Bibliography

Ayala, Francisco. "Biological Evolution: Facts and Theories." Presentation at Pontifical Gregorian University, Rome, March 3, 2009.
Bateson, Gregory. *Steps to an Ecology of Mind*. New York: Ballantine Books, 1972.
Chaisson, Eric. *Cosmic Evolution: The Rise of Complexity in Nature*. Cambridge, MA: Harvard University Press, 2001.
Clayton, Philip. *God and Contemporary Science*. Grand Rapids, MI: Eerdmans, 2004.
Clayton, Philip, and Paul Davies, eds. *The Re-emergence of Emergence: The Emergentist Hypothesis from Science to Religion*. Oxford: Oxford University Press, 2006.
Cody, George, Nabil Z. Boctor Robert Hazen, Jay Brandes, Harold Morowitz, and Hatten Yoder, Jr. "The Geochemical Roots of Autotrophic Carbon Fixation: Hydrothermal Experiments in the System Citric Acid, H_2O-(+/-FeS)-(+/-NiS)." *Geochimica et Cosmochimica Acta* 65 (2001): 3557-3576.
Cohen, Joel. "Irreproducible Results and the Breeding of Pigs." *Bioscience* 26 (1976): 391-394.
Corliss, John B., John A. Baross, and Sarah E. Hoffman. "A Hypothesis Concerning the Relationship between Submarine Hot Springs and the Origin of Life on Earth." *Oceanologica Acta* 4 (1981, Suppl.): 59–69.
Davies, Paul C. W. *The Mind of God: The Scientific Basis for a Rational World*. New York: Simon and Schuster, 1993.
Depew, David, and Bruce Weber. *Darwinism Evolving: Systems Dynamics and the Genealogy of Natural Selection*. Cambridge, MA: MIT Press, 1995.
Elsasser, Walter. "Acausal Phenomena in Physics and Biology: A Case for Reconstruction" *American Scientist* 57(1969): 502-516.
———. "A Form of Logic Suited for Biology?" In *Progress in Theoretical Biology, Vol. 6*. Edited by Robert Rosen, 23-62. New York: Academic Press, 1981.
Fiscus, Daniel. "The Ecosystemic Life Hypothesis II: Four Connected Concepts." *Bulletin of the Ecological Society of America* 83 (2002): 94-96.
Gödel, Kurt. "On Formally Undecidable Propositions of *Principia Mathematica* and Related Systems." Translated by Martin Hirzel. 2000, http://www.research.ibm.com/people/h/hirzel/papers/canon00-goedel.pdf.
Haught, John. *Is Nature Enough?: Meaning and Truth in the Age of Science*. New York: Cambridge University Press, 2006.
Janssen, Marco, and Elinor Ostrom. "Empirically Based, Agent-Based Models." *Ecology and Society* 11 (2006): 37.
Kauffman, Stuart. *At Home in the Universe: The Search for the Laws of Self-Organization and Complexity*. New York: Oxford University Press, 1995.
———. *Reinventing the Sacred: A New View of Science, Reason and Religion*. New York: Basic Books, 2008.
Lewin, Roger. "Why Is Development So Illogical?" *Science* 224 (1984): 1327-1329.

Lewontin, Richard. "Billions and Billions of Demons." *The New York Review* (January 9, 1997), 31.
Odum, Howard Thomas. *Environment, Power and Society*. New York: John Wiley and Sons, 1971.
Russell, Bertrand. *An Outline of Philosophy*. Oxford: Routledge, 2009.
Tiezzi, Enzo. *Steps towards an Evolutionary Physics*. Southampton, UK: WIT Press, 2006.
Ulanowicz, Robert Edward. "Utricularia's Secret: The Advantage of Positive Feedback in Oligotrophic Environments." *Ecological Modelling* 79 (1995): 49-57.
———. *Ecology, the Ascendant Perspective*. New York: Columbia University Press, 1997.
———. "Emergence, Naturally!" *Zygon* 42 (2007): 945-960.
———. *A Third Window: Natural Life beyond Newton and Darwin*. West Conshohocken, PA: Templeton, 2009.

Chapter Five
Ordinal Pluralism as Metaphysics for Biology

Lawrence Cahoone

Reductionism, by which I mean not the use of reductive explanations, but the view that *all* valid explanations of natural phenomena ought to be reductive, when applied to biology, is physicalist or materialist in its metaphysics. It takes material or physical components (not usually distinguished) and the rules governing their interactions to constitute nature or reality, and avoids positing entities or factors that cannot be understood as such. Hans Jonas argued that modern thought labors under a kind of bipolar disorder, imagining that there are at most only two kinds of beings, the material and the mental, the only question being their status and relation, for example, whether the mental can or cannot be explained by the material.[1] This dualism is *prima facie* inhospitable for biology, presenting it with a metaphysical dilemma: either living things are reducible to the material, or they must in some sense be mental. Reductionists take the former option. Some anti-reductionists—for instance, biosemioticians, who find sign-usage and intentionality in all life, and pan-psychists, who ascribe some form of mentality to all life—take the later. Whitehead's philosophy of organism is a particularly sophisticated example.

An alternative would be a *pluralism* that refuses the dualistic option. By this, I mean a perspective in which it is not assumed that all natural phenomena are, or are caused by, the physico-material, or the mental, but that these, along with the biological and perhaps other domains, are all distinctive types of phenomena in interaction, with causal arrows potentially pointing in any direction. If this view combines the notion of hierarchical dependence, accepting that the mental depends on the biological, the biological on the material, then it also accepts emergence, the fact of novel properties, and the necessity of non-reductive explanations at more complex levels of organization. Such a view would provide what philosopher William Wimsatt calls a "tropical rainforest" ontology—as opposed to Willard Van Orman Quine's famous ontological taste for "desert landscapes."[2] There is such a metaphysical language available: the "metaphysics of natural complexes" formulated by Justus Buchler, the American philosopher and scholar of Charles Sanders Peirce. Biologist Stanley Salthe has shown that Buchler's notion of "natural complexes" can be used to inform a scientific notion of complexity consonant with a hierarchical systems theory of nature.[3] We shall explore the advantages of such a meta-language for naturalism and natural science, including biology.

Buchler stipulated a principle of *ontological parity*, according to which nothing we can discriminate can be more or less real or genuine than anything else.[4] That is to say, he rejected entirely the various traditional philosophical distinctions between the "real," "true" (regarding things, not propositions), or "genuine," and the "apparent," "epiphenomenal," or "illusory." A fictional character, the possibility of my dying, the imaginary number i, and Heaven are all no less real than the computer keys under my fingers. Anything that can be discriminated, in any sense, is a "natural complex." Complexes can be physical objects, facts, processes, events, universals, experiences, institutions, numbers, possibilities, artifacts, and all their relations and properties and functions. For Buchler the qualifier "natural" signifies that there can be no discontinuous realms of complexes, no worldly versus transcendent complexes, while the noun "complex" means that nothing is simple or incapable of further analysis. Following Peirce, Buchler denies that anything is either utterly determinate or absolutely indeterminate, or that the traits of any complex can be exhausted.

Pluralism and parity require Buchler to endorse *ordinalism*. The question "what is real?" is transformed into, as John Herman Randall, Jr. put it, "*How* is something real?" or "In what orders of relation does it function?"[5] This is what replaces our usual distinction between the real and apparent. A fictional truck and the truck bearing down on me are equally real, one functioning in a literary order, another in the order of physical fact that includes my body. Every complex must be related to some other complexes—which is *not* to say related to all others, for irrelevance also obtains—hence is located in one or more contexts of relations or orders in which the complex functions and hence has an "integrity." Complexes and orders are either strongly or weakly related to a given complex—strongly, to its integrity, hence, an internal or constitutive relation—and weakly, to its breadth or scope in that order. A complex's identity, or in Buchler's terms its "contour," is the continuous relation between each of the complex's integrities and its total collection of integrities.[6]

Buchler's metaphysics is thus distributive, rather than collective; it characterizes *any*thing in its functional locale, not *every*thing together. Buchler denies that the Whole is a complex at all; there is no "order of orders."[7] We may, if we wish, use the phrase "innumerable orders" or the "provision of complexes" or some equivalent to refer to complexes and orders indifferently.[8]

Buchler's metaphysics is the closest thing we have to a metaphysics of any possible world. It is not presuppositionless or neutral with respect to all other perspectives, but it is the nearest thing to it, the least suppositional and the most neutral with respect to standard metaphysical problems. Buchler's metaphysics would apply equally well to a world of disembodied spirits, a quark-plasma, a set of Platonic forms, or an Olympus of Greek gods. But if Buchler's metaphysics fits many possible worlds, then it is equally true to say that *it does not pick out this world*. As far as we can reliably judge, lives, minds, selves, intentionality, and meanings respectively require organisms, matter, bodies, neurons and cultural objects. These relations of dependence are not symmetrical; living things presuppose atoms, but atoms do not generally presuppose living things. Buchler's metaphysics allows all sorts of facts and processes that either cannot or at least do not occur in our reality as far as we can tell. Despite Buchler's use of the qualifier in his term "natural complex" his metaphysics is not naturalistic in any strong sense.

But his scheme can provide us with an indispensable background language, allowing us to represent a naturalism *within* his pluralism, one which will still reap distinctive conceptual benefits from its inclusion in the latter. Four benefits are foremost. First, the acceptance of complexity, which in effect means fallibilism: we cannot ever assume analysis of a complex is at an end. The second is ordinalism (i.e., the fact that a complex functions and has integrities in multiple orders). Third, Buchler subtracts from the discussion of ontology questions of which strata are more real or genuine or fundamental. As noted, "real" as a comparative term is thrown out entirely. "Basic" or "fundamental" become ordinal terms, referring to the relative importance or scope of a complex's role in an order or system. Last is a methodological point: Buchler's metaphysics rejects the notion that the validity of a metaphysical analysis of a thing or order of things hangs on the valid characterization of the *most inclusive order* in which it participates, for example, the most comprehensive (the Whole), the most invariant (the Highest), the most elementary (Simples), or the most fundamental (Foundations). It is non-foundational, distributive, not collective, and local, not global. It does not stipulate the nature of the Whole.

Let us apply Buchler's terms to nature as understood by contemporary science. Regarding the term "complex," two qualifications are necessary. First, the scientific and Buchlerian use of complexity are different. For Buchler, "complex" does not admit of degree; this is because for him complexity refers to the actual and potential ordinal locations of a complex. In science, however, complexity means the organizational structure or information a system exhibits, hence the length of its shortest adequate description (in the Kolmogorov or algorithmic sense). Rather than using another term for physical complexity (e.g., "organizational complexity") in what follows, "complexity" will have its scientific meaning, under which things can be more or less complex, even if none can have zero complexity, there being *no* simples. Second, the last point may seem to be violated by elementary particles, or quarks and leptons. But this objection is not compelling. Even if quarks and leptons turn out to be the *most* elementary of components, the *least* complex, they are not "simples." Quarks only appear in clusters. Hence, they are constituted by relations to things outside of themselves—namely, to other quarks. That still leaves electrons. But electrons, like all quantum particles, are subject to non-locality or entanglement, meaning that their states are internally related to states of other entities. Quantum field theory conceives all particles as field excitations in an underlying ontology of fields, and whatever fields are, they are not simples.

Now, the natural sciences have a term that applies to a broad class of natural complexes: *systems*. Like "complex," "system" can be recursively applied to parts and wholes: a system is constituted in some sense by its component systems and by its relations to more encompassing systems of which it is a part. Without prejudging further analyses, we can say the pre-eminent objects of natural science are systems. We could *almost* say that "system" is a synonym for "entity," but we must reserve "entity" for narrower use. This does *not* mean that all natural complexes are systems. Properties and performances of systems, like temperature and velocity, or rotation and melting, are not. A hydrogen atom is a system, but hydrogen *per se* is not; it is a natural kind. Last, structures and processes such as a crystal lattice and the double-helix, or oxidation and eutrophication, are natural complexes but not systems. These can be conceived, respectively, as either traits of systems or kinds of systems. So, for the moment at least, we are orienting our consideration of any and all natural complexes in a neo-Aristotelian way, dividing beings or complexes into particular systems, second-order classifications of systems (kinds), and the traits of systems ("trait" including anything predicable of systems except kinds, hence all properties and performances, anything the system is, has, or does, in the present, past, or future). The flavor of this is still ordinal, however; anything discriminable is still a complex, a being, but our naturalistic analysis is picking out of all complexes the natural complexes ("nature" being not yet defined).

It may seem that in beginning with systems, and invoking Aristotle, we have adopted an entitative or "thing" ontology. We can avoid that result if, following Buchler's pluralism, we conceive many systems of nature—in fact, those we know most "robustly," or through multiple means of access—as simultaneously and co-primordially a set of lower-scale entities or *components*, a *structure* of relations among components, and a *process* of events that constitutes and maintains the structure. (*States* are snapshots of processes, or rather, system processes as Δt approaches zero.) We may say that *components (themselves systems), structure, and process exhibit ontological parity* (see Figure 5.1 below). A system equally *is* its parts, *is* a structure, and *is* a process.

The parity of parts, processes, and structures calls attention to something often unrecognized in philosophical discussions of reduction. What virtually all participants mean by the term "reduction" is "componential reduction" or reduction of wholes to parts. But the parity of entities, process, and structure suggests that to consider an individual or system or order a "process" or a "structure" is *also* reductionist, at least in spirit. For it so happens that some metaphysicians and cosmologists, in reaction against the historical supremacy of entitative metaphysics and componential reductionism, have attempted to conceive structure or process as the ultimate reality, replacing entities. From the perspective of ontological parity this is itself reductive in a broader sense. If there is no *a priori* reason to privilege entities there is also no *a priori* reason to privilege either processes and events or structures and relations. Absent a longer metaphysical discussion, we may at least note that in the robust orders of existence to which we have greatest and multiple means of access. Just as we find no entities that are not structured and undergoing some kind of process, we find no structures without something structured, and no processes without something undergoing the process. (In a more detailed discussion we would have to note that ensembles, like a volume of gas, have negligible structure and fields; for example, magnetic fields, have no parts.)

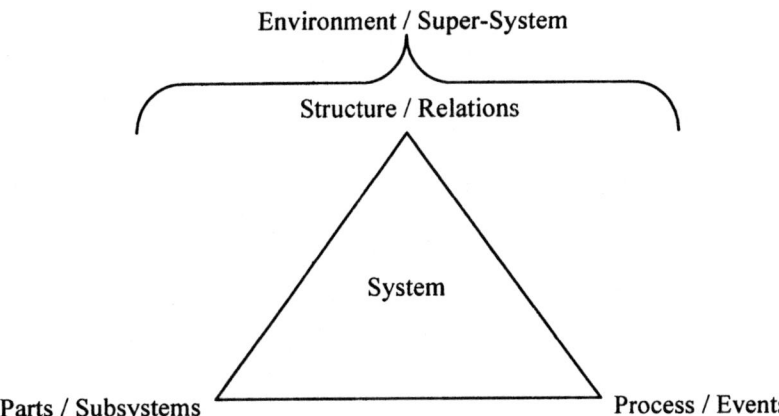

Figure 5.1: Systems as Natural Complexes

Now we turn to an empirical claim about nature, forwarded by British Emergentism in the 1920s, by hierarchical systems theorists, like Herbert Simon in the 1970s, and most recently Wimsatt and Stanley Salthe. Natural phenomena occur at *strata* or *levels* of scale, usually but not exclusively, levels of size, energy, rate of reaction and recovery time, for example, atomic, molecular, macromolecular, macroscopic, etc... The notion of levels is independent from the notion of emergence, but conforms to it. Wimsatt is able to define natural levels rather objectively as "peaks of regularity" or "local maxima of regularity and predictability in the phase space of different modes or organization of matter."[9] A level is ordered by "hierarchical part-whole composition relations." The range of entities with which an entity interacts is a non-arbitrary and informative fact about that entity; levels are collections or orders of interacting entities. Size or scale is a common, not always sufficient, indicator of level, for "size is . . . a robust indicator for many other kinds of causal interactions." Entities are generally at levels; levels are "where the entities are," ranges of scale where one finds the greatest "density of types" of entities. According to Wimsatt, "the identifications made between levels will of course mean that the same system will be found at a number of levels, if it has any reasonable degree of complexity, though it will of course be *a* system only at one level."[10] There are processes and phenomena between levels, but levels act as natural "attractors" for entities. They emerged during the evolution of the universe, perhaps selected by their achievement of stable equilibria. As such, they are real objects in the world; we perspectively select them because *that is where the explanations are.*

This is to say that higher-level properties commonly achieve multiple-realizability in, and dynamic autonomy from, lower levels.[11] Macro-state stability often rests on micro-state flux. It simply is the case that "the [relative] stability of macro-states . . . further entails that the vast majority of neighboring (dynamically accessible) micro-states map into the same or (more rarely) neighboring macro-states." Stable macro-states have to be "tuned" so that the chaos at lower levels within a numerically identical system does not lead to deviation-amplifying effects that would destroy macroscopic stability. While the macro-properties must be sensitive to *certain kinds* of micro-changes, "it is crucial that *most differences [at the micro-level] do not have significant [macro-level] effects most of the time.*"[12] This has a simple but powerful explanatory consequence: it necessitates that there cannot be purely micro-level explanations for most stable macro-level properties.

Wimsatt proposes that nature must exhibit compositional levels of natural phenomena, that is, "hierarchical division of stuff (paradigmatically but not necessarily material stuff) organized by part-whole relations, in which wholes at one level function as parts at the next (and all higher) levels."[13] He summarizes, "*levels of organization are a deep, non-arbitrary, and extremely important feature of the ontological architecture of our natural world, and almost certainly of any world that could produce, and be inhabited or understood by, intelligent beings.*"[14] Thus a major feature of nature is that it evolves hierarchically layered types of entities, structures, and processes of vastly differing scales, as the product of cosmic, chemical, biological, and cultural evolution.[15] This does not by itself indicate any "vital force" or intrinsic principle of design, nor determinism; it is presumably the contingent result of the favorable physical constants and force ratios characteristic of our universe. But it inevitably makes ours a world, not of one, but of multiple, mutually influencing inquiries. The distribution of the forms of inquiry is determined by objective features of nature; something about nature dictates that human inquiry cannot be one science, as positivism hoped, nor multiple unrelated sciences, as some postmodernists held (e.g., Lyotard). Nature's different kinds of things, and different strata or orders at which kinds obtain and interact, require a plurality of related sciences. The program of our inquiry is not the unity of *science*, but the integration of *sciences*.

Salthe distinguishes two types of hierarchies: scalar hierarchies relate nested systems of differing "extensional complexity" (e.g., size, speed) which have whole-part relations, whereas specification or subsumptive hierarchies relate types of systems of differing "intentional" or informational complexity that are simultaneously taxonomic and evolutionary.[16] Under the former I am a whole with parts and a part of more encompassing systems. In a specification hierarchy, on the other hand, the "focal" level is characterized by the greatest cumulative informational constraints and integrates the behavior of particular systems operating at lower levels as well, for example, the series [physical [material [biological [mental]]]]. The relations in this series are transitive; the biological, as more complex than the material, is also more complex than the physical. The latter are dependent on the former; the former are more widespread or inclusive and arise earlier in evolution. Salthe points out that there can be—and need be— no canonical listing of levels. The list is complex and variable. There are far more instances of emergence, far more componential constructions of wholes with novel properties, than we can enumerate. Hence, emergence is a necessary but insufficient criterion of strata. The criteria that legitimately may be used to distinguish strata include at least: scale of mass, energy, volume, or speed of processes; regularity peaks; scale of entification; compositional level, or what is a part of what whole; degree of complexity; emergent properties or performances; and description or explanation by distinctive scientific methods and concepts.

Fortunately, several of these criteria tend to cluster together. Regularity peaks are often entification peaks at scale. Emergent properties and performances, if they are distinctive enough, call for distinctive scientific methods and concepts. Thus the different sciences mark particularly robust points of emergence. Furthermore, a particular series of these sciences do correspond to rising complexity. Thus it is possible to highlight as "orders" of nature a specification hierarchy of phenomena studied by distinctive sciences, recognizing that there are many more local levels of emergent strata within them. The result is a short list of wide strata corresponding to a small subset of especially prominent emergent properties and sciences: the physical, material, biological, mental, and cultural. We can use Buchler's notion of order for these levels or strata. Levels are a kind of natural order.

We can now formulate our naturalism.[17] By "naturalism" I mean any view that accepts: (1) that the reality called "nature" is a set of systems whose members are open to interaction (which does not mean that everything in it has, does, or will interact with everything else); and (2) that the conclusions of the natural sciences are relevant to the metaphysics of nature. These are common to most naturalisms. But I will add three further claims to characterize my naturalism: (3) the orders listed above exhibit a hierarchy of dependence and emergence (which does not rule out "downward" causation); (4) the physical and the material are *two particular orders* of nature and do not encompass the whole of nature (nor is nature exhausted by the dualism of the physic-material and the mental); and (5) we need not and do not assume that everything that is or was or will be is natural, since we renounce the metaphysics of the Whole.

The first of these three additions implies emergence and hierarchy theory. The second indicates that a pluralistic naturalism, based in ordinalism and emergence, accepts in principle both an entity-pluralism and a property-pluralism, hence, a "tropical rainforest" ontology. The last means that a *local* naturalism is possible, in which we describe the most robustly accessible orders of reality, collectively called nature, as central. It *might* be exhaustive, but we do *not* presume it is or must be. Whether any complexes cannot be included in the orders of nature remains an open question. Given such a fallible, pluralist, and local starting point, I will argue that *naturalism is at least locally true*. However one might want to characterize the Whole, the Underlying, the Simple, or any candidate for the most inclusive order of all orders, it is arguable that anything we discriminate is locatable in the orders of nature. Whether this claim is true or false depends on whether we can discriminate complexes not locatable in the natural orders.

Now I will turn briefly to the orders themselves. The assumption that nature is physical is troublesome. First, the physical is not easy to define. If the material implies "matter," then the physical must be broader: electromagnetic fields, photons, and spacetime are not matter. Defining the physical as space-occupying is also problematic; fields are spatially extended, but do not exclude other fields from their space. Additionally, our yet-inadequate theories of Quantum Gravity hold that space itself emerges from non-spatial energy at the Planck scale (and/or Planck time). We may say the physical is the smallest (sub-atomic) components and widest environment (gravitation) of spacetime-energetic systems. If so, atomic matter emerges from the physical.

Second, to what extent is the physical primary, fundamental, or determinative of the other orders? Are the rules governing the electron more "fundamental" than the rules governing, say, eutrophication? I reply in the affirmative, but only in two relative senses. Ecological and biological processes are asymmetrically dependent on the existence of electrons: no electrons, no eutrophication, whereas the reverse does not hold. Also, physics is more extensive in scope than biology and ecology; the laws governing the free electron have something to say to all occurrences of electrons. But these rules neither comprehend nor determine eutrophication. In a pond undergoing eutrophication the changes in the behavior of electrons will have something to say, but it will be something *negligible*, because it will be swamped by the role dictated to the electrons by the atoms and molecules they inhabit and, at a higher level, by the pond and ecosystem in which they function. Eutrophication is not a process electrons can undergo. The enormous number of different electron configurations and energy states that underlie the region of a pond's phase space that signifies eutrophication will likely show no difference from the electronic configuration of other regions of that phase space. Physics has no general ontological or explanatory priority. The physical is one order of nature, not the whole of nature.

The material, or chemical, or atomic order emerges from the physical, and from it emerges the biological, organisms being a special kind of material system. Life means a massive leap in complexity. It is a set of processes, hence also a state, manifested by certain complex material individuals of characteristic structure. Like chemistry and unlike physics, biology has a natural smallest unit, the cell, and a set of natural kinds, species. For chemistry, they are the atom and elements, respectively. Biology's objects vary greatly in scale, but far less than those of physics; there a minimum scale, the bacterium, but the upper bound is strictly constrained—the largest organisms are of normal macroscopic size, or a bit bigger. But with biology also come larger ensembles—namely societies and ecosystems, as well as the Earth's biosphere itself. Neurologically complex, socially communicating animal species exhibit intentional performances, or "mind." These are studied by psychology, psychiatry, ethology, social psychology, neurology, and cognitive science. Mind is not solely a human phenomenon, but an animal phenomenon. It is supported by biological components, but not composed of them. Not minds *per se*, but minded organisms are components of societies and ecosystems. The mental activities they subtend or contain intentional objects, hence, are semantic. At the lowest levels, intentional events are internalist—emergent only upon brain and soma—but at more complex levels, externalist, emergent upon the organism's relation to environment as well. These properties are non-spatial but potentially causal traits of a biological entity in interaction with an environment. Last, we know of at least one species of minded animals that are *cultural* (i.e., that they are capable of autobiographical consciousness, sign construction, and taking the perspective of others, and hence, manipulating meanings in a way no other mind can do). In the cultural, as well as in the mental and ecological domains, causation becomes very complex, forming what Wimsatt calls "causal thickets."

This complexity hierarchy of orders (represented in Figure 5.2) is *itself* a robust hypothesis, attested by three different and powerful facts: the series of orders is characterized by increasing complexity; rational inquiry has found it necessary to develop special methods and concepts corresponding to this series; and contemporary science has strong evidence that this hierarchy roughly matches the temporal evolution of nature. Leaving aside the particulars of the relevant views, the British Emergentism of the 1920s, followed later by hierarchic systems theory, correctly perceived that stratified dependence and emergence is a central feature of nature. Cultural meanings, mental intentions, living organisms, material entities, and physical phenomena can all be located in a hierarchically organized, pluralistic nature. This is a fallible, contingent claim about whatever we encounter, including ourselves, located within an open-ended metaphysical language of complexes, not an *a priori* stipulation of the Whole. Such must be, as Wimsatt remarks, the working hypothesis for error-prone beings of finite reasoning capacity inhabiting a world of indefinitely great complexity.

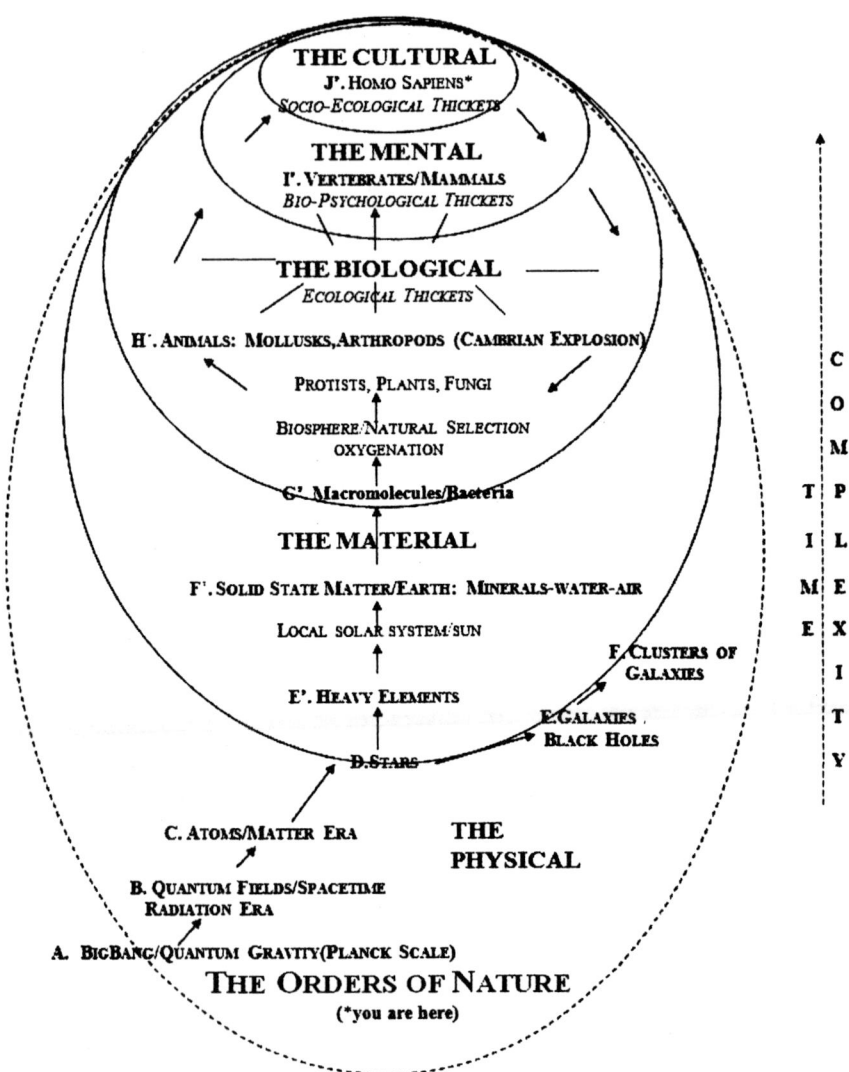

Figure 5.2: The Orders of Nature

Notes

1. Hans Jonas, *The Phenomenon of Life: Toward a Philosophical Biology* (New York: Harper and Row, 1966).
2. William C. Wimsatt, *Re-Engineering Philosophy for Limited Beings: Piecewise Approximations to Reality* (Cambridge, MA: Harvard University, 2007).
3. Stanley Salthe, *Evolving Hierarchical Systems* (New York: Columbia University Press, 1985).
4. Justus Buchler, *Metaphysics of Natural Complexes*, eds. K. Wallace and A. Marsoobian, with R. S. Corrington (Albany, NY: State University of New York Press, 1990).
5. John Herman Randall, *Nature and Historical Experience* (New York: Columbia University Press, 1958), 131.
6. Randall, *Nature and Historical Experience*, 22.
7. Buchler, "On the Concept of 'the World'," Appendix to *Metaphysics of Natural Complexes*.
8. Buchler, "Probing the Idea of Nature," Appendix to *Metaphysics of Natural Complexes*.
9. Wimsatt, *Re-Engineering Philosophy*, 249.
10. William C. Wimsatt, "Reductionsim, Levels of Organization, and the Mind-Body Problem," in *Consciousness and the Brain*, ed. G. G. Globus, G. Maxwell, and I. Savodnik (New York: Plenum, 1976), 242.
11. Wimsatt, *Re-Engineering Philosophy*, 217.
12. Wimsatt, *Re-Engineering Philosophy*, 218, emphasis mine.
13. Wimsatt, *Re-Engineering Philosophy*, 201, 213.
14. Wimsatt, *Re-Engineering Philosophy*, 203-204.
15. See Herbert A. Simon, "The Architecture of Complex Systems," *Proceedings of the American Philosophical Society* 162 (1962): 467-482; and Stanley Salthe, *Evolving Hierarchical Systems*.
16. Stanley Salthe, *Development and Evolution: Complexity and Change in Biology* (Cambridge, MA: MIT Press, 1993).
17. Lawrence Cahoone, "Emergence, Reduction, and Ordinal Physicalism," *Transactions of the Charles S. Peirce Society* 44, no. 1 (Winter 2008): 40-62.

Bibliography

Buchler, Justus. *Metaphysics of Natural Complexes*, edited by Kathleen Wallace and Armen Marsoobian, with Robert S. Corrington. Albany, NY: State University of New York Press, 1990.

———. "On the Concept of 'the World'." Appendix III in *Metaphysics of Natural Complexes*.

———. "Probing the Idea of Nature." Appendix IV in *Metaphysics of Natural Complexes*.

Cahoone, Lawrence. "Emergence, Reduction, and Ordinal Physicalism." *Transactions of the Charles S. Peirce Society* 44, no. 1 (Winter 2008): 40-62.

Jonas, Hans. *The Phenomenon of Life: Toward a Philosophical Biology.* New York: Harper and Row, 1966.
Mayr, Ernst. "Teleological and Teleonomic: A New Analysis." *Boston Studies in the Philosophy of Science* 14 (1974): 91-117.
Pittendrigh, Colin. "Adaptation, Natural Selection, and Behavior." In *Behavior and Evolution.* Edited by A. Roe and G. G. Simpson. New Haven: Yale University Press, 1958.
Randall, John Herman. *Nature and Historical Experience.* New York: Columbia University Press, 1958.
Salthe, Stanley. *Evolving Hierarchical Systems.* New York: Columbia University Press, 1985.
———. *Development and Evolution: Complexity and Change in Biology.* Cambridge, MA: MIT Press, 1993.
Simon, Herbert A. "The Architecture of Complex Systems." *Proceedings of the American Philosophical Society* 162 (1962): 467-82.
Wimsatt, William C. "Reductionsim, Levels of Organization, and the Mind-Body Problem." In *Consciousness and the Brain.* Edited by G. G. Globus, G. Maxwell, and I. Savodnik, 199-267. New York: Plenum, 1976.
———. *Re-Engineering Philosophy for Limited Beings: Piecewise Approximations to Reality.* Cambridge, MA: Harvard University, 2007.

Section 2:
Biosemiotics

Chapter Six

Why Do We Need a Semiotic Understanding of Life?

Jesper Hoffmeyer

The Problem of Agency

The prevailing story of life as presented by science through the last half century is generally known as neo-Darwinism, or perhaps better, "the modern synthesis." In the 1940s and 1950s, several major biological disciplines, systematics, plant biology, genetics, and paleontology, accommodated their theoretical structures to a generalized view of evolution based on population genetic models of evolution as developed in the 1930s by Ronald Fisher, J. B. S. Haldane, Sewall Wright, and others. What was noteworthy, however, was that developmental biology was missing from the modern synthesis, and questions of morphology or development were not thought to contribute anything important to the population genetically based approach. This occurred in spite of the warnings from, among others, geneticist Richard Goldschmidt that evolution is the result of heritable changes in development.[1] Instead the gene's double role as a causative factor in ontogeny and a causative factor in phylogeny was swept aside under the assumption that natural selection would not care about how the phenotypes were actually created but only about how good they were in the game of reproduction and survival. The gene became a unit of transmission, not of construction.

That evolutionary thinking would proceed along this track was not accidental. In spite of its reductionist propensities, science has never ascribed real constructive capacity to unitary agents. Such agents, whether atoms, genes, embryos or individuals were always understood as passive bodies to be steered by forces beyond their own control. As is well known, Darwin wished to be the "Newton of the grass blade" and although natural selection, as Darwin himself knew, is nothing like the austere gravitational forces, there nevertheless are parallels.[2] Most important, by posing natural selection as the essential mechanism of evolution, Darwin created a perfectly *externalist* theory, a theory that seeks to explain the internal properties of organisms, their adaptations, exclusively in terms of properties of their external environments, natural selection pressures.[3] True, the theory needed heritable variations to be present in the population, but this was not at the time seen as a cause to suspect any intrusion of constructive agents into the theoretical structure, for obviously such variations would occur simply as a result of chance fluctuations. To this day, mainstream biology ascribes variations solely to mutational events, and routinely plays down the generative role of developmental or ecological processes.

Another central pillar of the modern synthesis is *gradualism*. The population-dynamic formalism operates on the assumption of continuous and incremental genetic variation, and any non-gradualist events will therefore in principle be excluded from the horizon of the modern synthesis. An acceptance of sudden discontinuities would weaken the notion of natural selection as the major mechanism behind evolutionary change, and by implication also the strongly held belief that *macroevolution* (the large morphological changes seen between species, classes, and phyla) is nothing but an extension in time of *microevolution* (evolution within the species).

To these traditional biases of the modern synthesis we must add a more modern one: *gene-centrism*. Darwin did not know anything about genes, as we understand these units today (serious ambiguities remain in the modern gene concept, though),[4] but in the modern synthesis genes are seen as units of inheritance and as determinants of traits and variations in traits. In fact, the whole mathematical building raised by the modern synthesis would scramble should gene mutations prove not to be the basic unit of evolutionary change.

Until very recently, any attack on one of these three major pillars of evolutionary biology would have met with vigorous rejection, but the accumulation of new knowledge in several areas of biology, especially in molecular genetics, cell biology, developmental biology, and population genetics itself, has meant that critique cannot so easily be dismissed anymore. A re-evaluation of core presuppositions of the modern synthesis has thus become an urgent task.[5]

Of the three pillars supporting the modern synthesis, externalism is probably the more fundamental, but it also poses the strongest challenge to our intuitive experience of agency as an inherent property of life. Agency is used here in the sense of a capacity of a unit system to generate end-directed behaviors. Biologists of today embrace natural selection theory because it opens the doors for biology to become an exact science. At the same time, however, biology seems unable to get rid of teleological language that implicitly contradicts the non-agency premise of externalism and yet keeps popping up on every page of textbooks and even in scientific papers. The question is: why is the modern synthesis unable to explain life-processes without using teleological language? Why do we all feel that the creatures of this world, and humans in particular, possess agency?

When pressed, biologists may respond to such questions by alluding to our human propensity to project our own humanness upon the poor creatures of this world. We may think of them as agents, but they are not true agents. Rather, they are programmed by natural selection to act as if they were agents in their own right. When further pressed, the same explanation may be given even for human agency. We think we act out of a free will, but this again is only an illusion; instead, natural selection has formed us to experience a freedom of will, and in reality we are programmed by our genetic inheritance to feel that way. So, ultimately, the old Cartesian machine conception prevails—only, contrary to Descartes, the machine-thinkers of today do not think it necessary to equip humans with special *res cogitans*-like properties.

That many scientists and philosophers apparently feel confident that human intentional idioms do not denote anything real is due not the least, I suspect, to the successful fusion between information thinking and genetics that took place with the birth of molecular biology in the 1950s. The idea that selection operates on informational molecules, genes, intuitively seems to bridge the gap between soma and sema, between body and signification. What is selected is not only genes but "information." The very occurrence of "information" at the cellular level thus may be taken to support the idea that even high-level faculties such as cognition and feelings might some day in the future be derivable from molecular data.

The decisive step in this development was Francis Crick's formulation of the so-called *central dogma*, according to which the "flow of information" in a cell is unidirectional. Here information was thought of as *specifications* given by the sequence of bases in the DNA molecule, which in the process of transcription become expressed as a sequence of RNA and subsequently, in the process of translation, will come to specify a sequence of amino acids in a protein string. The point is, as Francis Crick himself expressed it, that: "once information has passed into protein it cannot get out again," i.e., the specificational process cannot run backwards.[6]

Now, fifty years later, the central dogma is far from undisputed. The problem is not only that "information" in many cases does indeed "run backwards"—at least indirectly—due to a variety of processes subsumed under the concept of "epigenetic inheritance."[7] The problem also is that the notion of information as something "flowing," "running," "being passed on," or "transferred" from one place to another presupposes that "information" is a thing or a property that can meaningfully be designated as being in time and space. But, if this were the case, it is hard to see why this "thing" or "property," information, should not simply be called a cause. Thus, for instance, patients suffering muscular dystrophy have a defective version of the gene for dystrophin, a protein, which in healthy persons, assures a smooth synchronization of the cell membrane with the interior of the muscle cells during muscle contraction. If this synchronization does not work properly, intracellular communication breaks down and a cascade of damaging processes sets in that ultimately yield pronounced myofiber necrosis as well as progressive muscle weakness and fatigability. What goes on in this case is a chain of well-known causal processes, and nothing is added to our understanding by ascribing the effect to a process of information flow.

And yet, something definitely seems to be added here. And that "something" is history. Genes are instructions handed down to present generations from the evolutionary past. But, considered as instructions, they are not passed around in the cell from one location to another—as little as the instructions you receive from a signpost telling you to stop the car is transported from your eye to your foot. What happens in both cases is something entirely different—namely, *an interpretation*. The cellular "machinery" interprets the genetic instructions in ways that are highly dependent on the context in which the cell is situated and initiates a set of suitable processes, and—likewise—the brain interprets the signpost and initiates suitable behaviors (activating appropriate muscles).

Genetic information (in the sense of Francis Crick) does not contain a key to its own interpretation. The key resides in the receptive system and is given by an interplay of developmental end environmental factors. Genes are not the only things handed down from one generation to the next. A highly complex cellular system, the egg, and an external environment are inherited by the organism. Offspring is not left unattended in haphazard ways, but typically placed in selected environments. Animals manufacture nests, burrows, webs, and pupal cases; plants decompose organic matter; and bacteria fix nutrients.[8] In addition to natural selection, we must also consider the processes of *niche construction*, as they were called by John Odling-Smee.[9] For decades, ecologists have realized that organisms do alter their environments in ecologically significant ways, now called *ecosystem engineering*.[10]

The modern synthesis and the central dogma builds on an understanding of the gene as hermetically sealed away from environmental influences, regulatory processes, trait associations, and interactions with the rest of the genome. But since the 1980s this "gene-in-a-bubble-approach," as Gregory Wray has termed it, has gradually been replaced by a much more interactive view.[11] Genetic background often has a strong effect on expressivity of a mutation, whether this is measured as an organismal trait or an intermediate phenotype such as gene expression. The implication is that the *information carried by a gene is highly context sensitive*.

Ninety percent of human genes, for instance, are prone to so-called alternative splicing, which means that the primary RNA transcript from the gene is split up into multiple segments that may then be reassembled in several different ways before being "translated" into protein by the ribosomes. As a consequence, the same gene may "specify" many different functional proteins depending on the context. Or, differently stated, the gene does not really specify the protein at all, rather it suggests a range of possible proteins to be constructed, and which of them will actually be produced depends on the local context in the tissue.

Other important context-dependent intervening factors are the presence or absence of a range of transcription factors, signaling molecules (signal transduction), and epigenetic modifications of the gene. The full complexity of the mechanisms whereby genotypes are related to phenotypes is only now beginning to be uncovered. As Marc Kirschner and John Gerhart observe,

> each spatial compartment of an embryo is defined by a small set of selector genes, which encodes transcription factors or signaling molecules that are expressed uniquely in that compartment. The selector gene can then "select" any other gene to be expressed or repressed in its compartment. This allows the core processes to be apportioned for usage in a specific compartment. For example, a process that causes cell proliferation may cause differentiation or cell death, or not being expressed, in a different compartment, depending on what else has been selected to occur there. The amplitude and timing of any response can easily be modulated.[12]

Recent developments in molecular genetics, cell biology, and developmental biology imply that an essential link is missing in the modern synthesis. The production of phenotypic variation depends on internal dynamics that cannot be determined by selection pressures from the environment alone. Instead, an interactive process must be considered, in which the activity of embryonic systems and whole organisms enter the equation. Or, in other words, agency cannot be accorded solely to the selective forces; individual entities exhibit agency.

There is a good reason why evolutionary biologists have shrunk away from taking this step, for it implies saying good-bye to externalism and it thus involves a break with deep-rooted preferences of science. By according agency to individual organisms, and even cells and embryos, a creative element is introduced in the world that has been forbidden ever since the Newtonian revolution. And the automatic explaining away of this agency by claiming it to be a product of natural selection is logically excluded in this case, since the whole reason why agency must be ascribed to organismic systems was that natural selection could not itself produce the variations upon which it acts. Or, in other words, without agency, there is no natural selection.

It is interesting that Darwin did not himself shy away from using a term like "striving" when talking about natural selection. In chapter three of *The Origin of Species*, he explicitly writes: "In looking at Nature, it is most necessary . . . never to forget that every single organic being may be said to be striving to the utmost to increase in numbers."[13] Darwin's book was about the origin of species, not the origin of life, and perhaps he was even content with assuming that "life, with its several powers, having been originally breathed by the Creator into a few forms or into one," as he expressed it in the very final paragraph of his great oeuvre. That organisms exhibited agency simply was so obvious to Darwin that he did not question it. So, although he managed to construct an externalist explanation for evolution, he was not a fundamentalist in his externalism, as were his followers in the twentieth century, who thought they could get rid of organismic agency by enthroning the gene and seeing organisms as passive derivatives of genotypes.

Biosemiotics is based on an understanding of agency as a real property of organismic life, a property that is ultimately rooted in the capacity of cells and organisms to interpret (whether consciously or unconsciously) events or states as referring to something other than themselves or, in other words, the capacity to interpret signs. These signs need not be emitted with a purpose of communication, in fact by far most signs are not part of a sender-receiver interaction but are simply important cues (internal or external) that organisms use to guide their activities. In the next section we shall explore this semiotic perspective of life.

Biosemiosis

The sign, in everyday parlance, is something that refers to something else, like the foam on the waves telling us that it's windy and perhaps even where the wind comes from. In Greek antiquity, natural signs, or *semeion* (for example, fragrances, sounds, animal tracks, thunderclouds, or fever), were distinguished from signs of human origin, *symbols* such as language signs. Augustine (354-430), however, merged the two sign types into one category of signs (*signum*) comprising "anything perceived which makes something besides itself come into awareness."[14] Augustine's definition is too narrow in its focus on perception since elements of awareness may well be signs also without being perceived, but Augustine nevertheless pointed to the core of the matter when he defined a *thing* as "what has so far not been made use of to signify something,"[15] implying that things may be signs, but need not be, and also that the essence of the sign is its formal relational character of evoking an awareness of something that it is not itself, thereby including the full triad of sign, object, and interpretant (here, the altered awareness). The evoking of such a triad is by no means exclusive for the workings of human awareness, but is rather, as was later realized, a purely logical relation to be established in any system capable of autonomous anticipatory activity—that is, all living systems.

Needless to say, semiotics has not always seen it this way. For most of the twentieth century, semiotics (semiology, as it was called by Ferdinand de Saussure) was conceived narrowly as a branch of linguistics, and human language was taken as model for sign systems in general. Animal signs such as alarm calls or courtship displays were seen as a degenerate kind of signs, if, that is, they were allotted a status of signs at all. Fortunately, a growing awareness of the true sophistication of animal cognition has meant that the traditional anthropocentric bias of the humanities, "the *reading humanness out of nature*," as the American philosopher Maxine Sheets-Johnstone has put it,[16] is now dwindling and the broader Peircean conception has become generally accepted.[17]

A sign is not, in the Peircean conception, a thing (material or mental); it is but rather a relation or a process that calls forth a relation. The Peircean sign unites the formal and the functional aspects of action, it is both *logical node* (i.e., a triadic relation), and *material process* (i.e., semiosis). And for this reason the Peircean sign cannot be decomposed into three dyads.[18] By convention the term for this process, the sign process, is *semiosis*, and the science that studies it is *semiotics*. The sign process implies that something, an event or a state, evokes the formation of an interpretant in a recipient system, an interpretant that somehow relates to the object of the sign in a way that reflects the sign's own relation to that object. Thus a human person sensing smoke may decide to call the firefighters or, if she is presently at the theater, she may decide to do just nothing, conceiving the smoke as a *symbolic sign* rather than *an index* for real fire. Sign processes are highly context-dependent, and a baby will of course not see smoke as an *index* pointing to fire, but may perhaps make an *iconic sign* of it, a sign that equates the smoke experience to other similar experiences.

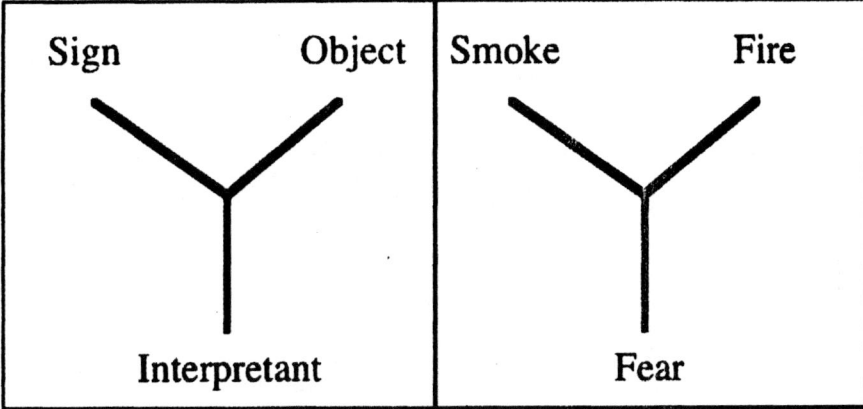

Figure 6.1: The Peircean concept of a sign as a triadic relation connecting a sign vehicle with an object through the formation of an interpretant in a receptive system.[19] In the right part is shown how smoke may act as a sign that evokes a sense of fear by making us aware of the risk of burning. For illustration we can think of a deer in the forest that is seized by alarm and flees away when exposed to smoke. This "being seized by alarm" is exactly what an interpretant is in the Peircean scheme. (Figure by author.)

Semiosis often assumes a web-like character where, for instance, the interpretant may release further sign processes by inducing the formation of new interpretants in the system itself or in other systems. For illustration, we can consider the process of male sex determination in humans. Growing human embryos encountering the Y-chromosome in their genomic setup normally "know" what to do about it—namely, constructing a male baby. Graphically we can depict this as a Peircean triadic sign process (as in Figure 6.1) in which the Y-chromosome occupies the position of the primary sign, maleness stands in the position of the object, and the embryonic reading of the chromosome is the interpretant. Looked upon in some more detail (Figure 6.2), what happens is that around the seventh week certain embryonic cells called epithelial sex cord cells for unknown reasons begin to express a gene (termed SRY) located at the Y-chromosome. This results in the production of a so-called testis-determining factor (TDF) and perhaps in the activation of the expression of a few other genes.[20] From there on, apparently, the male sex determination process is taken care of by these transformed cells.

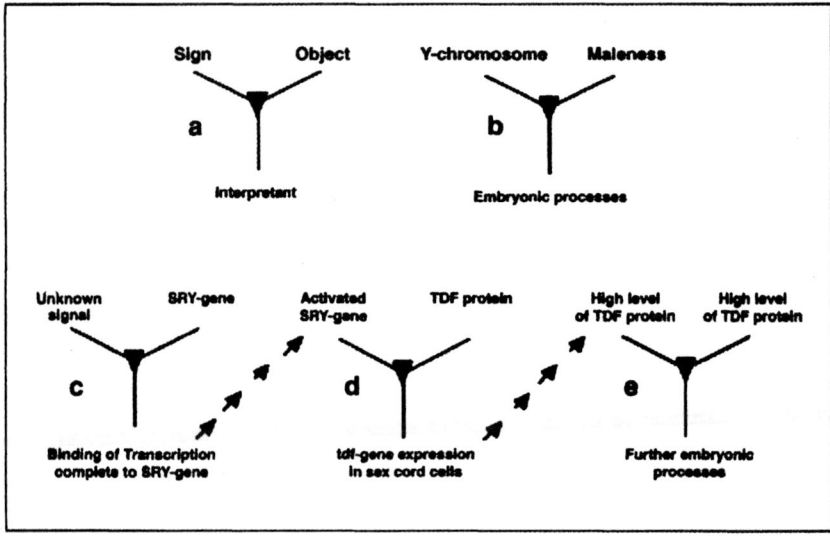

Figure 6.2: The semiotics of human male sex determination as triggered by a gene on the Y-chromosome, including the SRY gene expression in epithelial sex cord cells. (Figure by author.)[21]

What we see is that the organism acquires its male determination through a series of steps whereby semiotically competent cells "read the messages" made available to them in part from their internal genetic makeup, and in part from the external biochemical context set by a multitude of cues (molecular signs) derived from neighboring cells or from other embryonic tissues. These contextual cues are received at specific receptors located in the plasma membrane. Even though all these processes may sooner or later be fully characterized at the biochemical level, this will not by itself exhaust our need for explanation. For obviously, we are not dealing here with a haphazard mixture of biochemical processes, but with a delicately organized system of processes. What we really want to know is the logic of the organization of these biochemical processes, and this logic has to do with their developmental function.

In Figure 6.2 c-e is shown three chained semiotic steps in the matrix of male sex determination. First, an unknown biochemical event, perhaps the activation of a specific inducer, refers the epithelial sex cord cells to the SRY-gene (the object) by way of the binding of a transcription complex to the gene. Here the binding process occupies the position of an interpretant to the signal (c). This activates the SRY-gene to guide the formation of a new interpretant in the form of SRY-gene expression whereby testis-determining-factor (TDF) is produced (d). With increasing concentrations of TDF a level is reached such that further steps of embryonic processes are released in the chain of male sex determining semiosis (e). In each step, the interpretant becomes a sign in yet a new semiotic triad. Basically, the triadic nature of this functional logic derives from the fact that the expression of the SRY-gene is undertaken only because some unknown signal induces the cellular machinery to relate to the SRY-genes in a way which represents the historical relation of the unknown signal to those same SRY-genes. The sign points out a historically created relational logic of the macromolecular events. History is of essence in the semiotic analysis.

In a biosemiotic optic, the genome is best understood as a very sophisticated "inventory control system," a system that not only contains the specifications necessary to produce all the body's protein and RNA molecules, but also a number of switches and locks by which an agent may access the archives. Thus, if a given enzyme is needed in a tissue, cells will turn on the switch that opens the particular lock normally keeping the coding region of the responsible gene from being transcribed. When enough of the enzyme has been produced, the switch is turned off again. So, if we imagine this as a computerized system and a given enzyme is needed, say a peptidase, the cell will click down the menu for enzymes, pick the submenu for peptidases, and select a suitable one among the different kinds of peptidases on offer, depending on the kind of tissue and/or the concrete situation. The selection will then start off the process of operating the switches and keys corresponding to this particular gene (RNA splicing and other modifications included).

The most important factor that has been turned around in the semiotic description of this process relative to more traditional informational descriptions, is that agency here resides with the cell, the tissue, or the organism, not with the genetic system. Harvard geneticist Richard Lewontin expressed clearly why this is necessary when he stated that "genes do nothing" (it follows that the concept of "selfish genes" is outright nonsense).[22] Language use in modern biology is, to use Terrence Deacon's expression,[23] profused with hidden homunculi, and biosemiotics is needed precisely in order "to make explicit those assumptions imported into biology by such unanalyzed teleological concepts as function, adaptation, information, code, signal, cue, etc..., and to provide a theoretical grounding for these concepts."[24]

To say that cells interpret signs necessarily implies that such signs point the cell toward latent activities, whether now or later. The teleological character of semiosis cannot be explained away, and semiosis thus seems to break one of the deepest taboos of science, the taboo against what Aristotle called final causes. How can biosemiotics justify this forbidden step? There are at least two possible ways to answer this question. First, one may try to explain how agency and semiosis could arise in the course of processes that finally led to the formation of living systems. This possibility is addressed in a later section of this chapter. Second, one might point out that externalism has failed to produce a satisfactory theory of the evolution of life on Earth, and that biology is profused with hidden homunculi, as we said above. My view here is that we must stop this "*mummenschanz*" and admit defeat. Instead we should reevaluate the reasons why "final causation" is deemed impermissible in science.

Semiotic Causation

There are so many misunderstandings involved in this discussion that space doesn't allow a deeper discussion.[25] The most important of these misunderstandings is the belief that the introduction of final causes necessarily entails anthropomorphism. Peirce stated the misunderstanding in these words: "It is a widespread error to think that a *final cause* is necessarily a *purpose*. A purpose is merely that *form* of final cause which is most familiar to experience" (italics added),[26] or in other words, "purpose is the *conscious* modification of final causation" (italics added).[27]

That which in the Cartesian tradition is integrated into one concept, the concept of purposive, consciously conceived end causes (and which in a strict sense has only validity in the human world) is thus in Peirce's philosophy two things—one specifically human, and the other a general principle of emergent organization—that should not be confounded. Human purposes are only one special subcategory of the much broader category of final causes—and these, according to Peirce, are at play in any sort of goal-oriented activity in nature, as well as in culture. For, to Peirce, a final cause is simply the general form of any process that tends toward an end state (a finale):

> [A final cause is] . . . that mode of *bringing facts about* according to which a *general description of result* [read *law like system of regularity*] is made to come about, quite irrespective of any compulsion for it to come about in this or that particular way; although the means may be adapted to the end. The general result may be brought about at one time in one way, and at another time in another way. Final causation does not determine in what particular way it is to be brought about, but only that the result shall have a certain general character.[28]

When one is socialized into the Newtonian worldview—as are virtually all people today, whether they are conscious of it or not—then the idea of final causation is very difficult to accept. For the Newtonian view of causes is temporalized in a very thin, one-dimensional thread. That event A is the cause of event B means that first comes A and then comes B. That is to say, event A happens at time t1 and event B at time t2, where t2 > t1. According to this scheme of thinking, final causes imply that events that happen at time t1 are caused by an event that first takes place at time t2, which profoundly contradicts both scientific and everyday experience.[29]

In order to understand the idea of final causes, one must broaden one's conception of time so that it contains more than just one dimension. Event B is not just caused by the preceding event A, for the very fact that A takes place is already part of a pattern of events—a pattern that also includes occurrences of type B—which in the end obeys general laws of a more compelling kind. For example, the natural law that goes under the name of *the second law of thermodynamics* expresses an entirely universal condition pertaining to all known physical existence—namely, that the net amount of entropy will always increase with each spontaneous process that takes place in this universe. What this law determines, as Peirce observed, is not "in what particular way a given end is to be brought about, but only that the result shall have a certain general character."[30] The second law of thermodynamics can be understood as one of the final causes that must be part of any satisfactory explanation of evolutionary phenomena in general, including Darwinian selection.

This understanding of a final cause corresponds to the original Aristotelian view of *telos* as "that 'for the sake of which' something exists or occurs or is done." The philosopher T. L. Short has expressed it this way: "The stock example is familiar: It is the nature of an acorn to grow into an oak, not into a spruce tree or a butterfly. The final cause, then, is a potentiality whether or not actualized."[31] This, however, is not the view of *finality* that one normally finds in the literature. Most often, final causes are there linked with human (or at least human-like) intentions or desires about the future. But for Aristotle, such causes are in fact *efficient causes* that is, forces or energies that concretely make things happen in our familiar sense.[32]

While the acceptance of agency as an inherent property of living systems does contradict the Darwinian ambition for establishing an externalist theory of evolution (mirroring Newtonian ideals), it does not imply anthropomorphism in the narrow sense that ascribes specific human psychological properties to living systems in general. Since we are descendants from earlier life-forms, it would be absurd to claim all of our properties to be unique for the human species; thus we, like dogs, have two ears and a nose, but as Karl Popper noticed long ago, this fact is a result of homology (common descent) rather than analogy.[33] Likewise, agency is a common property of life shared by us as descendants from ancestor organisms that themselves possessed agency all the way back to the first organisms on Earth.

This raises the question of how agency and semiosis could arise in the first place on our planet. We shall turn to this problem next, but first we must disarm the latent confusion clinging to the concept of final causation. Since our acceptance of the reality of final causation does not imply any acceptance of the

Aristotelian metaphysics with its matter-form dualism and ideal of harmony, we shall prefer to call it semiotic causation and define it as *bringing about things under guidance of interpretation in a local context*.[34] By calling it semiotic causation we recognize our debt to Charles Peirce's understanding of final causes as ultimately based on randomness and irreversibility. Peircean final causation is a very different thing from Aristotelian final causation.

Semiotic causality cannot be reduced to efficient causality but, on the contrary, is dependent upon the workings of efficient causality, since interpretation, even in its most primitive modes, is of no need if it does not result in anticipatory action—and action unquestionably depends on efficient causality. Semiotic causality thus gives direction to efficient causality, while efficient causality gives power to semiotic causality. Their reciprocal relation is one of interdependence, not one of exclusion.

Origin of Agency, Semiosis, and Life

Elsewhere we have claimed that life, agency, and semiosis are co-existent and that the formation of a closed membrane space defining an inside-outside asymmetry must have been a decisive step on the path leading to the appearance of living systems.[35] Usually the origin-of-life problem is discussed in terms of how certain macromolecules may have first been formed under the conditions of the primitive Earth. This is of course a highly interesting and very important question, but we also need to understand a living system in its end-directed interaction with its environment. We therefore need (a) to define a spatial and structural separation between the system and its environment, or, in other words, we need an inside-outside asymmetry, and we also need (b) to have an inside representation of what is going on outside of the system. This last question may be understood as a measuring problem and, as pointed out by American biophysicist Howard Pattee, this leads us into an epistemological difficulty. The essential function of a measurement, Pattee says, "is to generate a computable symbol, usually a number, corresponding to some aspect of the physical system,"[36] and the problem is that

> a system cannot be measured without a measuring device, which itself is a physical system . . . but to function *as* a measuring device, it requires an observer's simplified description that is not derivable from the physical description. The observer must in effect choose what aspects of the physical system to ignore and invent those aspects that must be heeded. This selection process is a decision of the observer or organism, and cannot be derived from the laws.[37]

From this follows, Pattee says, that "we must define an *epistemic cut* separating the world from the organism or observer."[38] Pattee sees this epistemic cut as analogous—but by no means identical—to Niels Bohr's complementarity principle (analogous in the sense that the observation defines the observed reality which includes the observer, and thus the observation).

A biosemiotic reply to this claim would be to question the premise that a measurement must generate a "computable symbol." If this were indeed the case, it is hard to see how we could ever succeed in explaining the origin of life, for there seems to be no naturalistic way in which symbolic reference could jump out from scratch. On the other hand, it is possible to imagine that simple dynamic processes located in a membrane-enclosed complex chemical system with access to a source of energy-rich compounds might somehow have managed to self-organize into sets of repeatable cascades of reactions in response to distinct external challenges. Even a weakly stable correspondence between external cues and internal reaction cascades might suffice to initiate a semiosis-like process. In modern cells so-called signal transduction exhibits exactly this property. Receptor molecules stretching all the way through the cell membrane recognize and bind to specific molecular surfaces outside of the cell, and this binding subsequently causes changes at the inside end of the receptor molecule that serves to initiate cascades of internal processes in the cell. In semiotic terms this process is indexical rather than symbolic; the presence or (more correctly) concentration of the "recognized" external molecule "indicates" the necessity to the cell of a certain output, and hence of a chain of internal chemical reactions.

Now, for this to be a real case of semiosis, we need something more. However, we need a kind of time independence: the system must, so to say, exist for itself in order to extract usable "information"[39] from the environment. The way modern cells or organisms manage to do this is by their dependence on a hereditary system, the genome. Heredity may be seen as *semiotic survival*, survival as a digitally coded self-referential description contained in an analogically-coded cellular system (the egg) capable of interpreting the digital description as referring to a "reconstruction" of the organism. This way of persistence, called code duality[40] (i.e., the reshuffling back and forth between digital and analog representations) is the dynamic core of the life process.

Looked upon in this way, the appearance of life would require the establishment of a system capable of integrating two different referential systems; a digital *self-referential system* such as the genome, containing a complete set of descriptions of the sequential structure of the organism's macromolecules, and an analogically coded *other-referential system* containing the molecular tools for "recognizing" specific molecules or features of the environment and "translating" them to useful instructions to the cellular machinery. These are instructions that (through a process of signal transduction) enable the cell to command the cell's inventory control system, the genome, according to its own "survival" project.[41]

It may be objected that this line of argument puts the threshold for agency, semiosis, and life too high. Might it not be that simpler systems than whole cells could exhibit agency and semiosis, and, if so, would that not qualify as a living system? Stuart Kauffman and Philip Clayton have proposed a tentative five-part definition of what they call a "minimal molecular autonomous agent":[42] "Such a system should be able to reproduce with heritable variation, should perform at least one work cycle, should have boundaries such that it can be individuated naturally, should engage in self-propagating work and constraint construction, and should be able to choose between at least two alternatives."[43] Kauffman and Clayton claim this minimal molecular autonomous agent "to be about as minimal as one can get and still plausibly use the language of agency."[44] They furthermore—like Pattee—stress the necessity of agents to make distinctions between relevant and irrelevant acts and ponder how this fits into a biosemiotic organization. As far as I can see, the five requirements of the minimal molecular autonomous agent suggested here would hardly be fulfilled by any simpler system than a cell.

Terrence Deacon, on the other hand, has worked out a model of so-called "autogens"[45] that are far simpler than whole cells.[46] Autogens are self-assembling molecular structures that derive their individuality from a synergistic relationship between two kinds of self-organizing processes that reciprocally depend upon one another's persistence:

> An autocell is an autocatalytic set of molecules that produce one another and also produce molecules that spontaneously accrete to form a hollow container, analogous to the way virus capsules form. The molecular capsules that result will spontaneously enclose some of the nearby molecules of the autocatalytic set, keeping them together so that when the autocell is broken open autocatalysis will resume. Autocells are thus self-reconstituting, self-reproducing, and minimally evolvable.[47]

Autogens, however, are not yet full-blown living systems, since they lack several features that are generally considered criteria for being alive, such as the possession of the replicative molecules of RNA or DNA and differential survival through replications. But they do have the "necessary precursor attributes to *telos*, including individuality, functional interdependence of parts, end-directedness, a minimal form of representation, and a normative (evaluational) relationship to different environmental properties," claim Sherman and Deacon.[48] The autogen hypothesis is open to empirical test but, as Sherman and Deacon admit, such support for the autogen model has not yet been obtained. The model nevertheless does demonstrate that there is an unbroken continuity from thermodynamics to evolvability.

In Kull et al., we wrote:

> Often the emergence of life is seen as a sudden transition where the many properties defining life arise together or are tightly interconnected (like self-replication, autocatalysis, function, and cellularity). However, this appears to be both too simple and implausible. There is no simple dividing line where all the interconnected properties of living systems, as we know them, emerge. Instead we observe what we call a threshold zone, probably involving incremental

stages in which different component processes emerge. This is an open issue for further investigation and will probably develop into a fertile area for both molecular biology and biosemiotic research to contribute.[49]

While the concrete succession of stages on the path between lifeless matter and human agency is not at present clear, it seems safe to claim that it is a "stunning fact that the universe has given rise to entities that do, daily, modify the universe to their own ends" as Kauffman and Clayton have put it.[50]

Evolution and the Growth of Semiotic Freedom

Much confusion surrounds the relation between the twin concepts of perception and semiosis. The failure to see perception as a higher-order phenomenon based on the co-occurrence or interaction of multiple semiotic processes in the whole organism has often resulted in accusations of biosemiotics to imply panpsychism. To counter this misunderstanding, it is important to underline the evolutionary perspective that semiotic freedom (i.e., the capacity of a system to derive useful information from complex signs), is itself an evolved property and that proper psychological phenomena only appear in very late stages of evolution and in relatively few big brained species of animals, mostly mammals, whereas semiosis—the more parsimonious concept relative to perception—may, as we said, be taken to be coextensive with life.

To this day, millions of successful species persist that are very poorly equipped in semiotic terms. Thus, there are more than six hundred genera of grasses with an estimated total amount of ten thousand different species on our planet, and although even a grass plant necessarily must manage to adapt its growth under a number of factors such as gravity, humidity, porosity of the earth, and must also possess an elaborate internal system of cellular communication, grass plants nevertheless exhibit very restricted semiotic freedom. The evolution of a richer semiotic capacity is thus only one among many strategies available in the evolutionary game. But obviously, the ability to foresee important events or behaviors of others will be of advantage to all such species that depend for their survival upon a correct situated "reading" of decisive niche parameters. I have therefore suggested that the concept of an ecological niche should be extended to comprise also the semiotic niche—namely, "the totality of cues around the organism (or species) which the organism (or species) must necessarily be capable of interpreting wisely in order to survive and reproduce."[51] One important aspect of the semiotic niche concerns the semiotic interactions between individuals from different species. Most notably among these, of course, stands predator-prey relations but just as significant perhaps is the range of interactive behaviors connected to symbiotic relations. When symbiosis is seen as a semiotically controlled kind of interspecific interaction, it becomes evident that symbiosis encompasses a lot of subtle interaction forms that are not normally seen as belonging in this category.

An illustrative case is so-called *plant signaling*. Undamaged fava beans (*Vicia fabea*), for instance, immediately started to attract aphid parasites (*Aphidus ervi*) after having been grown in a sterilized nutrient medium in which aphid-infected fava beans had previously grown. The damaged beans thus had managed to signal their predicament through the medium to the undamaged beans, which then immediately started to attract aphid parasites, although no aphids were, of course, available for parasites to find.

The difference between this kind of *semiotic mutualism*, involving a delicate balance of interactions between many species, and symbiosis proper is one of degree rather than one of kind, and biosemiotics takes semiotic mutualism to be not only widespread in nature, but nearly ubiquitous. This implies that the relative *fitness* of changed morphological or behavioral traits become dependent on the whole system of existing semiotic relations that the species finds itself a part of and, accordingly, the firm organism-versus-environment borderline will be dissolved, and a new integrative level intermediate between the species and the ecosystem would have to be considered (i.e., the level of the *ecosemiotic interaction structure*).[52]

Clearly, this possibility becomes most interesting in cases where individual experience and learning enters the interaction pattern, as will often be the case in mammals and birds. Such learning might on occasion even subsume the evolutionary process, as is the case in human culture. Conversely, one might wonder if a relatively autonomous ecosemiotic interaction structure is precisely what is needed for learning to evolve in the first place. In this way, eventual increases in semiotic freedom (i.e., the capacity of a system to derive useful information from complex signs), will be prone to feed back into the evolutionary process by strengthening the advantages of possessing semiotic freedom. It follows that the evolutionary dynamics would possess an in-built tendency to invent and establish species exhibiting more and more *semiotic freedom* in the sense that their behavior would be increasingly underdetermined by the constraints of natural lawfulness and increasingly dependent on the *in situ* interpretative capacity of individual organisms.

Humanity's Place in Nature and Nature's Place in the Human

A semiotic understanding of animate nature will potentially influence science and culture in important ways. Above all, it will strengthen our human feeling of relatedness to the other creatures of this world and our belonging in the biosphere. The image of animals and plants as stupidly obedient slaves of simple survival schemes will dwindle and be replaced by an understanding of, and an admiration for, the marvelous semiotic interaction loops through which organisms pursue their interests. Living beings are not the senseless and ignorant machines that science has taught us they are, and in the long run this can well have profound implications for how we treat natural systems. The environmental movement was from the early beginnings with Rachel Carson's book *Silent Spring* as much motivated by the threat to the semiotic aspects of life, here bird song, as by the threat to more traditional ecological parameters such as food webs and nutrient cycles. Substituting the ecological niche concept with the concept of a semiotic niche may throw new light on interactive aspects of ecosystems that have not so far got sufficient attention. In medicine, we are going to see a reconceptualization of the body-mind separation at all levels. This does not so much lead toward a conception of the mind as "embodied" in the body, for this very expression already—falsely as seen from a semiotic point of view—ascribes a different kind of reality to the body and the mind. Rather, I have argued that the mental sphere does not consist of some mysterious subject matter at all, but must be understood as an interface through which our bodies have managed to connect to the world around them, physical as social.[53] The so-called psychosomatic aspect of diseases would be much easier to handle if the medical image of the body proper was a semiotic body and not just a biological machine.

We shall finish this overview of potential impacts of the semiotic understanding of nature by quoting the French molecular biologist and Nobel Laureate who in his influential book *Chance and Necessity*, from 1971, concluded that "if he accepts this message—accepts all it contains—then man must at last wake out of his millenary dream; and in doing so, wake to his total solitude, his fundamental isolation. Now he at last realizes that, like a gypsy, he lives on the boundary of an alien world. A world that is deaf to his music, just as indifferent to his hopes as it is to his suffering or his crimes."[54] The heroic "ethics of necessity" that Monod proposes in this book is a logical consequence of the scientific view of nature to which the book is dedicated. Introducing a semiotic view of nature opens for a return to a nature in which we belong. Monod's idea of the human being as "a gypsy at the edge of time" may finally be dismantled—to be replaced by a conception of human beings as embedded in the general biosemiosis of living nature. Human mind is not an alien element in the universe—but rather, an instantiation of evolutionary trends that penetrate the life sphere and that (I suspect) is deeply rooted in the general dynamics of the universe.

Notes

1. Richard Goldschmidt, *The Material Basis of Evolution* (New Haven: Yale University Press, 1940).

2. David L. Depew and Bruce H. Weber, *Darwinism Evolving: Systems Dynamics and the Genealogy of Natural Selection* (Cambridge, MA: Bradford/The MIT Press, 1995).

3. Peter Godfrey-Smith, *Complexity and Function of Mind in Nature* (New York: Cambridge University Press, 1998).

4. Eva M. Neumann-Held, "The Gene Is Dead—Long Live the Gene: Conceptualising the Gene the Constructionist Way," in *Developmental Systems, Competition and Cooperation in Sociobiology and Economics*, ed. P. Koslowsky (Berlin: Springer-Verlag, 1998), 105-137.

5. A recent summary is given in Massimo Pigliucci and Gerd Müller, eds., *Evolution: The Extended Synthesis* (Cambridge: The MIT Press, 2010).

6. Francis Crick, *What Mad Pursuit* (New York: Basic Books, 1988), 109.

7. Eva Jablonka and Marion J. Lamb, *Evolution in Four Dimensions: Genetic, Epigenetic, Behavioral, and Symbolic Variation in the History of Life* (Cambridge, MA: MIT Press, 2005); Eva Jablonka and Marion J. Lamb, "Transgenerational Epigenetic Inheritance," in *Evolution: The Extended Synthesis*, ed. M. Pigliucci, and G. Müller (Cambridge, MA: The MIT Press, 2010), 137-174; see also chapters 15 and 16 in this volume.

8. See Richard C. Lewontin, "Gene, Organism, and Environment," in *Evolution from Molecules to Men*, ed. D. S. Bendall (Cambridge: Cambridge University Press, 1983), 273-285; see also chapter 8 in this volume.

9. F. John Odling-Smee, "Niche Constructing Phenotypes," in *The Role of Behavior in Evoluton*, ed. C. H. Plotkin (Cambridge, MA: MIT Press, 1988), 72-132.

10. F. John Odling-Smee, "Niche Inheritance," in *Evolution. The Extended Synthesis*, ed. M. Pigliucci, and G. Müller (New York: MIT Press, 2010), 175-208.

11. Gregory Wray, "Integration Genomics into Evolutionary Theory," in *Evolution: The Extended Synthesis*, ed. M. Pigliucci, and G. Müller (New York: MIT Press, 2010), 100.

12. Mark Kirschner and John Gerhart, "Facilitated Variation" in *Evolution. The Extended Synthesis.* ed. M. Pigliucci, and G. Müller (New York: Cambridge University Press, 2010), 267.

13. Charles Darwin, *On the Origin of Species by Means of Natural Selection or the Preservation of Favored Races in the Struggle for Life* (1859; repr. London: J. M. Dent & Sons, 1971), 73.

14. John Deely, *Four Ages of Understanding: The First Postmodern Survey of Philosophy from Ancient Times to the Turn of the Twenty-first Century* (Toronto: Toronto University Press, 2001), 221.

15. Deely, *Four Ages of Understanding*, 221.

16. Maxine Sheets-Johnstone, *The Corporeal Turn: An Interdisciplinary Reader* (Charlottesville, VA: Imprint Academic, 2009), 125.

17. John Deely, *Intentionality and Semiotics* (Scranton PA: Scranton University Press, 2007); T. L. Short, *Pierce's Theory of Signs* (Cambridge, MA: Cambridge University Press, 2007).

18. The positions in the triad are not logically equivalent since the interpretant is related both to the sign and to the object but in such a way that the relation to the object is grounded in the sign's relation to that same object. The interpretant belongs to Peirce's category of Thirdness, whereas the sign is Firstness, and the object Secondness.

19. It may seem contradictory that "sign" is put in as one element in the sign-relation. As explained in the text the sign always presupposes the whole triadic relation

and technically speaking the term "representamen" or at least "sign vehicle" should have been used instead of sign. However, since everyday language uses the term "sign" as equivalent to the representamen as such, I have chosen to stick with it.

20. Scott F. Gilbert, *Developmental Biology*, 3rd ed. (Sunderland MA: Sinauer, 1991).

21. From "Genes, Development and Semiosis," in *Genes in Development: Rereading the Molecular Paradigm*, ed. E. M. Neumann-Held and C. Rehmann-Sutter (Durham, London: Duke University Press, 2006), 152-174.

22. Lewontin, "Gene, Organism, and Environment."

23. Terrence Deacon, "Shannon-Boltzmann-Darwin: Redefining Information, Part 1," *Cognitive Semiotics* 1 (2007): 123-148.

24. Kalevi Kull, Terrence Deacon, Claus Emmeche, Jesper Hoffmeyer, and Frederik Stjernfelt, "Theses on Biosemiotics: The Saka Convention," *Biological Theory* 4 (2009): 167-173.

25. For greater detail see Jesper Hoffmeyer, *A Legacy for Living Systems: Gregory Bateson as Precursor to Biosemiotics* (Dordrecht: Springer, 2008b).

26. Charles S. Peirce, *Collected Papers Vol. 1-8*, eds. Charles Hartshorne and Paul Weiss (Cambridge, MA: Harvard University Press, 1931), 1.211.

27. Peirce, *Collected Papers*, 7.366.

28. Peirce, *Collected Papers*, 1.211.

29. Perhaps more than anyone else, Robert Rosen has, from a mathematical foundation, seen that the Newtonian tradition rests upon a needlessly narrow concept of causality. Rosen defended a return to Aristotle's broader understanding of causality, but suggested the term "functional entailment" in place of "final cause," using functional in its biological sense. See Robert Rosen, *Life Itself: A Comprehensive Inquiry into the Nature, Origin, and Fabrication of Life* (New York: Columbia University Press, 1991).

30. Peirce, *Collected Papers*, 1.211.

31. T. L. Short, "Darwin's Concept of Final Cause: Neither New Nor Trivial," *Biology and Philosophy* 17 (2002): 326.

32. The vital force of Hans Driesch then, paradoxically, was a strange kind of an efficient cause; but ascribing *telos* to an efficient cause is a confusion of categories, and the ridicule that became the fate of vitalism is thus quite justified it would seem. See Short, "Darwin's Concept of Final Cause."

33. Karl Popper, *A World of Propensities* (Bristol: Thoemmes Antiquarian Books, 1990), 30.

34. Jesper Hoffmeyer, "Semiotic Scaffolding of Living Systems," in *Introduction to Biosemiotics*, ed. M. Barbieri, 149-166 (Dordrecht, Holland: Springer, 2007).

35. Jesper Hoffmeyer, "Surfaces inside Surfaces: On the Origin of Agency and Life," *Cybernetics & Human Knowing* 5 (1998): 33-42; Jesper Hoffmeyer, *Biosemiotics: An Examination into the Signs of Life and the Life of Signs* (Scranton and London: University of Scranton Press, 2008), 31-38.

36. Howard H. Pattee, "The Physics of Symbols and the Evolution of Semiotic Controls," in *Proceedings of the 1996 International Workshop on Control Mechanisms for Complex Systems,* ed. M. Coombs (Redwood City, CA: Addison-Wesley, 1997), http://informatics.indiana.edu/rocha/pattee/pattee.html.

37. Pattee, "The Physics of Symbols."

38. Pattee, "The Physics of Symbols."

39. Here I use the term "information" in its everyday language sense of "message" or "news."

40. Jesper Hoffmeyer and Claus Emmeche, "Code-Duality and the Semiotics of Nature," in *On Semiotic Modeling*, ed. M. Anderson, and F. Merrell (New York: Mouton de Gruyter, 1991), 117-166.

41. The term "survival" is here used as a *post facto* characterization of the project. Survival is not as such a part of the "sake for which" a living system acts, but rather the result of the process when looked upon in retrospect from the outside.

42. Stuart Kauffman and Philip Clayton, "On Emergence, Agency, and Organization," *Biology and Philosophy* 21 (2005): 501-521.

43. Kauffman and Clayton, "On Emergence," 505.

44. Kauffman and Clayton, "On Emergence," 517.

45. For example, see chapters 13 and 14 of this volume; Deacon refers to "autocells" elsewhere.

46. Terrence Deacon, "Reciprocal Linkage between Self-organizing Processes Is Sufficient for Self-Reproduction and Evolvability," *Theoretical Biology* 1 (2006): 136-149.

47. Jeremy Sherman and Terrence Deacon, "Teleology for the Perplexed: How Matter Began to Matter," *Zygon* 42 (2010), 873.

48. Sherman and Deacon, "Teleology for the Perplexed," 873.

49. Kull et al., "Theses on Biosemiotics," 168.

50. Kauffman and Clayton, "On Emergence," 504.

51. Jesper Hoffmeyer, *Signs of Meaning in the Universe* (Bloomington, IN: Indiana University Press, 1996), 94

52. Hoffmeyer, *Biosemiotics*, 198-202.

53. Jesper Hoffmeyer, "A Biosemiotic Approach to Health," in *Signifying Bodies. Biosemiosis, Interaction and Health*, ed. S. Cowley et al. (Braga: The Faculty of Philosophy of Braga, Portuguese Catholic University, 2010), 21-41.

54. Jacques Monod, *Chance and Necessity* (New York: Knopf, 1971), 172-173.

Bibliography

Crick, Francis. *What Mad Pursuit*. New York: Basic Books, 1988.

Darwin, Charles. *On the Origin of Species by Means of Natural Selection or the Preservation of Favored Races in the Struggle for Life*. 1859. Reprint, London: J. M. Dent & Sons, 1971.

Deacon, Terrence. "Reciprocal Linkage between Self-organizing Processes Is Sufficient for Self-Reproduction and Evolvability." *Theoretical Biology* 1 (2006): 136-149.

———. "Shannon-Boltzmann-Darwin: Redefining Information, Part 1." *Cognitive Semiotics* 1, (2007): 123-148.

Deely, John. *Four Ages of Understanding. The First Postmodern Survey of Philosophy from Ancient Times to the Turn of the Twenty-first Century*. Toronto: Toronto University Press, 2001.

———. *Intentionality and Semiotics*. Scranton PA: Scranton University Press, 2007.

Depew, David L., and Bruce H. Weber. *Darwinism Evolving: Systems Dynamics and the Genealogy of Natural Selection*. Cambridge, MA: Bradford / The MIT Press, 1995.

Gilbert, Scott F. *Developmental Biology*, 3rd ed. Sunderland MA: Sinauer, 1991.

Godfrey-Smith, Peter. *Complexity and Function of Mind in Nature*. New York: Cambridge University Press, 1998.

Goldschmidt, Richard. *The Material Basis of Evolution*. New Haven: Yale University Press, 1940.
Hoffmeyer, Jesper. *Signs of Meaning in the Universe*. Bloomington, IN: Indiana University Press, 1996.
———. "Surfaces inside Surfaces: On the Origin of Agency and Life." *Cybernetics & Human Knowing* 5 (1998): 33-42.
———. "Genes. Development and Semiosis." In *Genes in Development: Rereading the Molecular Paradigm*, edited by E. M. Neumann-Held and C. Rehmann-Sutter, 152-174. Durham, London: Duke University Press, 2006.
———. "Semiotic Scaffolding of Living Systems." In *Introduction to Biosemiotics*, edited by M. Barbieri, 149-166. Dordrecht: Springer, 2007.
———. *Biosemiotics: An Examination Into the Signs of Life and the Life of Signs*. Scranton and London: University of Scranton Press, 2008a.
———. *A Legacy for Living Systems: Gregory Bateson as Precursor to Biosemiotics*. Dordrecht: Springer, 2008b.
———. "A Biosemiotic Approach to Health." In *Signifying Bodies: Biosemiosis, Interaction and Health*, edited by S. Cowley et al., 21-42. Braga: The Faculty of Philosophy of Braga, Portuguese Catholic University, 2010.
Hoffmeyer, Jesper, and Claus Emmeche. "Code-Duality and the Semiotics of Nature." In *On Semiotic Modeling*, edited by M. Anderson and F. Merrell, 117-166. New York: Mouton de Gruyter, 1991.
Jablonka, Eva, and Marion J. Lamb. *Evolution in Four Dimensions: Genetic, Epigenetic, Behavioral, and Symbolic Variation in the History of Life*. Cambridge, MA: MIT Press, 2005.
———. "Transgenerational Epigenetic Inheritance." In *Evolution: The Extended Synthesis*, edited by M. Pigliucci and G. Müller, 137-174. Cambridge, MA: The MIT Press, 2010.
Kauffman, Stuart, and Philip Clayton. "On Emergence, Agency, and Organization." *Biology and Philosophy* 21 (2005): 501-521.
Kirschner, Mark, and Gerhart John. "Facilitated Variation." In *Evolution: The Extended Synthesis*, edited by M. Pigliucci and G. Müller, 253-280. New York: Cambridge University Press, 2010.
Kull, Kalevi, Terrence Deacon, Claus Emmeche, Jesper Hoffmeyer, and Frederik Stjernfelt. "Theses on Biosemiotics: The Saka Convention." *Biological Theory* 4 (2009): 167-173.
Lewontin, Richard C. "Gene, Organism, and Environment." In *Evolution from Molecules to Men*, edited by D. S. Bendall, 273-285. Cambridge: Cambridge University Press, 1983.
———. "The Dream of the Human Genome," *The New York Review of Books* (May 28, 1992): 31-40.
Monod, Jacques. *Chance and Necessity*. New York: Knopf, 1971.
Neumann-Held, Eva M. "The Gene Is Dead—Long Live the Gene: Conceptualising the Gene the Constructionist Way." In *Developmental Systems, Competition and Cooperation in Sociobiology and Economics*, edited by P. Koslowsky, 105-137. Berlin: Springer-Verlag, 1998.
Odling-Smee, F. John. "Niche Constructing Phenotypes." In *The Role of Behavior in Evolution*, edited by C. H. Plotkin, 72-132. Cambridge, MA: MIT Press, 1988.
———. "Niche Inheritance." In *Evolution: The Extended Synthesis*, edited by M. Pigliucci and G. Müller, 175-208. New York: MIT Press, 2010.

Pattee, Howard H. "The Physics of Symbols and the Evolution of Semiotic Controls." In *Proceedings of the 1996 International Workshop on Control Mechanisms for Complex Systems*, edited by M. Coombs. Redwood City, CA: Addison-Wesley, 1997, http://www.ssie.binghamton.edu/pattee/semiotic.html.

Peirce, Charles S. *Collected Papers, Vols. 1-8*. Cambridge, MA: Harvard University Press, 1931.

Pigliucci, Massimo, and Gerd Müller, eds. *Evolution: The Extended Synthesis*. Cambridge, MA: The MIT Press, 2010.

Popper, Karl. *A World of Propensities*. Bristol: Thoemmes Antiquarian Books, 1990.

Rosen, Robert. *Life Itself: A Comprehensive Inquiry into the Nature, Origin, and Fabrication of Life*. New York: Columbia University Press, 1991.

Sheets-Johnstone, Maxine. *The Corporeal Turn: An Interdisciplinary Reader*. Charlottesville, VA: Imprint Academic, 2009.

Sherman, Jeremy, and Terrence Deacon. "Teleology for the Perplexed: How Matter Began to Matter." *Zygon* 42, (2010): 873-901.

Short, T. L. "Darwin's Concept of Final Cause: Neither New Nor Trivial." *Biology and Philosophy* 17, (2002): 323-340.

———. *Peirce's Theory of Signs*. Cambridge, MA: Cambridge University Press, 2007.

Wray, Gregory. "Integration Genomics into Evolutionary Theory." In *Evolution: The Extended Synthesis*, edited by Massimo Pigliucci and Gerd Müller, 97-116. Cambridge: The MIT Press, 2010.

Chapter Seven
The Irreducibility of Life to Mentality: Biosemiotics or Emergence?

Lawrence Cahoone

Biosemiotics has been an important research program in theoretical biology for several decades. Developed in different ways by Howard H. Pattee and Thomas Sebeok, it draws inspiration both from the cybernetic approach to self-copying systems inaugurated by John Von Neumann and the phenomenological study of "*umwelten*" of Jacob Von Uexküll. One of its major recent expositors has been Jesper Hoffmeyer. Biosemiotics regards life as semiotic or sign-using. In addition to the particular biological phenomena that it has interpreted, there are two prominent philosophical sources of its appeal. First, by claiming that living things employ a means, or function in a dimension, that is absent from the non-living physico-chemical orders of nature, it constitutes one form of biological anti-reductionism, opposed to those who seek to explain all biological phenomena chemically, physically, or as the manifestation of a purely genetic determinism. Second and more broadly, if it is right, biosemiotics would automatically make mentality, consciousness, or intentionality continuous with other biological phenomena, thereby narrowing, if not erasing, the "explanatory gap" between body and mind. In this brief chapter, I cannot evaluate biosemiotics systematically, nor the breadth of Hoffmeyer's work, but only raise some philosophical questions about the role of biosemiotics as an alternative to reductionism in biology.

Hoffmeyer's central claim is that "life is based entirely on semiosis, on sign operations."[1] Most fundamental is genetic coding itself, in which RNA and DNA molecules become "messages," as Pattee put it, and the organism contains an internal self-representation.[2] This constitutes, Hoffmeyer claims, "a root form of interpretative activity . . . where the specifications carried in the genetic code . . . become transcribed, translated, read, or used by the cellular apparatus."[3] Then there are a variety of semiotic activities of the cell or organism. In cases of downward causation, complex systems impose parameters on simpler composing systems that "interpret" those parameters as signs. Any organismic process which selects, identifies, or classifies a phenomenon for a differential response takes the phenomenon *as* a sign of something, hence is interpretive. There is "semethic interaction" in which organisms capitalize on signs of other species (e.g., heterotrophs that use their light sensitivity to feed on light-sensitive autotrophs). Cells and organisms exhibit intentionality, for "they can differentiate between phenomena in their surroundings and react to them selectively, as though some were better than others."[4]

Then there is animal communication and even dissembling, along with animal mind, hence internal representation of environment and soma. Processes undergone by lineages in evolution are themselves semiotic. Adaptation is a "semiotic process," for which "the term 'natural selection' is in fact confusing and misleading."[5] In explaining the evolution of the killdeer, a bird that fakes a broken wing to distract predators from its nest, he writes, "the evolutionary process has somehow made an anticipation in the sense that it has managed to genetically instantiate a general rule . . . based on the outcome of myriad individual cases. . . . [T]o make a general rule out of single cases logically seems to come close to what interpretation actually means."[6] Ultimately, Hoffmeyer ascribes selfhood, intentionality, subjectivity, and "semiotic freedom" to all living things. Life can only be explained by historical, triadic semiotic processes—using Peirce's triadic definition of signs—not the dyadic causal processes envisioned by physics and chemistry, with their natural laws, determinism, and an entitative ontology that fails to recognize the causal reality of relations and collectivities, such as "swarms" or "superorganisms" (see chapters 8 and 10 in the present volume).

Now, if one accepts a non-reductionistic, and hence pluralistic, approach to nature (to be described below), one can agree with Hoffmeyer that: relations or structures, and processes or events, are just as real as systems and their components; nature, at all levels of scale, exhibits determinacy *and* indeterminacy; life is an emergent property of some complex material systems, and manifests traits that cannot be reductively explained; ensembles or collections are sometimes the proper level of causal analysis; and there is no good reason to restrict mind and intentionality to human beings. Nevertheless, the question is whether the best method to ensure these recognitions is to make *life itself semiotic*? If the claim of biosemiotics is merely that "since evolution did in fact produce language-using organisms (us), and language is an enriched version of symbol-use, we should expect to find examples of proto-symbolism in simpler living systems," then it is probably right, just as living cells were probably preceded by "autocells" or pre-biotic self-replicating chemical systems.[7] But finding semiosis to underlie all the distinctive features of living systems, and constitute the difference between life and nonlife, is another matter.

For Hoffmeyer, following Peirce, a sign is some X which has a referential relation R to, or is *about*, some object O, such that X and R provoke an organism to produce some Y which likewise relates to O by R, that is, Y also refers to

O. In Peirce's language, this means Y is an interpretant (see Figure. 6.1). Interpretants must themselves be signs, which for Peirce meant they must be mental, and it seems the entity for which X is a sign must *know* it is a sign. In the usual sense, a sign is something *constructed* to serve that sign-function. Peirce's primary example of an indexical sign, the weathervane, is still artificial, even if its relation to its object is causal. In his late work, Peirce did distinguish three kinds of interpretants, the emotional, or a feeling, the energetic, or an action, and the logical, or an idea.[8] But these are clearly intentional or mental phenomena. Biosemiotics, however, must employ signs and interpretants that are not mental and in which the entity for which X is a sign need not *know* that it, or its interpretant, is a sign.

Now, the ribosome does seem to *read* RNA. It seems legitimate to regard this as some kind of proto-semiosis. There is a reading of the information in a nucleotide sequence, and a construction of a matching sequence of amino acids. This is transcription (by mRNA) and translation (by tRNA), but is it *interpretation*? If so, it is a very mechanical kind of interpretation, like a thermostat reading temperature or a computer translating Morse code. Furthermore, does the cell interpret the RNA or does the ribosome? If the latter, does this expand semiosis beyond the living, since we must ask, is a ribosome alive? If semiosis is definitive of life, and nothing but life, then it should be ascribable only to the living unit, not its chemical subsystems. In any case, is the amino acid product a sign? If not, then is it still the interpretant?

All organisms are sensitive. Often this sensitivity translates a surface contact into an internal electro-chemical message, neural or not, that leads to a motor response. But while selective, is the message, or the response, an *interpretation*? When a worm's muscles fire to lurch away from an undesirable gradient, so the subsystem muscles undergo downward causation from the system that is the worm, are the muscles reading or encountering a sign? Selective response seems not enough to qualify as interpretation, for even non-living systems selectively respond to a fraction of their environment's phenomena. The molecules of a ferromagnetic material "select" a common orientation out of many degrees of freedom in response to a drop below its Curie temperature. Does a block of ice "interpret" an increase of local kinetic energy as heat and melt in response?

There is another, more complex biological phenomenon that produces what we call mentality or experience. Neurological firing patterns can be said to "read" other neurological firing patterns, the result being felt experience of pain, hunger, fear, or sensory imaging. This may produce, as Antonio Damasio (whom Hoffmeyer cites approvingly in his 2010 article) claims, an ongoing "core conscious" mapping of the environment and its relation to somatic states.[9] Consciousness or mentality may then be said to be emergent from this complex reading of one brain state by another. Now, like transcription and translation, this could be claimed to be a proto-semiotic process of a more complex sort. However, the "reading" is performed by the brain, not the organism, and produces mind, rather than mind being the "agent" of the reading. Again it is not clear how this arguably proto-semiotic reading is interpretation.

Sophisticated organisms learn and communicate with conspecifics by special signaling behaviors. Here again it seems legitimate to speak of a kind of proto-semiosis. In some sense, the dog's growling at another dog is "about" its state or social status or its impending aggressive behavior, and for Pavlov's dogs the bell is arguably "about" the food. These are cases of intentionality, and we may say that mental activity is intrinsically intentional. But are these cases of semiosis? George Herbert Mead refused to call the growl semiotic; it is a gesture, but not a "significant" gesture. The latter requires that the gesturer respond to its own emitted gesture as the other would, which requires taking the standpoint of the other.[10] For Mead, the gesture can be about its object only if the signaler views it objectively, which is to say, socially. This arguably requires what Damasio calls "autobiographical" or "extended" consciousness, the solely human consciousness in which the ongoing awareness of somatic state in relation to environment is read or mapped onto an historical self who "owns" it. This may mean that a requirement for a fully functioning human self is the ability to take the perspective of the other on the organism, and on "parts" of the organism (i.e., components of the flow of its own experience). Hoffmeyer writes that any system is semiotic if it relates, within itself, a self-referential description to an object-referential description. But does the fact that the organism contains a genetic self-description mean that the organism is self-referring, hence has a "self"? For Damasio, a human patient during an asognosic seizure negotiating a crowded lobby is *core* conscious of her body in relation to surroundings but precisely lacks self-reference or self-consciousness. Her self has been turned off. Selves seem rare in nature.

In ascribing semiosis to all life the biosemiotic approach to the fundamental problem of how matter relates to life, and both to mind, is at least akin to the pan-psychism of Whitehead and before him, Peirce. For Whitehead, the ultimate constituents of reality, actual occasions, are vital and mental, exhibiting a developmental cycle and "prehension" of all other occasions. Peirce believed the life of protoplasm must include a kind of rudimentary mentality. Hoffmeyer suggests that the biosphere lies within the "semiosphere." But the genetic code, adaptation, sensation, animal communication, the experience of minded creatures, and self-reference are very different processes. If they satisfy the definition of semiosis, they do so in very different ways.

Two correlated notions seem essential to the biosemiotic analysis: signs and agency. The notion that the evolving organism is a kind of "hypothesis," that it is "about" its environment, that it trades in and represents information, can only apply to living things in distinction from complex non-living systems *if* one ascribes some kind of supra-chemical agency to the organism. "Anticipation," Hoffmeyer writes, "is a semiotic activity in which a sign is interpreted . . . like the dark cloud that warns us of an approaching thunderstorm."[11] But the cloud is only a sign *for* some being that takes it as a sign. The usual sense of sign makes the status of sign dependent on the kind of agency we ascribe to the system for which it functions as a sign. So we may say the cloud is a sign for the navigator, but not her ship. Signs are signs *for some agent*. So biosemiotics must think of all organisms as interpreting agents. Does semiosis enable and explain this distinctive agency, or rather does something about the organism's agency enable and explain the semiosis?

It seems possible to avoid reductionism in biology, to recognize the distinctiveness of life from the objects of chemistry and physics, without ascribing intentionality, sign use, interpretation, or mentality to all living things. One approach would be simply to assert living things manifest emergent properties, meaning properties not subject to reductionist explanation in terms of physicochemical parts. Hoffmeyer, and other anti-reductionists, often use the term "emergent" or "emergence," and implicitly make semiosis emergent from the physico-chemical. But they seek a means or model that can account for all the emergent properties of all phylogenetic orders of organisms, like semiosis.

One sophisticated notion of emergence and its metaphysical implications comes from the work of philosopher of biology William Wimsatt, who argues that emergence and reduction are not in conflict. Reduction (more precisely, componential reduction) is an attempt to explain a system's properties or performances through the properties or performances of its parts. Emergence is a name for the fact that some properties and performances of systems are not reductively explainable, but are instead explained either phenomenologically (not in the Husserlian sense) or functionally. Entities at a scale, call it N, are explained phenomenologically by interactions among comparable entities at N. We explain the dent in the fender by the impact of the other car. When, on the other hand, we explain some trait of the N-system by its being selected within some supra-system of level N+1, that is a functional explanation, which explains through reference to a more encompassing system engaged in a process that serves to select the property or performance in question due to the latter's role within the larger system. Biological evolution explains an organism's genotype through the relations of its ancestors' phenotype to their environment.

For Wimsatt, "a reductive explanation of a behavior or property of a system is one that shows it to be mechanistically explicable in terms of the properties of and interactions among the parts of the system."[12] In scientific practice reduction explains only *some* properties or performances of a system on the basis of a *perspectival decomposition* of the system (i.e., a particular way of cutting it into parts), by using an *idealized model* of the parts and/or their interactions, resting on or employing significant *approximations*. According to Wimsatt, the endpoint of a complete reduction, which would justify the claim that a system property is "nothing but" its part properties, are those cases in which the system properties are *aggregations* of part properties or performances, which requires four conditions: *intersubstitutability* (invariance of the system property under rearrangements of parts), *qualitative similarity* under scale change (larger or smaller but with the same properties), *re-aggregativity* (invariance of the system property under decomposition and re-composition), and *linearity* (the absence of feedback).[13]

Wimsatt points out that this is rare. It holds only when there are no complex relations among the parts. Mass is one of the few properties of physical systems which is just the aggregation of the same property of the components. But this is not true for as seemingly simple a quantity as volume, for in some chemical reactions volume changes. Wimsatt makes the interesting suggestion that aggregativity tracks the conservation laws of physics, that the properties, which are the subject of conservation laws—mass, energy, charge, spin—are those whose values are indeed invariant in all interactions. This shows how fundamental, and yet how narrow, the band of aggregative properties is. Reduction is crucial but incomplete: *reduction explains something about almost everything, but everything about almost nothing*. Reduction nevertheless remains a crucial component of our explanatory practice. As noted, *something* about a system is almost always explained by reduction. When and where this works, it is tremendously simplifying, and generates the most context-independent entification. So we will keep trying to find it by decomposing systems and idealizing their components' interactions in just such a way as is likely to yield workable reductions. But most of the time we will find that the reduction must be supplemented by non-reductive explanations.

Note that the attempt to ascribe complex system processes and relations to the components does not save reductionism. If those "relational properties" of the parts are *themselves* determined and explained by the location of a part in the whole system, then we have now turned round to explain the parts by the whole, which is fine, but it is no longer a reductive explanation. To include such properties in the parts is to miss the point of reduction, which is to substitute a simpler problem for a complex one, to explain properties of complex wholes as the result of simple interactions among relatively isolated parts, relying only on background interaction rules which would govern those parts *if they existed outside* this particular whole. If the whole is the product of complex, context- or environment-sensitive relations among these components, then the whole must be referenced to explain those relational properties. Hence, the whole is being explained by parts being explained by the whole.

The multiplicity of non-coincident perspectival decompositions is particularly important in biology. In analyzing a piece of granite we can produce different decompositional maps or diagrams of its parts or regions, based distinctively on chemical composition, thermal conductivity, electrical conductivity, density, tensile strength, etc... In the non-biological realm these different maps are likely to divide up the system into parts with what Donald Campbell called "coincident boundaries."[14] *Not* so when we turn to organisms. When we produce decompositions based on anatomical organs, cell types, developmental gradients, types of biochemical reactions, and cybernetic flow, boundary coincidence across decompositions *disappears*. Each decomposition produces its own unique map of the "parts" of the organism. Furthermore, we find "cross-perspectival" causal interactions, or increased "interactional complexity" *between* these perspectival decompositions.[15]

The implications for biological explanation are straightforward. Wimsatt argues that functional explanations are necessary to biology, they are causal explanations, they are inherently teleological—or as I will say below, teleonomic—hence have an ineliminable reference to *purpose*. That is, "purpose and teleology [teleonomy] are correlative concepts."[16] None of this implies any necessary reference to consciousness, subjective agency, mental intentionality, or divine design. Wimsatt formalizes function as $F[B(i), S, E, P, T] = C$, meaning that "according to theory T, a function F of behavior B of item i in system S in environment E relative to purpose P is to do C."[17] The P-variable, purpose, is essential for it picks out criteria for an "intentionally" defined class of state-descriptions of the system and its environment," that is, a set of conditions of which $B(i)$ "promotes the attainment." Furthermore, there is feedback by which $B(i)$, S, and E are themselves partly the product of P as it fits into a system understood by T.[18] The functional consequence is "causally responsible for the selection and presence of the functional entity. . . ." It answers the "why" question, "giving a causal answer to a causal question."[19] If this view is correct, then, "only by denying that evolution has occurred as a result of selection processes or that it is of any scientific interest could teleology be eliminated from biology."[20]

There are many emergent properties of biological existence, from macromolecules to self-replication to communication and societies. But we may comment on one of the most central emergent traits of organisms. As the philosopher Hans Jonas writes, "Profound singleness and heterogeneousness within a universe of homogeneously interrelated existence mark the selfhood of organism."[21] The existence of living things is based on their ability to maintain boundaries from the environment and control their internal chemistry. "The point of life itself," he writes, is "its being self-centered individuality, being for itself and in contraposition to all the rest of the world, with an essential boundary dividing inside and outside—notwithstanding, nay, on the very basis of the actual exchange."[22] This is a kind of autonomy, although without any anthropomorphic notion of freedom. There must be *identity*, or else there is no "auto" to be the "nomos," the law or governor, of its activity. The organism remains a unity, with an inside and outside, and the difference *matters*. But the coin of autonomy has another side. Jonas writes, "The privilege of freedom carries the burden of need and means precarious being. . . . [L]iving substance . . . has taken itself out of the general integration of things in the physical context, set itself over against the world, and introduced the tension of 'to be or not to be' into the neutral assuredness of existence."[23] Simply, organisms *need* and *die*. Precisely because they are so distinct, they are precarious. Unfulfilled need means death, the loss of inner activity leading to merger with the environment. Death is the price at which life comes. Life is "hazardous being" because it forms an identity that can and will end. Death is complete, rather than controlled, openness to environment.

Here we can use Maturana's and Varela's notion of "autopoiesis" or auto-making. They meant the term to "convey the central feature of the organization of the living, which is autonomy . . . what takes place in the dynamics of the autonomy proper to living systems."[24] We can follow them here without accepting their notion of organismic closure or lack of interaction with the external world. The organism is auto-making and auto-reproducing, but without a "self" or the kind of agency or mentality characteristic of highly intelligent animals. Life is the state of: (a) a particular bounded material entity; (b) manifested in ceaseless activity (homeostasis, growth, and behavior in environment); (c) in order to achieve and maintain a complex, largely (not completely) fixed and inherited typical structure (much of it common to a type of organism or species); (d) as part of a lineage (life only comes from life); (e) in material and energetic exchange with an environment. (The "particularity" listed first does not gainsay symbiosis or parasitism. All life is cellular, and every cell is a bounded entity.) The *prima facie* evidence for life is an entity engaged in complex auto-maintaining activity; death is cessation of activity and/or loss of boundary with physical-chemical environment. So auto-control, auto-*nomos*, auto-*poiesis* do correctly identify one core distinctiveness of live organisms, as long as we understand they are only achievable through interaction with, and past selection by, the environment. We can call this a kind of agency, *biotic* agency, distinct from both intentional or *mental* agency (e.g., of some animals), and *autobiographical* agency or selfhood (of humans).

As noted, the only way for this hierarchical form and identity to be maintained is through ceaseless activity. Jonas writes, "the organism has to keep going, because to be going is its very existence."[25] But this remarkable intensity of activity requires a novel means of control. Ernst Mayr used "teleonomic" to refer to living but un-minded systems, leaving "teleological" for behavior guided by mental intentions.[26] He took the concept from Pittendrigh, who had written, "a physiological process or a behavior that owes its goal-directedness to the operation of a program can be designated as 'teleonomic.'"[27] The program is genetic and supplemented by feedback systems, so that all processes are "guided by a program, and they depend on the existence of some endpoint or goal which is foreseen in the program regulating the behavior." As Mayr wrote, in biology we do not merely say the Wood Thrush migrates in the fall and *thereby* escapes the northern winter. Such a statement misses a crucial bit of information: the Wood Thrush migrates *in order to* escape the winter.[28] That is *why* the behavior has been naturally selected and maintained.

Now, purposes are ends, and imply value, or to put it simply, they imply good and bad (which is *not* to say a moral sense). As the philosopher of biology Elizabeth Baeten remarks, there is value at the least wherever there is *evaluation*. The ecological philosopher Holmes Rolston has developed a general theory of the role of value in living things. First and most obviously, organisms value themselves, their states, and sometimes their offspring. They act so as to maximize their survival and/or reproductive chances. They value preferred states of their own organism, pursued homeostatically. They have *norms*; they are healthy or sick, living or dying, succeeding or failing. Second, they value parts and states of the environment; they turn toward the sun, stretch their roots for water, expand in search of resources, chase prey, seek secure locations. This value is *relative, but objective*, not subjective: the presence of certain carbohydrate-hungry beings means a potato has food value. Seed-bearing plants develop fruits because they scatter their plant's seeds in the feces of animals *because* they have value for the animals who eat them. Third, value *circulates* through an ecosystem. When a predator eats its prey, or a herbivore eats vegetation, they are "capturing the value" of the organism eaten. Fourth, *the species* evaluates, not as a collective agent, but as a "form" that presses its case genetically. The developing bird is *supposed to* have two wings. Rolston thus rightly characterizes the organism as a "valuational system."[29] Part of what makes it survive is that it evaluates in the ways it does; it is genetically coded to make those evaluations and embody certain norms. Valuing is one of its crucial activities; what it values can mean life or death. The valuational system itself evolves, in that populations can be selected by environment because of their distinctive preferences.

The emergence of life, with its inherent autopoiesis and teleonomic purposiveness, can be accounted for without the use of semiosis as its definitive essence. That several elements of life present a proto-semiotic relation, a kind of "aboutness," is importantly true, but this does not mean semiosis or intentionality are necessary or sufficient conditions of life, or that they explain the kind of agency and teleonomy of living things. Just as we must avoid reduction of the biological to the physical and chemical, we ought to avoid explaining life by surrogates for mentality. In the emergent, hierarchically structured orders of nature, there is no mind without life, but there is copious life without mind. The relation of life and intentionality does not appear symmetrical: intentionality presupposes life, life does not presuppose intentionality.

Notes

1. Jesper Hoffmeyer, *Signs of Meaning in the Universe* (Bloomington, IN: Indiana University, 1997), 24.
2. Howard H. Pattee, "How Does a Molecule Become a Message?" in *Communication in Development*, ed. Anton Lang (New York and London: Academic Press, 1969), 1-16.
3. Jesper Hoffmeyer, "A Biosemiotic Approach to the Question of Meaning," *Zygon* 45, no. 2 (2010): 372-373.
4. Hoffmeyer, *Signs*, 48.
5. Hoffmeyer, *Signs*, 22.
6. Hoffmeyer, "A Biosemiotic Approach," 371.
7. Liz Stillwaggon and Louis J. Goldberg, "Biosymbols: Symbols in Life and Mind," *Biosemiotics* 3 (2009): 17-31.

8. Charles Sanders Peirce, *Collected Papers, Volume Five*, ed. Charles Hartshorne and Paul Weiss (Cambridge, MA: Harvard University, 1931), 5.475-6.

9. Antonio Damasio, *The Feeling of What Happens: Body and Emotion in the Making of Consciousness* (Orlando, FL: Harvest, 2000).

10. George Herbert Mead, *Mind, Self, and Society: From the Standpoint of a Social Behaviorist*, ed. Charles Morris (Chicago: University of Chicago Press, 1962).

11. Hoffmeyer, "A Biosemiotic Approach," 375.

12. William Wimsatt, *Re-Engineering Philosophy for Limited Beings: Piecewise Approximations to Reality* (Cambridge, MA: Harvard University Press, 2007), 275.

13. Wimsatt, *Re-Engineering Philosophy*, 281.

14. Donald Campbell, "Blind Variation and Selective Retention in Creative Thought as in Other Knowledge Processes," *Psychological Review* 67, no. 6 (1960), 380-400.

15. Wimsatt, *Re-Engineering Philosophy*, 182-186.

16. William Wimsatt, "Teleology and the Logical Structure of Function Statements," *Studies in the History and Philosophy of Science* 3, no. 1 (1972), 12.

17. Wimsatt, "Teleology and the Logical Structure," 32.

18. Wimsatt, "Teleology and the Logical Structure," 39-40.

19. Wimsatt, "Teleology and the Logical Structure," 70.

20. Wimsatt, "Teleology and the Logical Structure," 66-67.

21. Hans Jonas, *The Phenomenon of Life: Toward a Philosophical Biology* (Chicago: University of Chicago, 1966), 83.

22. Jonas, *The Phenomenon of Life*, 79.

23. Jonas, *The Phenomenon of Life*, 4.

24. Humberto Maturana and Francisco Varela, *Autopoiesis and Cognition: The Realization of the Living* (Dordrecht, Holland: Springer, 1991), xvi.

25. Jonas, *The Phenomenon of Life*, 126.

26. Ernst Mayr, "Teleological and Teleonomic: A New Analysis," *Boston Studies in the Philosophy of Science* 14 (1974): 91-117.

27. Colin Pittendrigh, "Adaptation, Natural Selection, and Behavior," in *Behavior and Evolution*, ed. Anne Roe and George G. Simpson (New Haven: Yale, 1958), 390-416.

28. Mayr, "Teleological and Teleonomic," 152.

29. Holmes Rolston III, *Environmental Ethics: Duties to and Values in the Natural World* (Philadelphia: Temple University, 1988).

Bibliography

Campbell, Donald. "Blind Variation and Selective Retention in Creative Thought as in Other Knowledge Processes." *Psychological Review* 67, no. 6, 1960, 380-400.

Damasio, Antonio. *The Feeling of What Happens: Body and Emotion in the Making of Consciousness*. Orlando, FL: Harvest, 2000.

Hoffmeyer, Jesper. *Signs of Meaning in the Universe*. Bloomington, IN: Indiana University Press, 1997.

———. "The Semiotic Body." *Biosemiotics* 2 (2008): 169-190.

———. "A Biosemiotic Approach to the Question of Meaning." *Zygon* 45, no. 2 (2010): 367-390.
Jonas, Hans. *The Phenomenon of Life: Toward a Philosophical Biology*. Chicago: University of Chicago, 1966.
Maturana, Humberto, and Francisco Varela. *Autopoiesis and Cognition: The Realization of the Living*. Dordrecht, Holland: Springer, 1991.
Mayr, Ernst. "Teleological and Teleonomic: A New Analysis," *Boston Studies in the Philosophy of Science* 14 (1974): 91-117.
Mead, George Herbert. *Mind, Self, and Society: From the Standpoint of a Social Behaviorist*. Edited by Charles Morris. Chicago: University of Chicago, 1962.
Pattee, Howard H. "How Does a Molecule Become a Message?" In *Communication in Development*, edited by Anton Lang, 1-16. New York and London: Academic Press, 1963.
Peirce, Charles Sanders. *Collected Papers, Volume Five*. Edited by Charles Hartshorne and Paul Weis. Cambridge, MA: Harvard University Press, 1931.
Pittendrigh, Colin. "Adaptation, Natural Selection, and Behavior." In *Behavior and Evolution*, edited by Anne Roe and George Gaylord Simpson, 390-416. New Haven: Yale, 1958.
Rolston, Holmes. *Environmental Ethics: Duties to and Values in the Natural World*. Philadelphia: Temple University Press, 1988.
Stillwaggon, Liz, and Louis J. Goldberg. "Biosymbols: Symbols in Life and Mind." *Biosemiotics* 3, no. 1 (2009): 17-31.
Wimsatt, William. "Teleology and the Logical Structure of Function Statements." *Studies in the History and Philosophy of Science* 3, no.1 (1972).
———. *Re-Engineering Philosophy for Limited Beings: Piecewise Approximations to Reality*. Cambridge: Harvard University, 2007.

Section 3:
Homeostasis, Thermodynamics, and Symbiogenesis

Chapter Eight
Biology's Second Law: Homeostasis, Purpose, and Desire

J. Scott Turner

The title of this volume asks us to go "beyond mechanism," which invites the obvious question of where, precisely, should we go? The invitation is motivated by a long-standing question which has arisen about life, in one form or another, literally for millennia: is life a qualitatively special phenomenon, or is it not? The study of life now finds itself at the end of more than a century of triumphant scientific progress dedicated to the proposition that the answer to this question is "No." By clarifying the chemical and physical properties of living systems to their uttermost details, the scope of this scientific long march has illuminated the phenomenon of life at scales ranging from the atomic to the planetary, from the span of nanoseconds to encompassing the entire history of life on Earth. Our understanding of the mechanisms of life is now so comprehensive that it is easy to conclude that the question is settled, and that there is no place fruitful for biology to go "beyond mechanism."

Yet the question cannot be settled, because even the most perfect understanding of the mechanisms of life does not disprove the proposition that life is special. There are many marvelous phenomena in the universe, and mechanisms are undoubtedly at work in them all. A cauliflower is a marvelous phenomenon, as is a cumulus cloud. There is a striking phenomenology in the regularity of an array of Bénard cells in a heated dish, and also in the regularity of termite mounds distributed across an African savanna. A thought is certainly a unique phenomenon, and so too is the twinkling of a sparkler in a child's hand. For each comparison I have posed, we have no difficulty identifying one as a phenomenon of life, and the other as not. Why is this so? The possibility exists, of course, that the distinction is illusory, and that we are just very good at fooling ourselves into thinking that there is something deeply special about cauliflowers, termite mounds, and thoughts. But what if the special qualities we perceive are, in fact, special? If there is someplace "beyond mechanism" for biologists to go, mapping the road there will have to grapple with just what it is that we imagine life's special quality to be.

In this essay, I will argue that life does indeed have a special quality, two of them in fact, that taken both together may provide the basis for fully coherent theory of life. The first, and obvious quality, is that life evolves and it does so by a process that is unique to living systems—namely, Darwinian natural selection. I will refer to this as Biology's First Law. Biology's First Law explains much, including the necessity for heritable memory and a means for these memories to be realized in well-functioning repositories that some call vehicles. Yet, Biology's First Law is inadequate by itself. It has no credible explanation for the origin of life.[1] In its present form, it cannot explain the evolution of life before there were genes, nor can it explain how genes came to be in the first place. Its explanations for important phenomena like biological design, adaptation, cognition and consciousness are embarrassingly tautological.[2]

I propose that those gaps are filled in, and that the path is opened to the coherent theory of biology we seek, by joining Biology's First Law with what I will call Biology's Second Law. This is the phenomenon of homeostasis, first proposed by Darwin's contemporary, the great French physiologist Claude Bernard, and embodied in his famous dictum: "*La fixité du milieu intérieur est la condition d'une vie libre et indépendante*" ("The constancy of the internal environment is the condition for a free and independent life"). Homeostasis is a mainstay of physiological thought, but by itself stands mostly as a dogmatic assertion about how life works in the present: it sheds little light on how such a phenomenon could have come into being, or why such a condition, and not some other, should characterize living systems.

Darwin's and Bernard's insights offer complementary visions into the nature of life. Through the twentieth century, these visions have developed mostly as Non-Overlapping Mechanistic Magisteria (NOMMA), to mangle Stephen Jay Gould's well-turned phrase.[3] While the NOMMA have had an important place in the scientific development of biology in the twentieth century, their continuing division is now proving to be a frank obstacle to the emergence of a fully coherent understanding of the phenomenon of life.

The V-Word

The overlap between the NOMMA is slight: one can pursue elegant studies in physiology without ever bringing evolution into it, and evolutionists have long been agnostic about the actual workings of life. Where they overlap is a deep suspicion of vitalism. Both Darwin and Bernard are lauded, in their own ways, for purging the science of life of vitalist obscurantism, and for opening the door for "real" science (i.e., mechanistic and law-driven science), in their respective realms.

There is a whiff of whiggish revisionism in both stories, however, which relies largely on a caricature of vitalist philosophy as it was developing in the late eighteenth and early nineteenth centuries.[4] Both Darwin and Bernard drank strongly from this vitalist spring, but for historical reasons peculiar to each, this became philosophically inconvenient to the emerging NOMMA that came to divide biology during the twentieth century. For Darwinism, the shroud began to come down in the late nineteenth century "eclipse of Darwinism," as Peter Bowler has described this period,[5] followed by the co-opting of classical Darwinism by its genetic opponents into its modern gene-centric form, neo-Darwinism. Meanwhile, Bernard's inconvenient idea, homeostasis, was cloaked in a discreet silence for nearly a half century following Bernard's death.[6] Despite honorific nods to the concept, mostly in textbooks of physiology, respectable eyes remained mostly averted well into the twentieth century: a recent biographical essay on Bernard does not even mention the term.[7]

It is not my intention to disparage the caricature, far from it, for it provided an intellectual shelter that had enormous salutary consequences for both evolutionary biology and physiology. For evolutionary biology, it led to the neo-Darwinist synthesis, arguably one of the greatest intellectual achievements of all time. For physiology, it paved the way to the spectacular emergence of molecular biology. But a caricature, even a widely accepted and fruitful one, is nevertheless still a caricature. Unveiling the caricature shows both Darwin and Bernard, not as vanquishers of nineteenth century vitalism, but actually its vindicators.[8] Understanding why this is so will help to constructively break down the barriers that currently divide biology into its NOMMA.

The vitalism of caricature is the vitalism of the *vis essentialis*, the notion that there is an ineffable substance that confers life upon what would otherwise be dead matter. That form of vitalism effectively died in the eighteenth century, however, at the hands of academic physicians who responded to the Cartesian rise of mechanism with a vigorous flowering of debate among the various European faculties of medicine.[9] These were driven by the conviction that living things were more than mere mechanisms, that there was indeed something special about life, and by the desire to clarify just what life's special quality could be.[10] These schools of thought ranged from the school of "Newtonian medicine," based mostly in Leiden, which sought laws governing health and disease that could compare to Newton's laws governing matter and motion; to the "Organismist" school, centered in Montpellier, that took what might now be described as a "holistic" approach to medicine, viewing the body as an indivisible system with an integrated functional identity. What united these various schools of medical thought was a consensus that what distinguished life was not a vital stuff that animated dead matter, but a unique form of process, a form of negotiation between many subsidiary "little lives" that constituted a whole organism. How these "little lives" were packaged was a matter of disagreement. The Organicist school in Paris, for example, pointed to organs as the little lives, citing as evidence the ability of organs removed from the body to continue functioning for a time. Berlin's Cellulist school, in contrast, identified cells as the ultimate "little lives." No matter how the "little lives" were packaged, though, the existence of a whole, well-functioning and adaptively responsive organism was *prima facie* evidence of their successful accommodation to one another. The important question to ask, then, was how the accommodation was negotiated.

The vitalism of the *vis essentialis*, or "metaphysical vitalism" as it came to be called, survived in this ferment for a time, with some proposing a *vis mediatrix* that would replace the *vis essentialis* as the vital stuff.[11] This could not stand, however, because one could see organism-like behavior also in collections of organisms, like bee colonies (a remarkable eighteenth century anticipation of the superorganism idea) that presumably could not share a "mediating stuff" between their disconnected bodies. From observations like these, there emerged early on a division in vitalist thought, between the metaphysical vitalists on the one hand, and "scientific vitalists" on the other, who focused on the transactions and processes that made living things uniquely special. The vitalism that is mocked in caricature is the straw man of metaphysical vitalism, which was not taken seriously by anyone after the eighteenth century. The scientific vitalism that replaced it, however, significantly shaped biology throughout the nineteenth century. Cuvier's principle of correlation of parts and his "conditions of existence," for example, explicitly addressed the problem of how the myriad "little lives" negotiated their way to a coherent organism.[12] Lamarck's notions of adaptive and complexifying forces drew from the same well.[13] Similarly, Pasteur was a vigorous defender of the doctrine that life could only arise from other life, making him a quintessential vitalist.[14] And the same can be said of Claude Bernard and homeostasis: his outstanding experimental work, like Cuvier's groundbreaking work in comparative anatomy, was aimed at working out the mechanisms underlying life's special quality: the processes of negotiation between parts to produce a coherent whole. Seen in this light, Bernard's principle of homeostasis was in fact not the refutation of scientific vitalism, but its culmination.[15]

And what of Darwin? It is worth remembering that Darwin's formative years and his early thinking about evolution were steeped in the phenomenon of adaptation, which itself has deep roots in both metaphysical and scientific vitalism.[16] The marvelous adaptive contrivances of nature's "little lives," written about so vividly by William Paley, and that were a mainstay of natural theology and the argument from design, were evidently an inspiration to Darwin, who took pains to include Paley's *Natural Theology* in the sparse baggage he was allowed on HMS Beagle.[17] The influence of Darwin's grandfather, Erasmus (a notorious Lamarckian), the medical profession of his father, Robert, and his own sporadic medical training (at the feet of his scientific mentor in Edinburgh, Robert Grant) would have immersed Charles in the bath of scientific vitalism. It is also noteworthy that Darwin's mature thoughts about heredity were strongly influenced by Lamarck's notions of use and disuse, and Cuvier's correlation of parts, both of which derived directly from the scientific vitalist tradition, even to the point that Darwin himself promulgated an explicitly Lamarckian theory of heredity. Even though Darwin backed away somewhat from adaptationism in his later thinking,[18] it remained the core question for many of his scientific critics. To the extent that Darwin sought a natural law to explain adaptation, though, his theory of natural selection is as much a statement of core principles of scientific vitalism as was Bernard's theory of homeostasis.

The Oversold Triumph of Mechanism

In the late nineteenth and early twentieth centuries, this vigorous tradition of vitalist thought was replaced nearly completely by the materialist revolution that swept through Western scientific and political thought. In the evolutionary sciences, this came to the fore primarily in material theories of heredity, like the particulate theory of the gene.[19] In physiology, it emerged as chemistry replacing homeostasis as a foundational principle.[20] In the sciences, this shift in philosophy was driven by powerful advantages in experimental technique that advanced the materialist cause against the opposing scientific vitalists, who were left, for the most part, to fall back on vague principle, appeals to reason, and ad hoc "systems" of empirical "laws" that could point to no coherent principle to justify their claims.

This rise of materialism ushered in what is arguably biology's golden age, but it is important to realize that this Elysium did not rest entirely on a scientific foundation. Scientific materialism arose in a complex and confusing stew of personal rivalries and larger social and political trends, which included the rise of materialist (i.e. "scientific") principles of history, politics and sociology, progressivism, of technocratic theories of governance, and the rise of viable socialist and Marxist political movements during this period.[21] It would be a fallacy to say that these political developments drove the scientific, or that the scientific drove the political. However, it would be fair to say that the scientific, social, and political spheres were all shaped together by a peculiar mutualism. The triumph of mechanism was both a scientific triumph and a political vindication, and this confluence of interests is largely what drives the whiggish tendencies I mentioned above. Among the most significant has been the tendency to overstate the impact of scientific results when they justify the intertwined interests of the mutualists. Reference has already been made to one aspect of this tendency for over-reach: the argument, which fails even the most elementary test of logic, that demonstration of mechanism in a living system invalidates the proposition that life has a special quality. Two other examples are more specifically instructive, for they contain the roots of the various internal contradictions that permeate modern biology, which, I propose, underscores the necessity for a Second Law of Biology. The first concerns the significance of August Weismann's famous experiment at the dawn of the neo-Darwinist synthesis, on germ-line versus somatic inheritance.[22] The second example comes from neo-Darwinism's seeming culmination, the formulation of the famous Central Dogma of Molecular Biology.[23]

August Weismann's experiment took place in the context of the late-nineteenth century crisis of Darwinism.[24] This was a time of mounting skepticism over the validity of the Darwinian idea, expressed succinctly in the famous question posed by the American paleontologist Edward Drinker Cope: Darwin may have an explanation for the origin of species, but no explanation for the origin of fitness (i.e., adaptation).[25] Evolutionary biology then parsed roughly into two broad camps: the Developmentalists, who were neo-Lamarckian in outlook, in that they were focused on mechanisms of adaptation as drivers of evolution (Cope himself was a prominent neo-Lamarckian) and the Geneticists, who for their part, were concerned more with mechanisms of heredity. Of the two, the Geneticists were more openly hostile to Darwinism, looking to mutation as a principal driver of evolution. Oddly, the Developmentalists regarded themselves as the keepers of the Darwinian flame, largely because of Darwin's own adaptationist roots and his explicitly Lamarckian speculations on heredity.

In his experiment, Weismann elegantly asked a simple question: would the repeated amputation of tails and subsequent breeding of the mutilated mice produce a congenital shortening of tails in their lineages? No such modification occurred, of course. This simple result has been regarded as pivotal to resolving the crisis of Darwinism, for two reasons. First, by showing that modifications to the soma did not produce heritable variation in a lineage, the experiment is supposed to have verified Weismann's theory of heritability solely through the germ-line. Second, by verifying his theory of germ-line segregation, Weismann is supposed to have discredited Lamarckism experimentally, once-and-for-all, and in any of its varieties.[26]

Weismann's experiment did neither, of course. What Weismann was testing was not Lamarckism itself, but a particular mechanism of Lamarckian inheritance—namely, Darwin's gemmule theory of inheritance.[27] It did not explore any other supposed mechanisms of Lamarckian inheritance. Certainly, it did not explore mechanisms that went beyond mutilation to produce an acquired character. Furthermore, Weismann did not test Lamarck's basic premise that physiological adaptation in one generation would produce heritable adaptation across generations. Finally, Weismann's experiment did not illuminate in any way the generality of germ-line segregation. The emergence of a differentiated germ-line in embryonic development is complicated and varies widely among different kinds of animals—in some it occurs with the first cleavage division, while in others it never occurs.[28] The very idea of germ-line segregation is also inapplicable to taxa, like plants, fungi, and bacteria, that do not have differentiated germ lines or sexual reproduction. Serial mutilation of a lineage of mice does not explore these complications. Weismann himself was appropriately modest in how he interpreted his experiment, but not so others: his experiment has been cited as one of the most significant in the history of biology.[29] The interesting question is why Weismann's experiment became freighted with such grandiose claims.

On physiology's side of the NOMMA, a similar philosophical debate played out over the surprisingly persistent issue of spontaneous generation, or "spontogenesis" as Darwin termed it, a debate that ranged well into the last quarter of the nineteenth century.[30] The lines in the spontogenesis debate roughly paralleled the materialist-scientific vitalist divide that roiled the contemporaneous crisis of Darwinism.[31] Spontogenesis opponents tended to be scientific vitalists, while its supporters were more materialist, and the disagreement played out over numerous venues. One of these involved competing theories for the origin of disease. On the one side were advocates of the promulgated germ theory, championed by Pasteur and Lister, among others. The germ theory was vitalist in outlook. Disease was clearly associated with the presence of microbes, but it was not clear whether these were the cause or the consequence of pathology. On the other side were the opponents of vitalism who championed the the "zymotic theory" of disease. This doctrine held that disease was a derangement of body chemistry wrought by chemical agents called "zymes." These could arise spontaneously in putrefying conditions, as could the bacteria that kept showing up. This ran counter, to say the least, to the vitalist predilections of the germ theorists, who believed that bacteria could only arise from other bacteria, and could not arise de novo from putrefying flesh.

According to the myth, Louis Pasteur settled this argument with his famous gooseneck flasks and boiled broth, vindicating the germ theory by casting spontaneous generation into the wilderness. What is often left out of the hagiography is that Pasteur's experiments were a vindication of a vitalist doctrine—cells can only arise from other cells. Also rarely mentioned is that spontogenesis did not go quietly under Pasteur's assault, and that it was scientists, not obstinate metaphysicians, who pushed back the hardest. Central to this scientific challenge to Pasteur was the work of Henry Charles Bastian, a protégé of Thomas Huxley, who repeated Pasteur's experiments, but with different kinds of infusions, such as boiled cheese or cucumbers.[32] With these experiments, Bastian was able to show, repeatedly, and with seemingly impeccable experimental technique, that microbes could originate even in infusions that had been boiled for many hours and impeccably sheltered from contamination from the outside. The fly in Bastian's infusions was the existence of heat-resistant strains of bacteria, the existence of which were only established by the physicist John Tyndall in 1877. It was this discovery, and not Pasteur's, that finally cratered the nineteenth century case for spontaneous generation.

Tyndall's experiment left the vitalists on top, but their triumph did not last long. Where the neo-Lamarckians were left gasping in the aftermath of Weismann, and with nothing to fall back on, the spontogenesist's strong grounding in chemistry and experiment gave them a foundation to regroup and eventually to recast their materialist stance into explorations of the nature of metabolism and biological catalysis. Thus was borne the nascent field of physiological chemistry, hived off from the Pasteurian (vitalist) root of microbiology, and which gave the defeated (materialist) spontogenesists a congenial home. There they wandered in the wilderness for a time, looking to explain biological catalysis as a kind of self-assembled form of crystalline protein known as cyclols.[33] By the 1930s, pioneering crystallographic work on the structure of proteins revealed both a surprising specificity of amino acid sequence combined with a striking structural asymmetry that blew the cyclol concept completely apart. Our modern conception of enzymes rose from this wreckage, but the high complexity of proteins now cast the question of heredity into a new light. This was a major impetus for the discovery of the material basis of heredity.

The story that follows is well-known, and need not be recounted here, except to note two salient points. First, the confluence of physiological chemistry and genetics so perfectly combined several fundamental questions with the ability to experimentally answer them that it swept all of biology into its wake, including the erstwhile redoubts of Pasteurian vitalism. Second, the culmination of this trend came in 1958 with Francis Crick's statement of the Central Dogma of Molecular Biology (CDMB).

Like Weismann's experiment, the Central Dogma stands as one of the great over-interpretations in modern biology.[34] By itself, it makes a simple statement about the relationship between nucleotide-encoded sequence information in DNA and amino-acid encoded sequence information in proteins. As the Weismann experiment was, the Central Dogma is an elegant encapsulation of this limited scientific question, and as Weismann himself was, Crick was modest in the meaning he attached to it, regarding it essentially as a "brash" summary of limited scope that has been misinterpreted the more broadly it has been applied.[35] Among these broad applications of the CDMB was the way it seemed to recast Weismann's theory of germ-line inheritance into a molecular form. Where Weismann would say that it is impossible for changes acquired during an organism's lifetime to feed back onto transmissible traits in the germ line, the CDMB now added that it was impossible for information encoded in proteins to feed back and affect genetic information in any form whatsoever, which was essentially a molecular recasting of the Weismann barrier. As we have come to know more about the nature of gene expression, the metabolic interaction of DNA with the cell environment in which it resides, and even the nature of the gene itself, the generality of these assertions has been fatally weakened.

Thus, twenty-first century biology finds at its heart two supreme ironies. Dating from the dawn of the twentieth century is the irony of the emergence of genetics, a fundamentally anti-Darwinian doctrine, as the modern keeper of the Darwinian flame, with the arguably more Darwinian (albeit neo-Lamarckian) adaptationists left in the cold. The second irony dates from the CDMB, which completed the capture of biology by materialists who arguably do not believe in life, at least not as a phenomenon that has any dimension that cannot be reduced ultimately to chemistry. From these unresolved ironies spring many odd things about modern biology, which, for the most part, stand unremarked or unnoticed.

Take, for example, the study of life's origins, which is characterized by two irresolvable issues.[36] The first, bequeathed to us by the anti-vitalist spontogenesists, is that the origin of life is exclusively a chemistry project, even to the point that we can speak unabashedly of creating artificial life despite there being no credible theory for a chemical origin of life that does not require large teams of scientists and multimillion dollar research budgets to pull off.[37] The second is the bequest of the anti-Darwinian geneticists, who equate the origin of life with the origin of the gene. This logic must necessarily view the origin of life as an event, rather than as an evolutionary process itself, largely because the evolution of life before there were genes is simply inconceivable.[38] Life on Earth may indeed have arisen as an event, but presently the most likely scenario for that is the seeding of the planet by extra-terrestrial inocula.[39] Panspermia is not, however, a theory for the origin of life *per se*.

Biology's Second Law

Which brings me to my claim that homeostasis is biology's needed Second Law. I have written extensively about the phenomenon of homeostasis elsewhere, both in the context of the phenomenon of animal-built structures[40] and in the origins of biological design,[41] so I will not reprise those discussions here. Rather, I wish to focus on two questions. First, is there something about the phenomenon of homeostasis that resolves the conundra inherent in a strictly materialist approach to life and its evolution? Second, is homeostasis a phenomenon that is incident from (and hence subordinate to) the principle of natural selection, so that it explains nothing beyond what could adequately be explained by Biology's First Law—Darwinian natural selection—alone?

The first question rephrases the one I posed in my introduction: is there something qualitatively special about life? For most of the history of Western philosophy, life's special quality was thought to reflect some underlying purposeful nature—whether it be a disembodied ideal like the *telos*, or an inherent purposefulness like the *physis*. The philosophical issue that roils modern biology turns on whether this purposefulness is illusory or actually exists. The question can be resolved if there exists a scientifically credible theory of purpose. Materialists would say there is no such theory. Homeostasis, I argue, provides that theory. Although the principle of homeostasis was first offered by Bernard as a physiological aphorism, its principal virtue today is that it provides a conceptual framework for living systems that is independent of details of scale and mechanism. Let me begin by restating Bernard's dictum in the language of open-system thermodynamics: *homeostasis is a dynamic, persistent and specified state of thermodynamic disequilibrium that does work to resist perturbations to it.*

At its most general, the disequilibrium that characterizes a living system takes the form of a low-entropy stream of matter that continually degrades to equilibrium with the surroundings in accordance with the Second Law of Thermodynamics. For such a system to persist, work must be done to feed into the pool a stream of matter and to impose specified order on it, so as to offset the loss of matter and order to the higher entropy surroundings. Thus, a living system is marked by two flows of matter: a thermodynamically-favored flux (TFF) of matter up an entropy gradient (which liberates heat in the process), and a physiological flux (PF) of matter down an entropy gradient, which mobilizes free energy to do entropy-reducing work. In the context of cellular systems, the orderliness takes the specific form of a catalytic milieu that biases the flows of matter and energy in a way that sustains the specified disequilibrium. The orderliness can take other forms in other contexts, such as the organism, such as potential energy differences in heat content (temperature), solute concentration, or oxidation potential.

This thermodynamic reformulation of homeostasis goes beyond Bernard's original conception in some key ways. The first provides a fundamental contextualization of physiology: a living system implies some boundary that divides environments and that couples both the TFF and PF to the entropy gradient across it. The second extends physiology's reach. Where Bernard envisioned homeostasis as a property of environments within living boundaries (*milieu intérieur*), principles of mass and energy conservation demand that any ordering of matter inside an adaptive boundary inevitably impacts matter outside the boundary as well. In short, physiology (and by implication homeostasis) is both intensive and extensive. This is the phenomenon of "extended physiology," which makes both physiology and homeostasis scalable and nestable. The concept of extended physiology recasts somewhat the common idea that living systems are net entropy exporters, the usual implication being that living systems necessarily export entropy to their immediate surroundings. Extended physiology pushes the boundary for net entropy export outwardly, because homeostatic systems at any particular scale of organization can benefit by nesting themselves within new adaptive boundaries that can bring the formerly uncontrolled external environment under homeostatic control. The most obvious example of such nested homeostatic systems is the organism, with its multitudinous epithelial interfaces. The nesting can extend further, producing "extended organisms," of which social insect colonies provide the most dramatic examples, or up the scale to Clementsian ecosystems,[42] or even ultimately to the planetary scale, making the Universe, with its nearly infinite capacity to absorb it, the ultimate entropy sink.[43] Third, this reformulation of homeostasis provides a link to evolutionary process that Bernard (who was not an evolutionist) did not envision. Specifically, it posits a physiological definition of evolutionary fitness: temporal persistence of a living system. By this logic, living systems that are characterized by more robust homeostasis are more likely to persist in the face of unpredictable perturbations than systems that are not so robust. These I have termed "persis-tors," to distinguish them from the more conventional neo-Darwinian idea of replicators.[44] This leads to a fourth interesting connection between physiology and evolution: a persistor, because it biases future function similarly to the way genes do, can represent an alternate form of hereditary memory that can be subject to natural selection, just as memory that resides in sequence information in nucleic acids can be.

This reformulation of homeostasis justifies Biology's Second Law, for it provides what I argue is a sound theory for reintroducing purposefulness to biology. Let us consider an example. The problem of purposefulness arises strikingly in the problem of biological design: why are organisms constructed so well to perform their functions? Modern evolutionism has traditionally explained biological design as the consequence of selection of "good function genes." Design, which implies purposefulness and intentionality, is therefore illusory. Because the connection from gene to function is unspecified, though, this has led modern evolution into a problematic tautology: good function genes are those that specify good function. As I have described elsewhere, homeostasis provides a way out of the tautology, through the agency of what the computer scientist Cosma Shalizi has termed Bernard machines, agents that create adaptive boundaries that create and sequester new environments and impose homeostasis on them. This establishes a credible scientific paradigm for restoring biological design to the real phenomenon that it is. It also offers a credible response to the rejoinder that design implies a discredited intentionality, which by association discredits design as a real phenomenon. The logic of Bernard machines offers a rational model for intentionality being the inverse of cognition, which is also a form of homeostasis. Among the radical implications of this argument is that all living systems are imbued with frank (not illusory) intentionality.

What of evolution? Can homeostasis play a role there? Neo-Darwinism's commitment to selection of the gene as the sole driver of evolution has come under challenge recently, most prominently from developmental biology,[45] and somewhat more tentatively from the emerging field of epigenetics.[46] For the most part, these challenges have sought to constrain themselves within the mechanistic umbrella. I argue here that the most profound challenge to gene selectionism comes from the explicit intentionality that homeostasis brings to the evolutionary process. One can point, for example, to new modes of embryonic development, such as gastrulation, that lead to the evolution of entirely new body plans. One can very profitably cast developmental events like this as a problem of gene selection, by searching for and understanding the dynamics of the nucleotide sequences that influence embryogenesis. This approach does not ask what is arguably the more basic question: what is the benefit of, say, gastrulation itself? I have argued elsewhere,[47] that new types of embryonic development can be regarded as newly emerging systems of Bernard machines that produce entirely novel forms of existence. This puts homeostasis, not nucleotide sequence, as the principal driver of body plan evolution, an idea which now admits the explicit intentionality of homeostasis to the evolutionary process. Many interesting philosophical issues follow from this.

To the notion that homeostasis could be a significant driver of evolution, two common counter-arguments are commonly made.[48] The first is to grant that homeostasis, while certainly an important component of phenotype, can only arise from the selection of the genetic specifiers of it. The second counter-argument follows the logic of the Weismann barrier: physiology (homeostasis) is a property of transient "vehicles" while evolution, by definition, transcends the vehicle. The only credible means for transcending generations is the gene. Therefore, the only object of natural selection can be the gene. The first counter-argument subordinates homeostasis to the gene. The second treats homeostasis as irrelevant to evolution.

Both arguments rely upon a conception of the gene as an object, as a replicator, a notion that is growing increasingly at variance with new understanding of gene expression and its relationship to the cellular milieu in which it operates.[49] Arguably, sequence information in nucleic acids is better viewed as one component of a complex network of process, all components of which potentially can (and do) feed back on, and influence, the others. Each feedback loop complicates the mapping of nucleotide sequence onto phenotypic trait, even to the most immediate mapping onto RNA transcripts or translated proteins. Thus, a gene is more a process than an object, which makes it part of the broader physiological—homeostatic—system in which it resides. The ultimate logic of treating the gene in this way is actually to undercut the Central Dogma, because the gene is no longer the sole form of hereditary memory, nor is its function solely the repository of hereditary memory. In a homeostatic system, fitness accrues most strongly to living systems that can bias future function most reliably—persistence. DNA nucleotide sequence is one means of biasing future function, but it is embedded in, and is influenced by, a multitude of other means that living systems have of biasing the future. This could take the form of post-translational persistence of particular suites of enzymes (proteomes), to long-lived modifications of physical environments, as in structures built by social insects.[50] One can easily envision circumstances where nucleotide sequences are indeed the strongest biasers of future function, and that selection will involve these forms of hereditary memory. Homeostasis as Biology's Second Law means that it is now at least conceivable that other forms of hereditary memory can also significantly shape evolution.

This brings the discussion to the second question I posed: must not homeostasis necessarily be subsumed under the selection of the genes that specify it? This question is illustrated fruitfully by the recently promulgated idea of niche construction (NC) as a purportedly new (or at least neglected) evolutionary process.[51] The logic of NC springs from Richard Lewontin's famous conception of the "triple helix" (genome, organism, and environment), which asserts that modification of the environment by organisms is as important a feature of selection as effects of the environment upon the organism are.[52]

Niche Construction

Niche construction theory is fruitfully understood by tracing its intellectual lineage back to its roots, which starts with Sewall Wright's idea of "adaptive landscapes" and continues through G. Evelyn Hutchison's notion of the "ecological niche."[53] Wright's adaptive landscapes were a conceptual metaphor to resolve what he perceived as the principal shortcoming of Ronald Fisher's essentially atomistic approach to genetic natural selection: it was not sufficiently appreciative of biology, in particular adaptation.[54] By defining a theoretical "space" that plotted fitness of all possible genotypes in a population with respect to some function, "peaks" in fitness could be identified. Adaptation through natural selection was thus explained by lineages of organisms "climbing" these fitness peaks. Hutchinson's ecological niche idea was a refinement on this concept, with fitness (or some proxy of it) now being an explicit function of environmental conditions.[55] These conditions could be abiotic (as in a range of possible environmental temperatures) or biotic (shadiness, presence of predators, competing trophic guilds, etc...), but in any circumstance, there should be a particular suite of conditions where an organism functioned best. By combining all possible environmental parameters, one could identify a multi-dimensional space that localized those conditions that were most congenial to the species. This was the species' ecological niche, where organisms would most likely exist, and toward which lineages would naturally evolve. The niche concept has meanwhile expanded to include particular functional roles in ecosystems, such as a "predator niche" or even such things as a "flying predator niche," where they are now termed "adaptive states" localized in a larger "adaptive state space." Other variations on this theme can be found throughout the ecological and evolutionary literature, and they have seemingly impressive explanatory power. The adaptive state space concept, for example, has been employed to explain why wheeled animals have never evolved:[56] the adaptive state of wheeled animals may exist theoretically, but cannot be accessed because there is no conceivable route through adaptive state space for any imagined lineage to navigate there.[57]

In its classic conception, the ecological niche is reconciled with gene selectionism by treating the niche as an environmental "selective filter": the offspring of each generation represents a broad set of genetically determined tests of conformity to the ecological niche. Those individuals that conform well survive, reproduce, and are selected for, and those that do not are selected against. Niche construction departs from this rather gloomy adapt-or-die dynamic by allowing organisms to manipulate environments to suit themselves, essentially constructing their own ecological niches, and so, in some sense controlling the selective milieus they inhabit.[58]

Commensurate with the adaptive space, niche construction theory was formulated as a means to reconcile neo-Darwinian gene selection with the obvious ecological fact that organisms modify their environments. This gene-selectionist logic is illustrated by a simple scenario.[59] Suppose there is an environmental resource, RE, that is essential to the functioning of an organism, O. If O can modify an environment, E, so that RE becomes more readily available to O, this will increase the fitness of lineages that do so compared to lineages that do not, and hence select for the genes G[O] that specify the ability to construct environments.

Niche construction theory, like the theory of the niche itself, is a very powerful metaphor for explaining a broad suite of evolutionary phenomena. For example, because NC leaves unspecified the precise boundary between environment and organism, it brings scalability to gene selection: it could easily treat the evolution of the organism, for example, with its multitudinous internal environments, as constructed niches for the differentiated tissues that build them.[60] Mimicking Darwinism's own agnosticism on the matter, NC is not tied too closely to particular mechanistic details of heredity, which allows it to accommodate both cultural as well as genetic evolution,[61] and to comfortably incorporate persistently modified environments as a form of "ecological memory" that could influence selection.

In the end, though, NC cannot explain adaptation except in the same way that adaptive landscapes and the ecological niche do—through a metaphor that essentially sneaks Platonic idealism in through the back door. Consider the logic behind the argument that flying predators exist because there exists a "flying predator niche" in an adaptive state space that will be filled by any genome that happens to sit near enough to evolve into it. In this argument, the niche is the fitness equivalent of the Linnaean species archetype: a disembodied ideal state toward which real living things gravitate. This implicit essentialist legacy survives undiminished in current niche construction theory, and it can only explain adaptation by introducing a surreptitious purposefulness. It is my claim that homeostasis provides a more constructive way to bring purposefulness in than the implicit Platonic idealism of niche construction theory. This does not diminish the philosophical challenge, of course, for it introduces the notion that flying predators, for example, exist because a lineage of non-flying predators in some sense wanted to fly and could make good on the desire.

Conclusion

I conclude by returning to my original question: where "beyond mechanism" should biology go? The question is perhaps best rephrased slightly: how should we look at mechanism? This is because the history of biology is only partly a history of discovery, it is also a history of perception. For most of our intellectual history, we have viewed the phenomenon of life through a lens of purposefulness. For a time, we swapped that lens for the polaroid filter of mechanism, which allowed us to see many things lurking beneath the purposeful glare that had previously blinded us. We have now explored those mechanistic depths to the point where we have forgotten there was a glare at all, never mind a reason for it. The challenge for modern biology, then, is to find a new set of lenses that now allow us to see again what is life's most fundamental quality: its frank purposefulness, even its striving and evolutionary desire. We have no idea currently what those lenses might be. I propose that homeostasis, as Biology's Second Law, may fit the bill.

Acknowledgments

This essay is an abstract of a forthcoming book tentatively titled *Biology's Second Law: Evolution, Purpose and Desire*. The John C. Templeton Foundation has generously supported the writing of this work, including a stint as a Visiting Scholar at Cambridge University under the generous hospitality of Prof. Simon Conway Morris and St. John's College.

Notes

1. Robert Shapiro, *Origins: A Skeptics Guide to the Creation of Life on Earth* (London: Heinemann, 1986); Iris Fry, *The Emergence of Life on Earth. A Historical and Scientific Overview* (New Brunswick, NJ: Rutgers University Press, 2000); Eric Smith, "Before Darwin: How the Earth Went from Lifeless to Life," *The Scientist* 22, no. 6 (2008): 32-41.

2. Michael R. Rose and Laurence D. Mueller, *Evolution and Ecology of the Organism* (Upper Saddle River, NJ: Pearson Prentice Hall, 2006).

3. Stephen J. Gould, "Nonoverlapping Magisteria," *Natural History* 106 (1997): 16-22.

4. Sebastian Normandin, "Claude Bernard and an Introduction to the Study of Experimental Medicine: 'Physical Vitalism,' Dialectic, and Epistemology," *Journal of the History of Medicine and Allied Sciences* 62, no. 4 (2007): 495-528.

5. Peter J. Bowler, *The Eclipse of Darwinism. Anti-Darwinian Evolution Theories in the Decades around 1900* (Baltimore, MD: The Johns Hopkins University Press, 1983).

6. Charles G. Gross, "Claude Bernard and the Constancy of the Internal Environment," *The Neuroscientist* 4, no. 5 (2008): 380-385.

7. Harry Bloch, "François Magendie, Claude Bernard, and the Inter-relationship of Science, History and Philosophy," *Southern Medical Journal* 82, no. 10 (1989): 1259-1261.

8. Normandin, "Study of Experimental Medicine."

9. Charles T. Wolfe, "Introduction: Vitalism without Metaphysics? Medical Vitalism in the Enlightenment," *Science in Context* 21, no. 4 (2008): 461-463.

10. Dana Ullman, "The Philosophical and Historical Roots of the Holistic Approaches to Health: A Review of the First Three Volumes of 'Divided Legacy' by Harris L. Coulter, Ph.D." *Homeopathic Educational Services*, http://www.homeopathic.com/Articles/Introduction_to_Homeopathy/A_review_of_the_first_three_volumes_of_Divid.html; Tobias Cheung, "Regulating Agents, Functional Interactions, and Stimulus-Reaction-Schemes: The Concept of 'Organism' in the Organic System Theories of Stahl, Bordeu, and Barthez," *Science in Context* 21, no. 4 (2008): 495-519.

11. Cheung, "Regulating Agents."

12. John O. Reiss, *Not by Design: Retiring Darwin's Watchmaker* (Berkeley: University of California Press, 2009).

13. Richard W. Burkhardt, *The Spirit of System. Lamarck and Evolutionary Biology* (Cambridge, MA: Harvard University Press, 1995).

14. Ernst Mayr, *The Growth of Biological Thought: Diversity, Evolution, and Inheritance* (Cambridge, MA: Belknap/Harvard University Press, 1982).

15. Normandin, "Study of Experimental Medicine."

16. Michael Ruse, *Darwin and Design. Does Evolution Have a Purpose?* (Cambridge, MA: Harvard University Press, 2003); Timothy Shanahan, *The Evolution of Darwinism: Selection, Adaptation and Progress in Evolution* (Cambridge, MA: Cambridge University Press, 2004); Michael Ruse, "Purpose in a Darwinian World," in *The Deep Structure of Biology: Is Convergence Sufficiently Ubiquitous to Give a Directional Signal?* (West Consohocken, PA: Templeton Foundation Press, 2008), 178-194.

17. Charles Darwin, *Charles Darwin: His Life Told in an Autobiographical Chapter and in a Selected Series of His Published Letters* (New York: D Appleton and Company, 1893).

18. Peter Bowler, *Fossils and Progress: Paleontology and the Idea of Progressive Evolution in the Nineteenth Century* (Sagamore Beach, MA: Science History Publications, 1976); Bowler, *The Eclipse of Darwinism*.

19. Evelyn F. Keller, *The Century of the Gene* (Cambridge, MA: Harvard University Press, 2000).

20. Robert E. Kohler, *From Medical Chemistry to Biochemistry. The Making of a Biomedical Discipline* (Cambridge, Cambridge University Press, 1982).

21. Zhores A. Medvedev, *The Rise and Fall of T.D. Lysenko* (New York: Columbia University Press, 1969); Cynthia E. Russett, *Darwin in America. The Intellectual Response 1865-1912* (San Francisco: W H Freeman and Company, 1976); Marek Kohn, *A Reason for Everything. Natural Selection and the English Imagination* (London: Faber and Faber, 2004); Piers J. Hale, "Of Mice and Men: Evolution and the Socialist Utopia. William Morris, H.G. Wells, and George Bernard Shaw," *Journal of the History of Biology* 43 (2010), 17-66.

22. Alexander Petrunkevitch, "August Weismann," *Systematic Zoology* 13, no. 4 (1964): 215-219.

23. Francis Crick, "Central Dogma of Molecular Biology," *Nature* 227 (1970): 561-563; Francis Crick, "On Protein Synthesis," *Symposia of the Society for Experimental Biology* 12 (1958): 138-163.

24. Bowler, *The Eclipse of Darwinism*.

25. Edward D. Cope, *The Primary Factors of Organic Evolution* (Chicago, IL: The Open Court Publishing Company, 1896).

26. Mordecai Gabriel and Seymour Fogel, eds. *Great Experiments in Biology* (Englewood Cliffs, New Jersey: Prentice-Hall, 1955).

27. Scott F. Gilbert and David Epel, *Ecological Developmental Biology: Integrating Epigenetics, Medicine, and Evolution* (Sunderland, MA: Sinauer Associates, 2008).

28. Helen V. Crouse, "The Controlling Element in Sex Chromosome Behavior in Sciara," *Genetics* 45 (1960): 1429-1443; Chikara Furusawa and Kunihiko Kaneko, "Origin of Multicellular Organisms as an Inevitable Consequence of Dynamical Systems," *Anatomical Record* 268, no. 3 (2002): 327-342; Elena de la Casa-Esperón and Carmen Sapienza, "Natural Selection and the Evolution of Genome Imprinting," *Annual Review of Genetics* 37, no. 1 (2003): 349-370.

29. Gabriel and Fogel, *Great Experiments in Biology*.

30. James E. Strick, "Spontaneous Generation," in *Encyclopedia of Microbiology, 2nd Edition Vol. 4*. (New York: Academic Press, 2000), 364-376; James E. Strick, *Sparks of Life: Darwinism and the Victorian Debates over Spontaneous Generation* (Cambridge, MA: Harvard University Press, 2000).

31. Mayr, *The Growth of Biological Thought*.

32. Strick, *Sparks of Life*.

33. John T. Edsall and David Bearman, "Historical Records of Scientific Activity: The Survey of Sources for the History of Biochemistry and Molecular Biology," *Proceedings of the American Philosophical Society* 123, no. 5 (1979): 279-292.

34. Crick, "Central Dogma."

35. Richard Dawkins, "Extended Phenotype—But Not Too Extended: A Reply to Laland, Turner, and Jablonka," *Biology and Philosophy* 19, no. 3 (2004): 377-396.

36. Shapiro, *Origins: A Skeptics Guide;* Fry, *The Emergence of Life on Earth*.

37. Anonymous, "Meanings of 'Life'," *Nature* 447 (2007): 1031-1032.

38. Alexander Cairns-Smith and Hyman Hartman, eds., *Clay Minerals and the Origin of Life* (Cambridge, MA: Cambridge University Press, 1986); Smith, "Before Darwin."

39. Fred Hoyle and N. Chandra Wickramasinghe, "The Case for Life as a Cosmic Phenomenon," *Nature* 332 (1986): 509-511.

40. J. Scott Turner, *The Extended Organism: The Physiology of Animal-Built Structures* (Cambridge, MA: Harvard University Press, 2000).

41. J. Scott Turner, *The Tinkerer's Accomplice: How Design Emerges from Life Itself* (Cambridge, MA: Harvard University Press, 2007).

42. Frank B. Golley, *A History of the Ecosystem Concept in Ecology* (New Haven, CT: Yale University Press, 1993).

43. James Lovelock, *Gaia. A New Look at Life on Earth* (Oxford: Oxford University Press, 1987).

44. Turner, *The Tinkerer's Accomplice*.

45. Gilbert and Epel, *Ecological Developmental Biology*.

46. Eva Jablonka and Marion J. Lamb, "Genic Neo-Darwinism: Is It the Whole Story?" *Journal of Evolutionary Biology* 11 (1998): 243-260; Eva Jablonka and Marion Lamb, *Evolution in Four Dimensions: Genetic, Epigenetic, Behavioral and Symbolic Variation in the History of Life* (Cambridge, MA: MIT Press, 2005).

47. Turner, *The Tinkerer's Accomplice*.

48. Dawkins, "Extended Phenotype—But Not Too Extended."

49. Jablonka and Lamb, *Evolution in Four Dimensions*.

50. Turner, "Extended Phenotypes and Extended Organisms," *Biology and Philosophy* 19, no. 3 (2004): 327-352.

51. F. John Odling-Smee, Kevin N. Laland and Marcus W. Feldman, *Niche Construction: The Neglected Process in Evolution* (Princeton, NJ: Princeton University Press, 2003).
52. Richard Lewontin, "Gene, Organism, and Environment," in *Evolution from Molecules to Men*, ed. D. S. Bendall (New York: Cambridge University Press, 1983), 594; Richard Lewontin, *The Triple Helix: Gene, Organism and Environment* (Cambridge, MA: Harvard University Press, 2000).
53. George E. Hutchinson, "Concluding Remarks," in Cold Spring Harbor Symposia on Quantitative Biology, *Population Studies: Animal Ecology and Demography* 22 (1957): 415-427.
54. William B. Provine, "The R. A. Fisher-Sewall Wright Controversy," in *The Founders of Evolutionary Genetics. A Centenary Reappraisal*, ed. Sahotra Sarkar (Dordrecht: Kluwer Academic Publishers, 1992), 201-229.
55. Odling-Smee, Laland, and Feldman, *Niche Construction*.
56. Stephen J. Gould, "Kingdoms without Wheels," *Natural History* 90, no. 4 (1981): 42-48.
57. For an interesting counterargument, see Michael LaBarbera, "Why the Wheels Won't Go," *American Naturalist* 121 (1982): 395-408.
58. Odling-Smee, Laland, and Feldman, *Niche Construction: The Negeleted Process in Evolution*; J. Scott Turner, "Homeostasis, Complexity and the Problem of Biological Design," in *Explorations in Complexity Thinking: Pre-Proceedings of the 3rd International Workshop on Complexity and Philosophy*, ed. K. A. Richardson and P. Cilliers (Mansfield, MA: ISCE Publishing, 2007), 131-147.
59. Kevin N. Laland, F. John Odling-Smee, and Marcus W. Feldman, "Evolutionary Consequences of Niche Construction and Their Implications for Ecology," *Proceedings of the National Academy of Sciences (USA)* 96 (1999): 10242-10247.
60. Kevin N. Laland, F. John Odling-Smee, and Scott F. Gilbert, "Evo-Devo and Niche Construction: Building Bridges," *Journal of Experimental Zoology* 310, no. 7 (2008): 549-566.
61. Kevin N. Laland, F. John Odling-Smee, and Marcus W. Feldman, "Niche Construction, Biological Evolution and Cultural Change," *Behavioral and Brain Sciences* 23 (2000): 131-175; Kevin N. Laland, F. John Odling-Smee, and Sean Myles, "How Culture Shaped the Human Genome," *Nature Reviews Genetics* 11 (2010): 137-148.

Bibliography

Anonymous. "Meanings of 'Life'." *Nature* 447 (2007): 1031-1032.
Bloch, Harry. "François Magendie, Claude Bernard, and the Inter-relationship of Science, History and Philosophy." *Southern Medical Journal* 82, no. 10 (1989): 1259-1261.
Bowler, Peter J. *Fossils and Progress: Paleontology and the Idea of Progressive Evolution in the Nineteenth Century*. Sagamore Beach, MA: Science History Publications, 1976.
———. *The Eclipse of Darwinism. Anti-Darwinian Evolution Theories in the Decades around 1900*. Baltimore, MD: The Johns Hopkins University Press, 1983.
Burkhardt, Richard W. J. *The Spirit of System. Lamarck and Evolutionary Biology*. Cambridge, MA: Harvard University Press, 1995.

Buss, Leo W. *The Evolution of Individuality*. Princeton, NJ: Princeton University Press, 1987.
Cairns-Smith, Alexander, and Hyman Hartman, eds. *Clay Minerals and the Origin of Life*. Cambridge, MA: Cambridge University Press, 1986.
Cheung, Tobias. "Regulating Agents, Functional Interactions, and Stimulus-Reaction-Schemes: The Concept of "Organism" in the Organic System Theories of Stahl, Bordeu, and Barthez." *Science in Context* 21, no. 4 (2008): 495-519.
Cope, Edward D. *The Primary Factors of Organic Evolution*. Chicago: The Open Court Publishing Company, 1896.
Crick, Francis. "Central Dogma of Molecular Biology." *Nature* 227 (1970): 561-563.
———. "On Protein Synthesis." *Symposia of the Society for Experimental Biology* 12 (1958): 138-163.
Crouse, Helen V. "The Controlling Element in Sex Chromosome Behavior in Sciara." *Genetics* 45 (1960): 1429-1443.
Darwin, Charles. *Charles Darwin: His Life Told in an Autobiographical Chapter and in a Selected Series of His Published Letters*, edited by Francis Darwin. New York: D Appleton and Company, 1893.
Dawkins, Richard. "Extended Phenotype—But Not Too Extended: A Reply to Laland, Turner, and Jablonka." *Biology and Philosophy* 19, no. 3 (2004): 377-396.
de la Casa-Esperón, Elena, and Carmen Sapienza. "Natural Selection and the Evolution of Genome Imprinting." *Annual Review of Genetics* 37, no. 1 (2003): 349-370.
Edsall, John T., and David Bearman. "Historical Records of Scientific Activity: The Survey of Sources for the History of Biochemistry and Molecular Biology." *Proceedings of the American Philosophical Society* 123, no. 5 (1979): 279-292.
Fry, Iris. *The Emergence of Life on Earth. A Historical and Scientific Overview*. New Brunswick, NJ: Rutgers University Press, 2000.
Furusawa, Chikara, and Kunihiko Kaneko. "Origin of Multicellular Organisms as an Inevitable Consequence of Dynamical Systems." *Anatomical Record* 268, no. 3 (2002): 327-342.
Gabriel, Mordecai L., and Seymour Fogel, eds. *Great Experiments in Biology*. Englewood Cliffs, NJ: Prentice-Hall, 1955.
Gilbert, Scott F., and David Epel. *Ecological Developmental Biology: Integrating Epigenetics, Medicine, and Evolution*. Sunderland, MA: Sinauer Associates, 2008.
Golley, Frank B. *A History of the Ecosystem Concept in Ecology*. New Haven, CT: Yale University Press, 1993.
Gould, Stephen J. "Kingdoms without Wheels." *Natural History* 90, no. 4 (1981): 42-48.
———. "Nonoverlapping Magisteria." *Natural History* 106 (1997): 16-22.
Gross, Charles G. "Claude Bernard and the Constancy of the Internal Environment." *The Neuroscientist* 4, no. 5 (2008): 380-385.
Hale, Piers J. "Of Mice and Men: Evolution and the Socialist Utopia. William Morris, H. G. Wells, and George Bernard Shaw." *Journal of the History of Biology* 43 (2010): 17-66.
Hoyle, Fred, and N. Chandra Wickramasinghe. "The Case for Life as a Cosmic Phenomenon." *Nature* 332 (1986): 509-511.
Hutchinson, George E. "Concluding Remarks." Cold Spring Harbor Symposia on Quantitative Biology, *Population Studies: Animal Ecology and Demography* 22 (1957): 415-427.
Jablonka, Eva, and Marion J. Lamb. "Genic Neo-Darwinism: Is It the Whole Story?" *Journal of Evolutionary Biology* 11 (1998): 243-260.

———. *Evolution in Four Dimensions: Genetic, Epigenetic, Behavioral and Symbolic Variation in the History of Life*. Cambridge, MA: MIT Press, 2005.

Keller, Evelyn F. *The Century of the Gene*. Cambridge, MA: Harvard University Press, 2000.

Kohler, Robert E. *From Medical Chemistry to Biochemistry. The Making of a Biomedical Discipline*. Cambridge, MA: Cambridge University Press, 1982.

Kohn, Marek. *A Reason for Everything. Natural Selection and the English Imagination*. London: Faber and Faber, 2004.

LaBarbera, Michael. "Why the Wheels Won't Go." *American Naturalist* 121 (1982): 395-408.

Laland, Kevin N., F. John Odling-Smee, and Marcus W. Feldman. "Evolutionary Consequences of Niche Construction and Their Implications for Ecology." *Proceedings of the National Academy of Sciences (USA)* 96 (1999): 10242-10247.

———. "Niche Construction, Biological Evolution and Cultural Change." *Behavioral and Brain Sciences* 23 (2000): 131-175.

Laland, Kevin N., F. John Odling-Smee, and Scott F. Gilbert. "Evo-Devo and Niche Construction: Building Bridges." *Journal of Experimental Zoology* 310, no. 7 (2008): 549-566.

Laland, Kevin N., F John Odling-Smee, and Sean Myles. "How Culture Shaped the Human Genome." *Nature Reviews Genetics* 11 (2010): 137-148.

Lewontin, Richard. "Gene, Organism, and Environment." In *Evolution from Molecules to Men*, edited by D. S. Bendall. New York: Cambridge University Press, 1983.

———. *The Triple Helix: Gene, Organism and Environment*. Cambridge, MA: Harvard University Press, 2000.

Lovelock, James E. *Gaia. A New Look at Life on Earth*. Oxford: Oxford University Press, 1987.

Mayr, Ernst. *The Growth of Biological Thought: Diversity, Evolution, and Inheritance*. Cambridge, MA: Belknap/Harvard University Press, 1982.

Medvedev, Zhores A. *The Rise and Fall of T.D. Lysenko*. New York: Columbia University Press, 1969.

Normandin, Sebastian. "Claude Bernard and an Introduction to the Study of Experimental Medicine: 'Physical Vitalism,' Dialectic, and Epistemology." *Journal of the History of Medicine and Allied Sciences* 62, no. 4 (2007): 495-528.

Odling-Smee, F. John, Kevin N. Laland, and Marcus W. Feldman. *Niche Construction: The Neglected Process in Evolution*. Princeton, NJ: Princeton University Press, 2003.

Petrunkevitch, Alexander. "August Weismann." *Systematic Zoology* 13, no. 4 (1964): 215-219.

Provine, William B. "The R. A. Fisher-Sewall Wright Controversy." In *The Founders of Evolutionary Genetics. A Centenary Reappraisal*, edited by Sahotra Sarkar 201-229. Dordrecht: Kluwer Academic Publishers, 1992.

Reiss, John O. *Not by Design: Retiring Darwin's Watchmaker*. Berkeley, CA: University of California Press, 2009.

Rose, Michael R., and Laurence D. Mueller. *Evolution and Ecology of the Organism*. Upper Saddle River, NJ: Pearson Prentice Hall, 2006.

Ruse, Michael. *Darwin and Design: Does Evolution Have a Purpose?* Cambridge, MA: Harvard University Press, 2003.

———. "Purpose in a Darwinian World." In *The Deep Structure of Biology: Is Convergence Sufficiently Ubiquitous to Give a Directional Signal?*, edited by Simon Conway Morris, 178-194. West Consohocken, PA: Templeton Foundation Press, 2008.

Russett, Cynthia E. *Darwin in America. The Intellectual Response 1865-1912*. San Francisco: W H Freeman and Company, 1976.

Shanahan, Timothy. *The Evolution of Darwinism: Selection, Adaptation and Progress in Evolution*. Cambridge, MA: Cambridge University Press, 2004.

Shapiro, Robert. *Origins: A Skeptics Guide to the Creation of Life on Earth*. London: Heinemann, 1986.

Smith, Eric. "Before Darwin: How the Earth Went from Lifeless to Life." *The Scientist* 22, no. 6 (2008): 32-41.

Strick, James E. "Spontaneous Generation." In *Encyclopedia of Microbiology, 2nd Edition Vol. 4*, edited by Martin Alexander et al., 364-376. New York: Academic Press, 2000.

———. *Sparks of Life: Darwinism and the Victorian Debates over Spontaneous Generation*. Cambridge, MA: Harvard University Press, 2000.

Turner, J. Scott. *The Extended Organism: The Physiology of Animal-Built Structures*. Cambridge, MA: Harvard University Press, 2000.

———. "Extended Phenotypes and Extended Organisms." *Biology and Philosophy* 19, no. 3 (2004): 327-352.

———. "Homeostasis, Complexity and the Problem of Biological Design." In *Explorations in Complexity Thinking: Pre-Proceedings of the 3rd International Workshop on Complexity and Philosophy*, edited by K. A. Richardson and P. Cilliers, 131-147. Mansfield, MA: ISCE Publishing, 2007.

———. *The Tinkerer's Accomplice: How Design Emerges from Life Itself*. Cambridge, MA: Harvard University Press, 2007.

Ullman, Dana. "The Philosophical and Historical Roots of the Holistic Approaches to Health: A Review of the First Three Volumes of 'Divided Legacy' by Harris L. Coulter, Ph.D." *Homeopathic Educational Services*, http://www.homeopathic.com/Articles/Introduction_to_Homeopathy/A_review_of_t he_first_three_volumes_of_Divid.html.

Wolfe, Charles T. "Introduction: Vitalism without Metaphysics? Medical Vitalism in the Enlightenment." *Science in Context* 21, no. 4 (2008): 461-463.

Chapter Nine
"Wind at Life's Back"—Toward a Naturalistic, Whiteheadian Teleology: Symbiogenesis and the Second Law

Dorion Sagan and Lynn Margulis

Without evidence, examples, and empirical checking, science can devolve into hyper-conceptualization, mere theory loosely or unconnected to reality. This chapter briefly examines two areas in which abundant data supports a new theoretical understanding of life as a naturally purposeful phenomenon: thermodynamics and natural selection.

In the first case, the inherently telic tendency of energy to spread informs life, whose living systems measurably produce more entropy, that is, reduce more gradients and delocalize more concentrated sources of energy than would be the case without them, informs life at all levels.[1] Before making abstruse abstractions, inventing new terminology, or announcing new research programs, this basic fact has to be realized. Metabolizing, growing, reproducing organisms are natural, energy-spreading, gradient-perceiving, and gradient-reducing systems. The essence of the phenomenon described by the Second Law as an increase in entropy is not theoretical, but the simple, easily verified observation that energy, if not hindered, spreads.[2] This does not necessarily entail an increase in disorder, comparatively meaningless at the molecular level because of the huge number of states of any collection of atoms above absolute zero, since naturally complex nonliving and living systems organize continuously spread energy into their surroundings. The association of entropy with "disorder" has been traced to a common language summary by Ludwig Boltzmann in his *Lecture on Gas Theory* after four hundred pages of dense mathematics.[3] Entropy, a ratio and statistical measure, should not be reified, but recognized as a measure of energy spread; like an odometer that does not drive a car, entropy does not drive anything, but reflects a naturally telic process in which energy moves from being concentrated to spread out in space. This applies to sound waves emitted from the cone of a loudspeaker, to hot coffee equilibrating with the temperature of a surrounding room, and to life as cells, organisms, ecosystems, and the Earth System. And while it is true that life uses energy to build up sturdy macromolecules, and that it has means of self-repair, it is crucial to notice that the cumulative effect of such structures and subsystems is not to impede but to exacerbate the spread of energy. To take just one example, most of solar energy used by plants goes not into growth but into evapotranspiration, in which latent heat is transferred to raindrops and released in the atmosphere. Experiments with low-flying airplanes show that old growth forests are cooler than twenty-five-year-old plantations and that both are cooler than grasslands, which are cooler than deserts; thermal satellite measurements show that Borneo and the Amazon jungle are as cool in the height of summer as Siberia is in mid-winter. These data at first might be interpreted as suggesting that life has reversed the spread of energy predicted by the Second Law by trapping energy in the molecules of life's growing organization, and thus preventing heat's spread. In fact, there is a more compelling explanation: given the temperature difference between the hot sun and cool space, the coolness of ecosystems represents a transport of heat—a spread of energy—further out. The situation is similar to a refrigerator, which stays cool but produces excess heat around it to do so. The view we get, and it is as basic as it is important, is that life is a partner process in the phenomenon described by the Second Law. At the same time, the tendency for energy if not hindered to spread (a simple statement that encapsulates the essence of the observations described by the Second Law of thermodynamics) does *not* mean that entropy is maximized: immediate maximization of entropy or energy spread would entail, for example, you and this book bursting into flame. But that this does *not* happen allows the naturally telic energetic process to continue, allowing for more gradients to be reduced and energy to be spread *in the long run.* Organisms have low specific entropy production, but since they are open, producing waste and entropy as heat around them, as systems they are completely complicit with the Second Law, vectorizing and accelerating its predicted effects. The bottom line here, based on data, is that life can be regarded as a manifestation of the energy spread, telically tending toward equilibrium, predicted by the Second Law.

In the second case, data and observation increasingly show the importance of symbiogenetics—the merging of organisms, metabolism, and genomes—in providing the raw material for natural selection. Indeed, these two themes are connected—thermodynamics is a kind of "wind at the back" of evolution by the natural selection of symbiogenesis: economies of scale, granted by successful symbiotic (and social) mergers, become more effective means of stably spreading energy in conformity with thermodynamics' Second Law.

Whitehead identified science as a combination of the lucid Greek genius for bold speculation and experimentation. He noticed, nontrivially, an anti-intellectual strain in science: it is the continuous checking against nature, the emphasis on the empirical, the evidentiary, and the experimental, that keep scia historically dependent thermodynamic, rather than mathematically idealized and reversible, dynamic pathway of development. This broader metabolic understanding of living matter—including the formal perspectival shifts and practices of science and technology—as energetic process provides us with a richer perspective and epistemological backdrop from which to understand the evolutionary process. Against this backdrop, moreover, empirical observations and analyses lend credence to a view of evolution as driven more by thermodynamically directed symbiogenesis than random mutation.

Accumulation of Random Variations or Symbiogenesis?

The term "symbiosis," coined by Heinrich Anton de Bary in 1879, refers to the living together of differently named organisms. We can be more specific and define symbiosis as a physical alliance between two organisms that involves touching for the greater part of the lives of the involved organisms. The term is often used loosely to refer to fleeting encounters and alliances, but the truly generative symbiotic interactions in nature tend to be endosymbiotic—the living inside ("endo") of one organism by another. Symbiogenesis refers to the evolution of living novelties by this process: new behaviors, organelles, tissues, organisms, species, and higher taxa. Symbiogenesis emerges from prolonged contact and eventual metabolic and genetic merger.

Symbiogenetics is our coinage for the study of such mixed life-forms. Human beings, like all animals, plants, fungi, and non-bacterial microbes (protoctists), evolved by symbiogenesis. Genetic analyses confirm with extremely high probability that our cells have combined bacterial genomes, including those of respiring bacteria (the ancestors to mitochondria) and sulfide gas-producing fermenting bacteria (archaea). Mitochondrial ancestors, ingested but never completely digested by swimming protists, became trapped. Three protoctist lineages evolved to become single-celled and colonial organisms in the Kingdom Protoctista, which today includes amoebae, ciliates, and other microbes exclusive of the bacteria. After early nucleated cells ingested but failed to digest cyanobacteria, algae evolved. Some macroscopic forms, such as seaweeds, which are not true plants in part because they do not form embryos, evolved from smaller algae.[25] Colonial protoctists evolved to become animals, seaweeds (some huge), and fungi, including mushrooms. Lineages that had added, through symbiogenetic acquisition, photosynthetic bacteria to the mix, evolved to become brown, red, and green algae, the latter of which evolved to become members of the plant kingdom. Heterogenomic mixtures, symbiogenetic acquisition among bacteria, laid the groundwork for all familiar life-forms.

Although the symbiotic origin of the cells that evolved into plants, animals, and fungi is now an accepted fact taught in biology texts, it was initially rejected in western Europe and the United States. Today biologists accept that symbiogenesis was important in early evolution because gene sequences in mitochondria and chloroplasts are so similar to those in certain free-living bacteria (oxygen-breathing and cyanobacteria, respectively). This modern acceptance follows much persuasive early work by Russian and other researchers, currently becoming more available in the West.[26] The earlier researchers, working before the exciting discovery of DNA's role in replication, argued not on the basis of genetic sequence likenesses, but on the grounds of morphological similarity, cytoplasmic inheritance, and the bacteria-like reproductive behavior of organelles such as mitochondria and chloroplasts.

Despite the hard-won recognition of symbiogenesis as the key process that led to the animal-style cell with gene-containing mitochondria (inherited from bacteria) in the cytoplasm outside the nucleus, symbiogenesis remains significantly underappreciated as an evolutionary mode. This underacceptance or marginalization, however, reflects not lack of evidence for symbiogenesis but rather institutional and academic scientific practices. The prevailing "thought-collective" of mainstream genocentric biology[27] flatly contradicts the ecological approach of symbiogenesis. In contrast to belief, evidence suggests that a saltational process, symbiogenesis, is considerably more important than gradual mutation in driving evolutionary change. But the symbiogenesis we attempt to present here in its thermodynamic context does not mesh well with neo-Darwinian assumptions and the general picture that gene frequencies in populations account for evolutionary change. When zoologist Richard Dawkins was given several specific examples of symbiogenesis in Oxford, England, for example, rather than accepting them as rich empirical evidence for evolution, Dawkins said he was not interested in hearing any other "anecdotes."[28] Considering that this was at a celebration for 150 years of Darwinism, one would think that he would be interested in such real evidence of evolutionary change. But the tenaciousness of the Fleckian "thought-style" often trumps the spirit of scientific inquiry; as Kuhn, who read but did not cite Fleck, argues, scientific revolutions do not necessarily embrace evidence that contradicts or does not fit into established theories; more often theories change suddenly, through a process of schism or even mortal attrition of the leading advocates of a prevailing thought-style.[29]

Dawkins, although he dismissed symbiogenetic examples as anecdotes, was unable, when requested, to provide a single example of speciation by gradual accumulation of mutations. In addition, the videorecorded exchange was moderated by a member of Dawkins' thesis committee, physiologist Denis Noble, who has long been critical of genetic essentialism.[30] Thus a picture emerges of the prevailing theory that has little direct evidence to support its main contention that evolution proceeds by gradual accumulation of genetic mutations, and a less popular theory which provides numerous examples of a real mode of evolutionary speciation, one which has already been admitted by neo-Darwinism to account for crucial early major evolutionary transitions, but which is still dismissed as marginal.

Alive Inside: Interspecies Encounters and Merged Beings

Examples of symbiogenesis as an evolutionary mode are not confined to bacteria presumably eating but not digesting or, alternatively, infecting but not killing one another. The thermodynamic-metabolic systems that we call organisms continue to exchange matter, energy, and information. Occasionally in the act of eating or interacting, they acquire whole genomes that fatefully alter their identity and destiny. Sometimes new metabolic skills, resistance to extreme conditions, expansions of behavioral repertoires, and other abilities are acquired, making a "new and improved" organism. Moreover, these materially significant exchanges are not confined to bacteria infecting and eating and then symbiogenetically merging with one another billions of years ago in the primeval muds. Rather, they happen and have been observed to happen, including in laboratory settings. There is extensive evidence for symbiogenesis not only among modern bacteria, protoctists, and animals,[31] but also between different phyla of animal species, leading to metamorphosing marine invertebrates.[32]

Various forms of bacteria in the genomes of species of insects have been shown to alter their gender, and to make them incapable of interbreeding, thus speciating. Bacterial genomic acquisition by amoebae has been seen and shown in the laboratory to induce speciation of *Amoeba proteus*. The acquisition of ambient photosynthetic genomes of cyanobacteria and algae is not just theoretically postulated, but actually observed in the laboratory to induce speciation. Species of animals, for example, certain snails and flatworms, are virtually identical except for their possession of green symbionts. The presence of algae in tissues of these animals allows them to bask in sunlight and live without eating. Photosynthetic snails include *Plachobranchus sp.*, *Tridacna*, giant clams, and marine worms (e.g., *Convoluta sp.*). These green variant species clearly owe their existence to the symbiogenetic acquisition of photosynthetic symbionts. Such "green animals" derive a clear benefit (not having to eat) from the presence of their internal symbionts, the plastids of algae or whole cyanobacteria, which in some cases for example, *Convoluta roscoffensis*, allow them to go through their adult lives with mouths closed. The photosynthetic and nonphotosynthetic worm species *Convoluta convoluta* and *Convoluta roscoffensis*, living side by side on the same Channel islands allows inference of the cause of the speciation event: the addition of photosynthesis to animal cells relieve the animals of need to seek food.

Another whole class of productive interspecies beings is represented by luminous bacteria, such as *Photobacter sp* or *Vibrio fischeri*. These glowing symbionts have merged with marine animals. They are found as "light organs" in flashlight fishes and in bobtail squid. The females of some deep-sea anglerfishes use their bacteria-appropriated lights as lures to catch other fish.

John Werren from the University of Rochester in New York has shown that nestled within the chromosomes of some parasitic wasps are genes from *Wolbachia*, a genus of bacterium. Many different types of insects undergo gender change due to the presence of these bacteria. Antibiotics can make separate species of jewel wasps interbreed again.

While we do not doubt that low rate of accumulation of random mutations is a possible means of speciation, such speciation has not been observed in plants or animals and thus should not be unquestioningly taught as the main means of speciation. Ironically, of arguably the two most convincing traditional Darwinian examples of inherited change—namely, of Darwin's "descent with modification" coupled with his "natural selection," one turns out to be false and the other symbiogenetic. The first case is the literally colorful example of industrial melanism in moths, which is still included in many textbooks today. The idea is that polluted (soot-darkened) trees in industrial England gave randomly variant dark moths an adaptive advantage against their fellows, leading to statistically significant increase in genes conferring darkness and eventual speciation in surviving moths. Despite the plausibility of such evolution by lucky additional camouflage against the background of newly dark tree trunks, ultimately enough to form a new species or subspecies in less than a century via accumulation of random genetic differences in moth populations, historical sleuthwork reveals the example to be manufactured. For example, to illustrate neo-Darwinian orthodoxy, dark moths were allegedly pinned to dark tree trunks prior to photographing them to provide pictures for textbooks.[33]

The second case is even more intriguing. When one of us asked noted evolutionary biologist Niles Eldredge for his best documentation of animal speciation in the laboratory, the field, or the fossil record, he gave the example of Dobzhansky's fruit flies. Here is another canonical textbook example that touts natural selection by accumulation in variant populations of random mutations, this one in a laboratory in two years. Theodosius Dobzhansky, the Russian-born Columbia University geneticist and professor, exposed homogenous populations to low and to elevated temperatures. He picked out for further breeding the ones that grew faster and better under progressively warmer and colder laboratory conditions. After some two years "hot flies" were unable to successfully fertilize "cold" ones. The "hot" and "cold" flies were reproductively compatible amongst themselves, but the genetic crosses (hot x cold or cold x hot flies) were infertile. The two separated populations of the different populations now conformed to the traditional zoological definition of new animal species. Populations of *Drosophila paulistorium* had been reproductively and geographically isolated, and were now only able to breed with their own kind.

Here symbiogenesis seems almost to be thumbing its nose at prevailing orthodoxy. However, Wolfgang Miller of the University of Vienna Medical School, Austria, found that the "cold-fertile fly population" had retained a bacterial population widely distributed in certain tissues whereas the "hot-fertile flies" had been "cured" of that same set of bacteria. In other words, the presence or absence of the same bacterium apparently correlates with the speciation event. No documentation of accumulation of random genetic mutation was shown. Rather the investigators selected for the presence of bacterial genomes in fly tissue.

Jeon's X-Bacteria

Again, Kwang Jeon of the University of Tennessee in his laboratory in the 1970s observed newly arrived unidentified bacteria inside a newly acquired batch of amoebae. He called these "X-bacteria."[34] Some 100,000 per cell were perpetuated from a few that invaded his laboratory-grown amoebae. Most were killed, but by tending to the ill amoebae, he selected moribund survivors and with time saved those relatively resistant to the bacterial infection. After five years of careful nurture, he found the amoebae with the "pathogenic" X-bacteria had undergone a permanent species transition. The most notable survivors had retained 40,000 X-bacteria per cell. Jeon showed experimentally that these internal bacteria were now absolutely required for the nuclei of the transformed amoebae to survive.

The species-changing abilities of symbionts complicate and correct the too-linear notion that illness is simply caused by a named "germ" microbe that is uniquely responsible for disease. Fungi and bacteria are themselves the main sources of our antibiotics, and many, if not most, of our medicines are derived from plants. Although not themselves metabolic, inherited viruses in sheep have been identified as crucial for attachment of maternal placenta. As in the Jeon example, an organism associated with illness or the cause of disease can evolve to become necessary and part of the infected organism's self, at which point it is no longer infected but evolutionarily altered and in time speciated. Far from being simple diseases medicine must always eradicate to make us healthy and whole, bacterial and viral genomes are us. Indeed, it now appears that most of the human genome is composed of viral DNA.[35]

Thus to relegate symbiogenesis à la Dawkins to the status of evolutionary anecdotes, important only for the ancient origin of cytoplasmic organelles, requires both willful ignorance of the evidence and a healthy dash of rhetorical subterfuge. Whether or not Neo-Darwinist evidence achieves empirical validation, symbiogenesis appears to be central to evolution. Neo-Darwinism's mathematizations may be theoretically valid, but unless and until they can be corroborated by empirical observation, they are hardly deserving of the level of exclusive professional reverence they now unscientifically enjoy.

Toxoplasma, Cats, and Mind-Altering Symbionts

Wallowing in Aristotle's playground of evidence rather than Plato's garden of ideas provides richer fertilizer for theory. Organisms are not Platonic abstractions but real bodies with perception and sensation geared to the ambient energy gradients and material resources that sustain them. They should be understood in a thermodynamic context. Thus, for example, the hungry organisms that feed on photosynthesizers that then end up inside their transparent skins or translucent cells were not acquired accidentally, but because they were sought as food. This naturalistic teleological rapprochement contrasts with the neo-Darwinist mantra of randomness: in seeking mates and food, organisms regularly fuse. When they merge, in feeding or fertilization, animals may pick up new genomes, thereby helping themselves, intentionally or unwittingly, to evolve.

Food-seeking itself is a gradient-driven process, as food represents a chemical concentration of energy that can be spread to fuel growth. The acquisition of foreign symbionts changes not only the form of organisms, but their behaviors, in ways that may or may not enhance the survival of either or both parties. We rightly consider cholera, caused by a bacterium that can lead to epidemics (such as recently in Haiti), a disease, but the disease organism spreads via explosive diarrhea in unsanitary conditions. Similarly, bacteria and viruses may be transmitted by sneezes that expel genes from one organism even as the mode of expulsion tends to inoculate others. As with Jeon's diseased amoebae that came to require their former pathogens, what is a fatal disease at the individual level can lead to evolutionary transformation at the population level. The recognition of our mitochondria as former bacteria with their own DNA—ancient pathogens of our microbial ancestors—but now essential to healthy bodies and, indeed, impaired in sick ones[36]—requires a re-evaluation of medical epistemology, as does the integration into each of us of multiple lineages of organisms with separate evolutionary histories.[37]

While we fear intruders as disease organisms, they are also part of our natural defenses against illness-causing organisms. Probiotics are popularly known for their health properties in displacing less friendly gut bacteria. More deeply, the once-alien mitochondria are now deeply integrated into our immune system's intracellular deployment of oxygen and nitrogen compounds to destroy pathogens. Neo-Darwinism problematizes the individual, but the animal body itself is a multi-level evolved community. Bodies are not stable entities of a singular taxonomic kind—biological instantiations of some Platonic Idea. Rather, they are necessarily open to productive flux.

The bacteria identified in Dobzhansky's flies as a "mycoplasm" (cell-wall-less bacterium) are now recognized to be in the genus *Wolbachia*. *Wolbachia* of different kinds are nearly ubiquitous inhabitants of insect tissues. They bring in suites of genes for metabolic and reproductive features as they establish symbioses, often permanent, in arthropods. Because life is an open thermodynamic system, as well as an open informational one, genomic transfer is rampant. And identifiable new behaviors and physiologies come into being along the continuum from pathology to symbiogenesis. Human gut microbiota are not simply hangers-on, but influence the timing of maturation of our intestinal cells, our internal nutrient supplies and distribution, our blood vessel growth, our immune systems, and the levels of cholesterol and other lipids in our blood.[38] Partly because of the presence of neurons in the mammalian intestinal tract, and the communication between the gut and the brain, the presence of various types of bacteria alters not just body, but mood. Lab work with *Campylobacter jejuni* showed that this bacterium causes increased anxiety in mice, whereas the soil bacterium *Mycobacterium vaccae*, cheers them up. In people, it has been suggested that yogurt microbiota, including *bifidobacteria*, improves a sense of well-being.

Toxoplasmosis provides a striking example of new modes of behavior conferred upon animals due, not to "themselves," but to genomes that have partly become them.[39] *Toxoplasma gondii* is a protist (i.e., a "protozoan," a member of the *Protoctist* Kingdom in the five-kingdom taxonomy of life).[40] Toxoplasma infection has long been known to be a serious problem for pregnant mothers who may contract it from kitty litter or infected meat. From the mother, Toxoplasma moves to the fetus, often devouring it and leading to a miscarriage. *Toxoplasma gondii* sexually reproduces in members of the *Felidae* family, notably inside domestic cats. Toxoplasmosis alters animal behavior in bizarre if sometimes subtle ways. Infected mice, usually deathly afraid of cats, lose their fear of cats when their brains become infected with Toxoplasma. More bizarre still, they become sexually attracted to cat urine. Large numbers of humans are also infected with Toxoplasma. Even though the possessor of the protist genome may not know it is there, it confers statistically significant behavioral differences on him or her. These depend on gender. Toxoplasma infection in men correlates with enhanced risk-taking, jealousy, and asocial behavior. Even if bereft of obvious symptoms, men who carry Toxoplasma are less likely to be found attractive to women relative to controls. Women with Toxoplasma by contrast are more likely to be judged outgoing, friendly, conscientious, and promiscuous. An index of the risk-taking behavior is provided by the fact that males in car and motorcycle accidents are more likely to test Toxoplasma-positive.[41] The action of the strange genome is interesting as well. Toxoplasma makes enzymes (tyrosine hydroxylase and phenylalanine hydroxylase) that alter brain levels of the neurotransmitter dopamine. Dopamine is a neurotransmitter involved in mood, attention, sociability, and sleep. Cocaine and amphetamines work in large part by blocking the reuptake of dopamine in the brain. Dysfunctional dopamine regulation is theorized to be linked to schizophrenia, and several antipsychotic drugs target dopamine receptors. Up to a third of the world population is thought to be infected with Toxo, with an estimated infection rate of almost 90 percent in France. In short, there are significant statistically verified behavioral variances on the basis of infection of this one kind of protist, and Toxoplasma is only one of probably hundreds that influence human behavior.

Multiple Symbiogenetic Agency versus Neo-Darwinist Hyperbole

These examples of the effects of Toxoplasma on behavior indicate that our notions of unique human agency are at best incomplete and at worst woefully mistaken. Spinoza (1677) argued that our faith in our own free will is an illusion, attributable to our lack of knowledge of the actual causes of our own behavior.[42] In other words, that we do not know the causes of our behavior does not mean that it is "free" in the sense of not caused. Much of our would-be agency, which we perceive to issue smoothly from the ego, will, or "I" (but which is probably a kind of oversimplifying biological interface),[43] appears to owe much to heretofore-undetected symbiogenetic actors.

In a causal or even probabilistic world that does not uniquely imbue human beings with freedom, our behaviors are part of a vast causal weave that includes the cycling of matter "to" reduce gradients or spread energy. There is little reason to believe that we alone, among the vast interconnected parts of the cosmos, are impervious to the causal and/or probabilistic nexus we are happy to use in explaining the behaviors of nonhuman processes. Yet our human tendency is to identify unique, rather than mixed causes or correlations so as to explain our behavior and existence. Of these putative sole causative factors the "I"—the ego—remains the most tenacious and egregious. From a thermodynamic perspective, the most inclusive level of purposeful activity is that of the cycling system as a whole, which acts intelligently or with unconscious purpose to degrade ambient gradients. Multiple agents aside from the individual animal act to degrade gradients, and sometimes do so in conflicting way. We can see the problem of mixed agency in the relationship of individual humans, not only to the microbes within them, but to the corporate and national structures of which they are a part, which influence them to perform activities they would not "on their own." The lure of money, for example, makes people work for organizations, corporations, or the state, that have their own persistence that depends upon people as a whole, but in which specific individuals are replaceable. Like two-faced Janus, the individual, who attributes his action to himself, is co-determined by organisms within and external factors without, in which money stands as a symbol of exchange for materials thermodynamically cycling in more-than-human gradient-reducing systems.

Neo-Darwinism seriously underestimates the complexity of this mixed agency, however. Attempting to rip the veil of phenomena (where agency is associated with the individual), it loves to assume the gene is really in control and portray animal behavior as an epiphenomenon of the gene. Consider Bryan

Sykes, an accomplished geneticist who predicts that in 120,000 years men may be gone. Only females in the human species will remain, Sykes argues in *Adam's Curse: A Future without Men* (2004).[44] This view is based on the idea of selfish genes (never mind that genes have no selves properly speaking, but only ever exist as functioning wholes inside cells, the minimal living selves). The energy producers of human cells (i.e., mitochondria), are found outside the nucleus with its spaghetti-strand chromosomes. Not only do these extranuclear structures have their own genes, but they are inherited in our species only through the mother's egg cell. The male's DNA-containing mitochondria, found in the tails of sperm, do not enter into the fertilized egg that grows into the infant's embryo.

The basis of Sykes wild prognostication is that, for the DNA of female mitochondria, reproducing through males is a complete waste of time. Your mother's mitochondrial genes might produce more copies of themselves and their DNA if they could be passed directly to daughters. According to such thinking, "male homosexuality . . . is a subtle plan by [the female genome] not only to get rid of Y-chromosomes but to help itself at the same time." But no sooner does Sykes ascribe teleological powers to women's mitochondrial genes than he takes it back: "Genes don't plan that far ahead,"[45] he says and, later in the book, "Genes are blind and have no concept of the future."[46] But if so, then the female mitochondrial genome does not have a "subtle [homoerotic] plan . . . to get rid of Y-chromosomes [and] help itself at the same time."[47]

Admittedly this is a rather flagrant example of selfish gene rhetoric. But often extremes make the case more clearly, and the case here is that even its users know that the agential rhetoric of genocentric neo-Darwinism is hopelessly awry. It should be obvious by now that agency is a more complicated, and labile affair. As theorist Richard Doyle has shown and environmental writer Michael Pollan has popularized, plants can be granted agency in narratives of plant-animal interaction. In the words of Doyle, for example, plants have "outsourced their sexuality"[48] to animals. Plant colors, tastes, and textures work to seduce us into propagating them; that the spit-outability of seeds aids in such a process, and that in certain cases, such as commercial corn, whose husks are so thick that they must be stripped by humans if they are to reproduce, we have become joined to both food plants and technology due to the in-growth of recently evolved, mutually reinforcing life cycles.[49] Also, fungi can tentatively be granted agency due to the psychophysical attractants of their alkaloids, by turns poisonous and nutritious, which by turns attract and repel animals, as well as get them to cultivate them and introduce them into religious ceremonies.[50] True agency cannot be located in the brain, which stands to the process of thinking like an engine does to driving, which is not in the engine but necessary for it to take place. Mammal brains, as Toxoplasma shows, are open to "control" (however unconscious) by foreign genomes. While we do not necessarily know the source of our behavior, we are quick to convince ourselves that the locus of our behavior originates entirely with us—we are adept at first-person narratives.

Finally, with regard to multispecies influences, all of evolution can be, as it were, turned on its head, seen from the perspective of a pan-microbial agency, one which portrays cows as forty-gallon tanks for temperate storage of methanogenic and grass-digesting microbes. Trees are platforms for the solar exposure of cyanobacterial descendants in leaves. Animal, plant, and fungal life in general are instruments for the breeding of the ubiquitous mitochondria—descendants of respiring bacteria that are now holed up in but provide energy to the tissues of most all Earth's surface organisms.[51]

Despite the topsy-turvy truth of such stories, however, they are themselves provisional and can be deconstructed—in Derrida's sense of the term, meaning first to displace a mutually reinforcing conceptual hierarchy (here, putting microbes before men) but then, after an interim strategic period of reversing the usual hierarchy within a metaphysical pair,[52] moving on from such a hierarchy (in which both "opposed" terms are ultimately complicit) altogether. For Derrida in *Of Grammatology* (1967), the privileged member of the conceptual dyad was speech, which is historically granted primacy over writing. For us, it was granting newfound primacy to metabolically diverse bacteria, our evolutionary seniors, in order to compensate for the distorted puffery of traditional evolutionary biological narratives featuring vertebrates. In both cases, the reversal of the privileged term is only strategic, done in order to "deconstruct" the received wisdom of an overused conceptual pair. In both cases, the hierarchy itself is best not reversed so much as displaced.

Biology's Mistress: Back to Aristotle

In the end, true biological agency does not belong to any member of an ecosystem or autocatalytic cycle, but can be traced thermodynamically to the environment, which "wants" to degrade gradients and spread energy. Irrespective of natural selection's ability to fashion features that, in retrospect, remind us of a conscious human creator, thermodynamics provides a directional arrow which applies to living organisms as well as to complex nonliving systems; and it is here, in the tendency for energy to be degraded, and for nature to find ways to do so, that a kind of agency behind the separate parts of an ecosystem or interspecies alliance lies. "Teleology is like a mistress to a biologist," quipped British geneticist J. B. S. Haldane. "He cannot live without her but he's unwilling to be seen with her in public."[53]

Organisms do seem to be genuinely teleological. They exist in a world "to" do things and "for" certain purposes. Moreover, teleological descriptions cannot be rooted out of language, with which we describe life. Neo-Darwinists argue that there are purposeful features in organisms but they never came from any being (e.g., God) looking ahead, but must be explained by a mechanical principle (e.g., natural selection) which eliminates relatively non-functional forms, leaving an appearance of design. But this, in Ernst Mayr's term, is not teleology but teleonomy—purposeful organs result from natural selection, not consciousness. Leaving aside the *aporia* of evolutionary biologists themselves—where does their own "real" purposeful behavior come from, if not God or nature—we can certainly identify part of the problem Haldane alludes to—namely, of purpose being biologically taboo (see chapters 8, 13, and 14 in this volume). As evolutionary biologist Stephen Jay Gould (2002) points out, Aristotle was also already ahead of most modern scientists in his nuanced views of causality, of which purpose was only one kind. Gould explains that efficient causation on chemical and physical grounds is how biologists explain behavior and morphological features.[54] Chemical and physical descriptions of immediately antecedent events explain how organisms operate, and biologists call such explanations "proximate cause." "Ultimate cause" is reserved for purpose, and refers to evolutionary explanations; this is the only allowable way in which biologists can explain why behaviors exist.

Ironically, this truncated typology of cause is shallow compared to the thinking of Aristotle, who had, depending upon how one counts, a fourfold or fivefold theory of causality that could be applied to phenomena in order to explain them: efficient, material, formal, final, and first. Sometimes they overlapped. In modern terms, the basic difference between cause and purpose ("final cause" for Aristotle), is that cause explains a phenomenon in terms of antecedent phenomena, whereas purpose explains phenomena in terms of the need, desire, will, or imperative to arrive at a future state—and thus is usually associated with consciousness. In industrial mechanical science, which ironically evolved historically from looking at the universe as a machine created by God, scientific cause is almost wholly confined to efficient causation, immediate antecedents. Purpose, except, of course, for our own all-too-human desires, is inevitably disparaged as a kind of mystic reverse causation. True purpose does not exist in nature; it only exists in humans. Unfortunately, this philosophically shallow concept results from our Cartesian heritage. We separate people with real free will from the rest of the universe, which is a blind mechanism. In Dawkins' (1986) words,

> Natural selection, the blind, unconscious, automatic process which Darwin discovered, and which we now know is the explanation for the existence and apparently purposeful form of all life, has no purpose in mind. It has no mind and no mind's eye. It does not plan for the future. It has no vision, no foresight, no sight at all. If it can be said to play the role of watchmaker in nature, it is the blind watchmaker.[55]

A look at the imperative spreading of energy—namely, the production of entropy or reduction of gradients, however, shows that this is not the case. As Aristotle intuited, there is a prevailing, inherent, not-necessarily-conscious *telos* or entelechy—an internal *telos*—that is natural. And this naturalistic teleology is genuinely retrocausal without being supernatural. In fact, it would be the height of both anthropocentrism and supernaturalism to think that when a human being, driving on a highway, turns on his clicker to get into the right lane to stop at a rest stop to use a bathroom, that her purposeful behavior somehow violates the laws of nature. That such stopping-to-go-to-the-bathroom-type behavior is somehow of a theistic, Cartesian, or even uniquely human mental order, disconnected from other life-forms and their metabolic relation to finding and dispersing concentrated sources, even Dawkins might agree. Indeed, there is an entirely real sense in which a future state of equilibrium seems to act retrocausally on matter-energy configurations, giving them a naturalistic purpose. Hot streamers of air in a heated house "purposefully" attempt to equilibrate with the colder air outside, and their unconscious "efforts" to do so may even be misapprehended as ghosts by the gullible huddled in darkly creaking Victorian houses. Beyond the bankrupt rhetoric and philosophy of neo-Darwinism, teleology looks natural, not anthropomorphic. In fact, it is exactly the opposite of anthropomorphism: we are part of a teleological continuum of gradient-reducing, energy-spreading arrangements of cycling matter in regions of energy. Why would we expect to be excepted?

Naturally Teleological Systems

Despite the time-symmetrical mathematizations of ideal physics, real world phenomena exist in time. Like narratives, they have beginnings, middles, and ends. We remember the past not the future and real objects are not perfect and eternal, but subject to wear and tear or, in the case of living beings, aging and death. Organizations of matter are temporary, expeditious configurations that help bring the areas around them to equilibrium. A Victorian house may seem haunted but is a very simple, close-to-equilibrium system in which, nonetheless, streamers of hot air "try" to escape, leaving the heated house cold—bringing it to thermal equilibrium with its surroundings.

Organisms can be conceptualized as relatively far-from-equilibrium systems that have, with the infrastructure of DNA, genetic copying, and the operationally closed metabolism of biological identity, called "autopoiesis,"[56] evolved ways to reliably seek and expend ambient energy sources. Only two classes of usable energy exist for any life-form: light and chemical oxidation. Both generate heat that spreads. Heat energy, while it may warm organisms, is not ever concentrated for use in metabolism. The second law is not disobeyed.

Once they run out of energy for metabolism and materials for their construction, of course, life-forms are finished. They go the way of all flesh toward thermal and chemical equilibrium. Life, however, operates in a vast non-equilibrium universe. And unlike less persistently active complex systems, living matter has happened upon a crucial, genetically dependent expedient: life-forms make copies of themselves, their complex, energy-degrading structures. Evolution might thus be characterized as possessing a thermodynamic algorithm to continue the process of energy depletion by producing new effective bodies (i.e., offspring). Such, in any event, is our naturalistic reading of purpose. This is a view of purpose that diverges both from neo-Darwinism and with the prevalent form of theistic teleology against which neo-Darwinism has pitted itself in a kind of eternal catfight. But a thermodynamic naturalistic teleology is one which Aristotle himself might have found congenial. Rather than disparaging naturalistic teleology as primitive, or exemplary of an atavistic tendency toward metaphor and pathetic fallacy, one should consider human purposefulness, including unconscious purposeful processes, as rooted in nature. Ultimately, this non-human-exceptionalist stance is far more in keeping with the scientific spirit and the deep intuition that we are not things so much as emergent processes within an energy-rich cosmos. Rather than thinking ourselves, along with Descartes, as rare aspects of the divine mind mulling about in a senselessly extended body, why should we not consider human purposefulness as rooted in the thermodynamic teleology of nature? It seems to us that everything from conscious planning to evolutionarily accelerating technology[57] may be instantiations, if not epitomizations, of a natural thermodynamic tendency.

Although the longevity of its process is extended thanks to its capacity for accurate reproduction, living matter is typical of natural thermodynamic flow processes: dust devils, tornadoes, Bénard (convection) cells, Taylor vortices, Liesegang rings, Belousov-Zhabotinski reactions, and other long-lasting autocatalytic chemical reactions. Salt fingers in the ocean, Hadley cells in the atmosphere, the northward flowing Gulf Stream, "Whirlpool" downstream of Niagara Falls, the Great Red Spot of Jupiter, and other natural thermodynamic systems—all maintain themselves (insofar as they are "selves") by cycling matter as they stably degrade available energy that sustains their recognizable three-dimensional identities. Even storms, often named to better to keep track of them, and in prior times associated with willful wind gods, are active sites of energy degradation. While they last, they have identity with boundaries, they reduce pressure and temperature gradients in the atmosphere until their impressive stints of material cycling and natural tasks of entropy production are accomplished. Storms are natural non-random energy-driven processes. Here they are like life, although, significantly, few today feel that they require a divine explanation, least of all for their teleological tendencies to reduce gradients.

Geophysiologically Active Interspecies Association (Gaia)

Bacterial cells, growing fetuses, prairie and rainforest ecosystems, and Earth's entire biosphere (sometimes called "Gaia") are also examples of volumes or spaces where matter cycles "in order" to reduce ambient gradients in accord with the natural tendency of energy to spread. This "in order" can be viewed as genuinely "teleological" because nature "prefers" a state of equilibrium and routinely naturally "designs" complex systems to bring about such equilibrium if materials and energy are available. The notion of Gaia, that is, of Earth's biosphere as a self-regulating system, has been criticized severely by Neo-Darwinists and other would-be hardheaded thinkers for being atavistically teleological, metaphorical, and new age. (The acronym heading this section is a playful attempt to demythologize what is a real physical phenomenon, the life-associated metastabilty of the planetary environment). For, they wonder, how could selfish organisms conspire to regulate global environmental variables over periods of hundreds of millions of years? Why, they couldn't, and that is that.

But we cannot dismiss a phenomenon because of our inability to understand it. We do not know in detail exactly how specific molecules and atoms arrange themselves in whirling eddies to reduce a barometric pressure gradient, but we know it happens, that it is not random but organized, and that we do not have to attribute the process to divine intervention or enlightened air.

In the same way, without knowing all the details—although this is a very productive area of research, often called Earth System Science—there is abundant evidence for global thermoregulation, atmospheric regulation of chemical composition, and marine salinity regulation for multi-million-year periods. These regularities, moreover, keep perfectly with the tendency of complex systems—not just living ones, or only living ones being naturally selected in vast populations containing genes—to organize and lay regions of concentrated energy methodically to waste. The biosphere itself is a natural complex thermodynamic system, one that measurably degrades the thermal gradient between the hot sun and cold space. Although we do not know most details on how it does it, we do know that geo-physiologically active interspecies associations exist. Finding how the thermodynamic teleological imperative informs emerging patterns of global organization should be an area of research for evolutionary biologists, not a source of literal ignorance motivated by neo-Darwinism's desire to distance itself from religion-associated teleology.

A fact-based science demands that we confront, rather than avoid or deny the data, whether that data be evidence for evolution by symbiosis or evidence for natural teleology in evolution seen as a measurably trending, rather than as a largely random process. In the end, a human exceptionalist view that restricts truly teleological behaviors to the future-oriented desires of conscious human agents is anthropocentric, dualistically obsolete, anti-Spinozistic, and contrary to mounting scientific evidence calling into question the easy assumption of a uniquely human free will.[58] Our telic behaviors stem from those of nature.

Finally, life's teleological character can be marshaled to answer a question that has vexed neo-Darwinists—and set them back in their valiant struggle against anti-scientific creationists: if evolution is basically a random process, why does evolutionary history appear to have a direction? Denying that it has a direction will not help. But the phenomenon has a scientific, rather than anthropomorphic and creationist solution. Evolution's main trends—increase in number of individuals, species, and taxa; increase in bacterial and animal respiration efficiency; increase in number of cell types; and long-term increases, despite periodic setbacks from mass extinctions, in global biodiversity, captured energy, matter, information, networking, and net perceptivity, technology, and intelligence—measurably increase entropy production (gradient reduction) at local and global scales. These trends become comprehensible when we recognize that life is an open, natural energy flow system in which animal intelligence and microbial and plant craftiness help deploy stored energy to tap existing and new gradients. The abundant evidence for symbiogenesis does not rule out evolution by gradual accumulation of chance mutations, which may indeed have played a major role in the evolution of bacteria. But it is equally if not more clear that permanent encounters play a major role in evolutionary change. Symbiotic partnerships merge sensibilities and metabolisms as open thermodynamic systems synergize and improve their ability to access ambient gradients. The incorporation of human-built technology, while measurably imperiling global gradient reduction (as indicated by global heating), extends the potential for this ancient sensorily-aided process of location and depletion of concentrated energy reserves. Pope Pius XII argued that the Second Law (simplistically assumed to mandate disorder rather than to also favor the production of energy-degrading systems) offers proof of God, because He could miraculously flout the laws of nature to produce life. Neo-Darwinist philosopher Daniel Dennett, albeit in a less grandiose fashion, makes the same mistake, saying that life-forms "defy" the Second Law.[59] But the complex systems—including the symbiogenetic collectives of human beings, food plants, and industrial and informational technologies we call civilization—that enable energy spread are ultimately natural. Contrary to common understanding, life does not defy but accelerates the effects of the Second Law. And contrary to neo-Darwinist orthodoxy, life is clearly purposeful.

Acknowledgments

The authors thank Eric D. Schneider, Frank L. Lambert, Steven Shavel, Tori N. Alexander, and Bruce "Bruno" Clarke for suggestions, provocations, discussions, and feedback leading to the preparation of this essay.

Notes

1. The empirical data summarized in this essay that relates thermodynamics to life is detailed in Eric D. Schneider and Dorion Sagan, *Into the Cool: Energy Flow, Thermodynamics, and Life* (Chicago: University of Chicago Press, 2006).
2. Frank L. Lambert (b. 1918), a Professor Emeritus of Occidental College, Los Angeles, has, since 1999, produced publications that have led 29 chemistry textbooks, reaching approximately 450,000 students (to date), to discard "disorder" as a description of entropy. See, for example, Frank L. Lambert, "Disorder—A Cracked Crutch for Supporting Entropy Discussions," *Journal of Chemical Education* 79 (2002a), 187-192; Frank L. Lambert, "Entropy Is Simple, Qualitatively," *Journal of Chemical Education* 79 (2002b): 1241-1246; and Frank L. Lambert, "Disorder in Thermodynamic Entropy," http://entropysite.oxy.edu/boltzmann.html, retrieved June 22, 2012. His websites include secondlaw, shakespeare2ndlaw, and entropysite, all with the suffix oxy.edu; the latter containing multiple links to relevant publications at a variety of levels.
3. In *Lectures on Gas Theory* (New York: Dover Publications, 1898), 443, Ludwig Boltzmann writes that the world, "or at least most of the parts of it surrounding us are initially in a very ordered—therefore very improbable—state. When this is the case, then whenever two or more small parts of it come into interaction with each other, the system formed by these parts is also initially in an ordered state and when left to itself it rapidly proceeds to the disordered most probable state."

But, as Frank J. Lambert points out in "The Conceptual Meaning of Thermodynamic Entropy in the 21st Century," *International Research Journal of Pure & Applied Chemistry* 1, no. 3 (2011): 67, http://www.sciencedomain.org/issue.php?iid=82&id=7, physicist Max Plank did the actual calculation (inscribed on Boltzmann's tombstone) to show the equation for energy change. As Lambert writes, "If liquid water at 273 K, with its $10^{1,991,000,000,000,000,000,000,000}$ accessible microstates [quantized molecular arrangements] is considered 'disorderly,' how can ice at 273 K that has $10^{1,299,000,000,000,000,000,000,000}$ accessible microstates be considered 'orderly'?"

4. Alfred North Whitehead, *Science and the Modern World*, Lowell Lectures 1925 (New York: The New American Library, 1962), 15.
5. Lynn Margulis and Michael J. Chapman, *Kingdoms and Domains: An Illustrated Guide to the Phyla of Life on Earth* (San Diego: Elsevier, 2010).
6. Alla Katsnelson, "Microbe Gets Toxic Response. Researchers Question the Science behind Last Week's Revelation of Arsenic-Based Life," *Nature News*, December 2010, http://www.nature.com/news/2010/101207/full/468741a.html.
7. Lynn Margulis and Dorion Sagan, *What Is Life?* Foreword by Niles Eldredge (Berkeley, CA: University of California Press, 2000).
8. Harold J. Morowitz, *Personal Communication at Science and Literature: Narratives of Discovery*, Key West Literary Seminar, 2001.
9. Freeman Dyson, *Origins of Life: Revised Edition* (Cambridge, MA: Cambridge University Press, 1999).
10. Dennis Noble et al., "Homage to Darwin, Evolution Debate with Dawkins, Margulis, Brasier, Bell, Chaired by Denis Noble," Balliol College, Oxford University, archived debate, 2010, http://www.voicesfromoxford.com/homagedarwin_part1.html.
11. Vladimir I. Vernadsky, *The Biosphere: Complete Annotated Edition* (1926; repr., New York: Springer, 1998).

12. Dorion Sagan and Jessica Hope Whiteside, "Gradient-Reduction Theory: Thermodynamics and the Purpose of Life," in *Scientists Debate Gaia: The Next Century* (Cambridge, MA: MIT Press, 2004), 173-186.

13. Eric D. Schneider and James J. Kay, "Life as a Manifestation of the Second Law of Thermodynamics," *Mathematical and Computer Modeling* 19 (1994): 25-48; Frank Lambert, "Characterization-Description of the Second Law as Energy Spread Rather Than Production of Entropy," 2010, http://entropysite.oxy.edu.

14. Jeffrey S. Wicken, *Evolution, Thermodynamics and Information: Extending the Darwinian Program* (New York: Oxford University Press, 1987).

15. Schneider and Sagan, *Into the Cool*.

16. Sagan and Whiteside, "Gradient-Reduction Theory.".

17. Richard Dawkins, *The Extended Phenotype* (New York: Oxford University Press, 1982); W. Ford Doolittle, "Is Nature Really Motherly?" *The Coevolution Quarterly* 29 (Spring 1981): 58-63.

18. James Lovelock, *Gaia: A New Look at Life on Earth* (New York: Oxford University Press, 1979).

19. Stuart Kauffman, *Investigations* (New York: Oxford University Press, 2000).

20. Schneider and Sagan, *Into the Cool*.

21. Harold J. Morowitz, *Energy Flow in Biology: Biological Organization as a Problem in Thermal Physics* (New York: Academic Press, 1968).

22. Andrey Lapo, *Traces of Bygone Biospheres* (San Francisco: Synergetic Press, 1988).

23. Humberto R. Maturana and Francisco J. Varela, *Autopoiesis and Cognition* (Dordrecht, Holland: D. Reidel, 1980).

24. Iris Fry, *The Emergence of Life on Earth: A Historical and Scientific Overview* (New Brunswick, NJ: Rutgers University Press, 2000).

25. Margulis and Chapman, *Kingdoms and Domains*.

26. Liya Nikolaevna Khakhina, *Concepts of Symbiogenesis: A Historical and Critical Study of the Research of Russian Botanists* (New Haven, CT: Yale University Press, 1992); Jan Sapp, *The New Foundations of Evolution: On the Tree of Life* (New York: Oxford University Press, 2009); Boris M. Kozo-Polyansky, *Symbiogenesis: A New Principle of Evolution* (1924; repr., Cambridge, MA: Harvard University Press, 2010).

27. Ludwik Fleck, *Genesis and Development of a Scientific Fact* (Chicago: University of Chicago Press, 1981).

28. Dennis Noble et al., "Homage to Darwin."

29. Thomas S. Kuhn, *The Structure of Scientific Revolutions* (Chicago: University of Chicago Press, 1996).

30. Dennis Noble, *The Music of Life: Biology beyond Genes* (New York: Oxford University Press, 2008).

31. Lynn Margulis and Dorion Sagan, *Acquiring Genomes* (New York: Basic Books, 2002a); Kozo-Polyansky, *Symbiogenesis*.

32. Frank Ryan, *Metamophosis: The Beautiful Mystery* (White River Junction, VT: Chelsea Green Publishing, 2011).

33. Judith Hooper, *Of Moths and Men: An Evolutionary Tale. The Untold Story of Science and the Peppered Moth* (New York: W. W. Norton & Company, 2003).

34. Kwang W. Jeon, "Amoeba and x-Bacteria: Symbiont Acquisition and Possible Species Change," in *Symbiosis as a Source of Evolutionary Innovation*, eds. Lynn Margulis and René Fester (Cambridge, MA: MIT Press, 1991), 118-131.

35. Christopher Carter, "The Human Genome Is Composed of Viral DNA," *Natureprecedings*, August 2010, http://dx.doi.org/10.1038/npre.2010.4765.1.

36. Heinrich Kremer, *The Silent Revolution in Cancer and AIDS Medicine* (Philadelphia, PA: Xlibris Corporation, 2008).

37. Jessica Hope Whiteside and Dorion Sagan, "Medical Symbiotics," in *Chimeras and Consciousness: Evolution of the Sensory Self*, ed. L. Margulis, C. Asikainen, and W. E. Krumbein (Cambridge, MA: MIT Press, 2011), 207-218.

38. Valerie Brown, "Bacteria 'R' Us." *Miller-McCune*. December 2, 2010, http://www.miller-mccune.com/science-environment/bacteria-r-us-23628/.

39. Robert Sapolsky, "Toxo: A Conversation with Robert Sapolsky," *Edge* (2009), http://www.edge.org/documents/archive/edge307.html#tc.

40. Margulis and Chapman, *Kingdoms and Domains*.

41. Jaroslav Flegr, Jiří Klose, Martina Novotná, Miroslava Berenreitterová, and Jan Havlícek, "Increased Incidence of Traffic Accidents in Toxoplasma-Infected Military Drivers and Protective Effect RhD Molecule Revealed by a Large-Scale Prospective Cohort Study," *BMC Infectious Diseases* 9, no. 72 (2009), 72.

42. Richard McKeon, *The Philosophy of Spinoza: The Unity of His Thought* (Woodbridge, CT: Ox Bow Press, 1987).

43. See Tor Nørretranders, *The User Illusion: Cutting Consciousness Down to Size* (New York: Viking Press, 1998).

44. Bryan Sykes, *Adam's Curse: A Future without Men* (New York: W.W. Norton & Company, 2004).

45. Sykes, *Adam's Curse*, 104.

46. Sykes, *Adam's Curse*, 275.

47. Sykes, *Adam's Curse*, 273.

48. Richard M. Doyle, *Darwin's Pharmacy: Sex, Plants, and the Evolution of the Noösphere* (Seattle: University of Washington Press, 2011).

49. Margulis and Sagan, *Microcosmos: Four Billion Years of Microbial Evolution* (Berkeley, CA: University of California Press, 2002b).

50. Margulis and Sagan, *Microcosmos*.

51. Margulis and Sagan, *Microcosmos*.

52. Jacques Derrida, *Positions* (London: Continuum Publishing Company, 2004); Jacques Derrida, *Of Grammatology* (1967; repr., Baltimore: Johns Hopkins University Press, 1974).

53. Haldane "mistress quote" cited Mayr, Ernst, 1989. *Toward a New Philosophy of Biology: Observations of an Evolutionist,* 63.

54. See Stephen Jay Gould, *The Structure of Evolutionary Theory* (Belknap Press of Harvard University Press, 2002), 626-627.

55. Richard Dawkins, *The Blind Watchmaker* (New York: W.W. Norton & Company, 1986).

56. Maturana and Varela, *Autopoiesis and Cognition*.

57. Kevin Kelly, *What Technology Wants* (New York: Viking Press, 2010).

58. Sam Harris, *Free Will* (New York: Free Press, 2012).

59. Daniel Dennett, *Darwin's Dangerous Idea* (New York: Simon & Schuster, 1995).

Bibliography

Boltzmann, Ludwig. *Lectures on Gas Theory*. New York: Dover Publications, 1898.
Brown, Valerie. "Bacteria 'R' Us." *Miller-McCune*. December 2, 2010, http://www.miller-mccune.com/science-environment/bacteria-r-us-23628/.
Carter, Christopher. "The Human Genome Is Composed of Viral DNA: Viral Homologues of the Protein Products Cause Alzheimer's Disease and Others via Autoimmune Mechanisms." *Natureprecedings*. August 2010, http://dx.doi.org/10.1038/npre.2010.4765.1.
Dawkins, Richard. *The Extended Phenotype*. New York: Oxford University Press, 1982.
———. *The Blind Watchmaker*. New York: W. W. Norton & Company, 1986.
Dennett, Daniel. *Darwin's Dangerous Idea*. New York: Simon & Schuster, 1995.
Derrida, Jacques. *Positions*. London: Continuum Publishing Company, 2004.
———. *Of Grammatology*. Translated by Gayatri Chakravorty Spivak, 1967. Reprint, Baltimore: Johns Hopkins University Press, 1974.
Doolittle, W. Ford. "Is Nature Really Motherly?" *The Coevolution Quarterly* 29 (Spring 1981), 58–63.
Doyle, Richard M. *Darwin's Pharmacy: Sex, Plants, and the Evolution of the Noösphere*. Seattle: University of Washington Press, 2011.
Dyson, Freeman. *Origins of Life: Revised Edition*. Cambridge, MA: Cambridge University Press, 1999.
Fleck, Ludwik. *Genesis and Development of a Scientific Fact*. Translated by F. Bradley. Chicago: University of Chicago Press, 1981.
Flegr, Jaroslav, Jiří Klose, Martina Novotná, Miroslava Berenreitterová, and Jan Havlíček. "Increased Incidence of Traffic Accidents in Toxoplasma-Infected Military Drivers and Protective Effect RhD Molecule Revealed by a Large-Scale Prospective Cohort Study." *BMC Infectious Diseases* 9, no. 72 (2009): 72.
Fry, Iris. *The Emergence of Life on Earth: A Historical and Scientific Overview*. New Brunswick, NJ: Rutgers University Press, 2000.
Gould, Stephen Jay. *The Structure of Evolutionary Theory*. Belknap Press of Harvard University Press, 2002.
Harris, Sam. *Free Will*. New York: Free Press, 2012.
Hoffmeyer, Jesper. *Biosemiotics: An Examination into the Signs of Life and the Life of Signs*. Edited by Donald Favareau. Scranton, PA: University of Scranton Press, 2009.
Hooper, Judith. *Of Moths and Men: An Evolutionary Tale: The Untold Story of Science and the Peppered Moth*. New York: W. W. Norton & Company, 2003.
Jeon, Kwang, W. "Amoeba and x-Bacteria: Symbiont Acquisition and Possible Species Change." In *Symbiosis as a Source of Evolutionary Innovation*, edited by Lynn Margulis and René Fester, 118–131. Cambridge, MA: MIT Press, 1991.
Katsnelson, Alla. "Microbe Gets Toxic Response. Researchers Question the Science behind Last Week's Revelation of Arsenic-Based Life." *Nature News*. December 2010, http://www.nature.com/news/2010/101207/full/468741a.html.
Kauffman, Stuart. *Investigations*. New York: Oxford University Press, 2000.
Kelly, Kevin. *What Technology Wants*. New York: Viking Press, 2010.
Khakhina, Liya Nikolaevna. *Concepts of Symbiogenesis: A Historical and Critical Study of the Research of Russian Botanists*. New Haven, CT: Yale University Press, 1992.

Kirksey, Eben, and Stefan Helmreich. "The Emergence of Multispecies Ethnography." *Cultural Anthropology* 25, no. 4 (2010): 545-575.
Kozo-Polyansky, Boris M. *Symbiogenesis: A New Principle of Evolution*. Edited by Lynn Margulis and Victor Fet. 1924. Reprint, Cambridge, MA: Harvard University Press, 2010.
Kremer, Heinrich. *The Silent Revolution in Cancer and AIDS Medicine*. Philadelphia, PA: Xlibris Corporation, 2008.
Kuhn, Thomas S. *The Structure of Scientific Revolutions*. Chicago: University of Chicago Press, 1996.
Lambert, Frank L. "Disorder: A Cracked Crutch for Supporting Entropy Discussions." *Journal of Chemical Education* 79 (2002a), 187-192.
———. "Entropy Is Simple, Qualitatively," *Journal of Chemical Education*, 79 (2002b): 1241-1246.
———. "Characterization-Description of the Second Law as Energy Spread Rather Than Production of Entropy (A Difficult-to-Picture Mathematical Ratio Mistakenly Conflated with "Disorder") Has Been Pioneered by Chemist and Occidental Professor Emeritus Frank Lambert, Whose More Teachable and Accurate Understanding of the Second Law Has Been Adapted in the Majority of Recently Printed Chemistry Textbooks," 2010. http://entropysite.oxy.edu.
———. "The Conceptual Meaning of Thermodynamic Entropy in the 21st Century," *International Research Journal of Pure & Applied Chemistry* 1, no. 3 (2011), 67.
———. "Disorder in Thermodynamic Entropy," http://entropysite.oxy.edu/boltzmann.html. Accessed June 22, 2012.
Lapo, Andrey. *Traces of Bygone Biospheres*. San Francisco: Synergetic Press, 1988.
Lovelock, James. *Gaia: A New Look at Life on Earth*. New York: Oxford University Press, 1979.
Margulis, Lynn, and Michael J. Chapman. *Kingdoms and Domains: An Illustrated Guide to the Phyla of Life on Earth*. San Diego: Elsevier, 2010.
Margulis, Lynn, John O. Corliss, and Michael Melkonian. *Handbook of Protoctista: The Structure, Cultivation, Habitats and Life Histories of the Eukaryotic, Microorganisms and Their Descendants, Exclusive of Animals, Plants, and Fungi: A Guide to the Algae, Ciliates, Foraminifera, Sporozoa, Water Molds, Slime Molds, and the Other Protoctists*. Sudbury, MA: Jones and Bartlett Publishers, 2012, in press.
Margulis, Lynn, and Dorion Sagan. *What Is Life?* Foreword by Niles Eldredge. Berkeley, CA: University of California Press, 2000.
———. *Acquiring Genomes: A Theory of the Origins of Species*. New York: Basic Books, 2002a.
———. *Microcosmos: Four Billion Years of Microbial Evolution*. Berkeley, CA: University of California Press, 2002b.
Mayr, Ernst, *Toward a New Philosophy of Biology: Observations of an Evolutionist*. Cambridge, MA: Harvard University Press, 1989.
Maturana, Humberto R., and Francisco J. Varela. *Autopoiesis and Cognition*. Dordrecht, Holland: D. Reidel, 1980.
McFall-Ngai, Margaret. "Origins of the Immune System." In *Chimeras and Consciousness: Evolution of the Sensory Self*, edited by Lynn Margulis, C. A. Asikainen, and W. E. Krumbein, 199-206. Cambridge, MA: MIT Press, 2010.
McKeon, Richard. *The Philosophy of Spinoza: The Unity of His Thought*. Woodbridge, CT: Ox Bow Press, 1987.

Morowitz, Harold J. *Energy Flow in Biology: Biological Organization as a Problem in Thermal Physics*. New York: Academic Press, 1968.

———. *Personal Communication at Science and Literature: Narratives of Discovery*. Key West Literary Seminar, 2001.

Noble, Dennis. *The Music of Life: Biology beyond Genes*. New York: Oxford University Press, 2008.

Noble, Dennis, et al. "Homage to Darwin, Evolution Debate with Dawkins, Margulis, Brasier, Bell, Chaired by Denis Noble," Balliol College, Oxford University, archived debate, four parts, 2010, http://www.voicesfromoxford.com/homagedarwin_part1.html.

Nørretranders, Tor. *The User Illusion: Cutting Consciousness Down to Size*. New York: Viking Press, 1998.

Ryan, Frank. *Metamorphosis: The Beautiful Mystery*. White River Junction, VT: Chelsea Green Publishing, 2011.

Sagan, Dorion. "Introduction: Umwelt after Uexküll." In Jacob von Uexküll, *A Foray into the Worlds of Animals and Humans*, translated by Joseph D. O'Neil, afterword by Geoffrey Winthrop-Young. Posthumanities Series edited by Cary Wolfe, 1-34. Minneapolis: University of Minnesota Press, 2010.

Sagan, Dorion, and Jessica Hope Whiteside. "Gradient-Reduction Theory: Thermodynamics and the Purpose of Life." In *Scientists Debate Gaia: The Next Century*, edited by Stephen H. Schneider, James R. Miller, Eileen Crist, and Penelope J. Boston, 173-186. Cambridge, MA: MIT Press, 2004.

Sapolsky, Robert. "Toxo: A Conversation with Robert Sapolsky," *Edge* (2009), http://www.edge.org/documents/archive/edge307.html#tc.

Sapp, Jan. *The New Foundations of Evolution: On the Tree of Life*. New York: Oxford University Press, 2009.

Schneider, Eric D., and James J. Kay. "Life as a Manifestation of the Second Law of Thermodynamics." *Mathematical and Computer Modeling* 19 (1994): 25-48.

Schneider, Eric D., and Dorion Sagan. *Into the Cool: Energy Flow, Thermodynamics, and Life*. Chicago: University of Chicago Press, 2006.

Spinoza, Benedict. "Part 1, Concerning God, Prop. 17." In *The Ethics of Benedict de Spinoza* (1677), translated from the Latin by R. H. M. Elwes (1883). MTSU Philosophy WebWorks Hypertext Edition, http://frank.mtsu.edu/~rbombard/RB/Spinoza/ethica1.html#Prop.%20XVI

Stengers, Isabelle. *Cosmopolitics I*, translated by Robert Bononno. Minneapolis: University of Minnesota Press, 2010.

Sykes, Bryan. *Adam's Curse: A Future without Men*. New York: W. W. Norton & Company, 2004.

Vernadsky, Vladimir I. *The Biosphere: Complete Annotated Edition*. Translated by D. B. Langmuir. 1926. Reprint, New York: Springer, 1998.

Werren, John H. and David W. Lehlin. "Curing Wolbachia Infections in Nasonia (Parasitoid Wasp)." *Cold Spring Harbor Protocols* 5312, 2009.

Whitehead, Alfred North. *Science and the Modern World*. Lowell Lectures 1925. New York: The New American Library, 1962.

Whiteside, Jessica Hope, and Dorion Sagan. "Medical Symbiotics." In *Chimeras and Consciousness: Evolution of the Sensory Self*, edited by Lynn Margulis, Celeste Asikainen, and Wolfgang E. Krumbein, 207-218. Cambridge, MA: MIT Press, 2011.

Wicken, Jeffrey S. *Evolution, Thermodynamics and Information: Extending the Darwinian Program*. New York: Oxford University Press, 1987.

Chapter Ten
Of Termites and Men: On the Ontology of Collective Individuals

Brian G. Henning

It is likely that upon observing the effortless turning and looping ballet of a flock of pigeons or school of fish you have asked yourself the question "How do they do it?"[1] As Brian Partridge noted in a *Scientific American* essay from the 1980s:

> [This] question occurs naturally to anyone watching a school of silversides moving slowly over a reef in clear tropical waters. Hundreds of small silver fish glide in unison, more like a single organism than a collection of individuals. The school idles along on a straight course, then wheels suddenly; not a single fish is lost from the group. A barracuda darts from behind an outcropping of coral and the members of the school flash outward in an expanding sphere. The flash expansion dissolves the school in a fraction of a second, yet none of the fish collide. Moments later the scattered individuals collect in small groups; ultimately the school re-forms and continues to feed, lacking perhaps a member or two.[2]

Or consider the ostensibly simple act of a honeybee foraging for nectar as Bert Hölldobler and E. O. Wilson describe it in their 2009 collaboration *Superorganism*:

> Although simple in appearance, the act is a performance of high virtuosity. The forager was guided to this spot by dances of her nestmates that contained symbolic information about the direction, distance, and quality of the nectar source. To reach her destination, she traveled the bee equivalent of hundreds of human miles at bee-equivalent supersonic speed. She has arrived at an hour when the flowers are most likely to be richly productive. Now she closely inspects the willing blossoms by touch and smell and extracts the nectar with intricate movements of her legs and proboscis. Then she flies home in a straight line. All this she accomplishes with a brain the size of a grain of sand and with little or no prior experience.[3]

Finally, consider the complex forms of social organization achieved by African driver ants:

> Viewed from afar, the huge raiding column of a driver ant colony seems like a single living entity. It spreads like the pseudopodium of a giant amoeba across 70 meters or so of ground. . . . As the column emerges, it first resembles an expanding sheet and then metamorphoses into a treelike formation, with the trunk growing from the nest, the crown an advancing front the width of a small house, and numerous branches connecting the two. The swarm is leaderless. . . . These predatory feeder columns are rivers of ants coming and going. The frontal swarm, advancing at 20 meters an hour, engulfs all the ground and low vegetation in its path, gathering and killing all the insects and even snakes and other larger animals unable to escape. After a few hours, the direction of the flow is reversed, and the column drains backward into the nest holes.[4]

How indeed are such coordinated efforts possible? How can each of these simple-brained and seemingly independent individuals achieve such impressive acts of coordination, communication, and collaboration? Witnessing such performances, it is understandable why in the late nineteenth and early twentieth century some researchers believed that the corporate behavior of flocks of birds, schools of fish, and colonies of ants, bees, and termites involved some undiscovered form of telekinesis.

Organism, Superorganism, Mechanism

Contemporary research has instead revealed that swarms of birds, fish, and insects are in fact leaderless systems more akin to a single living organism than a mere collection of individuals.[5] The school, flock, and colony, it turns out, has just as much right to the title "individual" as does the solitary fish, bird, bee, ant, or termite.[6] Indeed, the degree of unity achieved by some societies of social insects—such as army ants, weaver ants, termites, and honeybees—is so great that many sociobiologists characterize them as "superorganisms." As one prominent researcher, Thomas Seeley, notes, "A colony of honey bees, for example, functions as an integrated whole and its members cannot survive on their own, yet individual honey bees are physically independent and closely resemble in physiology and morphology the solitary bees from which they evolved. In a colony of honey bees two levels of biological organization—organism and superorganism—coexist with equal prominence."[7]

The term "superorganism" was brought to prominent use in sociobiology in 1910 by William Morton Wheeler, who noted the striking similarities between, on the one hand, caste and division of labor in social insect colonies and, on the other hand, the functioning of cells and organs in individual organisms.[8] Individual members of a colony function in much the same way that individual cells do in the human body. For instance, just as particular cells in the body specialize and collectively perform certain functions within the body, particular ants or bees are members of specific "castes" that perform specific tasks, such as reproduction, defense, and food distribution. This isomorphism between an individual organism and a superorganism is nicely captured in the following table from Hölldobler and Wilson.

Organism	Superorganism
Cells	Colony members
Organs	Castes
Gonads	Reproductive castes
Somatic organs	Worker castes
Immune system	Defensive castes: alarm-defense communication; colony recognition labels
Circulatory system	Food distribution, including regurgitation between nestmates (trophallaxis), distribution of pheromones, and chemical cues
Sensory organs	Combined sensory apparatus of colony members
Nervous system	Communication and interactions among colony members
Skin, skeleton	Nest
Organogensis: growth and development of the embryo	Sociogenesis: growth and development of the colony

Figure 11.1: From *The Superorganism: The Beauty, Elegance, and Strangeness of Insect Societies* by Bert Hölldobler and Edward O. Wilson. Copyright © 2009 by Bert Hölldobler and Edward O. Wilson. Used by permission of W. W. Norton & Company, Inc.

What is all the more amazing is that this intense coordination of behavior is, in fact, "self-organizing"; there is no leader. By following very simple algorithms or decision-rules, colonies collectively achieve feats unthinkable by the individuals of which they are comprised, such as finding the shortest path to food, selecting a suitable nest site, defending the nest from invaders, maintaining a narrow range of optimal nest temperature, allocating workers to different tasks, distributing food.[9] "Nothing in the brain of a worker ant represents a blueprint of the social order" Hölldobler and Wilson write.

> There is no overseer or 'brain caste' who carries such a master plan in his head. Instead, colony life is the product of self-organization. The superorganism exists in the separate programmed responses of the organisms that compose it. The assembly instructions the organisms follow are the developmental algorithms, which create the castes, together with the behavioral algorithms, which are responsible for moment-to-moment behavior of the caste members.[10]

"Thus," Hölldobler and Wilson continue, "a distributed colony intelligence is created greater than the intelligence of any one of the members, sustained by the incessant pooling of information through communication."[11] It is the emergence of a "distributed colony intelligence" or what many researchers call "swarm intelligence" that makes such complex, integrated behavior possible.[12] As a collective individual, they achieve forms of social organization rivaled only by humans.[13]

These findings introduce many interesting and important metaphysical issues. Understanding how millions of insects can coordinate their behavior so closely that they become a single, collective individual introduces fascinating problems regarding individuality, identity, the boundary between living and non-living, the origin of societies, and perhaps a key to the evolutionary origins of consciousness itself. The philosophical question which seems most immediately pressing is, "How are we best to *explain*, not just describe, the emergence and maintenance of these forms of order?"

For their part, many scientists continue to use the metaphor of mechanism to describe the order of social insects. For instance, despite the fact that they describe insect societies as "emergent" forms of social order that arise through the collective "decision making" of individual insects, at times Bert Hölldobler and E. O. Wilson describe the colony as "a growth-maximizing machine"[14] composed of "cellular automata"[15] whose operations can be described by the language of physical and computer sciences.[16] Thomas Seeley is even more explicit in his use of the mechanistic model. "In choosing a nest site," he writes, "building a nest, collecting food, regulating the nest temperature, and deterring predators, a honey bee colony containing a queen resembles a smoothly running machine in which each part contributes to the efficient operation of the whole." Indeed, he goes even further, arguing that "It should be very revealing, and at most only slightly misleading, to view a honey bee colony as an integrated biological machine that promotes the success of the colony's genes."[17] Seeley's view is representative of both of the dominant trends within modern biology: molecular biology and neo-Darwinism.

As the physiological ecologist Scott Turner perceptively notes, the former, molecular biology, has "relentlessly pursued an understanding of life as a mechanism, as a special case of chemistry, physics, and thermodynamics."[18] The latter, neo-Darwinism, has come to focus exclusively on the transmission of genes, as is perhaps best represented by Richard Dawkins' "extended phenotype." An unintended consequence of these two trends, Turner notes, is the gradual disappearance of the very notion of an organism. For molecular biologists, "the organism itself has become, at best, an unwelcome distraction from the fascinating cellular and molecular business at hand" and for the neo-Darwinist, "the organism has become essentially an illusion, a wraith obscuring the 'real' biology of the genes, bound together in a conspiracy to promote the genetic interests of its members."[19] Much of Anglo-American philosophy is, for its part, largely in keeping with this account. Despite what I take to be its limited explanatory force, some version of mechanistic physicalism is so widely accepted among a certain segment of philosophers that it scarcely requires defense.

However, as the philosophers Alfred North Whitehead, Charles Sanders Peirce, William James, John Dewey, Henri Bergson, Pierre Teilhard de Chardin, and others so forcefully argued in response to an earlier generation of physicalists, the mechanistic metaphor cannot adequately do justice to the reality of living, evolving, striving, emoting, beings connected in interdependent social relations. Indeed, despite the fact that their own language betrays them at times, there is strong evidence to suggest that Hölldobler and Wilson perceive the inadequacy of the mechanistic model.

> Watched for only a few hours, a colony of social insects might be interpreted as consisting of automata driven with the same uniform set of decision rules. But that is far from the case. Each member of the colony is distinct in some manner or other that affects its behavior. Each has a mind of its own. By mind we do not mean a reflective, self-aware, wide-roaming consciousness of the human kind, but rather a cognitive consciousness built with a relatively complex brain that can store information from all its sensory modalities (taste, smell, touch, sight, and sound) as well as some memory of the events it has experienced during its short life.[20]

Though comparatively simple, an individual insect is not an interchangeable machine part, nor is it "a simple automaton."[21] Even on Hölldobler and Wilson's account, the algorithms upon which individual insects make their decisions are not rigidly deterministic, they are rather "central tendencies."[22]

If we are to be adequate to the beauty and dynamism of these complex societies of individuals, if we are to explain, and not just describe, how these patterns of behavior emerge and are perpetuated, we need a more adequate model of the relationships between individuals. We need, as it were, an explanatory framework for describing how social order can emerge. I will argue that, although not fully adequate, Alfred North Whitehead's "philosophy of organism" provides a more adequate conceptual system for explaining the ontological status of collective individuals and thereby points beyond the current hegemony of reductive molecular biology and neo-Darwinism.

Philosophy of Organism

Alfred North Whitehead (1861-1947) developed his "philosophy of organism" in opposition to two historical trends: the long-standing tradition of substance ontology, particularly as it came to be defined in the modern era by Rene Descartes, and the early twentieth century trend toward what Whitehead called "scientific materialism," of which physicalism is the contemporary heir.[23] According to Whitehead's organic view of individuality, there are no discrete individuals (or independent substances) mechanistically determined by absolute laws of nature. Although the mechanistic metaphor has been wildly successful, it ultimately does not do justice to the complex interrelations between individuals. We ought, Whitehead presciently insisted, to abandon the mechanistic metaphor in favor of the metaphor of organism, according to which individuals are determined by internal relations and nested within ever-expanding environments. "The only way of mitigating mechanism is by the discovery that it is not mechanism."[24]

On this organic model of individuality, the ontological fabric of the universe contains no true gaps.[25] Thus, the difference between, for instance, a wildflower and a boulder is ultimately found not in an appeal to different ontological kinds, but in the difference in the *degree* of "coordination" achieved by the occasions of which each is composed.[26] Whitehead writes, "The organic starting point is from the analysis of process as the realization of events disposed in an interlocking community. The event is the unit of things real. The emergent enduring pattern is the stabilization of the emergent achievement so as to become a fact which retains its identity throughout the process."[27]

The macroscopic objects which we experience (e.g., desks, bees, trees, rocks) are what Whitehead calls "nexūs" (the plural form of nexus) of actual occasions which are real, individual and particular "in the same sense" in which their constituent occasions are real, individual, and particular. Actually, to be more precise, entities such as bees and trees are particular types of nexūs which Whitehead refers to as "societies." While all societies are nexūs, not all nexūs are societies. For Whitehead, it is societies and nexūs, not actual occasions, which are the "things" that endure and that have adventures. For him, "The real actual things that endure are all societies. They are not actual occasions. It is the mistake that has thwarted European metaphysics from the time of the Greeks—namely, to confuse societies with the completely real things which are the actual occasions.[28]

On this view, a society is not an "aggregate" of "discrete," "externally related" beings held together in an "extrinsic unity." Rather, a society is a *socially ordered* nexus of *internally* related events that form an *intrinsic* unity. Societies are *not* mere collections, aggregates, or assemblages of entities to which the same class-name applies. This is the difference between a nexus and a society. Whereas a nexus is simply any real fact of togetherness, including extrinsic unities such as aggregate entities—for example, boulders—a society is a particular type of nexus which enjoys "social order." That is, a society's constituent occasions share a common, "defining characteristic" because of the conditions imposed upon them by their *internal* relatedness with previous members of that self-same society. Hence, contrary to aggregate entities, complex structured societies such as plants and animals are organic entities that, like systematic entities, are characterized by, as the philosopher Frederick Ferré puts it, "strong internal relations between parts that vary with one another and together perform a common function. The entity as a whole is what it is because of the [constitutive] interplay of these parts, and without them would cease to be an entity of that kind."[29]

All macroscopic individuality, on this reading, is a matter of order. If the degree of order is particularly high and the potential for novelty is introduced, then it is a living society. If it is higher still, it may be a personal society. Although a colony of weaver ants or honeybees may not have the same degree of intrinsic unity as a plant or animal, for instance, they are nonetheless real forms of togetherness with properties of their own. In this way, the organic model is better able to *explain* the unity of experience which perduring macroscopic individuals possess. A Whiteheadian organic model of individuality not only meets the challenge of providing an adequate account of the experienced unity of macroscopic individuals, but it does so with greater explanatory depth.

The Extended Organism

With this organic model of individuality in hand, consider the example of macrotermes colonies or African termites. As Scott Turner notes, opening a mound reveals "a capacious central chimney from which radiates a complex network of passages, connecting ultimately to an array of thin-walled tunnels that lie under the mound's surface like veins on an arm." Beyond the impressiveness of the construction, what is most surprising, Turner notes, is what you do not see—namely, termites. The mound, it turns out, is not built to house the millions of termites that continually maintain it. They live in a large spherical nest under the mound.[30] To understand the mound's purpose requires that we examine termites' dietary habits.

Termites, Turner explains, are unable to digest the bits of grass, bark, dead wood, and dung that they swallow. Instead, each species of termite cultivates a particular species of fungus that can break down the material into a digestable form. However, this digestive arrangement increases the oxygen requirement of the colony significantly, since the fungus requires five times the oxygen of the termites.[31] According to Turner,

This fungus, together with the bacteria and other soil microorganisms, raises the oxygen requirement to the amount needed by a cow. Indeed, ranchers in northern Namibia think of each termite ground mound as the equivalent of one livestock unit: each nest's foraging insects eat about the same quantity of grass as would one head of cattle. A cow buried alive would soon die without access to air, and so it is with a termite colony: without ventilation, it would suffocate.[32]

The mound, therefore, is not a residence or even a defensive structure, it is an external lung. By building the mound up vertically, the natural force of the wind exchanges the air through the network of capillary tunnels.[33] "Thus," Turner concludes, "the regulated environment, maintained by a constructed physiological organ—the mound—furthers the interests of both groups of inhabitants. The termite colony—insects, fungus, mound, and nest—becomes like any other body composed of functionally different parts working in concert and is ultimately capable of reproducing itself. Taken as a whole, the colony is an extended organism."[34] The subterranean nest is like the skin or skeleton of an organism, the fungus serves as its digestive system, the mound the respiratory system, various castes serve as the reproductive, sensory, immune, and nervous systems. Though a complete organic unity itself, a single termite is unintelligible apart from the collective organism of which it is a member. Indeed, as Turner himself notes at the end of his article, "Understanding the system requires thinking about the mound as not really an object but a process." "In the case of termite mounds," he continues, "the termites and fungi certainly qualify as living, but so does the mound, in a sense. After all, it does just what our lungs do for us. The primary difference is in perspective. For a human, what is inside the body is pretty clear, but for the termite colony, 'inside' includes the nest environment."[35]

Despite his critical approach, it is a mistake to infer that Turner is rejecting the modern synthesis. Indeed, he sees his position as complementary with, not contradictory to the gene-centric focus of molecular biology and its "extended phenotype." Turner's claim is not that evolutionary biology is incorrect, but that, by itself, it is inadequate. Whereas evolutionary biologists such as Richard Dawkins see the extended phenotype "as the extension of the action of genes beyond the outermost boundaries of an organism and asks how these extended phenotypes aid in the transmission of genes from one generation to the next,"[36] Turner's work sees the extended organism as the extension of the action of agents beyond the physical boundaries of an organism to include built structures and asks how these "extended organisms" might make evolution by natural selection possible and may in fact help explain the origins of life itself.

Turner's work with insects has led him to a much broader conclusion. In his 2007 book, *The Tinkerer's Accomplice*, he argues that "organisms are designed not so much because of natural selection of particular genes has made them that way, but because agents of homeostasis build them that way."[37] Indeed, he goes so far as to claim that "nothing about evolution makes sense except in light of the physiology that underpins it."[38] Turner contrasts his position with the dominant, gene-centric model in the following manner:

> Conventionally, Darwinian fitness is *thing*-based, measured in terms of replication of discrete things. In "traditional" Darwinism, for example, the replicate is the offspring, while to a Neo-Darwinist, it is the atom of heredity, the gene.... The fitter gene is the one whose bias reaches further into the future. A physiological process can also bias the future, and by this criterion could also qualify as heritable memory. In this instance, the forward reach in time is embodied in *persistence* of the process: how likely is it that the orderly stream of matter and energy that embodies the process will persist in the face of whatever perturbations are thrown at it? A fit process is therefore a persistent process: if a particular catalytic milieu, or a particular embodied physiology, can more persistently commandeer a stream of energy and matter than can another, the more persistent stream will be the fitter. Homeostasis, therefore, is the rough physiological equivalent of genetic fitness: a more robust homeostasis will ensure a system's persistence over a wider range of perturbations and further into the future than will a less robustly regulated system.[39]

As we see in Turner's contribution to the present volume, he argues that homeostasis, or the ability of an organism to maintain a stable internal environment, is the "second law" of biology, with natural selection being the first. The analogy here is with the laws of motion and just as the second law of motion is not reducible to or derivable from the first, he claims that homeostasis is a fundamental law of biology not reducible to or derivable from the first.

These scientific frontiers radically challenge both the substance and physicalist accounts of individuality. Individuals normally have clearly defined boundaries, a membrane that demarcates where they begin and end. Here we find that, as a single superorganism, the termite colony is extended in space and time, without clearly defined boundaries or a skin to define where the environment stops and the superorganism begins. Normally we would say that a single insect crawling on the ground is a proper individual. However, Turner's research shows that a single termite is no more an individual than a single cell in a petri dish solution. This research also muddles the usually sharp distinction between living and non-living. Here, inorganic soils, living insects, and fungus all constitute a single, collective individual. These built environments shape and determine the individuals that create them, often becoming a sort of external memory that shapes the evolutionary trajectory of the individuals that maintain them.

Recognizing that an entire colony—nest, mound, insect, and fungus—is a single, organic individual undermines the long-standing conception of individuals as discrete beings. As Turner puts it, "If the existence of physiological function is not dependent upon a clear partition of an organism from its environment, then there seems to be little reason to regard the organism as an entity discrete from its environment."[40] Whitehead's philosophy of organism provides a rich metaphysical basis for Turner's biological account of the "extended organism."

Returning to a holistic, organic model such as that developed by Whitehead provides an avenue for overcoming the mechanistic materialism that has come to dominate both science and philosophy throughout the late twentieth and early twenty-first century.

Earth System Science

It is worth noting, if only briefly, that these findings are not limited to social insects. One of the unexpected fruits of the unprecedented, worldwide scientific investigation of Earth's climate is the conclusion that our planet is not the lifeless rock it is normally taken to be. It is increasingly apparent that the Earth is a single, living system and must be studied as such. Indeed, this surprising conclusion is enshrined in the opening words of the 2001 "Amsterdam Declaration," signed by thousands of scientists at the European Geophysical Union, which states that "The Earth System behaves as a single, self-regulating system comprised of physical, chemical, biological, and human components."[41] Although there is still great controversy over what is meant by the term "self-regulating," this research is revealing a planet that functions as a dynamic whole.

As with Turner's work on termites, Earth System science is revealing that the reductive and mechanistic tendencies of neo-Darwinism is inadequate to account for the emergence of these forms of planetary-level homeostasis and self-regulation. As Richard Dawkins wrote in *The Extended Phenotype*, planetary-level homeostasis is not explicable via natural selection because it would "have all the notorious difficulties of 'group selection.'"[42] That is, it would be wide open to "cheats." "For instance," Dawkins writes, "if plants are supposed to make oxygen for the good of the biosphere, imagine a mutant plant which saved itself the costs of oxygen manufacture. Obviously it would outreproduce its more public-spirited colleagues, and genes for public-spiritedness would soon disappear."[43]

However, as Wilkinson,[44] Lenton,[45] and others have noted, planetary-level feedbacks and homeostatic regulation need only be *consistent* with natural selection, not be a *product* of it. For instance, Wilkinson argues that planetary level self-regulation could be the emergent result of "by-product mutualisms" similar to those found in population ecology.[46]

> In investment mutualisms both organisms provide some service to their partner at some cost to themselves, while in by-product mutualisms a waste product of one organism is used by its partner. Investment mutualisms are open to cheating (one partner could in theory reduce its investment while still taking the benefits).... However, many mutualisms are of the by-product type, in which there are no selective advantages to an organism's withholding its by-product. Indeed, if it were costly to prevent the partner from obtaining the by-product, then the subsequent fitness of a cheat would be lower than if it had cooperated

in supplying the by-product. . . . This avoids the criticisms of Dawkins, who, interestingly, used the example of oxygen production by plants, which is a by-product of oxygenic photosynthesis and thus not open to cheating.[47]

Dawkins and his physicalist philosophical friends' reductive emphasis on chemistry, physics, and thermodynamics is inadequate to explain the emergence of these system-level forms of regulation. A mechanistic ontology cannot make sense of the claim that the Earth System functions as "a single self-regulating system." Indeed, my claim is that traditional metaphysical accounts of individuality are unable to make sense of these non-traditional, but nevertheless real, forms of individuality.

I am suggesting that what is needed is a more robust metaphysical model that can make sense of systems of internally related, interdependent individuals that constitute integrated wholes with varying *degrees* of unity and identity. What is needed is a metaphysics that rejects absolute breaks between living and non-living, and between mental and physical, one that sees reality as a single, continuous whole. What is needed is a model that avoids static, inert conceptions of matter and recognizes the inherently dynamic, processive nature of reality. What is needed is a complex metaphysics that takes as its primary metaphor not inert bits of matter in a vast machine, but the metaphor of internally related organisms woven into systems of varying complexity such as developed by Alfred North Whitehead.

In the epilogue to *The Extended Organism*, Turner suggests that, although their explanatory success is indubitable, the main intellectual contributions of molecular biology and neo-Darwinism have already been achieved; their best insights are behind them. Molecular biology may have a rich future in developing new industrial applications, Turner writes, but "there is also nothing to come from it that will make us think about the world in a fundamentally different way." Similarly, as the proponents of neo-Darwinism are "engaged in endless rancorous debate over ever more arcane and abstract subjects" the field is becoming "scholastic" and "is now looking a bit frayed and dowdy."[48] If Turner is right that "the path to biology's next Golden Age will involve breaching the essentially arbitrary boundary between organisms and the environment, to create a biology that unifies the living and the inanimate worlds," there is good reason to believe that Whitehead's philosophy of organism will be an important part of the revolution.

Notes

1. Portions of this chapter originally appeared in the online magazine, *Global Spiral* under the title, "Swarms, Colonies, Flocks, and Schools: Exploring the Ontology of Collective Individuals," May 2009.

2. Brian L. Partridge, "The Structure and Function of Fish Schools," *Scientific American* 246, no. 6 (1982), 114.

3. Bert Hölldobler and E. O. Wilson, *The Superorganism: The Beauty, Elegance, and Strangeness of Insect Societies* (New York: Norton, 2009), 4.

As the prominent researcher Thomas Seeley noted in "The Honey Bee Colony as a Superorganism," *American Scientist* 77, no. 6 (1989), 550, "Within colonies there are various tappings, tuggings, shakings, buzzing, stroking, waggling, crossing of antennae, and puffings and streakings of chemicals, all of which seem to be communication signals. The result is that within a honey bee colony there exists an astonishingly intricate web of information pathways, the full magnitude of which is still only dimly perceived."

See also, Hölldobler and Wilson, *Superorganism*, 169:

> The essential element in the performance is the waggle run, or straight run; it is the middle piece of the figure-eight dance pattern, and it conveys the direction of the target during the outbound flight. Straight up on the vertical surface represents the direction of the sun the follower will see as she leaves the nest. If the target is on a line 40° to the right of the sun, say, the straight run is made 40° to the right of vertical on the comb.

And, also see Hölldobler and Wilson, *Superorganism*, 171: "The waggle dance conveys more information than just the direction of the outbound flight. The duration of the waggle run is correlated with the distance of the food site from the hive: farther away the site, the longer each waggle run takes. Circumstantial evidence suggests that the key element in the signal is the duration of the buzzing sound."

4. Hölldobler and Wilson, *Superorganism*, xx.

5. As Iain Couzin argues in "The Mob Rules." *BBC Wildlife* 22, no. 7 (2004), 38, "The organising principles employed by ants provide no evidence for leadership; in fact, they demonstrate that leadership is unnecessary to co-ordinate complex group behaviour. We now know that group behaviour may be co-ordinated by relatively simple interactions among the members of the group, a process termed 'self-organization'."

6. This is not to say that a flock of birds is as integrated as a colony of honeybees. As I will suggest, integration is a matter of degree.

7. Thomas D. Seeley, "The Honey Bee Colony as a Superorganism."

8. See Hölldobler and Wilson, *Superorganism*, 85. See also,

William Morton Wheeler, in his famous 1911 essay 'The Ant-Colony as an Organism,' brought the concept explicitly into sociobiology. 'The ant-colony is an organism,' he wrote, 'and not merely the analogue of the person.' The colony, Wheeler pointed out, has several diagnostic qualities of this status: (1) it behaves as a unit. (2) It shows some idiosyncrasies in behavior, size, and structure, some of which are peculiar to the species and others of which distinguish individual colonies belonging to the same species. (3) It undergoes a cycle of growth and reproduction that is clearly adaptive. (4) It is differentiated into 'germ plasm' (queens and males) and 'soma' (workers) (Hölldobler and Wilson, *Superorganism*, 10).

9. Peter Miller, "Swarm Theory," *National Geographic* 22, no. 1 (2007), 127-142.

As Sakata and Katayama put it in their article on ant colony defense systems,

> A colony of social insects, consisting of a large number of individuals, shows highly context-specific and well-organized behavior, similar to a sophisticated organism. . . . Within each colony, each individual gathers only a small part of the information necessary for decision-making, and makes only a limited re-

sponse. Sophisticated behavior of an organism often consists of responses of many subunits (e.g., organs and cells). A colony of social insects is an excellent model to observe the mechanism of information processing and decision-making through interactions among subunits with limited skills." (Hiroshi Sakata and Noboru Katayama, "Ant Defence System: A Mechanism Organizing Individual Responses into Efficient Collective Behavior," *Ecological Research* 16, no. 3 (2001): 395.

For a simple account of such rule following, see Hölldobler and Wilson 65:

Her decision rules can be stated as follows: (1) Not enough nectar collectors in the field? If yes, and if you also have immediate knowledge of a producing flower patch, perform the waggle dance. (2) Is the flower patch rich or the weather fine or the day early or does the colony need substantially more food? Perform the dance with appropriately greater vivacity and persistence. (3) Not enough active foragers to send into the field? Perform the shaking maneuver. (4) Not enough nectar processors in the hive to handle the nectar inflow? Perform the tremble dance. Hundreds of bees making such decisions more or less simultaneously yield the overall response of the superorganism.

10. Hölldobler and Wilson, *Superorganism*, 7.

11. Hölldobler and Wilson, *Superorganism*, 58-9.

12. Passino, Seeley, and Visscher give an excellent example of swarm cognition in their study of honeybee nest selection.

The swarm's distributed group memory is distinct from the internal neural-based memory of each individual bee. What is known by the swarm is actually *far more* than the sum of what is known by the individual bees, as the swarm's knowledge includes the information stored in the bees' brains *and the information coded in the locations of the bees and their actions*. No bee can know all the locations and activities for all other bees. But this information is coded at the swarm level and . . . is explicitly used in decision making ("Swarm Cognition in Honey Bees," *Behavioral Ecology and Sociobiology* 63, no. 3 (2008): 407, my emphasis.

13. See "The ants, bees, wasps, and termites are among the most socially advanced nonhuman organisms of which we have knowledge." Hölldobler and Wilson, *Superorganism*, xviii.

14. Hölldobler and Wilson, *Superorganism*, 114.

15. Hölldobler and Wilson, *Superorganism*, 53.

16. Hölldobler and Wilson, *Superorganism*, 58.

17. Seeley, "The Honey Bee Colony as a Superorganism," 549.

18. J. Scott Turner, *The Extended Organism: The Physiology of Animal-Built Structures* (Cambridge, MA: Harvard University Press, 2000), 2. See also chapter 8 in this volume.

19. Turner, *The Extended Organism*, 2.

20. Hölldobler and Wilson, *Superorganism*, 117. They continue, "Thus, while a brief surveillance of an insect colony may seem to disclose a confusing kaleidoscope of activity, longer periods of observation reveal that patterns are built from many quite individual minds linked by a high degree of organization. That amount of order is central to the colony's survival and reproduction" (119). "To conclude that the central tendencies of role change with aging have a genetic basis is not to imply that it is rigidly determined. Instead, the ensemble of genes that program the sequence of role changes have, like all such hereditary unites, a norm of reaction, the array of possible outcomes in physiology

and behavior determined by the interaction of the genes and the particular environment in which the development occurs (121).

21. Hölldobler and Wilson, *Superorganism,* 55.

22. Hölldobler and Wilson, *Superorganism,* 116. Although they would no doubt deny the connotation, the idea of a "central tendency" seems scarcely different than referring to final causality or a *telos*. As Turner argues in his chapter, there is real purpose in nature.

23. Although Whitehead called his thought the "philosophy of organism," it is often referred to as "process philosophy" because it conceives of reality in terms of dynamic pulses of energy, rather than static bits of matter.

24. Alfred North Whitehead, *Science and the Modern World* (New York: The Free Press, 1925), 76.

25. Alfred North Whitehead, *Process and Reality*, eds. David Ray Griffin and Donald W. Sherburne (1929; corr. ed., New York: Free Press, 1978), 110.

26. See Alfred North Whitehead's statement in *Adventures of Ideas* (New York: Free Press, 1933), 207,

It seems that, in bodies that are obviously living, a coördination has been achieved that raises into prominence some functionings inherent in the ultimate occasions. For lifeless matter these functionings thwart each other, and average out so as to produce a negligible total effect. In the case of living bodies the coördination intervenes, and the average effect of these intimate functionings has to be taken into account.

27. Whitehead, *Science and the Modern World*, 152.

28. Whitehead, *Adventures of Ideas*, 204.

29. Frederick Ferré, *Being and Value: Toward a Constructive Postmodern Metaphysics* (Albany: State University of New York Press, 1996), 337.

30. J. Scott Turner, "A Superorganism's Fuzzy Boundaries," *Natural History* 111, no. 6 (2002): 63.

31. See Turner, "A Superorganism's Fuzzy Boundaries," 63.

32. Turner, "A Superorganism's Fuzzy Boundaries," 63.

33. See Turner, "A Superorganism's Fuzzy Boundaries," 65.

The flow of the wind pushes air through the porous soil on the windward side and sucks it out on the leeward side, allowing the nest atmosphere to mix with fresh air from the outside world. This in itself is not surprising; lots of animals build structures that do similar things. What is remarkable is the pattern of ventilation: an in-and-out movement very similar to the way air flows into and out of our own lungs. In fact, what most distinguishes the action of the two 'organs' is that the termites' is powered by the ebb and flow of wind instead of by the contractions of muscle.

34. Turner, "A Superorganism's Fuzzy Boundaries," 66.

35. Turner, "A Superorganism's Fuzzy Boundaries," 67. For his part, Whitehead, writes in *Modes of Thought* (New York: Free Press, 1938), 21, that he is not even sure that we can define where the human organism begins and ends: "We think of ourselves as so intimately entwined in bodily life that a man is a complex unity—body and mind. But the body is part of the external world, continuous with it. In fact, it is just as much part of nature as anything else there—a river, or a mountain, or a cloud. Also, if we are fussily exact, we cannot define where a body begins and where external nature ends."

36. J. Scott Turner, *The Tinkerer's Accomplice: How Design Emerges from Life Itself* (Cambridge, MA: Harvard, 2007), 2.

37. Turner, *The Tinkerer's Accomplice*, 1.
38. Turner, *The Tinkerer's Accomplice*, 13.
39. Turner, *The Tinkerer's Accomplice*, 218-19.
40. Turner, *The Tinkerer's Accomplice*, 25.
41. James Lovelock, *The Vanishing Face of Gaia: A Final Warning* (New York: Basic Books, 2009), 179.
42. Dawkins, *The Extended Phenotype*, 236.
43. Dawkins, *The Extended Phenotype*, 236.
44. David M. Wilkinson, "Homeostatic Gaia: An Ecologist's Perspective on the Possibility of Regulation," in *Scientists Debate Gaia: A New Century*, ed. Stephen H. Schneider et al. (Cambridge, MA: MIT Press, 2004), 71-76.
45. Timothy M. Lenton, "Clarifying Gaia: Regulation with or without Natural Selection," in *Scientists Debate Gaia: A New Century*, ed. Stephen H. Schneider, et al. (Cambridge, MA: MIT Press, 2004), 15-25.
46. Wilkinson, "Homeostatic Gaia," 71.
47. Wilkinson, "Homeostatic Gaia," 73.
48. Turner, *Extended Organism*, 214.

Bibliography

Couzin, Iain. "Collective Minds." *Nature* 445, no. 15 (2007): 715-716.

———. "The Mob Rules." *BBC Wildlife*. 22.7 (2004): 36-40.

Croft, Darren, et al. "When Fish Shoals Meet: Outcomes for Evolution and Fisheries." *Fish and Fisheries* 4 (2003): 138-146.

Despland, Emma, Matthew Collett, and Stephen J. Simpson. "Small-Scale Processes in Desert Locust Swarm Formation: How Vegetation Patterns Influence Gregarization." *Oikos* 88, no. 3 (2000): 652-662.

Dawkins, Richard. *The Extended Phenotype: The Long Reach of the Gene*. New York: Oxford University Press, 1982 / 1999.

Ferré, Frederick. *Being and Value: Toward a Constructive Postmodern Metaphysics*. Albany: State University of New York Press, 1996.

Henning, Brian G. *The Ethics of Creativity: Beauty, Morality, and Nature in a Processive Cosmos*. Pittsburgh, PA: University of Pittsburgh Press, 2005.

———. "From Exception to Exemplification: Understanding the Debate over Darwin." In *Genesis, Evolution, and the Search for a Reasoned Faith*, edited by Mary Katherine Birge, Brian G. Henning, Rodica Soicoiu, and Ryan Taylor, 73-98. Winona, MN: Anselm Academic, 2011.

———. "Getting Substance to Go All the Way: Norris Clarke's Neo-Thomism and the Process Turn." *The Modern Schoolman* 81 (2004): 215-225.

Hölldobler, Bert, and E. O. Wilson. *The Superorganism: The Beauty, Elegance, and Strangeness of Insect Societies*. New York: W.W. Norton, 2009.

Hubbard, Simon, Petro Babak, Sven Th. Sigurdsson, Kjartan G. Magnússon. "A Model of the Formation of Fish Schools and Migrations of Fish." *Ecological Modeling* 174, no. 4 (2004): 359-374.

Lenton, Timothy M. "Clarifying Gaia: Regulation with or without Natural Selection." In *Scientists Debate Gaia: A New Century*, edited by Stephen H. Schneider et al., 15-25. Cambridge, MA: MIT Press, 2004.

Lovelock, James. *The Vanishing Face of Gaia: A Final Warning.* New York: Basic Books, 2009.
Miller, Peter. "Swarm Theory." *National Geographic* 22, no. 1 (2007): 127-142.
Passino, Kevin M., Thomas D. Seeley, and P. Kirk Visscher. "Swarm Cognition in Honey Bees." *Behavioral Ecology and Sociobiology* 63, no. 3 (2008): 401-414.
Partridge, Brian L. "The Structure and Function of Fish Schools." *Scientific American* 246, no. 6 (1982): 114-123.
Queller, David C., and Joan E. Strassman. "The Many Selves of Social Insects." *Science* 296, no. 5566 (2002): 311-331.
Sakata, Hiroshi, and Noboru Katayama. "Ant Defence System: A Mechanism Organizing Individual Responses into Efficient Collective Behavior." *Ecological Research* 16, no. 3 (2001): 395-403.
Seeley, Thomas D. "The Honey Bee Colony as a Superorganism." *American Scientist* 77, no. 6 (1989): 546-553.
Shaw, Evelyn. "The Schooling of Fishes." *Scientific American* 206, no. 6 (1962): 128-136.
Turner, J. Scott. *The Extended Organism: The Physiology of Animal-Built Structures.* Cambridge, MA: Harvard University Press, 2000.
———. "A Superorganism's Fuzzy Boundaries." *Natural History* 111, no. 6 (2002): 63-67.
———. "Gaia, Extended Organisms, and Emergent Homeostasis." In *Scientists Debate Gaia: A New Century*, edited by Stephen H. Schneider et al., 57-70. Cambridge, MA: MIT Press, 2004.
———. *The Tinkerer's Accomplice: How Design Emerges from Life Itself.* Cambridge, MA: Harvard, 2007.
Vabø, Rune, and Georg Skaret. "Emerging School Structures and Collective Dynamics in Spawning Herring: A Simulation Study." *Ecological Modeling* 214, nos. 2-4 (2008): 125-140.
Whitehead, Alfred North. *Science and the Modern World.* New York: Free Press, 1925.
———. *Process and Reality.* Edited by D. R. Griffin and D. W. Sherburne. 1929. Corrected edition, New York: The Free Press, 1978. Page references to corrected edition.
———. *Adventures of Ideas.* New York: Free Press, 1933.
———. *Modes of Thought.* New York: Free Press, 1938.
Wilkinson, David M. "Homeostatic Gaia: An Ecologist's Perspective on the Possibility of Regulation." In *Scientists Debate Gaia: A New Century*, edited by Stephen H. Schneider et al., 71-76. Cambridge, MA: MIT Press, 2004.

Section 4:
The Baldwin Effect, Behavior, and Evolution

Chapter Eleven
The Baldwin Effect in an Extended Evolutionary Synthesis

Bruce H. Weber

One consequence of developing an expanded and extended evolutionary synthesis informed by complex systems dynamics, as I described in chapter 1 of this volume, is that phenomena and/or theoretical approaches that were at best marginalized, if not outright rejected, by the Modern Evolutionary Synthesis can be revitalized and brought back within the discourse of evolutionary theory. As a case in point, I would like to consider the fate of notions of how behavior and mind might play a role in evolution, specifically the so-called "Baldwin Effect."

Darwin did consider the possibility that there might be some sort of mechanism by which a behavior that had survival value might become to be inherited.[1] But when August Weismann demonstrated the separation of somatic and germ line cells, it was inferred that such mechanisms, including genuine Lamarckian ones, were not possible.[2] This was taken to deprive evolutionary processes of the possibility of speeding up heritable change, the role of organismic agency, and in higher organisms a role for mind.[3] In 1896, James Mark Baldwin, Conway Lloyd Morgan, and Henry Fairfield Osborn independently published a proposal by which such phenomena could occur within a broadly Darwinian framework.[4] What has come to be known as "the Baldwin Effect" can be described as follows. If there is a stable, over generational time, selection pressure due to some aspect of the environment that sorts among non-heritable variations in behavior, and if that behavior can be transmitted by mimesis or other forms of learning, to the next generation, then, not only can there be selection for that behavior, but any genetic changes that make learning the behavior easier or that facilitate the execution of the behavior, will be favored. Thus the Baldwin Effect buys the lineage a breathing space for useful behavior until appropriate genetic support is selected.

The type of selection process Baldwin and Lloyd Morgan assumed was an all-or-nothing one.[5] In Fisher's model of selection as marginally changing gene frequencies each generation, the Baldwin Effect was still theoretically possible but was viewed as marginal at best.[6] Even so, as the Modern Evolutionary Synthesis was forged there was initial support for it by Julian Huxley.[7] But by the 1950s in the 'hardened' synthesis, Simpson saw the Baldwin Effect as causally inconsequential or trivially true at best, and Mayr argued that it should be discarded as it seemed to leave the door open to neo-Lamarckian mechanisms.[8] John Cobb suggests a possible deeper reason for this rejection—that purposeful behavior capable of being learned by con-specifics would imply that such organisms are not behaving in a mechanical or Newtonian manner.[9] However, after the Baldwin Effect was brought back to attention in the mathematical work of Hinton and Nowlan, who showed the usefulness of the Baldwin Effect in AI computer programs, interest was revived.[10] Indeed, Daniel Dennett, who used a computer-program metaphor of natural selection acting as an algorithm, was able to countenance the Baldwin Effect as an acceptable evolutionary mechanism.[11] Francisco Ayala has gone further, embracing the Baldwin Effect as enriching the causal power of natural selection thus enhancing modern evolutionary theory.[12]

Although defenders of the genocentrism of the Modern Evolutionary Synthesis, such as John Maynard Smith and Dennett, allow that the Baldwin Effect might play a role in Darwinian explanations, its status is still somewhat problematic.[13] Dennett goes out of his way to argue, using his metaphor in which the Baldwin Effect is a "crane" acting in a bottom-up manner rather than a "sky hook" that might act in a top-down fashion.[14] Robert Ulanowicz, however, offers an alternative metaphor of muscadine grapevine in which by natural processes top-down influence is a coherent concept.[15]

We can view the Baldwin Effect as belonging to a family of phenomena in which phenotypic processes are transmitted by learning or some other means in a broadly heritable way prior to genetic support for these adaptive activities. The notion of niche construction activity generalizes this concept to all organisms, not just those that can transmit learned behavior. Indeed, notions of niche construction proposed by Richard Lewontin and John Odling-Smee suggest that the behavior of organisms has not only causal agency but allows a process of co-adaptation of organisms and their environments.[16] It is plausible that the developmental plasticity explored by Susan Oyama, Paul Griffiths, Russell Gray, and Mary Jane West-Eberhard provides a further set of processes by which the Baldwin Effect could have causal consequences within the context of an expanded-extended evolutionary synthesis.[17] Terrence Deacon in particular has shown how, in the co-evolution of human brain and language, the Baldwin Effect processes could have played a major causal role, especially given the time frame for such evolution in our lineage, for humans to have evolved into a "symbolic species."[18]

Although the Baldwin Effect recently has been accepted by supporters of the Modern Evolutionary Synthesis in light of computational advances, it fits more comfortably in an expanded-extended evolutionary synthesis, informed by complex systems dynamics, that allows for selection to act at various levels and to interact with processes of development and self-organization.[19] Further, this shift thus focuses on processes more than objects or traits, and so has implications for process philosophy.[20] Such developments also can inform our continuing exploration of the process of emergence.[21]

Notes

1. Charles Darwin, *On the Origin of Species* (London: John Murray, 1859); Charles Darwin, *The Descent of Man* (London: John Murray, 1871).
2. August Weismann, *Essays on Heredity and Kindred Problems* (Oxford: Clarendon Press, 1891); August Weismann, "The All-Sufficiency of Natural Selection: A Reply to Herbert Spencer," *Contemporary Review* 64 (1893), 596-610.
3. David J. Depew, "Baldwin and His Many Effects," in *Evolution and Learning: The Baldwin Effect Reconsidered* (Cambridge, MA: MIT Press, 2003), 3-31.
4. James Mark Baldwin, "Heredity and Instinct," *Science* 3 (1896a): 438-441, 558-561; James Mark Baldwin, "Physical and Social Heredity," *American Naturalist* 30 (1896b): 422-428; James Mark Baldwin, "A New Factor in Evolution," *American Naturalist* 30 (1896c): 354, 422, 441-451, 536-553; Conway Lloyd Morgan, *Habit and Instinct* (London: Arnold, 1896a); Conway Lloyd Morgan, "Of Modification and Variation," *Science* 4 (1896b): 733-739; Henry Fairfield Osborn, "A Mode of Evolution Requiring Neither Natural Selection Nor the Inheritance of Acquired Characteristics," *Transactions of the New York Academy of Science* 15 (1896): 141-148.
5. Depew, "Baldwin and His Many Effects," 11.
6. Depew, "Baldwin and His Many Effects," 11.
7. Julian Huxley, *Evolution: The Modern Synthesis* (London: Allen and Unwin, 1942).
8. George Gaylord Simpson, "The Baldwin Effect," *Evolution* 7 (1953): 110-117; Ernst Mayr, *Animal Species and Evolution* (Cambridge, MA: Harvard University Press, 1963).
9. John Cobb, "Organisms as Agents in Evolution," in *Back to Darwin: A Richer Account of Evolution*, ed. John Cobb (Grand Rapids, MI: William B. Eerdmans Publishing Company, 2008), 215-241.
10. Geoffrey E. Hinton and Steven J. Nowlan, "How Learning Can Guide Evolution," *Complex Systems* 1 (1987): 495-502.
11. Daniel Dennett, *Consciousness Explained* (Boston: Little, Brown, and Company, 1991); Daniel Dennett, *Darwin's Dangerous Idea: Evolution and the Meanings of Life* (New York: Simon and Schuster, 1995); Daniel Dennett, "The Baldwin Effect: A Crane, Not a Skyhook," in *Evolution and Learning: The Baldwin Effect Reconsidered* (Cambridge, MA: MIT Press, 2003), 69-79.
12. Francisco Ayala, "The Baldwin Effect," in *Back to Darwin: A Richer Account of Evolution*, ed. John Cobb (Grand Rapids, MI: William B. Eerdmans Publishing Company, 2008), 193-195.
13. John Maynard Smith, "Natural Selection: When Learning Guides Evolution," in *Adaptive Individuals in Evolving Populations*, ed. R. K. Belew and M. Mitchell (Reading, MA: Addison-Wesley, 1996), 455-457.

14. Dennett, *Darwin's Dangerous Idea;* Dennett, "The Baldwin Effect."

15. Robert Ulanowicz, *A Third Window: Natural Life beyond Newton and Darwin* (West Conshohocken, PA: Templeton Press, 2009).

16. Richard Lewontin, "Genes, Organisms and Environment," in *Evolution From Molecules to Men* (Cambridge, MA: Cambridge University Press, 1983), 273-285; F. John Odling-Smee, "Niche-Constructing Phenotypes," in *The Role of Behavior in Evolution* (Cambridge MA: MIT Press, 1988), 73-132; F. John Odling-Smee, "Niche-Construction, Genetic Evolution, and Cultural Change," *Behavioural Processes* 35 (1995): 196-205; Paul E. Griffiths, "Beyond the Baldwin Effect: James Mark Baldwin's 'Social Heredity' Epigenetic Inheritance and Niche Construction," in *Evolution and Learning: The Baldwin Effect Reconsidered* (Cambridge, MA: MIT Press, 2003), 227-304.

17. Griffiths, "Beyond the Baldwin Effect"; Paul E. Griffiths and Russell D. Gray, "Developmental Systems and Evolutionary Explanations," *Journal of Philosophy* 91 (1994): 227-304; Paul E. Griffiths and Russell D. Gray, "Darwinism and Developmental Systems," in *Cycles of Contingency: Developmental Systems and Evolution* (Cambridge MA: MIT Press, 2001): 195-218; Susan Oyama, *The Ontogeny of Information: Developmental Systems and Evolution, 2nd Edition* (Durham, NC: Duke University Press, 2000); Susan Oyama, "Terms in Tension: What Do You Do When All the Good Words Are Taken?" in *Cycles of Contingency: Developmental Systems and Evolution* (Cambridge MA: MIT Press, 2001): 177-193; Susan Oyama, "On Having a Hammer," in *Evolution and Learning: The Baldwin Effect Reconsidered*, ed. Bruce H. Weber and David J. Depew, 169-191 (Cambridge, MA: MIT Press, 2003); Mary Jane West-Eberhard, *Developmental Plasticity and Evolution* (New York: Oxford University Press, 2003).

18. Terrence W. Deacon, *The Symbolic Species: The Co-Evolution of Language and the Brain* (New York: W. W. Norton and Company, 1997); Terrence W. Deacon, "Multilevel Selection in a Complex Adaptive System: The Problem of Language Origins," in *Evolution and Learning: The Baldwin Effect Reconsidered*, ed. Bruce H. Weber and David J. Depew, 81-106 (Cambridge, MA: MIT Press, 2003a); Bruce H. Weber, "Emergence of Mind and the Baldwin Effect," in *Evolution and Learning: The Baldwin Effect Reconsidered* (Cambridge, MA: MIT Press, 2003).

19. Weber, "Emergence of Mind and the Baldwin Effect," 309-326; Depew, "Baldwin and His Many Effects."

20. Ian G. Barbour, "Evolution and Process Thought," in *Back to Darwin: A Richer Account of Evolution*, ed. John Cobb (Grand Rapids, MI: William B. Eerdmans Publishing Company, 2008), 196-214; Ulanowicz, *A Third Window*.

21. Terrence W. Deacon, "The Hierarchic Logic of Emergence"; Weber, "Emergence of Mind: Untangling the Interdependence of Evolution and Self-Organization," in *Evolution and Learning: The Baldwin Effect Reconsidered*, ed. Bruce H. Weber and David J. Depew, 273-308 (Cambridge, MA: MIT Press, 2003b); Bruce H. Weber, "On the Emergence of Living Systems," *Biosemiotics* 2 (2009), 343-359; Bruce H. Weber, "Design and Its Discontents," *Synthèse* 178 (2011), 271-289.

Bibliography

Ayala, Francisco J. "The Baldwin Effect." In *Back to Darwin: A Richer Account of Evolution*, edited by J. Cobb, Jr., 193-195. Grand Rapids, MI: William B. Eerdmans Publishing Company, 2008.
Baldwin, James Mark. "Heredity and Instinct." *Science* 3 (March 20, April 10, 1896a): 438-441, 558-561.
———. "Physical and Social Heredity." *American Naturalist* 30 (1896b): 422-428.
———. "A New Factor in Evolution." *American Naturalist* 30 (1896c): 354, 422, 441-451, 536-553.
Barbour, Ian G. "Evolution and Process Thought." In *Back to Darwin: A Richer Account of Evolution*, edited by John Cobb, Jr., 196-214. Grand Rapids, MI: William B. Eerdmans Publishing Company, 2008.
Clayton, Philip. *Mind and Emergence: From Quantum to Consciousness*. New York: Oxford University Press, 2004.
Cobb, John. "Organisms as Agents in Evolution." In *Back to Darwin: A Richer Account of Evolution*, edited by J. Cobb, 215-241. Grand Rapids, MI: William B. Eerdmans Publishing Company, 2008.
Darwin, Charles. *On the Origin of Species by Means of Natural Selection or the Preservation of Favored Races in the Struggle for Life*. London: John Murray 1859 / Cambridge, MA: Harvard University Press, 1964.
———. *The Descent of Man*. London: John Murray, 1871.
Deacon, Terrence W. *The Symbolic Species: The Co-Evolution of Language and the Brain*. New York: W. W. Norton and Company, 1997.
———. "Multilevel Selection in a Complex Adaptive System: The Problem of Language Origins." In *Evolution and Learning: The Baldwin Effect Reconsidered*, edited by B. H. Weber and D. J. Depew, 81-106. Cambridge, MA: MIT Press, 2003a.
———. "The Hierarchic Logic of Emergence: Untangling the Interdependence of Evolution and Self-Organization." In *Evolution and Learning: The Baldwin Effect Reconsidered*, edited by B. H. Weber and D. J. Depew, 273-308. Cambridge, MA: MIT Press, 2003b.
Dennett, Daniel. *Consciousness Explained*. Boston: Little, Brown, and Company, 1991.
———. *Darwin's Dangerous Idea: Evolution and the Meanings of Life*. New York: Simon and Schuster, 1995.
———. "The Baldwin Effect: A Crane, Not a Skyhook." In *Evolution and Learning: The Baldwin Effect Reconsidered*, edited by B. H. Weber and D. J. Depew, 69-79. Cambridge, MA: MIT Press, 2003.
Depew, David J. "Baldwin and His Many Effects." In *Evolution and Learning: The Baldwin Effect Reconsidered*, edited by B. H. Weber and D. J. Depew, 3-31. Cambridge, MA: MIT Press, 2003.
Griffiths, Paul E. "Beyond the Baldwin Effect: James Mark Baldwin's 'Social Heredity' Epigenetic Inheritance and Niche Construction." In *Evolution and Learning: The Baldwin Effect Reconsidered*, edited by B. H. Weber and D. J. Depew, 193-215. Cambridge, MA: MIT Press, 2003.
Griffiths, Paul E., and Russell D. Gray. "Developmental Systems and Evolutionary Explanations." *Journal of Philosophy* 91 (1994): 227-304.

———. "Darwinism and Developmental Systems." In *Cycles of Contingency: Developmental Systems and Evolution*, edited by S. Oyama, P. E. Griffiths, and R. D. Gray, 195-218. Cambridge MA: MIT Press, 2001.

Hinton, Geoffrey E., and Steven J. Nowlan. "How Learning Can Guide Evolution." *Complex Systems* 1 (1987): 495-502.

Huxley, Julian. *Evolution: The Modern Synthesis*. London: Allen and Unwin, 1942.

Lewontin, Richard. "Genes, Organisms and Environment." In *Evolution from Molecules to Men*, edited by D. S. Bendall, 273-285. Cambridge, MA: Cambridge University Press, 1983.

Lloyd Morgan, Conway. *Habit and Instinct*. London: Arnold, 1896a.

———. "Of Modification and Variation." *Science* 4 (1896b): 733-739.

Macdonald, C., and G. Macdonald, eds. *Emergence in Mind*. New York: Oxford University Press, 2010.

Maynard Smith, John. "Natural Selection: When Learning Guides Evolution." In *Adaptive Individuals in Evolving Populations*, edited by R. K. Belew and M. Mitchell, 455-457. Reading, MA: Addison-Wesley, 1996.

Mayr, Ernst. *Animal Species and Evolution*. Cambridge, MA: Harvard University Press, 1963.

Odling-Smee, F. John. "Niche-Constructing Phenotypes." In *The Role of Behavior in Evolution*, edited by H. C. Plotkin, 73-132. Cambridge MA: MIT Press, 1988.

———. "Niche-Construction, Genetic Evolution, and Cultural Change." *Behavioural Processes* 35 (1995): 196-205.

Osborn, Henry Fairfield. "A Mode of Evolution Requiring Neither Natural Selection Nor the Inheritance of Acquired Characteristics." *Transactions of the New York Academy of Science* 15 (1896): 141-148.

Oyama, Susan. *The Ontogeny of Information: Developmental Systems and Evolution, 2nd Edition*. Cambridge, MA: Cambridge University Press, 1985 / Durham, NC: Duke University Press, 2000.

———. "Terms in Tension: What Do You Do When All the Good Words Are Taken?" In *Cycles of Contingency: Developmental Systems and Evolution*, edited by S. Oyama, P. E. Griffiths, and R. D. Gray, 177-193. Cambridge MA: MIT Press, 2001.

———. "On Having a Hammer." In *Evolution and Learning: The Baldwin Effect Reconsidered*, edited by B. H. Weber and D. J. Depew, 169-191. Cambridge, MA: MIT Press, 2003.

Simpson, George Gaylord. "The Baldwin Effect." *Evolution* 7 (1953): 110-117.

Ulanowicz, Robert, *A Third Window: Natural Life beyond Newton and Darwin*. West Conshohocken, PA: Templeton Press, 2009.

Weber, Bruce H. "Emergence of Mind and the Baldwin Effect." In *Evolution and Learning: The Baldwin Effect Reconsidered*, edited by B. H. Weber and D. J. Depew, 309-326. Cambridge, MA: MIT Press, 2003.

———. "On the Emergence of Living Systems." *Biosemiotics* 2 (2009): 343-359.

———. "Design and Its Discontents." *Synthèse* 178 (2011): 271-289.

Weber, Bruce H., and David J. Depew. "Natural Selection and Self-Organization: Dynamical Models as Clues to a New Evolutionary Synthesis." *Biology and Philosophy* 11 (1996): 33-65.

———. "Developmental Systems, Darwinian Evolution, and the Unity of Science." In *Cycles of Contingency: Developmental Systems and Evolution*, edited by S. Oyama, P. E. Griffiths, and R. D. Gray, 239-253. Cambridge MA: MIT Press, 2001.

Weismann, August. *Essays on Heredity and Kindred Problems.* Edited by E. Poulton et al. Oxford: Clarendon Press, 1891.
———. "The All-Sufficiency of Natural Selection: A Reply to Herbert Spencer." *Contemporary Review* 64 (1893): 309-338, 596-610.
West-Eberhard, Mary Jane. *Developmental Plasticity and Evolution.* New York: Oxford University Press, 2003.

Chapter Twelve
On the Ramifications of the Theory of Organic Selection for Environmental and Evolutionary Ethics

Adam C. Scarfe

Contemporary evolutionary theorists have been warming to the notion that genes can be "followers in evolution" and that "behavioral change often precedes and directs morphological change."[1] This warming to the notion that behavior can play an initiating role in evolution coincides with several developments in evolutionary theory, one being a surge of interest in the theory of organic selection, alternatively known as the "Baldwin Effect," over the past ten years. This chapter elucidates some of the main tenets of the theory of organic selection with a view to demonstrating how it can provide the basis for a novel environmental-evolutionary ethic that coheres with the inclusive evolutionary framework of the "extended synthesis"[2] that is emerging in biology.

The Theory of Organic Selection

The theory of "organic selection" was first postulated independently by evolutionary psychologist James Mark Baldwin (1861-1934), ethologist and psychologist Conway Lloyd Morgan (1852-1936), and paleontologist Henry Fairfield Osborn (1857-1935) in the mid-1890s. What was curious about this theory is that it provided a Darwinian explanation for the appearance of Lamarckian evolution. Precisely, the appearance of the inheritance of acquired characteristics could be explained with reference to the notion of natural selection working on, and in conjunction with, the behavioral selections of organisms. In turn, the theory of organic selection seemed to give behavior "its due" as a causal factor in the Darwinian evolutionary picture.

In *Development and Evolution* (1902), his most mature articulation of the theory of organic selection, Baldwin argued that organisms are agents in evolution and that natural selection operates "on the basis of what [organisms] do, rather than [merely] of what they are,"[3] in terms of their genetic makeups, their morphological structures, and their physiological features. For him, over the course of their lives, organisms within populations may appropriate, develop, and select novel behaviors or "good tricks"[4] by which they may procure what they need from their environment in order to survive—for example, new foraging behaviors, hunting strategies, and killing movements. Such novel good tricks may be arrived at through play, trial and error, the repetition of overproduced movements, spontaneous reflex action, chance, imitation, social learning, and by way of exploratory activities. Although the price of engaging in exploratory activities may impose a great energy cost on the organism, from a Baldwinian standpoint, they are especially important when a change in environmental circumstances requires a change in terms of behavioral habits, such as when an organism enters into a new environment and needs to exploit a new food source if it is to survive. Baldwin's analysis of the origin of behavioral novelties and new mental structures focuses on the fluctuating interplay of habit, by which he means "the repetition of what is worth repeating, and the conserving of this worth" and selective accommodation—namely, the selective "adaptation of the organism to new conditions, so that it secures, progressively, further reactions, which at an earlier stage would have been impossible."[5] The theory of organic selection places into question the ideas that function is necessarily determined by form and that the behavior of organisms, conceived of as a constitutive part of the organism's phenotype, is either fully subject to genotypic control or is fully determined by the environment. Rather, the theory of organic selection emphasizes that in the contexts of their processes of adaptation to the environment and of their modification of it, over the course of their lives, organisms may discover truly novel ways of employing their morphologies, and that behavior feeds back as a causal factor impacting upon heredity, thereby "prescribing the direction of evolution."[6]

According to Baldwin, since any advantage, even the most minute, is magnified in the struggle for existence, if a particular behavior or movement becomes requisite for survival, even provisionally, then individuals having phenotypic traits which serve to accentuate, amplify, or perfect the performance of the new behavior will be selected for. On the contrary, members of the population who are in similar circumstances that either do not or cannot obtain a degree of proficiency in respect to the novel behavior will more likely be eliminated. In subsequent generations of the variety, natural selection will ensure that the phenotypic characteristics of such organisms will evolve in a manner that is directed or channeled by the good trick. In other words, natural selection will favor phenotypic variations in the same direction that amplify the ability to perform the behavior and to perfect it. Baldwin attributes the term "orthoplasy" to "the general fact that evolution has a directive determination through organic selection."[7]

The combination of organisms selecting their behaviors and the resulting evolutionary channeling via natural selection is what is known more generally as the "Baldwin Effect."[8]

The Architects of the Modern Synthesis Critically Assess the "Baldwin Effect"

The architects of the modern synthesis were, in general, dismissive of the importance of the theory of organic selection as a widespread "mechanism" of evolutionary change. In 1942, Julian Huxley (1887-1975) emphasized that the theory of organic selection had been "unduly neglected"[9] by the evolutionists of his time, and he pointed out several instantiations of it which involved lice (*Pediculus*), the fruit fly (*Drosophila melanogaster*), the apple fly (*Rhagoletis pomonella*), the apple sucker (*Psylla mali*), the ermine moth (*Hyponomeuta padella*), and on other insects.[10] Yet, he still seemed to attribute to it the status of a "special case" rather than as a set of processes operating widely throughout nature.[11]

In 1953, George Gaylord Simpson (1902-1984) termed a version of the theory "the Baldwin Effect."[12] Simpson concluded that it was a "fully plausible" theory which could "be taken as a hypothesis subject to investigation," given that the "complex sequence of events" or "processes" it describes were "all known to occur separately."[13] However, Simpson downplayed it as having "little concrete ground for the view that it is a frequent and important element in adaptation."[14] He thought the examples that had been provided in support of it provided "possible" evidence for it, but they seemed "too trivial to establish that [its] role [is] universal or even [one that was] particularly important."[15]

In light of Simpson's critical assessment, in 1963, Ernst Mayr (1904-2005) dismissed the scientific legitimacy of the "the Baldwin Effect" outright. He claimed that it did not offer a "reconciliation between Lamarckism and Darwinism," as the title of Osborn's seminal paper had suggested,[16] because it was "Lamarckism pure and simple."[17] He also took issue with the "unproven claims" that had been given in support of it. Exemplifying the neo-Darwinist outlook, Mayr expressed that the theory of organic selection had "no validity"[18] as an explanatory framework because its originators did not duly emphasize that behavioral selection, plasticity, and rigidity are phenotypic traits that are controlled by the organism's genotype. His point is that behavior does not somehow float freely and independently apart from genotypic control. As such, Mayr concluded that it was "desirable to discard [the theory of organic selection] altogether."[19] That said, in later work, Mayr indicated that he was open to the notion that behavior played a crucial role in speciation.[20]

While being downplayed and criticized by the architects of the modern synthesis, over the last ten years, the theory of organic selection has seen a great surge in terms of interest and attention paid to it.[21] This interest is a testament to the validity of the underlying logic of the theory, even though there is a lack of widespread empirical data fully confirming the set of processes it identifies, as pervading nature. This interest is also the by-product of a suspicion that something has been missed due to the restrictiveness of neo-Darwinian categories and reductionist focus. Offering an explanatory "crane" rather than a "skyhook,"[22] the theory of organic selection is a fully falsifiable hypothesis. Research concerning it will continue to be at the forefront of the less genocentric, more inclusive and comprehensive, "extended synthesis" that is today gathering steam in evolutionary biology.

Examples of the Baldwin Effect

A lack of empirically verifiable examples has been a major criticism of the claim that the Baldwin Effect is a pervasive factor in evolution. As reasons for this lack, defenders of the theory point to the difficulties in observing events that involve novel shifts in behavior which lead to evolutionary changes in populations, the need to integrate multiple disciplines in carrying out such research,[23] and the deep time scales the theory deals with. Here, I provide three examples of the theory. The first example is hypothetical, and it serves to highlight the underlying logic of the theory of organic selection. The other two are based in recent empirical studies. These three examples serve to illustrate how a behavioral modification can initiate and can channel out the subsequent course of evolution in a population via natural selection, as Baldwin had described.

A hypothetical, although plausible, example that fits well with Baldwin's own descriptions of the scenario of organisms moving into a new environment is the evolution of marine iguanas (*Amblyrhynchus cristasus*), which are the only sea-going lizard species in the world. Marine iguanas live on the barren, volcanic islands of the Galapagos. While their origins are debated, it is believed that marine iguanas, a distinct species, are genetic descendants of other land-based iguana species that live in the tropical rainforests of South America, and may have arrived in the Galapagos via driftwood hundreds of thousands of years ago.[24] Finding themselves in a new, harsh, barren island environment where there was little to eat, the castaway iguanas may have engaged in exploratory activities, leading to the novel behavioral habit of diving into the water to feed almost exclusively on macrophytic marine algae,[25] procured by subtidal and intertidal foraging.[26] Like other iguana varieties, marine iguanas have a nasal gland that enables them to excrete and to remove salt from their bodies, in maintaining electrolyte balance. But they rely on this excretion system more than other species due to the fact that they digest large amounts of salt while underwater. In comparison with other iguana species, marine iguanas have evolved a blunter snout with which to effectively scrape algae off of rocks, and a horizontally flatter, longer, and more muscular tail than other iguana varieties, which enables them to propel themselves through the water. In comparison with other iguana varieties, they have larger hooked claws, which enable them to cling to rocks when exiting the water. Also, due to the fact that they forage in cold water, they must regulate their body temperatures. Marine iguanas are physiologically able to shunt blood away from the surface of their bodies to conserve heat, and they can reduce their heart rate and metabolism. They also warm up on land by huddling together gregariously[27] and basking in the sun on rocks near the shore for thermoregulation purposes. The marine iguana's unique dark coloration assists in this process. Darwin called them "imps of darkness" during his visit to the Galapagos on the Beagle. Marine iguanas suffered hardships in the 1980s when higher water temperatures affected the growth of undersea algae. This hypothetical example serves to demonstrate the logic behind the notion that a behavioral change in response to the selection pressures endemic to moving into a new environment can give rise, via natural selection, to morphological and physiological changes in a population that amplify the requisite behavioral habit.

A concrete example of the Baldwin Effect is highlighted in a 2006 study carried out by Jonathan Losos, entitled "Rapid Temporal Reversal in Predator-Driven Natural Selection,"[28] which is based on an experiment involving ground dwelling brown anole lizards (*Anolis sagrei*) on small islands in the Bahamas. Changes in the morphologies of the brown anole lizards were observed over several generations after the controlled introduction of curly-tail lizards (*Leiocephalus carinatus*), a larger predator, into their environment. Over the first six months of the experiment, brown anole lizards with longer legs were selected for because they had a faster running speed that assisted them to get away from the curly-tails. However, later, many of the brown anole lizards in the population seemed to adopt a new behavioral strategy that prevented predation more effectively. They became more arboreal and began perching on high branches where the bulky curly-tails could not give chase. Six months later, brown anole lizards with shorter legs, which better enabled them to climb and to reach their sanctuary, giving them an advantage over their longer-legged fellows who were poor climbers, were selected for. The study demonstrates how a behavioral shift preceded a morphological change, the type of change that we most readily associate with the term "evolution."[29]

Another example is provided by a 2009 study by Alexander Badyaev[30] who claims that Baldwin Effect processes are likely present in the colonization of North America by house finch (*Carpodacus mexicanus*) populations from 1850 to the present. Over this period, house finches expanded their range from a geographic strip running from southern Oregon to Oaxaca, Mexico, across the entire continent, occupying more diverse habitats. Faced with the harshness of climates and lower temperatures found in some of the newly inhabited regions (e.g., in Montana), and the fact that male house finch juveniles are more sensitive to colder temperatures than females, as an adaptive maternal strategy, breeding females altered their determinations of the order and timing of egg-laying as well as the length of incubation, thereby offsetting potential survival imbalances between males and females. Such sex-specific maternal resource allocation, affecting offspring growth, resulted in wide, locally advantageous divergences in sex-size dimorphism when comparisons were made between populations. In addition, females that biased their ovulation sequence more precisely in response to environmental circumstance maximized fledging success in comparison with females that did not. Hence, in subsequent generations, natural selection would favor those females whose physiological and hormonal constitutions accommodate the amplification of these maternal effects, thereby channeling out the course of subsequent evolutionary development in future generations of the population. Subsequent observations were made in respect to similar maternal effects in house finch populations in Arizona, in relation to seasonal nest mite (*Pellonyssus reedi*) infestations.

Organisms as "Loci of Valuative-Selective Activity"

The theory of organic selection illustrates how organisms, in the context of their adaptation to the environment, may be considered to have a degree of agency in evolutionary processes. It shows that organisms not only are "objects of selection"[31]—namely, the units upon which natural selection acts and whose fitness is tested, but they are also active *subjects* of selection.[32] Organisms are active developers, valuators, and selectors of their "structures of activity"[33] (i.e., the behavioral habits by which they adapt to, and modify, the natural environment—of which they are a composing part). In response to selection pressures and to the exigencies of the environment, such habits can also be broken up in favor of new structures of activity. And, as outlined above, according to Baldwin, processes of behavioral selection, in conjunction with natural selection, may channel out the direction of variations in subsequent generations of a population. They may do so no matter how limited the range of possible behavioral alternatives that are open to the organism, in consideration of the particular selection pressures the organism faces, as well as the capacities of its original morphological structure or of its physiological features in performing the novel behavior. Indeed, if the later Mayr is right that "almost invariably, a change in behavior is the crucial factor initiating evolutionary innovation,"[34] behavioral selections ought to be taken as a chief causal factor in speciation.

As active selectors of their behavioral habits in the context of their environments, organisms can be agents of predatory selection,[35] habitat selection,[36] host selection,[37] sexual selection,[38] community (or social) selection,[39] group selection,[40] kin selection,[41] artificial selection,[42] conservation selection,[43] psychosocial selection,[44] etc... and play active roles in the total process that is natural selection. In so far as selection processes at various biotic levels (micro- through macro-) and operating on various units is the efficient cause of evolutionary processes, organisms can be described as participating appendages in, or "loci" of, natural selection. This is because their agency in respect to these diverse selective activities impacts on the survivability of other organisms—namely, on whether the latter are preserved or eliminated in the struggle for existence. While Darwin personified nature as an "active scrutinizer of variations"[45] in describing natural selection, he rightly rejected the notion that nature, as a whole, is a conscious valuative and selective agency. But in so far as being an agent of selection infers valuation and a degree of mentality in doing so, organisms (at least in our contemporary evolutionary epoch) may alternatively be described as "loci of valuative[-selective] activity,"[46] a notion that emphasizes their status not only as objects of selection, but as subjects of selection. In advancing the notion that organisms in our contemporary evolutionary epoch are loci of selective activity, I do not intend for this concept to be employed in the "nothing but" sense, for organisms are not simply reducible to it.

The notion of organisms being loci of valuative-selective activity corresponds with some aspects of biosemiotic theory (see chapters 6 and 7), but also with Alfred North Whitehead's (1861-1947) philosophy of organism. Whitehead maintained that organismic experience is constituted by operations of valuation and selective appropriation (what he calls "prehensions"[47]) across a continuum of levels of creaturely experience: from physical feelings, to mental feelings, to consciousness. Like Darwin, who postulated that human mentality only differs from non-human mentality in terms of degree,[48] Whitehead emphasized that non-human organisms have a degree of mentality. This view runs counter to the 16th to 17th century Newtonian-Hobbesian-Cartesian belief that organisms are passive objects, devoid of feeling, whose functioning is fully determined by a set of mechanistic causal laws acting upon them. At the same time, Whitehead avoided the anthropocentrism that is generally involved in the employment of the term "mentality." While for Whitehead physical feeling was pervasive in nature, consciousness, which is marked by intense operations of selectivity—namely, of discrimination, comparison, negation, elimination, and judgment—occurs rarely, even in respect to the experience of so-called "higher organisms."[49]

As loci of valuative-selective activity, while no single organism or set of organisms "speaks for" the total "complex of related entities"[50] that constitutes nature, from a Baldwinian perspective, organisms are orthoplasic influences, making decisions,[51] be they non-conscious, proto-conscious, or conscious, which not only serve to help chart the evolutionary trajectory of their own species via natural selection, but which impact on the evolutionary destinies of other organisms. In this respect, for humankind, the theory of organic selection takes on great significance for environmental and evolutionary ethics.

Non-Reductionistic Critical Pan-Selectionism: An Evolutionary-Environmental Ethic

The theory of organic selection serves to accentuate the notion that the immediate actions of human beings have far-reaching evolutionary consequences over the long term, not only for themselves, but also for most life-forms on the planet. Today, humanity's all-out emphasis on instrumental reason and technology are the chief ways by which it exerts its selective powers and "directs its attack on the environment,"[52] thereby enabling us to procure what we desire from it; to cultivate, to prune, and to modify our habitat (e.g., by means of selecting out "undesirable" organisms in favor of "desirable" ones), expressing our valuations in doing so. In comparison with the selective activities of other organisms, *avant garde* evolutionary theorists Jablonka and Lamb (2005) point out that

> without doubt, humans are the major selective agents on our planet, and have carried out the most dramatic reconstruction [usually destruction] of environments. Today, in addition to changing plants and animals by artificial selection, humans can alter the genetic, epigenetic, and behavioral state of organisms by direct genetic, physiological, and behavioral manipulation.[53]

In similar fashion, Wcislo (1989) states that "many examples of human behavior show how schemes of artificial selection have profoundly influenced evolutionary change in domesticated plants and animals, and in their pests."[54] The morphological, physiological, and behavioral effects of artificial selection on non-human organisms have long been studied.[55] In parallel fashion, environmental ethics concerns the impact of human selective activity on wild nature as such, given that humankind's dominant position essentially assures it a form of artificial selection over all life-forms on the planet.

While nature is a complex interdependent system in which biological processes pass through the physical and conceptual boundaries erected and postulated by human beings, humankind's selective activities have impacted most, if not all, geographical regions and ecosystems of the world. The theory of organic selection highlights how decisions made by human beings, which advance their narrow short-term aims and which stem from a reductionist focus, may unwittingly alter the long-term evolutionary trajectories of the organisms living within them. And they may do so in adverse and in unpredictable (i.e., non-targeted) ways—for example, in manners favoring invasive species that may pose threats to native species and create imbalances in ecosystems;[56] or in releasing genetically modified strains of organisms into the environment which may cross-pollinate with, and outstrip, non-GM crops;[57] or in introducing systemic pesticides to render agriculture "more efficient," but which are a prime suspect in the effort to understand the causes of colony collapse disorder in honeybee populations;[58] or in developing genetically engineered and farmed salmon for human consumption, but unwittingly infecting wild stocks with infectious anemia, sea lice, and other diseases as a result;[59] or in pursuing modes of development that depend exclusively on the selected (and now widespread) habit of burning fossil fuels, thereby contributing to global warming and climate change.

The need for human beings to reflect critically on their selective activities is exemplified, in an analogous way, by a 2002 study on predatory selection, entitled "Multilevel Selection and the Evolution of Predatory Restraint,"[60] that was carried out by Joshua Mitteldorf et al. and involved computer models. The study postulates that predators are dependent for their survival on the sustainable renewal of populations of its prey. As a locus of selective activity, predators that do not exhibit restraint, and instead go to excess in nourishing themselves over the short term by consuming and depleting entire populations of its prey— perhaps overpopulating itself in doing so—may outstrip the means of their own existence. Such self-regulation is perhaps a trait that has been selected for, the truly successful predator being one that is neither deficient in providing the means of its own subsistence nor excessive by outstripping the very fact of its dependence on other organisms for its own subsistence.

Obviously, this study offers to us a metaphor that expresses why human beings would be wise to reflect critically on their selective activities and to embrace more sustainable ways of providing for themselves. The term "sustainable development" infers a similar Aristotelian "golden mean" between the deficiency of mere subsistence and the excess of overdevelopment. Sustainable development means that development should be kept at levels that do not exceed the capacity of the biosphere to absorb its effects (e.g., pollution). It also means limiting the consumption of renewable resources to a rate that is consistent with their natural replenishment, and limiting the use of non-renewable resources that takes into account the needs of future generations. Rather than hunting, for example, the highly desirable Bluefin Tuna (*Thunnus thynnus*) to extinction over a single generation,[61] using industrial-grade trawlers and nets, human beings would be wise to reflect critically on their selective activities and to engage in sustainable fishing practices, so as to be able to pass on the bounty of this precious resource to future generations.

There is a great need for human beings to reflect on their selective activities in general. The ecological imperative involves promoting what Conrad Hal Waddington (1905-1975) called "biological wisdom"[62] as well as values and lifestyles which are reflective of the realization of organic interdependence,[63] which will inform our selective activities. The realization of organic interdependence means an understanding of the fact that human beings are dependent on other organisms and on the biosphere as a whole for their long-term wellbeing. Stepping back and reflecting critically on the notion that humankind's selective activities may impact heavily on the evolutionary destinies of our own species as well as those of non-human organisms, is the meaning of the holistic environmental-evolutionary ethic I have elsewhere called "non-reductionistic, critical pan-selectionism."[64]

Having the theory of organic selection as its basis, non-reductionistic critical pan-selectionism holds that organisms, by way of their selective activities, are causally efficacious participants in respect to the eliminations and preservations belonging to the total process that is natural selection. It places natural selection into context with the emphasis on organic interdependence that is commonly advanced by environmental ethicists, integrating them together. Emphasizing the need for a phase of holistic reflection in the research processes of science and technological development, it challenges selections made solely on the basis of narrow, reductionist perspectives and short-term instrumental aims. It seeks to challenge eugenics programs,[65] selective cloning, genetic manipulation, gender-selective abortion, and other selective activities and discriminative practices. That said, organisms cannot somehow escape from operations of selectivity, so as to attain a state of "selective neutrality" where there are no consequences for other organisms. However, in respect to human beings, given the potential for our selective activities to have a long-term impact on our ecological life-support system, non-reductionist critical pan-selectionism emphasizes the need for critical moral deliberation in relation to our selective activities in general, with a view to mitigate our "selective footprints" through environmentally-evolutionarily ethical *praxis*.

The Theory of Organic Selection and Criticisms of Naturalistic Accounts of Morality

Given that Darwinian natural selection enforces the notion that biological reality is, as characterized by Herbert Spencer (1820-1903), constituted by a relentless struggle in which only the fittest survive,[66] evolutionary ethicists have long been puzzled in respect to the question of how to maintain a coherent, naturalistic sense of morality. In attempting to keep the harsh biological realities Darwin described from infiltrating human culture and society, in 1893, T. H. Huxley (1825-1895) stated, "Let us understand, once and for all, that the ethical progress of society depends, not on imitating the cosmic process, still less in running away from it, but in combating it."[67] And in response to the Spencerian notions that what has evolved via natural selection is good and that competition should not be dampened in society in favor of helping those of lower rank, G. E. Moore (1873-1958) sought to erect a firewall between natural selection as an 'is' and how we 'ought' to behave, by invoking his naturalistic fallacy.

More recently, it has perhaps largely been because of the apparent incompatibility between the realities of biotic competition and a morality that has its basis in the recognition of biological interdependence, as well as the lack of an adequate solution to this apparent incompatibility, that contemporary environmental ethicists have tended to steer clear of evolutionary biology altogether.[68] However, as has been articulated above, the theory of organic selection can provide the basis of an ethic that resolves these difficulties. This is because in elucidating the role of behavior in the evolutionary process, it serves to point the way as to how one may integrate the notions of natural selection and biological interdependence together. In addition, non-reductionistic critical pan-selectionism can withstand a central criticism that has plagued other naturalistic accounts of morality.

Influenced by Utilitarian ethical theory, Darwin attempted to deal with the apparent logical incompatibility between biotic competition and morality in *The Descent of Man* (1871). He postulated that while organisms are engaged in a struggle against other organisms for survival, central components of what is deemed "morality"—namely, human sympathy, altruism, and cooperativeness, could be explained as advantageous traits that have been selected for over the course of evolutionary time. Such traits would have been enforced through communal approbation, such that they would become progressively habitual, instinctual, and heritable. In explaining the origins of morality, Darwin reduced it to a by-product of the evolutionary mechanism of "community" or "social" selection, whereby in the evolutionary past, groups and tribes that were galvanized together by common symbols, life-meanings, expectations for behavior, and senses of obligation toward others outcompeted other tribes. Membership in tribes thus organized heightened survival value, enabling individuals to call on others to help ward off predators and enemies, and to have access to their skills, goods, and labor. In addition, traits that promoted peaceful social living served to maximize the potential for reproductive success. Darwin writes that in respect to humanity and to other species "those communities, which included the greatest number of the most sympathetic members, would flourish best and rear the greatest number of off-spring."[69] Darwin was also open to the idea that animals operating gregariously in packs and social insects exhibit something like a proto-morality *qua* community selection.[70] Today, in evolutionary biology, the concept of "inclusive fitness" incorporates the fitness contributions of cooperative group members toward the individual's capacity for reproductive success.

Friedrich Nietzsche (1844-1900) was one of the first philosophers to recognize the decisiveness of Darwin's account of evolution and its colossal implications for metaphysics, epistemology, and for ethics—namely, that it "necessitate[d] a radical overhaul of traditional metaphysics and ethics."[71] In fact, Darwin's articulation of the theory of natural selection was a necessary condition for Nietzsche's proclamation of the death of God, his emphasis on the will-to-power, and his confrontation with meaninglessness, nothingness, anxiety, and absurdity. To be sure, in *The Phenomenon of Life* (1966), Hans Jonas claims that "Nietzsche's [confrontation with] nihilism and his attempt to overcome it are demonstrably connected with the impact of Darwinism. The will to power seemed the only alternative left if the original essence of man had evaporated in the transitoriness and whimsicality of the evolutionary process."[72] However, while Nietzsche appropriated from Darwin's theses heavily in his early work, in his more mature writings such as *On the Genealogy of Morality* (1887) he was highly critical of Darwin's highly reductionist account of the evolution of morality.

From a Nietzschean perspective, Darwin's reductionist account of the origin of morality concerned the origination of what he viewed as an "all-too-human," "animal herd," or "slave" morality,[73] having its basis in mass conformation to established norms, thus making human behavior predictable so as to ensure the organization and the efficient functioning of human communities. More importantly, not only for Nietzsche was the reductionist form of morality Darwin had described, based in heightening survival value and maximizing the prospect of reproductive success, representative of a herd morality, it was ethical egoism,[74] and further, it was nihilism. While tribes and societies can often operate in "immoral ways"[75] (conceived of in the non-naturalistic sense) that may be communally approved of, for Nietzsche, the sovereign individual who, in "pursuit of creative self-affirmation,"[76] deviates critically from their codes and instead creates his or her own values, is more apt to be eliminated in the struggle for existence than preserved. Deviating from Darwin, Nietzsche developed an alternative naturalism that saw life as the clash of individual wills-to-power, beyond the former's emphasis on survival and reproductive success. He wrote, "A living thing seeks above all to discharge its strength-life itself is will to power; self-preservation is only one of the indirect and most frequent results."[77]

Given that the intent of this chapter is to argue that the theory of organic selection is the basis for a novel evolutionary-environmental ethic, I must respond briefly to the imminent question of how it addresses Nietzsche's criticism of Darwin's account of the origin of morality. To do so, I will refer to several affinities between Nietzsche's highly individualist philosophy and the theory of organic selection. First, the theory of organic selection highlights the importance of the individual specimen as a source of creative behavioral modifications and the selective valuation of novel good tricks, in helping to chart the evolutionary destiny not only of their own species, but also of others. Daniel Dennett's "pole diagrams" in *Darwin's Dangerous Idea*, which he employs to explain the Baldwin Effect, illustrate how novel good tricks may be appropriated from the individuals who developed them and disbursed more widely among members of a population;[78] this would have interested Nietzsche greatly.

Second, the theory of organic selection emphasizes the type of valuative-selective agency on the part of the individual that can operate critically against "immoral" communities. In addition, non-reductionistic critical pan-selectionism, emphasizing the need to reflect critically on our selective activities, can stand in contrast to the type of group-thinking that seems to pervade Darwin's naturalistic account of morality. That said, it would be foolish to think that we can jettison a concern for the consequences in terms of survival and reproductive success from our moral deliberations, and still act ethically. It would also be foolish to jettison concerns for the long-term evolutionary consequences our immediate actions have on other organisms in light of the fact that according to the theory of organic selection, we are orthoplasic influences on our species and on other organisms. The point is to see that morality may include, but cannot be reduced to, that which maximizes survival value and heightens the prospects of reproductive success. Rather, acting morally necessitates a multi-perspectival outlook, weighing diverse criteria and working out their antagonisms. Morality is also to be viewed as being a product of evolution, but as evolving or *in process*, given the exigencies of world problems such as overpopulation that did not exist previously, as well as world events, relationships, and new pressures placed on us by our environment. Furthermore, leading moral theories have been selected for (in the intellectual sense) in our current evolutionary epoch, and they may also be decentered in favor of new ones. But the fact that morality has a temporal aspect does not necessitate the relegation of ourselves to moral relativism.

Third, while the theory of organic selection makes no prescription for what is desirable in terms of evolutionary change, the ethic that may be derived from it is centrally concerned for the long-term trajectory of evolutionary change, and not merely concerned for heightening survival value and maximizing the potentiality for reproductive success. In moral deliberation, the concerns for how future generations will live and behave, for who and what our offspring will become, for what opportunities they will have during their lives, for their health and well-being, for what sort of world that they will live in, and for the overall quality of the environment that they will inherit, are as important as the concern for putting them into existence in the first place.[79]

Pointing to the evolutionary importance of both ethical reflection and of education, our selective activities in the here and now sow the seeds for what will transpire further down the evolutionary track (e.g., whether we leave to our descendants an ecologically depleted planet and/or render various species extinct). As such, because the theory of organic selection addresses Nietzsche's chief criticism of Darwin's account of the origin of morality, he might have been more impressed with the Baldwin Effect than not. After all, as partially embracing a quasi-Lamarckian perspective, Nietzsche was greatly interested in the notion of a willed evolution, and in what today amounts to *transhumanism*. While the Baldwin Effect most certainly does not legitimate such Nietzschean hopes, and the evolutionary-environmental ethic of non-reductionist critical pan-selectionism I have advanced in this chapter challenges them, the theory of organic selection does point to the existence of a degree of agency on the part of organisms in evolutionary processes. From these considerations, the emergence of the theory of organic selection in evolutionary biology, as well as of some of the other new frontiers of research in biology that are contributing to an "extended synthesis," will open up new doors and windows for philosophers to explore, and to inform us about, the natures of life and the human condition.

Notes

1. Mary Jane West-Eberhard, *Developmental Plasticity and Evolution* (New York: Oxford University Press, 2003), 157, 24.
2. Massimo Pigliucci and Gerd B. Müller, eds., *Evolution: The Extended Synthesis* (Cambridge, MA: MIT Press, 2010), 3.
3. James Mark Baldwin, *Development and Evolution* (New York: The Macmillan Company, 1902; New York: Elibron Classics, 2005), 117.
4. Daniel Dennett, *Darwin's Dangerous Idea* (New York: Simon & Schuster, 1995), 77.
5. James Mark Baldwin, *Mental Development in the Child and the Race* (New York: The Macmillan Company, 1894; Whitefish, MT: Kessinger Publishing, 2006), 161.
6. Julian Huxley, *Evolution: The Modern Synthesis* (Cambridge, MA: MIT Press, 2010), 524.
7. Baldwin, *Development and Evolution*, 142.
8. Erika Crispo, "The Baldwin Effect and Genetic Assimilation: Revisiting Two Mechanisms of Evolutionary Change Mediated by Phenotypic Plasticity," *Evolution* 61, no. 11 (2007): 2469.
9. J. Huxley, *Evolution: The Modern Synthesis*, 524.
10. J. Huxley, *Evolution: The Modern Synthesis*, 304-305, 304, 296, 296-297, 297.
11. J. Huxley, *Evolution: The Modern Synthesis*, 523-524. The assessment that it was a "special case" in the sense that it could be downplayed may be subject to interpretation.
12. George Gaylord Simpson, "The Baldwin Effect," *Evolution* 7, no. 2 (June 1953): 110, 112.

Recently, Patrick Bateson in "The Adaptability Driver: Links between Behavior and Evolution," *Biological Theory* 1, no. 4 (2006): 344, and in Bateson and Gluckman's *Plasticity, Robustness, Development, and Evolution* (Cambridge, MA: Cambridge University Press, 2011), 108, has objected to Simpson's term, "the Baldwin Effect," as an accurate name for the theory. He attributes the origin of the theory to Douglas Spalding (1873), rather than to Baldwin, Morgan, and Osborn. Morgan's research followed Spalding's, and Baldwin cites Spalding. Bateson characterizes the theory as "a behavioral driver in evolution" where there is "a mismatch between an organism's phenotype and its environment." As such, he seeks to rename the theory the "adaptability driver."

On this note, according to Robert J. Richards in *Darwin and the Emergence of Evolutionary Theories of Mind and Behavior* (Chicago, University of Chicago Press, 1987), 490-502, while Baldwin "never overtly claimed priority" for its discovery, nor did anything "intentionally deceitful" to assert his claim over it, he did the best job of the three in controlling the language surrounding discussion of the theory. Also, Baldwin repeatedly alluded to the genesis of the idea in his work, and published the others' papers as fully-acknowledged appendices in his work. For these reasons, Baldwin's name was largely given priority in the advancement of the notion in the scholarly and scientific culture, until it was dubbed the "the Baldwin Effect" by Simpson.

Although most scholars seem to embrace it, not only does it characterize the theory as simply Baldwin's discovery, but the word "effect" implies a focus on the net evolutionary result of the phenomenon, thereby neglecting Baldwin's original reference to the role of organisms in making individual accommodations as an indirect cause of the direction of the evolutionary processes. This, I believe, implies a mechanistic interpretation of the theory. Therefore, I prefer to return to Baldwin's term, "organic selection." This stance in terms of terminology is also helpful in deciphering between Baldwin's original and authentic formulations of the theory, some of its more contemporary neo-Baldwinian interpretations, as well as even looser, more popular interpretations which use the term to describe pretty much any quirk or wrinkle that is at odds with the standard version of evolutionary theory.

13. Simpson, "The Baldwin Effect," 115, 112.
14. Simpson, "The Baldwin Effect," 115.
15. Simpson, "The Baldwin Effect," 114, my additions.
16. Henry Fairfield Osborn, "A Mode of Evolution Requiring Neither Natural Selection Nor the Inheritance of Acquired Characteristics," *Transactions of the New York Academy of Science* 15 (1896): 141-142.
17. Ernst Mayr, *Animal Species and Evolution* (Cambridge, MA: Belknap Press of Harvard University Press, 1963), 610.
18. Mayr, *Animal Species and Evolution*, 612.
19. Mayr, *Animal Species and Evolution*, 611, my addition.
20. See William T. Wcislo, "Behavioral Environments and Evolutionary Change," *Annual Review of Ecology and Systematics* 20 (1989): 142.
21. For example, see the following sources: John M. Broughton and D. John Freeman-Moir, eds. *The Cognitive Developmental Psychology of James Mark Baldwin* (Norwood, NJ: Ablex Publishing Corporation, 1982); Robert J. Richards, *Darwin and the Emergence of Evolutionary Theories of Mind and Behavior*; Geoffrey E. Hinton, and Steven J. Nowlan, "How Learning Can Guide Evolution," *Complex Systems* 1 (1987): 447-454; Henry Plotkin, *Evolutionary Thought in Psychology: A Brief History* (Malden, MA: Blackwell Publishing, 2004); Wcislo, "Behavioral Environments and Evolutionary Change"; Gilbert Gottlieb, *Individual Development and Evolution: The Genesis of Novel Behavior* (New York: Oxford University Press, 1991); Daniel Dennett, *Consciousness Explained* (New York: Little, Brown, and Company, 1991); Dennett, *Darwin's Dangerous Idea*; Terrence W. Deacon, *The Symbolic Species* (New York: W. W. Norton & Company, Inc., 1997); Bruce H. Weber and David J. Depew, eds. *Evolution and Learning: The Baldwin Effect Reconsidered* (Cambridge, MA: MIT Press, 2003); West-Eberhard, *Developmental Plasticity and Evolution*; Eva Jablonka and Marion Lamb, *Evolution in Four Dimensions: Genetic, Epigenetic, Behavioral, and Symbolic in the History of Life* (Cambridge, MA: MIT Press, 2006); Bateson, "The Adaptability Driver"; Bateson and Gluckman, *Plasticity, Robustness, Development, and Evolution*. (Cambridge, MA: Cambridge University Press, 2011); Crispo, "The Baldwin Effect and Genetic Assimilation"; Adam Scarfe, "James Mark Baldwin with Alfred North Whitehead on Organic Selectivity: The 'Novel' Factor in Evolution," *Cosmos and History: The Journal of Natural and Social Philosophy* 5, no. 2 (2009): 40-107; Alexander V. Badyaev, "Evolutionary Significance of Phenotypic Accommodation in Novel Environments: An Empirical Test of the Baldwin Effect," *Philosophical Transactions of the Royal Society: Biological Sciences*, 364 (2009): 1125-1141.
22. Dennett, *Darwin's Dangerous Idea*, 77-80.
23. In "Evolutionary Significance of Phenotypic Accommodation in Novel Environments: An Empirical Test of the Baldwin Effect," *Philosophical Transactions of the Royal Society: Biological Sciences*, 364 (2009), 1126, Alexander Badayev asserts that "demonstrating the Baldwin effect requires an integration of approaches from developmental biology, physiological ecology and evolutionary ecology and an empirical system

in which one can observe organisms adapting to changing environments of variable recurrences."

24. Richard Dawkins, "The Turtle's Tale," *The Guardian*, (February 26, 2005), http://www.guardian.co.uk/books/2005/feb/26/scienceandnature.evolution.

25. Martin Wikelski, Chris Carbone, and Fritz Trillmich, "Lekking in Marine Iguanas: Female Grouping and Reproductive Strategies," *Animal Behavior* 52 (1996), 581.

26. Maren N. Vitousek, Dustin R. Rubenstein, and Martin Wikelski, "Galapagos Marine Iguana: Natural and Sexual Selection on Body Size Drives Ecological, Morphological, and Behavioral Specialization," in *Lizard Ecology: The Evolutionary Consequences of Foraging Mode* (Cambridge, MA: Cambridge University Press, 2007), 491.

27. Martin Wikelski, "Influences of Parasites and Thermoregulation on Grouping Tendencies in Marine Iguanas," *Behavioral Ecology* 10, no. 1 (1999): 22.

28. Jonathan Losos, Thomas W. Schoener, R. Brian Langerhans, and David A. Spiller, "Rapid Temporal Reversal in Predator-Driven Natural Selection," *Science* 314, no. 5802 (November 2006): 1111.

29. Typically, what is of concern to biologists in charting the evolution of a species, is genetic, morphological, and physiological selection and change over time—namely, changes in terms of the organism's genotype and phenotype, and not change in terms of its *ethotype* (i.e., of its behavioral constitution). While Baldwin emphasized behavior playing an initiating role in "evolution," he assumed this mainstream meaning of the term when outlining how it channeled out the course of morphological change in a population via natural selection. However, if we are to arrive at a more comprehensive definition of the terms "species" and "speciation," rather than being lumped in with other phenotypic characteristics, behavioral commonalities and differences among members of a population as well as behavioral change, in and of itself, could be included.

30. Badyaev, "Evolutionary Significance of Phenotypic Accommodation in Novel Environments." This study is also referred to in Bateson and Gluckman, *Plasticity, Robustness, Development, and Evolution*, 111-112. Also see: Alexander V. Badyaev and Kevin P. Oh, "Environmental Induction and Phenotypic Retention of Adaptive Maternal Effects," *BMC Evolutionary Biology* 8, no. 3 (2008): 1-10.

31. Ernst Mayr, "The Objects of Selection," *Proceedings of the National Academy of Science* 94 (March 1997): 2091.

32. The notion that organisms are not only objects of selection but are subjects of selection is part and parcel of Baldwin's blurring of the distinction between natural selection and organic selection. One criticism in relation to the theory of organic selection involves the apparent conflation of two distinct meanings of the term "selection," and that Baldwin commits the fallacy of equivocation in his writings. However, in response to this criticism, in *Development and Evolution*, Baldwin takes up a query from W. H. Hutton about the meaning of the word "selection." Baldwin distinguishes between the two different senses of the term and recognizes both. First, he agrees with Hutton's suggestion that "selection means the act of picking out certain objects from a number of others, and it implies that these objects are chosen for some reason or other" (168-169). At the same time, he preserves the Darwinian meaning of the term. Distinguishing between the two senses, Baldwin suggests that "there is only one thing to do, that is to recognize the two general uses of the term" as well as the ways in which they overlap.

33. Alfred North Whitehead, *Science and the Modern World* (1925; repr., New York: The Free Press, 1967), 187.

34. Ernst Mayr, *Toward a New Philosophy of Biology: Observations of an Evolutionist* (Cambridge, MA: Harvard University Press, 1988), 408, quoted in Wcislo, "Behavioral Environments and Evolutionary Change," 142.

35. See S. I. Nishimura, "A Predator's Selection of an Individual Prey from a Group," *Biosystems* 65, no. 1 (February 2002): 25-35.

36. See Douglas W. Morris, "Adaptation and Habitat Selection in the Eco-Evolutionary Process," *Proceedings of the Royal Society (Biological Sciences)* 10, no. 1098 (May 2011): 1-11.

37. See Andrew Delmar Hopkins, "Economic Investigations of the Scolytid Bark and Timber Beetles of North America," *U.S. Department of Agriculture Program of Work for 1917* (1916), 353; Andrew B. Barron, "The Life and Death of Hopkins' Host-Selection Principle," *Journal of Insect Behavior* 14, no. 6 (November 2001): 725-737; S. Bradleigh Vinson, "Host Selection by Insect Parasitoids," *Annual Review of Entomology* 21 (1976): 109-133; Kyrre L. Kausrud, Jean-Claude Grégoire, Olav Skarpaas, Nadir Erbilgin, Marius Gilbert, Bjørn Økland, and Nils C. Stenseth, "Dead or Alive! Host Selection and Population Dynamics in Tree-Killing Bark Beetles," *PLoS ONE* 6, no. 5 (2011): 1-13.

38. See Charles Darwin, *The Descent of Man* (Amherst, NY: Prometheus Books, 1998), from Part II, Chapter VIII: "Sexual Selection," 214ff.

39. See Charles Darwin, *The Descent of Man*, from Part I, Chapters IV: "Comparison of the Mental Powers of Man and the Lower Animals" to V: "On the Development of the Intellectual and Moral Faculties during Primeval and Civilized Times," 100-151.

40. See V. C. Wynne-Edwards, *Evolution through Group Selection* (London: Blackwell Scientific Publications, 1986).

41. See William Donald Hamilton, "The Genetical Evolution of Social Behavior: I and II," *Journal of Theoretical Biology* 7 (1964): 1-52; William Donald Hamilton, "Discriminating Nepotism: Expectable, Common, Overlooked," in *Kin Recognition in Animals*, ed. D. J. C. Fletcher and C. D. Michener (New York: Wiley, 1988), 417-487.

42. See Charles Darwin, *The Origin of Species* (1859; repr., New York: Penguin Books, 1968), Ch. 1: "Variation under Domestication," 71-100.

43. Here, I mean conservation of wildlife and habitats, in which some organisms are selected (protecting desirable animals) and other organisms are eliminated (e.g., predators) via human choice. Such measures include population control for the perceived long-term good of the ecosystem and/or to support human subsistence and development (e.g., culling seals that are purported to be depleting fish stocks, culling unwanted birds living near airports, etc...).

44. See Julian Huxley, *Evolutionary Humanism* (1964; repr., Amherst, MA: Prometheus Books, 1992). Huxley argues that the evolution of the human species has gone beyond its purely biological phase and that it has embarked upon a psychosocial phase of evolution. He states,

> The selective mechanism which determines what elements shall be incorporated and what rejected in the system of traditions, and so decides between alternative courses of cultural evolution, is primarily psychological or mental, involving human awareness instead of human genes, and directed towards the satisfaction of felt or imagined needs, instead of merely trending towards the survival of the more biological fit: further, it operations only within the framework of human societies. We may call it psychosocial selection. (33)

He goes on to suggest that,

Man's evolution is not biological but psychosocial: it operates by the mechanism of cultural tradition, which involves the cumulative self-reproduction and self-variation of mental activities and their products. Accordingly, major steps in the human phase of evolution are achieved by breakthroughs to new dominant patterns of mental organization, of knowledge, ideas and beliefs—ideological instead of physiological or biological organization. (76)

Later, Julian Huxley emphasizes that the destiny of humankind "is to be the sole agent for the future evolution of this planet" (77); to be "the controller or [selective] agent of future evolution on earth" (84, my addition); "to be responsible for the whole future of the evolutionary process on this planet" and "to steer it in the right direction" (121); "to be the spearhead and creative agent of the overall evolutionary process on this planet" (246).

For Huxley, this selective agency was not limited to be wielded in relation to non-human life-forms but on humanity itself. Huxley was a champion of eugenic selection, or "euselection" not only to stave off the Malthusian implications of global human overpopulation, but also to improve the species. While eugenic selection can be conceived of the selection of man by man at the genetic level, Huxley attempted to distinguish it from artificial selection. While Huxley believed that "successful psychosocial evolution demands a variety of gifts, temperaments and talents" (135), "it would be a good thing if the numbers of the too abnormal and the too defective could be reduced, those of the more intelligent and more gifted increased; and perhaps one day eugenics will get busy on this" (135).

45. In *The Origin of Species*, Darwin writes that the eliminative operations of natural selection are "daily and hourly scrutinizing, throughout the world, every variation, even the slightest; rejecting that which is bad, preserving and adding up all that is good; silently and insensibly working, whenever and wherever opportunity offers, at the improvement of each organic being in relation to its organic and inorganic conditions of life" (133).

46. Bruce Morito, *Thinking Ecologically: Environmental Thought, Values, and Policy* (Halifax: Fernwood Publishing, 2002), 143, my addition.

47. Whitehead, *Science and the Modern World*, 69.

48. See Darwin, *The Descent of Man*, especially 77-88, under the title "Comparison of the Mental Powers of Man and the Lower Animals."

49. Bruce Morito essentially sums up this Whiteheadian point in *Thinking Ecologically*, as follows:

Valuational activity may be amenable to rational deliberation, but it is not fundamentally the result of it. . . . Coupling the idea that values can be unconscious with the idea that they are set antecedently to rational activity, we can claim that underlying all valuational activity are judgments, or quasi-judgments, that an organism makes in determining the distinctions between suitable and unsuitable activities, good and bad situations. These primitive judgments direct the organism to be attracted to some objects and repelled by others, to perform certain functions and avoid others. When an organism is attracted to one object rather than another, it is drawn to it because of some property or set of properties it is judged to have and that will satisfy a need. If the object does not satisfy a need, the organism will likely not be drawn to it again; or, if the object is of negative value (e.g. pain-causing, noxious) the organism will learn to avoid that object in the future. Valuations are, among other things, motivations of organisms directing them to adapt. . . . This analysis of values presupposes a process ontology, that is to say, values are processes, not objects.

Further, the notion of loci of valuative-selective activity implies that organisms, including ourselves, are not independent or isolated units, but "members of an ecological community" (112-113).

50. Alfred North Whitehead, *The Concept of Nature* (1919; repr., Cambridge, MA: Cambridge University Press, 1995), 13.

51. In *Process and Reality* (1929; corrected ed., New York: The Free Press, 1978), Alfred North Whitehead qualifies that the word "decision" does not necessarily "imply conscious judgment, though in some 'decisions' consciousness will be a factor. The word is used in its root sense of a 'cutting off'" (43).

52. Alfred North Whitehead, *The Function of Reason* (Boston: Beacon Press, 1969), 8.

53. Jablonka and Lamb, *Evolution in Four Dimensions*, 241.

54. Wcislo, "Behavioral Environments," 138.

55. For example, see Konrad Lorenz, *Evolution and Modification of Behavior* (Chicago: University of Chicago Press, 1965); Dmitri K. Belaev, "Domestication of Animals," *Science Journal* (UK) 5 (1969): 47-52; I. Tiemann and G. Rehkämper. "Effect of Artificial Selection on Female Choice among Domesticated Chickens *Gallus gallus* f.d.," *Poultry Science* 88 (2009): 1948-1954; Lyudmilla Trut, Irina Oskina, and Anastasiya Kharlamova, "Animal Evolution during Domestication: The Domesticated Fox as a Model," *Bioessays* 31 (2009): 349-360; Márta Gàcsi, Paul McGreevy, Edina Kara, and Ádám Miklósi, "Effects of Selection for Cooperation and Attention in Dogs," *Behavioral and Brain Functions* 5, no. 31 (2009): 1-8.

56. For example, see Thomas Lorenz, "The Effects of Interspecific Competition, Salinity, and Hurricanes on the Success of an Invasive Fish, the Rio Grande Cichlid (*Herichthys cyanoguttatus*)," *University of New Orleans Theses and Dissertations*, Paper 846, 2008, http://scholarworks.uno.edu/td/846.

57. For instance, see Martin Hoyle and James E. Cresswell, "The Effect of Wind Direction on Cross-Pollination in Wind-Pollinated GM Crops," *Ecological Applications* 17 (2007): 1234-1243.

58. For example, see Chensheng Lu, Kenneth M. Warchol, and Richard A. Callahan, "*In Situ* Replication of Honey Bee Colony Collapse Disorder," *Bulletin of Insectology* 65, no. 1 (2012): 99-106.

59. For example, see Bruce Cohen, *The Cohen Commission of Inquiry into the Decline of Sockeye Salmon in the Fraser River*, Public Hearings transcript, December 15, 2011, http://alexandramorton.typepad.com/Cohen%20Dec%2015.pdf.

60. According to Joshua Mitteldorf, David H. Croll, and S. Chandu Ravela in "Multilevel Selection and the Evolution of Predatory Restraint" in *Artificial Life VIII: Proceedings of the Eighth International Conference on Artificial Life* (Cambridge, MA: MIT Press, 2002),

There is abundant suggestive evidence from field and experimental studies indicating the reality of population self-regulation. Fruitflies and nematodes appear to suppress their fertility in response to crowding, even with abundant nutrition (Guarente and Kenyon 2000). Observations in the wild suggest that rabbits exhibit the same response (Bittner and Chapman 1981). Arctic caribou in a fragile tundra environment breed less frequently than animals of the same species further south (Wynne-Edwards 1962). When deer are plentiful, wolves kill more deer and consume less of each (Kolenosky 1972). And the accumulated anecdotal experience of wildlife managers has created in that culture a belief that predator populations self-regulate (Nudds 1987). For each of these examples, evidence is not clean enough to rule out explanations from individual self-interest, which are deemed theoretically more conservative. (147)

In addition, they state:

Abrams (2000) emphasizes that most animals are predators in that they depend for nourishment on the culling of animal or plant populations. Every living thing is part of a local ecosystem, and its short-term interest in maximally exploiting the ecosystem is balanced by the long-term interest of its relatives and its progeny in maintaining the productivity of that ecosystem. (151)

61. See Gonçal Marion, John Furtado, Liliana Proaño, Lucia Corridoni, Majeda Al Musalli, and Marina Blanca, "Overfishing and the Case of the Atlantic Bluefin Tuna," *3rd UPC International Seminar on Sustainable Technology Development*, Technical University of Catalunya (June 2010): 1-15; Ana Maria Peñalosa, Govindran Jegatesen, José Llopis, Juliàn Patrón, Miguel Andreu Lozano, and Sara Alongi Skenhall, "Analysis of the Overfishing and Marine Ecosystem Degradation of Bluefin Tuna in the North Atlantic and Mediterranean Sea," *3rd UPC International Seminar on Sustainable Technology Development*, Technical University of Catalunya (June 2010): 1-15.

62. Conrad Hal Waddington, *The Ethical Animal* (New York: Atheneum, 1960), 23, 30.

63. Perhaps one of the best articulations of ecological interdependence is Whitehead's description in *Science and the Modern World*, 206, of the rich interrelatedness between organisms and elements found in a Brazilian rainforest, which I quoted in the introduction to this volume (44-45).

64. See Scarfe, "James Mark Baldwin with Alfred North Whitehead," especially 88-102. It must be noted that the term "pan-selectionism" should not be conflated with the same term that is employed in molecular biology, nor with Weismann and Wallace's views of the "all-sufficiency" of selection. See Stephen J. Gould, *The Structure of Evolutionary Theory* (Cambridge, MA: Harvard University Press, 2002), 198-203, 505.

65. Non-reductionistic critical pan-selectionism is highly critical of Huxley's endorsement of positive eugenic selection, or "euselection"—namely, the selection of man by man at the genetic level.

66. Herbert Spencer, *Principles of Biology, Volume 1* (London: Williams and Norgate, 1864), 444-445.

67. Thomas Huxley, *Evolution and Ethics and Other Essays* (1893; repr., New York: Barnes & Noble, 2006), 49.

68. To illustrate this point, we need but survey some of the classic work in the field of environmental ethics. Paul W. Taylor in "The Ethics of Respect for Nature," *Environmental Ethics* 3, no. 3 (Fall 1981) assumes that creatures are "teleological centers of life," without any reference to neo-Darwinian biology's apparent repudiation of teleology as an adequate notion with which to explain evolutionary processes. According to Dennett, in *Darwin's Dangerous Idea* (48-60) evolution is "algorithmic" rather than "teleological." Arne Naess was almost mute about the facts of evolution in advancing his Deep Ecological perspective. And, as Eco-Feminist, Ariel Salleh points out in "Deeper Than Deep Ecology: The Eco-Feminist Connection," in *Philosophy of Technology: The Philosophical Condition*, ed. R. C. Scharff and V. Dusek (Malden, MA: Blackwell Publishing, 2003), 483, environmental philosophies devalue themselves into a naïve set of prescriptions about how to "reconnect ourselves to the nature world," leaving us to drink the dec-

adent "karma-cola" of hippie-revivalist workshops on how to channel our bio-energy for personal fulfillment.

69. Darwin, *The Descent of Man*, 110.

70. In addition to Darwin's comments on social insects in *The Descent of Man*, another example of a proto-morality is exhibited in *The Origins of Morality: An Evolutionary Account* (New York: Oxford University Press, 2011), 133, where Dennis Krebs reports on Gerry Wilkinson's 1990 research on delayed reciprocity in vampire bats, who may regurgitate blood for starved members of their groups.

71. Dirk R. Johnson, *Nietzsche's Anti-Darwinism* (Cambridge, MA: Cambridge University Press, 2010), 25.

72. Hans Jonas, *The Phenomenon of Life: Toward a Philosophical Biology* (1966; repr., Evanston, IL: Northwestern University Press, 2001), 47, my addition.

73. See Friedrich Nietzsche, *Beyond Good and Evil*, trans. R. J. Hollingdale (1973; repr., New York: Penguin Books, 1990), §201-202, 122-126; §259-§260, 193-198.

74. Johnson, *Nietzsche's Anti-Darwinism*, 60. Nietzsche's analysis in the *Genealogy of Morality* reveals that in Darwin's account, the herd triumphs over "the strong" in and through "slave morality," which is the former's "own brand of will to power" (133). For these reasons, against Darwin, Nietzsche explored "the phenomenon of distinct 'moral' wills" (46) and he provided an alternative, multi-perspectival interpretation of life, conceived of as the interplay and/or the clash of individual wills to power rather than as a struggle for existence in which the "fit" are deemed to be those who have produced the most offspring. Rather, for Nietzsche, the goals of life include affirming the will-to-power (which, for him, is the will-to-life), the creation of one's own values, the self-founding of one's ownmost Being, and the domination of both others and the environment, all without pity for the "weaker."

75. One problem for evolutionary ethics in general is the tendency for proponents of neo-Darwinian accounts of the origin of morality to import non-naturalistic judgments and categories into their assessments of what is to be considered right and wrong. For example, in *The Origins of Morality*, Dennis Krebs writes "the function of morality is to induce people to strive to increase their inclusive fitness in certain (moral) ways, and not in other (immoral) ways" (258).

76. In *Foundations of Ethics*, (New York: Oxford University Press, 1939), Sir W. D. Ross stated, "Evolutionary theories [of ethics] do not seem to have offered us anything that can be accepted as a definition of 'right' or 'obligatory'" (21, my addition); "as we have seen, some evolutionary moralists have . . . identified 'right' with 'approved by the community'" (22); "Spencer's fundamental theory turns out to be universalistic Hedonism, or Utilitarianism; the apparently biological theory turns out to be really a psychological theory. And I believe this to be in the long run true of evolutionary ethics in general, so that it need not be examined as a separate form of theory regarding the ground of rightness" (59).

In conjunction with the Darwinian reduction of morality to a function of maximizing survival value and the prospect of reproductive success, we might contemplate the ramifications for ethics of Dawkins' more recent neo-Darwinian reduction of organisms to mere DNA-replicators. In "God's Utility Function," in *Scientific American* (November 1995), Dawkins claims that "life has no higher purpose than to perpetuate the survival of DNA" (81) and that "the true utility function of life, that which is being maximized in the natural world, is DNA survival" (83).

77. Nietzsche, "Beyond Good and Evil," in *Basic Writings of Nietzsche*. Translated and edited by W. Kaufmann (1967; repr., New York: The Modern Library, 2000), §13, 211.

78. Dennett, *Darwin's Dangerous Idea*, 78-79, figures 3.1 and 3.2. Also see Dennett, *Consciousness Explained*, 185-187, figures 7.1 and 7.2.

79. To be sure, in developing an evolutionary ethic that is not simply reducible to survival value and reproductive success, in *The Imperative of Responsibility: In Search of an Ethics for the Technological Age* (1979, 1984), Hans Jonas states that we have a duty not only to ensure "that there *be* a future mankind—even if no descendants of ours are among them" but also a duty "toward their *condition*, the quality of their life" (40).

Bibliography

Badyaev, Alexander V. "Evolutionary Significance of Phenotypic Accommodation in Novel Environments: An Empirical Test of the Baldwin Effect." *Philosophical Transactions of the Royal Society: Biological Sciences*, 364 (2009): 1125-1141.

Badyaev, Alexander V., and Kevin P. Oh. "Environmental Induction and Phenotypic Retention of Adaptive Maternal Effects." *BMC Evolutionary Biology* 8, no. 3 (2008): 1-10.

Baldwin, James Mark. *Mental Development in the Child and The Race: Methods and Processes*. New York: The Macmillan Company, 1894 and 1906. Reprint, Whitefish, MT: Kessinger Publishing, 2006. Page references are to the 2005 edition.

———. "A New Factor in Evolution." *American Naturalist* 30, (1896): 441-457, 536-554.

———. *Social and Ethical Interpretations in Mental Development*. New York: The Macmillan Company, 1897. Reprint, New York: Elibron Classics, 2005. Page references are to the 2005 edition.

———. *Development and Evolution: Including Psychophyiscal Evolution, Evolution by Orthoplasy, and the Theory of Genetic Modes*. New York: The Macmillan Company, 1902. Reprint, New York: Elibron Classics, 2005. Page references are to the 2005 edition.

———. *Darwin and the Humanities*. Baltimore: Review Publishing, 1909.

Barron, Andrew B. "The Life and Death of Hopkins' Host-Selection Principle." *Journal of Insect Behavior* 14, no. 6 (November 2001): 725-737.

Bateson, Patrick. "The Adaptability Driver: Links between Behavior and Evolution." *Biological Theory* 1, no. 4 (2006): 342-345.

———. "The Evolution of Evolutionary Theory." *European Review* 18, no. 3 (2010): 287-296.

Bateson, Patrick, and Peter Gluckman. *Plasticity, Robustness, Development, and Evolution*. Cambridge, MA: Cambridge University Press, 2011.

Belaev, Dmitri K. "Domestication of Animals." *Science Journal* (UK) 5 (1969), 47-52.

Broughton, John M., and D. John Freeman-Moir, eds. *The Cognitive Developmental Psychology of James Mark Baldwin*. Norwood, NJ: Ablex Publishing Corporation, 1982.

Cobb, John, ed. *Back to Darwin: A Richer Account of Evolution*. Grand Rapids: William B. Eerdman's Publishing Company, 2008.

Cohen, Bruce. *The Cohen Commission of Inquiry into the Decline of Sockeye Salmon in the Fraser River* Public Hearings transcript, December 15, 2011, http://alexandramorton.typepad.com/Cohen%20Dec%2015.pdf.

Crispo, Erika. "The Baldwin Effect and Genetic Assimilation: Revisiting Two Mechanisms of Evolutionary Change Mediated by Phenotype Plasticity." *Evolution* 61, no. 11 (2007): 2469-2479.

Darwin, Charles. *The Origin of Species by Means of Natural Selection*. New York: Penguin Books, 1859 / 1968.

———. *The Descent of Man*. Introduction by H. J. Birx, Amherst, NY: Prometheus Books, 1998.

Dawkins, Richard. "God's Utility Function." *Scientific American* (November 1995): 81-85.

———. "The Turtle's Tale," *The Guardian*, February 26, 2005, http://www.guardian.co.uk/books/2005/feb/26/scienceandnature.evolution.

Deacon, Terrence. *The Symbolic Species: The Co-Evolution of Language and the Brain*. New York: W. W. Norton & Company, Inc., 1997.

Dennett, Daniel. *Consciousness Explained*. New York: Little, Brown, and Company, 1991.

———. *Darwin's Dangerous Idea: Evolution and the Meanings of Life*. New York: Simon & Schuster, 1995.

Devall, Bill, and George Sessions. *Deep Ecology: Living as if Nature Mattered*. Layton, UT: Gibbs M. Smith, Inc., 1985.

Gàcsi, Márta, Paul McGreevy, Edina Kara, and Ádám Miklósi. "Effects of Selection for Cooperation and Attention in Dogs." *Behavioral and Brain Functions* 5, no. 31 (2009): 1-8.

Gould, Stephen J. *The Structure of Evolutionary Theory*. Cambridge, MA: Harvard University Press, 2002.

Gottlieb, Gilbert. *Individual Development and Evolution: The Genesis of Novel Behavior*. New York: Oxford University Press, 1991.

Hamilton, William Donald. "The Genetical Evolution of Social Behavior: I and II." *Journal of Theoretical Biology* 7 (1964): 1-52.

———. "Discriminating Nepotism: Expectable, Common, Overlooked." In *Kin Recognition in Animals*, edited by D. J. C. Fletcher and C. D. Michener, 417-487. New York: Wiley, 1988.

Hinton, Geoffrey E., and Steven J. Nowlan. "How Learning Can Guide Evolution." *Complex Systems* 1 (1987): 447-454.

Hopkins, Andrew Delmar. "Economic Investigations of the Scolytid Bark and Timber Beetles of North America." *U.S. Department of Agriculture Program of Work for 1917* (1916), 353.

Hoyle, Martin, and James E. Cresswell. "The Effect of Wind Direction on Cross-Pollination Wind-Pollinated GM Crops." *Ecological Applications* 17 (2007): 1234-1243.

Huxley, Julian. *Evolution: The Modern Synthesis (The Definitive Edition)*. Foreword by M. Pigliucci and G. B. Müller, 1942. Reprint, Cambridge, MA: MIT Press, 2010.

———. *Evolutionary Humanism*. Introduction by H. James Birx. 1964. Reprint, Amherst, MA: Prometheus Books, 1992.

Huxley, Thomas H. *Evolution and Ethics and Other Essays*. 1893. Reprint, New York: Barnes & Noble, 2006.

Jablonka, Eva, and Marion Lamb. *Evolution in Four Dimensions: Genetic, Epigenetic, Behavioral, and Symbolic in the History of Life*. Cambridge, MA: MIT Press, 2005.
Johnson, Dirk R. *Nietzsche's Anti-Darwinism*. Cambridge, MA: Cambridge University Press, 2010.
Jonas, Hans. *The Phenomenon of Life: Toward a Philosophical Biology*. 1966. Reprint, Evanston, IL: Northwestern University Press, 2001.
———. *The Imperative of Responsibility: In Search of an Ethics for the Technological Age*. Chicago: University of Chicago Press, 1984.
Kausrud, Kyrre L., Jean-Claude Grégoire, Olav Skarpaas, Nadir Erbilgin, Marius Gilbert, Bjørn Økland, and Nils C. Stenseth. "Dead or Alive! Host Selection and Population Dynamics in Tree-Killing Bark Beetles." *PLoS ONE* 6, no. 5 (2011): 1-13.
Krebs, Dennis L. *The Origins of Morality: An Evolutionary Account*. New York: Oxford University Press, 2011.
Lloyd Morgan, Conway. *Habit and Instinct*. New York: Edward Arnold, 1896. Reprint, Kessinger Publishing, 2007. Page references to 2007 edition.
Lorenz, Konrad. *Evolution and Modification of Behavior*. Chicago: University of Chicago Press, 1965.
Lorenz, Thomas. "The Effects of Interspecific Competition, Salinity, and Hurricanes on the Success of an Invasive Fish, the Rio Grande Cichlid (*Herichthys cyanoguttatus*)." *University of New Orleans Theses and Dissertations*. Paper 846, 2008, http://scholarworks.uno.edu/td/846.
Losos, Jonathan, Thomas W. Schoener, R. Brian Langerhans, and David A. Spiller. "Rapid Temporal Reversal in Predator-Driven Natural Selection." *Science* 314, no. 5802 (November 2006): 1111.
Lu, Chensheng, Kenneth M. Warchol, and Richard A. Callahan. "*In Situ* Replication of Honey Bee Colony Collapse Disorder." *Bulletin of Insectology* 65, no. 1 (2012): 99-106.
Marion, Gonçal, John Furtado, Liliana Proaño, Lucia Corridoni, Majeda Al Musalli, and Marina Blanca. "Overfishing and the Case of the Atlantic Bluefin Tuna." *3rd UPC International Seminar on Sustainable Technology Development*. Technical University of Catalunya (June 2010): 1-15.
Mayr, Ernst. *Animal Species and Evolution*. Cambridge, MA: Belknap Press of Harvard University Press, 1963.
———. *Toward a New Philosophy of Biology: Observations of an Evolutionist*. Cambridge, MA: Harvard University Press, 1988.
———. "The Objects of Selection." *Proceedings of the National Academy of Science* 94 (March 1997): 2091-2094.
Mitteldorf, Joshua, David H. Croll, and S. Chandu Ravela. "Multilevel Selection and the Evolution of Predatory Restraint." In *Artificial Life VIII: Proceedings of the Eighth International Conference on Artificial Life*, edited by R. Standish, M. Bedau, and H. A. Abbass, 146-152. Cambridge, MA: MIT Press, 2002.
Moore, G. E. *Principia Ethica*. London: Cambridge University Press, 1903. Reprint, Mineola, NY: Dover Publications, Inc., 2004. Page references to the Dover edition.
Morito, Bruce. *Thinking Ecologically: Environmental Thought, Values, and Policy*. Halifax: Fernwood Publishing, 2002.
Morris, Douglas W. "Adaptation and Habitat Selection in the Eco-Evolutionary Process." *Proceedings of the Royal Society (Biological Sciences)* 10, no. 1098 (May 2011): 1-11.

Nietzsche, Friedrich. *Basic Writings of Nietzsche*. Translated and edited by W. Kaufmann. 1967. Reprint, New York: The Modern Library, 2000.
———. *Beyond Good and Evil*. Translated by R. J. Hollingdale. Introduction by Michael Tanner. 1973. Reprint, New York: Penguin Books, 1990.
Nishimura, S. I. "A Predator's Selection of an Individual Prey from a Group." *Biosystems* 65, no. 1 (February 2002): 25-35.
Osborn, Henry Fairfield. "A Mode of Evolution Requiring Neither Natural Selection Nor the Inheritance of Acquired Characteristics." *Transactions of the New York Academy of Science* 15 (1896): 141-142.
———. "The Limits of Organic Selection." *American Naturalist* 31, no. 371 (November 1897): 944-951.
Peñalosa, Ana Maria, Govindran Jegatesen, José Llopis, Julián Patrón, Miguel Andreu Lozano, and Sara Alongi Skenhall. "Analysis of the Overfishing and Marine Ecosystem Degradation of Bluefin Tuna in the North Atlantic and Mediterranean Sea." *3rd UPC International Seminar on Sustainable Technology Development*. Technical University of Catalunya (June 2010): 1-15.
Pigliucci, Massimo, and Gerd B. Müller, eds. *Evolution: The Extended Synthesis*. Cambridge, MA: MIT Press, 2010.
Plotkin, Henry. *Evolutionary Thought in Psychology: A Brief History*. Malden, MA: Blackwell Publishing, 2004.
Richards, Robert J. *Darwin and the Emergence of Evolutionary Theories of Mind and Behavior*. Chicago: University of Chicago Press, 1987.
Richardson, John. *Nietzsche's New Darwinism*. New York: Oxford University Press, 2004.
Ross, W. David. *Foundations of Ethics*. Oxford: Clarendon Press, 1939.
Ruse, Michael, ed. *Philosophy After Darwin: Classic and Contemporary Readings*. Princeton, NJ: Princeton University Press, 2009.
Salleh, A. "Deeper Than Deep Ecology: The Eco-Feminist Connection." In *Philosophy of Technology: The Philosophical Condition*, edited by R. C. Scharff and V. Dusek, 480-484. Malden, MA: Blackwell Publishing, 2003.
Scarfe, Adam C. "James Mark Baldwin with Alfred North Whitehead on Organic Selectivity: The 'Novel' Factor in Evolution." *Cosmos and History: The Journal of Natural and Social Philosophy* 5, no. 2 (2009): 40-107.
Simpson, George Gaylord. "The Baldwin Effect." *Evolution* 7, no. 2 (June 1953): 110-117.
Spencer, Herbert. *Principles of Biology, Volume 1*. London: Williams and Norgate, 1864.
Taylor, Paul W. "The Ethics of Respect for Nature." *Environmental Ethics* 3, no. 3 (Fall 1981): 197-218.
Tiemann, I., and G. Rehkämper. "Effect of Artificial Selection on Female Choice among Domesticated Chickens *Gallus gallus* f.d." *Poultry Science* 88 (2009): 1948-1954.
Trut, Lyudmilla, Irina Oskina, and Anastasiya Kharlamova. "Animal Evolution during Domestication: The Domesticated Fox as a Model." *Bioessays* 31 (2009): 349-360.
Vinson, S. Bradleigh. "Host Selection by Insect Parasitoids." *Annual Review of Entomology* 21 (1976): 109-133.
Vitousek, Maren N., Dustin R. Rubenstein, and Martin Wikelski. "Galapagos Marine Iguana: Natural and Sexual Selection on Body Size Drives Ecological, Morphological, and Behavioral Specialization." In *Lizard Ecology: The Evolutionary Consequences of Foraging Mode*, edited by S. M. Reilly, L. D. McBrayer, and D. P. Miles, 491-507. Cambridge, MA: Cambridge University Press, 2007.

Waddington, Conrad Hal. *The Ethical Animal*. New York: Atheneum, 1960.
Wcislo, William T. "Behavioral Environments and Evolutionary Change." *Annual Review of Ecology and Systematics* 20 (1989): 137-169.
Weber, Bruce H., and David J. Depew, eds. *Evolution and Learning: The Baldwin Effect Reconsidered*. Cambridge, MA: MIT Press, 2003.
West-Eberhard, Mary Jane. *Developmental Plasticity and Evolution*. New York: Oxford University Press, 2003.
Whitehead, Alfred North. *The Concept of Nature*. 1919. Reprint, Cambridge, MA: Cambridge University Press, 1995.
———. *Science and the Modern World*. 1925. Reprint, New York: The Free Press, 1967.
———. *The Function of Reason*. 1929. Boston: Beacon Press, 1969.
———. *Process and Reality: Corrected Edition*, edited by D. R. Griffin and D. W. Sherburne. 1929. Corrected ed., New York: The Free Press, 1978. Page references to corrected edition.
Wikelski, Martin. "Influences of Parasites and Thermoregulation on Grouping Tendencies in Marine Iguanas." *Behavioral Ecology* 10, no. 1 (1999): 22-29.
Wikelski, Martin, Chris Carbone, and Fritz Trillmich. "Lekking in Marine Iguanas: Female Grouping and Reproductive Strategies." *Animal Behavior* 52 (1996): 581-596.
Williams, George C. *Adaptation and Natural Selection*. Princeton, NJ: Princeton University Press, 1966.
Wynne-Edwards, V. C. *Evolution through Group Selection*. London: Blackwell Scientific Publications, 1986.
Zimmerman, Michael E. *Contesting Earth's Future: Radical Ecology and Postmodernity*. Berkeley, CA: University of California Press, 1994.

Section 5:
Autogenesis, Teleology, and Teleodynamics

Chapter Thirteen
Teleology versus Mechanism in Biology: Beyond Self-Organization

Terrence Deacon and Tyrone Cashman

In the closing line of *On the Origin of Species*, Darwin cryptically acknowledges that the process of natural selection is insufficient to account for the core attributes of life. Thus he begins the poetic last line with: "There is grandeur in this view of life, with its several powers, having been originally breathed into a few forms or into one."[1] Implicitly he acknowledges that it is only subsequent to the existence of systems possessing these "several powers" that one is able to attribute to natural selection the ability to account for the many beautiful and wondrous forms of life. Natural selection does not explain the origin of these critical features of organism dynamics. They are its necessary initial conditions. So although Darwin's theory of natural selection is often credited with providing a thoroughly mechanistic and non-teleological account of the evolution of life, this cryptic caveat suggests that he was aware that natural selection could not provide a complete mechanistic account of the nature of organisms themselves. In the present chapter, our attempt is to throw light on those "several powers" left unexplained in Darwin's explicit theory—namely, the necessary capacities required for the emergence of a minimal "self" that can repair itself and reproduce, thus generating lineages that can then adapt to their environments due to the process of natural selection.

We argue that the recovery of a non-eliminative concept of information and a non-trivial concept of "selfhood" are both critical to a complete theory of the physics of biological organism. In recent decades, there have been a number of theoretical efforts to account for the atypical physical properties that constitute an organism, employing insights from dynamical systems theory (DST) and its biological counterpart, complex adaptive systems (CAS) theory. Most of these approaches have been informed by the pioneering work of Ilya Prigogine on far-from-equilibrium processes—in what he described as dissipative systems. The theory of dissipative systems helps to explain certain special features of organism thermodynamics. And the related concept of autocatalysis helps to explain aspects of the circularly generative property of life. But how exactly these principles contribute to organism dynamics is still a matter of intense debate. More importantly, we believe that DST principles are unable to account for certain critical features of biological selfhood. This limitation points to the need for the introduction of additional dynamical principles. Without an adequate concept of biological "selfhood," the referential and normative functions of biological information are left unexplained and adaptive function is left undefined. In this chapter we will provide a sketch of the conceptual shift that we believe is required to cross this last threshold to a more complete theory of organismic "selfhood."

A Biological Problem Reduced to an Engineering Problem

Long before the discovery of the structure of DNA and the so-called "central dogma" of molecular biology took shape, it was evident that genetics would inevitably need to be understood in terms of the storage and transmission of information. Hence, when the quantum physicist Erwin Schrödinger predicted in his 1944 book *What Is Life?* that the molecular basis for biological inheritance must be embodied in something like an aperiodic crystal, he was implicitly thinking in informational terms. This conception not only presaged the discovery of how DNA might work, but also implicitly employed a logical analysis that was later formalized in Claude Shannon's (1948) "The Mathematical Theory of Communication." Shannon's work formalized what came to be called information theory and provided an unambiguous way to measure the amount of data stored on a page of text, in a computer memory, or conveyed within a communication channel, such as an Internet connection. Originally developed as a tool for engineers endeavoring to analyze and quantify data processing, storage, and transmission, the mathematical theory of communication specifically excludes any consideration of referential relationships or significance. Or rather it simply assumes them to be independent and extrinsic considerations that can be ignored.

The term *communication* in the title of his "report" offers a hint of the special use of the term *information* that was to become the standard technical definition. Colloquially, information is presumed to be data that is "about" something else, but the term communication does not necessarily entail an interpretation of what if anything that is. Indeed, Shannon defines his problem in his introduction to the paper as follows: "The fundamental problem of communication is that of reproducing at one point either exactly or approximately a message selected at another point."[2] In this conception, communication is reduced to copying or conveying physical patterns. The power, as well as the limitation, of defining the problem in this way is that it dispenses entirely with issues of reference, function, and significance. The concept of information understood in this way is abstracted from its implicit functional or referential contexts and is reduced operationally to what amount to medium properties that can be precisely measured.

When applied to biological processes the Shannonian notion of information easily maps onto the various molecular genetic substrates and their correspondences to one another, but it brackets from discussion any reference to their functional and adaptive significance. These features can be ignored for the purpose of doing basic molecular biochemistry, but not when it comes to assessing functional adaptation, which is critical to evolutionary biology and physiology. This way of conceptualizing biological information processes has contributed to a major paradigm for modeling evolutionary processes. Thus the *replicator selection* logic originally caricatured in Richard Dawkins' 1976 book, *The Selfish Gene*, has contributed to a narrowing of the definition of biological evolution. For example, most modern introductory textbooks of biology and evolutionary theory define "evolution" as change in allele or gene frequency.[3] Notice that, besides reducing genetics to transcribed DNA sequences, this definition completely ignores any reference to function or adaptation, except in accounting for the differential copying of genes. Abstracted from this analysis are the processes responsible for the organism's growth and maintenance, the work required to copy (reproduce) its genetic information, and the means by which it responds to it environment.

Although vitalistic theories of life like that of Driesch (1908) and teleological theories of evolution like those of Bergson (1907), Berg (1922), and Teilhard de Chardin (1955) became anachronistic by the midpoint of the twentieth century in conjunction with the rise of molecular biology, this did not obviate the need to expand classic concepts of mechanistic causality to account for biological phenomena. Hence, as recently as fifty years ago, one of the main architects of the Modern Synthesis, George Gaylord Simpson, pointedly noted that "the study of organisms requires principles additional to those of the physical sciences [and yet] does not imply a dualistic or vitalistic view of nature." Simpson goes on to explain that the "'How?' [question] is the typical, and only meaningful question in the physical sciences. But biology can and must go on from there. Here, 'What for?'—the dreadful teleological question—not only is legitimate but also must eventually be asked of every vital phenomenon." He adds, "In biology, then, a second kind of explanation must be added to the first, or reductionist, explanation made in terms of physical, chemical, and mechanical principles. This second form of explanation, which can be called compositionist in contrast with reductionist, is in terms of the adaptive usefulness of structures and processes to the whole organism."[4]

Ernst Mayr decided that the "principles additional to those of the physical sciences" that biology required were not to be found in Simpson's proposed "compositionist explanation," but from programs (i.e., algorithms) within "the genetic code." In 1961, he proposed "a purely mechanistic purposiveness" to explain the appearances of teleology, for which he borrowed Pittendrigh's term, *teleonomy*.[5] Though originally understood in cybernetic (i.e., negative feedback) terms, in Mayr's more general definition, "a teleonomic process or behavior is one which owes its goal-directedness to the operation of a program."[6] Earlier he had asked,

> Where, then, is it legitimate to speak of purpose and purposiveness in nature and where is it not? To this question we can now give a firm and unambiguous answer. An individual who—to use the language of the computer—has been programmed can act purposefully. A bird that starts its migration, an insect that selects its host plant, an animal that avoids a predator, a male that displays to a female—they all act purposefully because they have been programmed to do so.[7]

During the last thirty years, the science of molecular genetics on the one hand, and the computational technologies and information sciences on the other, have each become dominant explanatory paradigms in the sciences. Their twin successes have fed on each other. This confluence of events in intellectual history has reinforced the teleonomic theory of organisms. In this context, Mayr's teleonomy no longer looks like "principles additional to those of the physical sciences."[8] A computer program seems purposeful to us, because computer programs are created by programmers with, and for, a human purpose. But the purpose is in human minds, not in the computer. Computing is just another blindly mechanistic physical process.

We contend that G. G. Simpson was right that one cannot understand biology in simple reductionist terms, even though such explanations work well for physics, chemistry, and computers. They are even explanatory for organism functions at certain levels. We agree with him that the question "What for?" is essential to any complete biological explanation. Furthermore, we agree that some sort of *compositionist* explanation must be found for the phenomena constitutive of living beings. But what are the components of the composition, and *how* must they fit together for the composite to be a living being?

Darwin's implicit assumption and Simpson's claim are almost certainly right. Individual organisms are goal-directed, and their functioning components exist "for" the role they play in supporting life and reproduction. This ultimately appeals to what Aristotle called "final causality," pointing to that "for the sake of which" something is done or happens. In Aristotle's way of thinking, when there is a beneficiary, there is also an end, a purpose—namely, a property that later generations called *teleology*. We would also add that for something to benefit from a dynamic process, that something must be, in at least a primitive sense, a *self*. So, what we require in order to explain the existence of this property of life is an explanation of how a biological self can emerge from non-self components.

Schrödinger's Unstated Hint

Erwin Schrödinger made two major points in his book, *What Is Life?* First, an organism must maintain itself persistently in a far-from-equilibrium state. Second, an organism must inherit, preserve, replicate, and transmit information in a molecular form, and this information must play the crucial role of specifying organismic "selfhood" in terms of its structure and functions. What Schrödinger only hinted at was that these two physically anomalous defining attributes of life might be intimately and necessarily interdependent—namely, that the property of functioning as information depends on a persistently far-from-equilibrium dynamic and that a system with this dynamical property necessarily depends on information. So what is ignored in teleonomic paradigms for modeling self-reproduction is the special self-directed nature of the physical *work* required to accomplish the construction of the organism (i.e., its development) and the reproduction of a replica sufficient to reinitiate this process. Unfortunately, the nature and special character of this form of work is ignored in theories of evolution that are limited to natural selection logic alone. For example, replicator selection explanations of biological evolution and Artificial Life (A-Life) simulations of evolutionary processes both bracket from consideration the actual physical work of component fabrication and their assembly into appropriately interdependent structures. Allowing this fundamental property of organic life to be merely assumed or supplied by an extrinsic mechanism (as in A-Life simulations) evades confronting this most fundamental physical property of life. Not only does the replication of genetic information require a system able to perform this copying task, the physical patterns taken to be genetic information inherit this property because of their contribution to this form-generating work.

Dynamical Systems, Dissipative Structures, and Self-Organization

A major advance over this overly simple engineering conception of life is provided by the incorporation of insights from DST. In many respects, this analysis addressed the first of Schrödinger's riddles concerning the physics of living processes: how organisms manage to maintain orderly structure and resist the ravages of the Second Law of Thermodynamics in a far-from-equilibrium dynamical condition. To summarize the critical insight to be drawn from this work in a single phrase, it demonstrates that *persistently far-from-equilibrium systems can spontaneously develop toward more orderly global organization as they are forced to more efficiently dissipate an incessantly imposed disturbance.* Thus, despite the ubiquitous tendency for such a system to become less ordered over time when left undisturbed, if a dynamic system such as a fluid or chemical process is incessantly perturbed (such as by continual heating or the constant introduction of new materials), it can at the same time develop additional internal large-scale regularities that re-establish a balance between the rate of perturbation and the rate that these perturbing influences are dissipated into a larger environment. By definition, such a system is both dynamic and open to inputs and outputs. But the form that this large-scale organization takes is not a direct reflection of the form of the perturbing influence but rather of the limitations on the options for dynamical interactions among the components of the system. For example, consider a whirlpool formed when trajectories of moving water molecules become progressively restricted due to their viscous interactions and the imposition of an asymmetric partial blockage of flow (e.g., by a rock in a stream). The term for this restriction of dynamical options is "constraint." In his 1962 seminal account of this use of the term W. Ross Ashby describes its relationship to organization as follows:

> Thus the presence of "organization" between variables is equivalent to the existence of *constraint* in the product-space of the possibilities. I stress this point because while, in the past, biologists have tended to think of organization as something extra, something *added* to the elementary variables, the modern theory, based on the logic of communication, regards organization as a restriction or constraint. The two points of view are thus diametrically opposed.[9]

In general, a constraint is that which reduces the possible degrees of freedom of a process. From the dynamical systems perspective, such a limitation is merely something less than random or chaotic and thus, conversely, some semblance of dynamical regularity or organization. This provides a way to define order in a way that does not depend on any comparison with an extrinsic or ideal model. At the same time, one can also re-describe the increase of entropy as an elimination of constraints. When a dissipative system generates order spontaneously, the generation of intrinsic dynamical constraints compensates for the external imposition of constraints at a faster rate than they can be spontaneously dissipated without this global regularity. This way of thinking about dynamical processes has given rise to an important generalization of the Second Law of Thermodynamics. The *Maximum Entropy Production* (MEP) principle states that a dynamical system will tend to organize itself in such a way that it maximizes the generation of entropy, given the available pathways for dissipation (i.e., within given constraints), even if it is persistently perturbed far from equilibrium. For instance, consider the regularity of flow that we recognize as the vortex structure of the whirlpool. This form of flow more efficiently and stably dissipates the perturbation of fluid movement through that location than would other less regular turbulent patterns of flow. The circular symmetry is not imposed from without. It is the result of a geometric principle by which this circular trajectory minimizes the total path-length taken by the average water molecule. In this sense the circular symmetry is a constraint on movement that arises intrinsically.

The intrinsic origin of the specific form that this regularity takes is what warrants calling such processes self-organized. The term "self" in this phrase does not of course implicate any conception of autonomy as used in describing organism or psychological selves. It merely designates that the form that this orderliness takes is more dependent on intrinsic than extrinsic form-generating influences. Thus a whirlpool produced by stirring a fluid in a circular manner is externally organized, in contrast to one that forms spontaneously. But the term *self*-organization can lead one to over-interpret this special sense of intrinsic causal influence and generalize it to the more thoroughly autonomous concept of selfhood that we recognize at work in organism dynamics. The point of this chapter is not to deny that this form of self-organization (which Deacon in his book, *Incomplete Nature* has renamed *morphodynamics* to avoid this conflation)[10] is relevant to explaining the way that precisely ordered dynamical regularities are generated and maintained in living organisms. The point is rather to show what is additionally required to generate the form of autonomous selfhood that is the fundamental principle defining life.

Another way to think of this mode of order-creation is that constant perturbation supplies work to generate this order. Just as it takes work to compensate for the way things spontaneously fall into messiness and to prevent our machines from eventually succumbing to accumulated damage, the incessant perturbation that produces self-organization is the source of work that helps to produce this order. In general, work is performed by introducing constraints into a spontaneous process. For example, the increase of entropy of hot gases expanding within a cylinder can be harnessed to do work if this expansion is confined to only one direction. The movement of a piston in the cylinder can be used to impose constraints on some other system, such as propelling a vehicle against the drag of friction or the pull of gravity. An organism's continuous generation of orderly processes and new structures thus requires work, and that work requires both an energy gradient and specific constraints to selectively channel its dissipation. The energy to drive this work must be derived from its environment (e.g., in the form of sunlight or food), but the constraints that channel it into specific forms of order-creation must be generated intrinsically. These intrinsic constraints arise from self-organizing processes within the organism. Whereas the spontaneous dissipation of an energy gradient results in a reduction of the capacity to do work, the creation of new kinds of constraints offers the potential to do new kinds of work. So, if novel constraints can be intrinsically generated under certain special (e.g., far-from-equilibrium) conditions, then novel forms of work can emerge. The exploitation of this capacity of constraints to engender further constraint is the ultimate basis of all forms of evolution.

Despite this critical role played by self-organization, we contend that—as with theories of organisms' reproduction and evolution that are mechanistic or teleonomic—those that are modeled on self-organized dynamics alone are nevertheless incompletely able to account for organism dynamics. In an insightful 1968 paper in *Science*, Michael Polanyi pointed out that a machine is a mechanism defined by "boundary conditions" (the set of initial constraints affecting possible trajectories of change) imposed by a designer through the choice of materials and the way they are allowed to interact.[11] This is as true of a computer as of an internal combustion engine. So its function is defined not so much by the dynamical laws that it realizes, but by the possibilities of dynamics that are not realized or are suppressed. Thus, frequently, a machine breaks down when these constraints are loosened in some significant way. That loss of function is due to an increase in the degrees of freedom (which may paradoxically lead to a complete cessation of its dynamics). Polanyi made a similar argument about living organisms by pointing out that there are no new laws of chemistry involved in life, only highly specific and complex constraints on the allowed chemistry constituting an organism. Polanyi's insight locates a critical difference between organisms and machines in the locus of these constraints. He notes that they are imposed extrinsically in the design of machines, but arise intrinsically in organisms imposed by their DNA. Although we agree with this extrinsic-intrinsic distinction, there are reasons to be more cautious in locating the source of living constraints in DNA.

Characterizing living processes in self-organizational terms is not new, even if thinking about it in terms of constraints is a modern formulation. The logic of this characterization was made explicit almost two centuries earlier by the philosopher Immanuel Kant. He distinguished organisms from machines with respect to two attributes: "An organized being is, therefore, not a mere machine. For a machine has solely motive power, whereas an organized being possesses inherent formative power, and such, moreover, as it can impart to material devoid of it—material which it organizes." He goes on to say that "this, therefore, is a self-propagating formative power, which cannot he explained by the capacity of movement alone, that is to say, by mechanism. The definition of an organic body is that it is a body, every part of which is there for the sake of the other (reciprocally as end, and at the same time, means)."[12] We can translate Kant's insight into these more modern terms by understanding "formative power" as the capacity of metabolic processes to generate new intrinsic components; "self-propagating" as something like self-replicating, which is of course one of the most basic defining characteristics of life; and "reciprocal ends and means" as the indirect co-creation of organism components each by the others.

Kant's effort was to account for the intrinsic teleological features of living processes in a way that is consistent with his Newtonian understanding of physical laws.[13] The modern problem of understanding how the physics of life diverges from inorganic physics and chemistry is in this respect a scientific reframing of Kant's problem. But, Kant's analysis betrays an epistemological framing of this relationship since the terms "ends" and "means" already assume a teleological framing. One way that this is often translated into more concrete terms is to understand this reciprocity of formative processes as a description of autocatalysis.[14]

Autocatalysis is a special case of catalysis in which a set of two or more molecular types catalyze the synthesis of each of the others comprising that set. This would, for instance, be the case if some molecule A catalyzes the production of molecule B, which catalyzes the production of molecule C, which catalyzes the production of molecule A, and so forth. The production of these molecules would be reciprocally self-amplifying (assuming that there is free energy available to perpetuate the process). Autocatalysis is a special case of self-organization, because it increases constraints locally (i.e., the asymmetric increase in catalyst concentration) in the process of maximizing the production of entropy into the larger environment. Catalysis more rapidly generates entropy and liberates the potential energy in molecular bonds than would occur in the absence of catalysis, thus exemplifying the MEP principle. *Auto*-catalysis is even more efficient because it progressively amplifies the catalytic components that facilitate this transition. The relevance of autocatalysis to living processes is exemplified by this reciprocity, since this is characteristic of the reciprocal synthesis of molecular components in cells. The geometrically increasing production of more of the same catalysts also appears superficially analogous to reproduction, although it is actually more analogous to growth processes (see below). This characteristic has led to autocatalysis being treated as the essential dynamic defining the living process.[15] This apparent interdependent production of molecular types has motivated the coining of the term *autopoiesis* (literally "self-forming" or "self-producing") to characterize this fundamental property of life.[16] The question remains whether component co-production of this sort is sufficient to constitute organismic "selfhood."

As Kant himself recognized, however, ascribing "means" and "ends" properties to these reciprocal catalytic processes is an externally imposed teleological interpretation. A purpose or end is not merely a consequent state of things. It is a general consequence that has a beneficiary. This defining feature of teleology goes back to Aristotle's characterization of what he described as "final cause." A final cause is that for the *sake* of which something is generated. So what is it that is the beneficiary of autocatalysis? Is it simply the autocatalytic process itself? The problem with this interpretation is that autocatalysis is not a distinct entity or a specific physical object, but rather merely a description of a type of chemical relationship involving a distributed set of interacting molecules. So there is no clear beneficiary. Even this general dynamical tendency itself is not benefitted because its very operation undermines the basis for its own persistence. The cyclic reciprocal nature of this process is only one of many possible descriptors that can be applied to such a collection of inter-reacting molecules. In this respect, both the autocatalytic set of molecules and the process of autocatalysis are merely abstract descriptions of molecular relationships, not distinct individuated physical entities. And although it is sometimes claimed (e.g., by Juarrero 1999) that autocatalysis comprises a "whole" which exerts a "top-down" causal influence (e.g., constraint) on the interactions of its catalytic "parts," exactly what constitutes the whole and the parts in autocatalysis is altered with each individual interaction. Indeed, even as they are synthesized, catalysts will diffuse away from the site of their production, so ultimately there is neither a consistent substrate nor a persistent locus. The one factor most relevant to autocatalytic persistence is the co-presence of substrates and catalysts, and the process of autocatalysis itself rapidly undermines this relationship.

Autocatalysis is not, then, fundamentally different from any typical chemical reaction. In both there is merely a repetition of similar chemical reactions generating increasing numbers of reaction products over time. And in both cases the conditions (constraints) that make any catalytic reaction possible are imposed extrinsically. They do not arise from anything one could describe as a self. There is no autocatalytic individual "self" that persists, only cycles of similar but individually distinct catalytic reactions. Moreover, this absence of individuation is inherited by any conception of organism framed in terms of autocatalysis even complexly interlinked autocatalytic processes such as Eigen's *hypercyles* exhibit this fundamental lack of intrinsic individuation.

So despite the self-promoting growth-like logic of autocatalytic processes, they have merely descriptive unity. The constituents of an autocatalytic system are intrinsically dissociated from one another. Arbitrarily distributed molecules interacting with each other in solution do not constitute a higher-order entity except as abstractly conceived. Moreover, the individual transformations that qualify as formative—the catalytic processes—are not in themselves self-organizing (i.e., form-generating in the sense of constraint-production), only the collective effect as a whole is self-organizing. With no particular unitary beneficiary—no self—for the sake of which these processes take place and no localizable source for the generation of these processes there is no actual (as opposed to merely descriptive) teleology. Lacking this, there can be nothing with respect to which some component process can be assessed as effective or ineffective, functional or non-functional, nothing to succeed or fail, live or die. As a candidate reinterpretation of Kant's description of what constitutes intrinsic teleology, autocatalysis fails for lack of a particular individuated locus that produces, and is benefitted by, this recursive production of molecules. In this respect, Kant's original assessment seems accurate—namely, that teleology in such cases is merely a "regulative" assessment, not a constitutive property.

Containment and Autopoiesis

While some accounts have considered autocatalysis to be sufficient in itself to qualify as *autopoiesis*,[17] others[18] consider that a minimal autopoietic molecular system requires that the autocatalytic process must be contained within an interdependently created, semi-permeable membrane. In an effort to remedy the diffuse nature of autocatalysis, Luisi (1993; 2003), Kauffman and Clayton (2006), and Thompson (2007), among others, have each argued that some form of closure and containment of an autocatalytic process is required before it can be understood as truly autopoietic. Introducing physical boundedness presumably answers the criticism that autocatalysis alone lacks a clear self-other distinction.

Thompson (2007) provides the most recent articulation of this enhanced model of autopoiesis. He describes a number of candidate molecular systems where an autocatalytic process is enclosed within a container.[19] Probably the most complete model envisions an autocatalytic process that is physically bounded within an amphiphilic vesicle, such as a bilayered lipid membrane.[20] In this model system, a lipid vesicle contains an autocatalytic set that both increases itself and additionally contributes to the growth of its lipid "container." The result is a physically bounded self-contained autocatalytic system able to grow. One important and essential complication is that autocatalysis can only occur in this system if the bounding membrane selectively enables catalytic substrates to selectively enter and waste products to be selectively excreted from the membrane. This is not a trivial complication, but assuming that it is achieved, growth of the whole complex can take place. It is further argued that continued growth will eventually cause the "cell" to become unstable and split, potentially forming two "daughter cells." Superficially, this appears analogous to reproduction, modeled after mitotic division of a simple living cell with autocatalysis substituting for its metabolism and genetics. Thompson (2007) agues that such a system exhibits three properties that must be present for molecular forms of autopoiesis: a reciprocal chemical reaction network, a semi-permeable boundary (generated by the contained chemistry), and autonomy (i.e., precise separation from the environment). Thompson (2011) thus describes this conception of autopoiesis as "the paradigm case of a system exhibiting self-creation and autonomy—the best understood and minimal case of an autonomous organization."[21] But let us examine this model system a bit more carefully to understand its properties.

Except for involving a chemical reaction and a boundary, how is such a system dynamically different from crystal growth that reaches a sufficiently large size so that it becomes unstable and breaks into two growing crystals? The shift to a lower energy of molecules being incorporated into a crystal lattice drives the process of growth, with the regularity of the lattice structure effectively acting like a catalyst which self-promotes lattice structure replication as new molecules accrete. When separated (e.g., by mechanical forces), the two new half crystals are individuated but do not lose this growth capacity. In both crystal growth and autopoietic growth, the creation of new individuals from one parent structure is a consequence of the physical properties of the materials at the point of breakage and growth. Molecular interactions occurring elsewhere in this molecular complex play no role in determining this individuation process. The creation of two individuals from one is thus an incidental physical consequence of structural growth and instability.

Like crystal separation, autopoietic individuation is not a disposition that is intrinsic to the systemic chemistry of the whole. Individuation also depends on the spontaneous re-annealing of the containing membrane and, consistent with this way of conceiving the process, Thompson (2007) echoes the views of Maturana and Varela (1987) in claiming that reproduction and evolvability are not features intrinsic to autopoiesis, but are processes that require additional complexity.[22] In this way, autopoiesis theory prioritizes growth and the replication of components over whole system reproduction. This inverts a critical biological relationship. Single cell organisms do not just break in two in order to reproduce, they generate duplicate autonomous systems (including duplicate genomes and cytoplasmic components) within a single membrane prior to initiating a process of walling each off from the other and then separating in space. Rather than growth, this generation of extra cellular components is more appropriately understood as development to a point where each potential daughter cell has all the components necessary to autonomously reiterate this same process.[23] Growth understood as mere multiplication of components is more characteristic of multicelled organisms, where mitotic cell multiplication is critical to developing a mature organism capable of reproducing.

This critique indicates that to distinguish simple catalyzed growth and separation from the end-directed dynamic of life we need to more carefully address two major questions. Most important is the question of what distinguishes the dynamical and physical autonomy that constitutes organismic "selfhood" from mere material boundedness, contiguity, or containment. This is coupled with a second question concerning the adequacy of the *Maximum Entropy Production* (MEP) principle to characterize the dynamics of life.

Autogenesis

These questions can be addressed by comparing autopoiesis as modeled above, with the concept of *autogenesis*[24] as developed by Deacon (2006a, 2006b, 2012) and by asking whether their respective forms of organization achieve what Stuart Kauffman (2000) calls "autonomous agency," by which he means the property of "acting on its own behalf."[25] Kauffman argues that this involves two requirements: it must propagate its organization to repair itself or produce a replica of itself and it must carry out at least one thermodynamic work cycle in the process. This latter requirement is what enables the process to be iterated indefinitely. In a subsequent work, Kauffman and Clayton (2006) also introduce the requirement of physical boundedness, which more explicitly localizes this property of agency. The phrase "acting on its own behalf" sets a useful standard with respect to which these model systems can be compared for their adequacy. The question is whether a given model's particular constellation of properties is sufficient to achieve self-directed agency. We argue that to meet the criterion of autonomous agency there must be a clearly defined individuated finite system that is both a source of work and the beneficiary of that work and which makes that system's existence / persistence more likely.

A minimal model system, called an *autogen* (or autocell), was proposed by Deacon to exemplify the distinctive features of autogenesis.[26] This model system incorporates a specific form of reciprocal coupling between autocatalysis and an enclosure-generating process, the latter of which is called *self-assembly*. It is similar to the most sophisticated exemplars of autopoiesis described above, but more explicitly addresses the critical property of self-individuation. Such a coupling can occur if one or more of the side products of autocatalysis tends to self-assemble as an enclosure, such as a polyhedron (like a virus shell) or a hollow tube (as in microtubule formation). This subtle augmentation helps to refocus attention on properties that are otherwise obscured or ignored in autocatalytic and autopoietic systems. This coupling of two distinct types of self-organized processes shifts the circularity of process up a level. It differs from the coupling of autocatalytic processes—or hypercycles—as described by Eigen and his colleagues[27] because the coupled self-organizing processes in autogenesis each generate very different forms of emergent constraint. In autogenesis, it is not just constituents that are joined in a reciprocally productive loop, but the *constraints* that each process generates, because each of these processes generates the boundary constraints that make the other process possible. Autocatalysis and self-assembly are in this respect reciprocally reinforcing self-organizing processes. They are not merely reciprocal phases of a self-organizing process.

This shift in the nature of the reciprocity toward the constraints generated rather than the materials, forms, or enclosure generated is a global property that exists independent of its specific material embodiment. The complementarity of the constraints generated by each component self-organizing process means that each provides certain critical extrinsic parameters that the other process depends on in order to persist. Thus, autocatalysis generates an accelerating increase in the local concentration of a few molecular forms. This is exactly the condition that a spontaneous self-assembly process requires. As molecules accrete to the edges of a self-assembling surface (e.g., viral shell or microtubule) their surrounding concentration is depleted until the growth process slows to a stop. But if there is a process (e.g., autocatalysis) that is continually generating an increasing number of these molecules locally, it will replenish those molecules that have accreted to the surface or have diffused away. Consequently, there will always be sufficient local raw materials for growth to proceed to an enclosed stage. Correspondingly, co-proximity (i.e., sufficient local concentration) of the reciprocal catalysts is necessary for autocatalysis, and the spontaneous diffusion of catalysts away from one another is a major impediment to the continuation of this process. The rate of autocatalysis must therefore be equal to or greater than the diffusion rate or the process will stop. But diffusion can be impeded by the growth of physical barriers to molecular movement. Such a boundary can be generated by self-assembly of a molecular sheet (e.g., a membrane), a polyhedron (e.g., a virus capsule), or a tube (e.g., a microtubule). So, if autocatalysis occurs in the vicinity of a rapidly growing molecular containment process all the necessary catalysts in the set will likely become enclosed. This can occur if autocatalysis generates a side-product that tends to self-assemble in one of these ways. Although enclosure will end autocatalysis, it will produce a bounded individual unit capable of re-initiating catalysis, and thereafter re-enclosure, if damaged in the vicinity of sufficient substrate molecules. *It is a unit structure that contains within itself the very set of constraints that are necessary and sufficient to re-create these same constraints in a new system in a supportive environment.*

A key feature of this model of autogenesis is that each component process (autocatalysis and self-assembly) is a different but complementary self-organizing process. When reciprocally bound together in this way what gets replicated are the constraints that are amplified by each as they individually maximize entropy production, not just the material components produced. Each of these self-organizing processes produces precisely those constraints that serve as permissive boundary conditions contributing to the operation of the other. But because eventual self-enclosure in a non-permeable container also limits the development of each component self-organizing process to a finite dissipated state, this increases the probability that the whole will reach a stable non-dynamic state before essential extrinsic resources are depleted. This then is a self-limiting constraint on entropy production. As a result, *this complementarity of permissive and limiting constraints effectively constitutes an additional higher-order constraint.* This higher-order constraint on the relationship between component constraint-generating processes is the locus of the potential to determine what constitutes an individuated system with the capacity to maintain, reconstitute, and reproduce a replica of itself. Indeed what ultimately gets reproduced is this constraint complex, not any specific physical object or collection. This higher-order constraint is thus the ultimate locus of selfhood—a property that is determined irrespective of any specific physical boundedness or material constitution. So in addition to constituent-co-creation and growth, as epitomized by the concept of autopoiesis, this co-generative co-constraining linkage of self-organizing process—which Deacon (2012) calls *autogenesis*—explains both how a molecular complex can be capable of spontaneously repairing itself when damaged and can generate a complete self-replica embodying these same properties when more completely dissociated. This provides a more stringent criterion for autonomy and points to a critical difference in focus as compared to theories of autopoiesis.

In simple terms, it is the difference between simple growth (accretion) with externally imposed parthenogenesis (splitting) and reproduction. Reproduction is an intrinsically whole-directed self-reconstitution process. Parthenogenesis in plants (e.g., strawberries) does not require reconstitution of whole organism self because all components (cells) already have the information necessary to "prescribe" the structure of a whole new individual. The "self" of a plant is already present (in potential) in its individual cells' genetics and cytoplasm. Cuttings of such a plant are autonomous individuals because they can reactivate this potential. Any of the autopoietic systems so far described (i.e., those comprised of a bounded and growing autocatalytic system) also becomes two individual systems by virtue of extrinsically induced parthenogenesis, but in this case, its components do not in themselves embody any whole-directed self-reconstitution process or information specifying what constitutes system completeness. The "damaged" structure merely re-anneals as two. This analogue of parthenogenesis, by which physical disruption is the determinant of the reproduction of new autopoietic individuals, is therefore an extrinsic property of autopoiesis. In contrast, external disruption is simply a loss of individual system integrity for an autogenic system. It is not a determiner of individuation, even though it may lead to reproduction. When the integrity of such a system is disrupted, catalysts and self-assembling molecules become dispersed and any semblance of structural individuality is lost. Individuation is instead determined by the way the component processes "respond" to disruption; individuation is a dynamic process not a structural property.

In response to extrinsic damage a dissociated autogenic complex initiates work that will specifically counter this disruption, and reintegrate the components into a non-dynamical structure again. The inert structure that results nevertheless possesses the *potential* to initiate exactly that form of work necessary to maintain this potential. This work of reintegration is a function of the complementarity of the constraints whose generation is potentiated by the way disruption enables the component self-organizing processes to be re-initiated. The physical work determined by these constraints does not merely grow the system larger, rather it specifies the unit-structure of what constitutes its autonomous individuality. Individuation is not a mere side effect of growth beyond a certain size in autogenesis (or in organisms in general). It is embodied implicitly in the system of constraints (or genetic information) irrespective of size or growth. In this way, autogenesis clearly exhibits a form of final causality, a specific disposition to achieve an end-state that is, circularly, what accounts for the existence of these co-dependent constraint-generating processes. Thus individuation appears to be distinguished in these two paradigms by an extrinsic-intrinsic dichotomy. Disruption of an autopoietic system individuates it by simply separating its material embodiment in space, whereas disruption of an autogen damages its individual integrity temporarily, but its embodied system of constraints initiate a process to re-individuate it, doing so immediately. Only an intrinsically embodied disposition to re-establish individual integrity after damage or to actively construct individuality in reproduction should qualify as autonomous agency.

Not being able to identify a clear final cause implicit in the above-described autopoietic model systems may explain why Maturana, Varela, Thompson, and others avoid invoking notions of teleology (and ultimately representation) as an important property. Indeed, Varela appears to have been as unwilling to consider notions of teleology and representation[28] as are most dynamical systems theorists.[29] In contrast, a form of final causality becomes the core property for defining selfhood, based on the autogenic analysis.[30] While this is a fundamental difference, it does not deny the possibility that something like these autopoietic model systems might be capable of evolving, but it would not involve global reorganization of the autopoietic dynamic, only inclusion of unprecedented components that may or may not be passed on in subsequent parthenogenic divisions (such as a local structural abnormality in a plant dividing by parthenogenesis). Indeed, Thompson (2007) argues that an autopoietic system need not exhibit either a tendency to reproduce or to evolve. In contrast, even a minimal autogenic system exhibits both capacities because during the open dynamic phase of reconstitution those acquired components that enter into effective autocatalysis or self-assembly have the capacity to pass on the changed constraints they introduce to all subsequent "progeny." Thus reproduction and evolvability are implicit in this kind of self-reconstitution dynamic.

The autogenic approach also addresses the second unmet challenge—explaining the nature of life's unusual thermodynamics—in a way that is not envisioned for autopoiesis. As noted above, self-organizing processes exemplify the MEP principle, in that their global order develops as a result of the way that these dynamical constraints aid the dissipation of imposed perturbations. But the constraint-generation reciprocity of an autogenic process ultimately takes advantage of MEP to eventually generate a non-dynamical consequence (an inert autogenic structure) that no longer generates an increase in entropy and thereby maintains the constraints necessary to re-initiate this dynamic again and again. This is what constitutes a work cycle: the employment of a constraint-generating process to produce the constraints necessary to repeat this incessantly. As a result, this circularity introduces a special case of MEP, since this property is defined within the context of boundary constraints. Autogenesis specifically replicates its own new boundary constraints, and so with each completed cycle of self-enclosure entropy production ceases before the potential to continue is exhausted. This produces what is essentially a *ratchet effect* in which full dissipation of constraints is never completed. Those that are newly generated by this process are thereby stabilized. This becomes the basis for the accumulation of new constraints over time—a necessary condition for evolution.

So this hypothetical molecular entity, which Deacon calls an autogen or autocell, embodies Kant's criterion in a way that he could not have conceived. It is not merely a self-propagating formative process, but *a reciprocally reinforcing system of self-propagating formative processes, which together form a cohesive integrated unit with the potential to actively maintain this unity of processes against dissociation.* Only this is sufficient to generate autonomous agency—a physical self that is the locus of an autonomous intrinsic end-directed dynamic, pointing to the meaning of the term *teleodynamics*.

Conclusion

Because they rapidly undermine the conditions that drive them, self-organizing processes are not sufficient to explain the coming-to-be of life-forms. Organisms are not merely complex self-organizing dissipative systems, nor is the addition of self-containment sufficient to generate true autonomy. Biological selfhood is not a function of physical enclosure or the circular production of all components from other components within the whole. Though these are common features of living organisms, it would be an error to attribute biological selfhood to these physical processes or their conjunction. Rather biological selfhood is vested in a particular configuration of the constraints embodied in such physical processes as these if they are collectively organized to produce and maintain autonomous individuation. This is a configuration among two or more constraint-generating (i.e., self-organizing) processes in which the critical boundary constraints of each are provided by the constraints generated by the other(s). In this way, autonomous individuation is not a function of any specific physical process, but rather of the multilevel system of constraints that maintain and reproduce themselves. Although this property appears to be implicitly subsumed within the concept of autopoiesis, and may in fact be exemplified in certain autopoietic model systems, a failure of this concept to shift out of dynamical relationships to focus on relations among the constraints that these dynamics embody prevents autopoietic theories from providing a fully comprehensive account of this essential property.[31]

This shifts the focus from a disposition to grow and reproduce to a disposition to produce a specific unrealized potential—an end, which is a type of condition rather than any specific bounded structure. The autonomous generation of precisely those constraints necessary to specify a complete dynamical system with a disposition to do this incessantly is the critical attribute of biological selfhood and its agency. It is, in effect, a disposition to maintain and potentially reproduce this same disposition indefinitely. The goal state that defines this disposition is even maintained *in potentia* after the integrity of the system is breached (as in a dissociated autogen). An intrinsic tendency to realize a general type of end state that is beneficial to the source of that tendency is the essence of final causality (i.e., teleology). It is what warrants calling this emergent disposition of life and mind "teleodynamics." And it is the defining attribute of biological selfhood.

Only with respect to such a self-reconstituting, self-individuating disposition can an actualized state of things be about another un-actualized state; i.e. be information *about* something. In this way, the origin of teleodynamics is the emergence of both self and semiosis at the same time. Ultimately, this is what constitutes the "several powers" that make possible the evolution of "endless forms most beautiful and most wonderful."[32]

Notes

1. By the sixth edition Darwin inserts "by the Creator" into this position, presumably to placate religious critiques and, as many have suggested, to reassure his devout wife who feared for his immortal soul.
2. Claude Shannon, "The Mathematical Theory of Communication," *Bell System Technological Journal* 27 (July and October 1948): 379–423, 623–656.
3. This way of simplifying and condensing the logic of neo-Darwinian theory remains the standard but has been extensively criticized in many contemporary texts. For an early critique see Brian Goodwin and Gerry Webster, "History and Structure in Biology," *Perspectives in Biology and Medicine* 25, no. 1 (1981): 39-62. And for a recent sympathetic critique see James Griesemer, "The Units of Evolutionary Transition," *Selection* 1 (2000): 67-80.
4. George Gaylord Simpson, "Biology and the Nature of Science," *Science* 139, no. 3550, (January 11, 1963): 87.
5. Ernst Mayr, "Cause and Effect in Biology," *Science* 10 (November, 1961): 1504.
6. Ernst Mayr, *Toward a New Philosophy of Biology: Observations of an Evolutionist* (Cambridge, MA: Harvard University Press, 1988), 45.
7. Mayr, "Cause and Effect in Biology."
8. Mayr, *Toward a New Philosophy of Biology: Observations of an Evolutionist.*
9. W. Ross Ashby, "Principles of the Self-Organizing System," in *Principles of Self-Organization: Transactions of the University of Illinois Symposium*, ed. Heinz Von Foerster and George W. Zopf, Jr. (London: Pergamon Press, 1962), 257.
10. Terrence W. Deacon, "Reciprocal Linkage between Self-Organizing Processes is Sufficient for Self-Reproduction and Evolvability," *Biological Theory* 1, no. 2 (2006a): 136–149; Terrence W. Deacon, "Emergence: The Hole at the Wheel's Hub," in *The Re-Emergence of Emergence*, ed. Philip Clayton and Paul Davies (Cambridge, MA: MIT Press, 2006), 111-150; Terrence W. Deacon, *Incomplete Nature: How Mind Emerged from Matter* (New York: W.W. Norton & Company, 2012).
11. Michael Polanyi, "Life's Irreducible Structure," *Science* 160, no. 3834 (1968): 1308–1312.
12. Immanuel Kant, *Critique of Judgment*, trans. J. H. Bernard (New York: Hafner Press, 1951). The attention given to Kant's *Critique of Judgment* by systems biologists interested in issues of self-organization seems to have begun in the early 1980s. Brian Goodwin and Gerry Webster discuss it in their "History and Structure in Biology" (1981). Complexity biologist, Stuart Kauffman, who was directly influenced by Goodwin's work, also presents Kant's description of organism as important to the understanding of causality in organisms in his 1993 book *Origins of Order*. (See also, Kauffman's Foreword to this volume.) Philosophers of biology have also focused on the relationship between a Kantian approach to teleology and a systems view of biology. For example, in 1985, philosopher, Alicia Juarrero, published a detailed philosophical analysis comparing Kant's concept of organism with the work of Ilya Prigogine on dissipative structures.
13. Alicia Juarrero-Roqué, in "Self-Organization: Kant's Concept of Teleology and Modern Chemistry," *Review of Metaphysics* 39 (September 1985): 131-132, italics in the original, notes, "Clearly Kant's claims are epistemological: final causation means for Kant only that *we* cannot think of organisms except as self-organized, not that organisms *are* self-organized. The emphasis on the limits of cognition, not ontology, is the reason

for the distinction between regulative and constitutive principles (and is what keeps the *Critique of Teleological Judgment* firmly inside Kant's critical work)."

14. Alicia Juarrero, *Dynamics in Action: Intentional Behavior as a Complex System* (Cambridge, MA: MIT Press, 1999).

15. For example, see: Manfred Eigen and Peter Schuster, *The Hypercycle—A Principle of Natural Self-Organization* (Heidelberg: Springer, 1979); Pier Luigi Luisi, "Defining the Transition to Life: Self-Replicating Bounded Structures and Chemical Autopoiesis," in *Thinking about Biology: Santa Fe Studies in the Sciences of Complexity, Lecture Notes, Volume 3*, edited by Wilfred D. Stein and Francisco J. Varela (New York: Addison-Wesley, 1993), 17-39; Pier Luigi Luisi, "Autopoiesis: A Review and a Reappraisal," *Naturwissenschaften* 90 (2003): 49-59; Juarrero, *Dynamics in Action: Intentional Behavior as a Complex System*; Evan Thompson, *Mind in Life: Biology, Phenomenology, and the Sciences of Mind* (Cambridge, MA: The Belknap Press of Harvard University Press, 2007).

16. See Humberto Maturana and Francisco Varela, *Autopoiesis and Cognition: The Realization of the Living*, Boston Studies in the Philosophy of Science, Volume 42, ed. Robert S. Cohen and Marx W. Wortofsky (Dordrecht: D. Reidel Publishing Company, 1980); Humberto Maturana and Francisco Varela, *The Tree of Knowledge: The Biological Roots of Human Understanding* (Boston: Shambala Press, New Science Library, 1987).

17. For example, Juarrero-Roqué, "Self-Organization: Kant's Concept of Teleology and Modern Chemistry"; Juarrero, *Dynamics in Action: Intentional Behavior as a Complex System*; Stuart Kauffman, *Investigations* (New York: Oxford University Press, 2000).

18. Francisco Varela, *Principles of Biological Autonomy* (New York: Elsevier North Holland, 1979); Thompson, *Mind in Life: Biology, Phenomenology, and the Sciences of Mind*.

19. Thompson, *Mind in Life: Biology, Phenomenology, and the Sciences of Mind*, 111-116.

20. See Luisi, "Defining the Transition to Life: Self-Replicating Bounded Structures and Chemical Autopoiesis."

21. Evan Thompson, "Mind in Life and Life in Mind," paper presented at the eSMCs Summer School, *The Future of the Embodied Mind*, San Sebastián, Spain, September 8, 2011 from the PowerPoint™ presentation of the lecture.

22. In "Mind in Life and Life in Mind," 167, Thompson explains this as follows: "Although autopoiesis and reproduction go hand in hand in living cells, there is a logical asymmetry between the two. Reproduction presupposes autopoiesis, but autopoiesis does not necessarily entail reproduction, for a system can be self-producing according to the autopoiesis criteria without being capable of reproduction."

23. See Griesemer, "The Units of Evolutionary Transition."

24. The neologism "autogenesis" has been used previously to mean various things in biology over the years. These include spontaneous generation, abiogenesis, and the emergence of eukaryotic cells. See, for example, Csanyi and Kampis' "Autogenesis: The Evolution of Replicative Systems." In this chapter, following Deacon ("Reciprocal Linkage," "Emergence," *Incomplete Nature*) we are using it for a very precise type of reciprocally interdependent interaction between different kinds of self-organizing process.

25. Kaufmann, *Investigations*.

26. Deacon, "Reciprocal Linkage between Self-Organizing Processes Is Sufficient for Self-reproduction and Evolvability."

27. See Eigen and Schuster, *The Hypercycle—A Principle of Natural Self-Organization*; Manfred Eigen and Ruthild Winkler-Oswatitsch, *Steps toward Life: A Perspective on Evolution* (New York: Oxford University Press, 1992).

28. Thompson, *Mind in Life: Biology, Phenomenology, and the Sciences of Mind*, 453, note 8.

29. For example, see the following dynamical systems approaches: J. A. Scott Kelso, *Dynamic Patterns* (Cambridge, MA: MIT Press, 1995); Tim van Gelder, "The Dynamical Hypothesis in Cognitive Science," *Behavioral and Brain Sciences* 21, no. 5 (1998): 615-665; Walter J. Freeman and Christine A. Skarda, "Representations: Who Needs Them?," in *Brain Organization and Memory Cells, Systems, & Circuits*, ed. James L. McGaugh, Norman Weinberger, and Gary Lynch (New York, Oxford Science Publications, 1992), 375-380.

30. Deacon, *Incomplete Nature*.

31. In a critical re-assessment of the concept of autopoiesis, Kepa Ruiz-Mirazo and Alvaro Moreno, in "Basic Autonomy as a Fundamental Step in the Synthesis of Life," *Artificial Life* 10: (2004): 238, come closest to the criteria here described as autogenesis by shifting the focus to the constraints involved. They argue that:

> Although the phenomenon of self-organization always involves the generation and maintenance of a global (or high-level) pattern or correlation that constrains the (low-level) dynamics of the components of the system, in standard dissipative structures this occurs only provided that the system is put under the appropriate boundary conditions. If those (externally controlled) conditions are changed (in particular, if the input of matter or energy is outside a certain range), the self-organizing dynamic vanishes. Therefore, there is an important difference between the typical examples of "spontaneous" dissipative structures and real autonomous systems: in the former case, the flow of energy and/or matter that keeps the system away from equilibrium is not controlled by the organization of the system (the key boundary conditions are externally established, either by the scientist in the lab or by some natural phenomenon that is not causally dependent on the self-organizing one), whereas in the latter case, the constraints that actually guide energy/matter flows from the environment through the constitutive processes of the system are endogenously created and maintained. (238)

Notice however, that this corrective does not also delineate what is additionally necessary to prescribe an autonomous individuation process, as described for autogenesis.

32. From the last phrase of the last line of Darwin's *On the Origin of Species*.

Bibliography

Ashby, W. Ross. "Principles of the Self-Organizing System." In *Principles of Self-Organization: Transactions of the University of Illinois Symposium*, edited by Heinz Von Foerster and George W. Zopf, Jr., 255-278. London: Pergamon Press, 1962.

Berg, Lev Semenovich. *Nomogenesis: Or, Evolution Determined by Law*. 1922. Reprint, Cambridge, MA: MIT Press, 1969.

Bergson, Henri. *Creative Evolution*. Translated by Arthur Mitchell. 1907. Reprint, New York: Henry Holt and Company, 1911.

Csanyi Vilmos, and György Kampis. "Autogenesis: The Evolution of Replicative Systems." *Journal of Theoretical Biology* 114, no. 2 (1985): 303-21.

Darwin, Charles. *On the Origin of Species by Means of Natural Selection or the Preservation of Favored Races in the Struggle for Life.* 1859. Reprint, London: J. M. Dent & Sons, 1971.

Dawkins, Richard. *The Selfish Gene.* New York: Oxford University Press, 1976.

Deacon, Terrence W. "Reciprocal Linkage between Self-Organizing Processes Is Sufficient for Self-reproduction and Evolvability." *Biological Theory* 1, no. 2 (2006a): 136-49.

———. "Emergence: The Hole at the Wheel's Hub." In *The Re-Emergence of Emergence*, edited by Philip Clayton and Paul Davies, 111-150. Cambridge, MA: MIT Press, 2006.

———. *Incomplete Nature: How Mind Emerged from Matter.* New York: W.W. Norton & Company, 2012.

Driesch, Hans Adolf Eduard. *The Science and Philosophy of the Organism, Volumes 1 and 2.* London: Adam and Charles Black, 1908.

Eigen, Manfred, and Peter Schuster. *The Hypercycle—A Principle of Natural Self-Organization.* Heidelberg: Springer, 1979.

Eigen, Manfred, and Ruthild Winkler-Oswatitsch. *Steps Toward Life: A Perspective on Evolution.* New York: Oxford University Press, 1992.

Freeman, Walter J., and Christine A. Skarda. "Representations: Who Needs Them?" In *Brain Organization and Memory Cells, Systems, & Circuits*, edited by James L. McGaugh, Norman Weinberger, and Gary Lynch, 375-380. New York, Oxford Science Publications, 1992.

Goodwin, Brian, and Gerry Webster. "History and Structure in Biology." *Perspectives in Biology and Medicine* 25, no. 1 (1981): 39-62.

Griesemer, James. "The Units of Evolutionary Transition." *Selection* 1 (2000): 67-80.

Juarrero, Alicia. *Dynamics in Action: Intentional Behavior as a Complex System.* Cambridge, MA: MIT Press, 1999.

Juarrero-Roqué, Alicia. "Self-Organization: Kant's Concept of Teleology and Modern Chemistry." *Review of Metaphysics* 39 (September 1985): 107-135.

Kant, Immanuel. *Critique of Judgment.* Translated by J. H. Bernard. New York: Hafner Press, 1951.

Kauffman, Stuart. *Origins of Order: Self-Organization and Selection in Evolution.* New York: Oxford University Press, 1993.

———. *Investigations.* New York: Oxford University Press, 2000.

Kauffman, Stuart, and Philip Clayton. "On Emergence, Agency, and Organization." *Biology and Philosophy* 21, no. 4 (2006): 501-521.

Kelso, J. A. Scott. *Dynamic Patterns.* Cambridge, MA: MIT Press, 1995.

Luisi, Pier Luigi. "Defining the Transition to Life: Self-Replicating Bounded Structures and Chemical Autopoiesis." In *Thinking about Biology: Santa Fe Studies in the Sciences of Complexity, Lecture Notes, Volume 3*, edited by Wilfred D. Stein and Francisco J. Varela, 17-39. New York: Addison-Wesley, 1993.

———. "Autopoiesis: A Review and a Reappraisal." *Naturwissenschaften* 90 (2003): 49-59.

Maturana, Humberto, and Francisco Varela. *Autopoiesis and Cognition: The Realization of the Living.* Boston Studies in the Philosophy of Science, Volume 42, edited by Robert S. Cohen and Marx W. Wortofsky. Dordrecht: D. Reidel Publishing Company, 1980.

———. *The Tree of Knowledge: The Biological Roots of Human Understanding.* Boston: Shambala Press / New Science Library, 1987.
Mayr, Ernst. "Cause and Effect in Biology." *Science* 10 (November, 1961): 1501-1506.
———. "Footnotes on the Philosophy of Biology." *Philosophy of Science* 36, no. 2, (June 1969): 197-202.
———. *Toward a New Philosophy of Biology: Observations of an Evolutionist.* Cambridge, MA: Harvard University Press, 1988.
Neumann, John von, and Arthur W. Burks. *Theory of Self-Reproducing Automata.* Champaign-Urbana, IL: University of Illinois Press, 1966.
Pittendrigh, Colin S. "Adaptation, Natural Selection, and Behavior." In *Behavior and Evolution,* edited by A. Roe and George Gaylord Simpson, 390-416. New Haven, CT: Yale University Press, 1958.
Polanyi, Michael. "Life's Irreducible Structure." *Science* 160, no. 3834 (1968): 1308–1312.
Prigogine, Ilya. *Introduction to the Thermodynamics of Irreversible Processes.* Chicago: Thournes, 1955.
Prigogine, Ilya, and Isabelle Stengers. *Order out of Chaos: Man's New Dialogue with Nature.* New York: Bantam Books, 1984.
Ruiz-Mirazo, Kepa, and Alvaro Moreno. "Basic Autonomy as a Fundamental Step in the Synthesis of Life." *Artificial Life* 10: (2004): 235-259.
Shannon, Claude. "The Mathematical Theory of Communication." *Bell System Technological Journal* 27 (July and October 1948): 379–423, 623–656.
Schrödinger, Erwin. *What Is Life? The Physical Aspect of the Living Cell.* Cambridge, UK: Cambridge University Press, 1944.
Simpson, George Gaylord. "Biology and the Nature of Science." *Science* 139, no. 3550, (January 11, 1963): 81-88.
Teilhard de Chardin, Pierre. *The Phenomenon of Man.* 1955. Reprint, New York: Harper and Row, 1961.
Thompson, Evan. *Mind in Life: Biology, Phenomenology, and the Sciences of Mind.* Cambridge, MA: The Belknap Press of Harvard University Press, 2007.
———. "Mind in Life and Life in Mind." Paper presented at the eSMCs Summer School, *The Future of the Embodied Mind,* San Sebastián, Spain, September 8, 2011 (from the PowerPoint™ presentation of the lecture).
van Gelder, Tim. "The Dynamical Hypothesis in Cognitive Science." *Behavioral and Brain Sciences* 21, no. 5 (1998): 615-665.
Varela, Francisco. *Principles of Biological Autonomy.* New York: Elsevier North Holland, 1979.
Weber, Andreas, and Francisco J. Varela. "Life after Kant: Natural Purposes and the Autopoietic Foundations of Biological Individuality." *Phenomenology and the Cognitive Sciences* 1 (2002): 97-125.
Zeleny, Milan, "Autopoiesis, A Paradigm Lost?" In *Autopoiesis, Dissipative Structures, and Spontaneous Social Orders,* edited by Milan Zeleny, 3-43. Boulder, CO: Westview Press, 1980.

Chapter Fourteen
Teleodynamics: A Neo-Naturalistic Conception of Organismic Teleology

Spyridon Koutroufinis

Introduction: *Philosophy of Biology* and *Biophilosophy*

The philosophy of biology is a discipline that was founded in the early 1970s. The best-known representatives of this discipline, which has become established especially in the United States, are theoretical biologists and philosophers.[1] Because of their commitment to the metaphysics of today's biosciences, many of the philosophers of biology and their adherents make up a mere subset—even if the most efficacious one at present—within a philosophic tradition that has existed since antiquity and will hereafter be referred to as "biophilosophy."[2] There are two reasons why I suggest making a distinction between philosophy of biology and biophilosophy, considering the former as included in the latter. First, in contrast to philosophy of biology, which does not allow reflections about matter and causality that go beyond the basic metaphysical framework dictated by today's mainstream biology, biophilosophy allows for questioning in relation to the metaphysical assumptions underlying research in biology and for exposing their limitations. Second, considering biophilosophy to be the broader field, metaphysically-speaking, allows for the pointing out of the relevance of the works of philosophers like Aristotle and Kant for contemporary bioscience, without characterizing them as "philosophers of biology." This is the case since the term "biology" was introduced in the very beginning of the nineteenth century when this discipline was founded. The borders, however, between philosophy of biology and biophilosophy are fluid.

The reasons for why the philosophy of biology seems to be unable to transcend the metaphysical framework of mainstream biology can be explained through the history of one branch of biology—namely, theoretical biology. This discipline began in the early twentieth century and aimed to develop a philosophically consistent foundation for biology. But in the 1920s, mathematical models of population dynamics were developed that initiated the systematic mathematization of theoretical biology, which began in earnest in the 1930s, with the works of Ludwig von Bertalanffy. With the development of theories of non-linear dynamic systems and the derivative concepts of self-organization, chaos, and complexity, theoretical biology became a mathematical discipline. As a result, the range of topics considered became much more limited.

In today's institutes of theoretical biology, it is mainly mathematical models, and computer simulations, of processes issuing from evolutionary theory, developmental biology, ecology, neurobiology, and epidemiology that are tested. In other words, it seems that only the branch of theoretical biology, which can be traced back to von Bertalanffy, has survived. In this way, the transformation of theoretical biology into a mathematical discipline has largely led to a lack of theoretical reflections about the foundations of biology as a scientific discipline. Today, the resulting void is increasingly being filled by philosophy of biology, which addresses topics involving the concepts of natural selection and adaptation; the concept of the gene; the meaning of teleology, purpose, and function; the relationship between micro- and macroevolution; the nature of biological species; and the conflict between evolution and theism. The wide range of topics it addresses provides a demonstration of the fact that the borders between theoretical biology and philosophy of biology are fluid. This also explains the very close connection between the philosophy of biology and the metaphysics of mainstream biology.

Due to its long development since classical antiquity, biophilosophy is able to exhibit a diverse range of naturalistic understandings of life in general, and of the organism in particular, which transcend the physicalistic metaphysics of most biologists. One central problem for the biosciences, namely, whether or not the concept of teleology has any legitimacy in biological explanation—a notion which has been discussed extensively since antiquity—exemplifies the central philosophical differences between the major representatives of the philosophy of biology and the more metaphysically open-minded considerations of the nature of life carried out by the biophilosophers. Yet, even inside current philosophy of biology, there are widely varying understandings of teleology. In this chapter, I will show how a recently suggested model of pre-cellular organization introduces a new form of teleology that has important consequences for our understanding of biological naturalism.

On Different Approaches to Teleology in the Philosophy of Biology

In the last few decades, several biologists and philosophers of biology have claimed that organisms may be considered teleological entities, spurring on a movement that is often celebrated as the renaissance of teleological thinking. I describe this movement as "neo-teleologism." I have coined this term based on the notion of "neo-teleology," which summarizes a strategy of biological explanation that has been supported by several philosophers of biology.[3] Neo-teleological explanations are based on the idea of natural selection.[4] In this chapter, I employ "neo-teleology" and similar terms in a more broad sense that includes three different understandings of biological teleology that appeared successively since the 1940's (see below). However, the issue of teleology in contemporary biology is a very contentious and complicated subject that has been tainted by many misunderstandings. It is doubtful whether opponents are arguing over the same thing.[5]

Aristotle is the biophilosopher who is mentioned most often by philosophers of biology. The interpretation, and subsequent acceptance or rejection, of central concepts in Aristotelian teleological theory is a reliable landmark for recognizing the metaphysical foundation of different authors' understandings of teleology. Philosophers of biology tend to reject all forms of universal teleology or pan-teleology. In other words, they reject the Aristotelian, Platonic, or Leibnizian conception that the whole cosmos is a finally aligned totality.[6] Neo-teleologism, however, is confined to a specific biological variety of teleological reasoning—the "special"[7] or "regional"[8] teleology which merely refers to single living bodies and not to global phenomena like the evolution of species or of the cosmos. Special teleology is again subdivided into "inner" and "external" teleology, the former emphasizing the growth of the entire organism, its elements, and their functional role, and the latter ascribing "utility for something else"[9] to the organism as a whole. Philosophy of biology recognizes only special, inner teleology.

In the long history of biophilosophy, a great deal of attention has been paid to special inner teleology. However, biophilosophers have not considered special inner teleology as the only kind of natural teleology. In biophilosophy, for more than two millennia, the term "*telos*" kept its double meaning of both "end," or rather "end-state," on the one hand, and "purpose," "aim," or "goal" on the other. In Aristotelian hylomorphism, the end-state of living processes is something both aimed at and purposed. As Hans Jonas has shown, Aristotle made a distinction between the "mere ending and internal 'end' of a movement."[10] In the first clearly formulated theory of teleology, as presented in *Physics* and *On the Soul*, the concepts of aim or goal, end and purpose denote inseparable aspects of one and the same thing: they designate essential elements of the "*eidos*" (εἶδος), or the form of the biological species to which a single living being belongs. But it is even more important to note that in Aristotelian teleology *these concepts presuppose the existence of a striving being*. Against the background of the last great metaphysical systems—created in the early twentieth century by William James, Charles Sanders Peirce, Alfred North Whitehead, Henri Bergson, etc...—Aristotle's concept of "striving" (*orexis*) may be interpreted in terms of *proto-mental* or *proto-experiencing* agency. According to pan-experientialistic metaphysics, entities at all levels of complexity are able to enjoy some degree of subjective experience. This is often misunderstood since we usually ascribe experience only to conscious beings. But, as Whitehead says, "consciousness presupposes experience, and not experience consciousness."[11] In other words: not all experience is conscious experience. Whitehead's explanations harmonize well with an important position in Aristotle's *Physics* where he very clearly accentuates that a striving entity is only rarely conscious of its acting. He states, "It is absurd to suppose that purpose is not present because we do not observe the agent deliberating. Art does not deliberate. If the ship-building art were in the wood, it would produce the same results by nature. If, therefore, purpose is present in art, it is present also in nature."[12] In other words, conscious action is only a seldom, special case of purposeful end-directed action. This interpretation of Aristotle's understanding of teleology confirms Mark Bedau's position that one may talk of teleology only in connection with entities to which one can ascribe values, since an experiencing being always strives to attain something that is experienced as something desirable.

In neo-teleological approaches to the philosophy of biology, the concept of *telos* is understood as end-directedness, but here "end" means the end-state of a material process that has been achieved by blind, deterministic, non-mental factors alone. Embryogenesis, physiological processes, the search for food, achieving a certain geographic position (e.g., in the case of migratory birds) and final acts of behavior (e.g., in the case of mating) are considered to be typical examples of end-directed processes.[13] According to Aristotle, in contrast, organismic end-directedness is the result of striving factors. If, as I suggest, Aristotle's understanding of striving be re-interpreted in pan-experientialistic terms, his concept of *telos* may be extended to include the idea that achieving a certain end requires that an experiencing being, even a proto-mental one, desires to achieve this end. In other words, the end-state has an intrinsic value for the striving being.

A major source of confusion in contemporary debates is due to the fact that it is easy to get the impression that the Aristotelian amalgamation of end and purpose has survived. The relevant literature is teeming with expressions like "purpose," "aim," and "goal." Less frequently there is talk of "purposiveness." This gives the impression of transcending the framework of physicalistic metaphysics of most biologists, which often makes proponents of mental teleology euphoric. Evidently, the crucial question that will reveal the nature of neo-teleologism is the actual meaning of "purpose" and "aim" in this approach. This question leads us to three milestones in the renaissance of teleology in the twentieth century.

First, in 1943, the founders of Cybernetics—Wiener, Rosenbleuth, and Bigellow—published the article "Behavior, Purpose and Teleology," in which they argued for the rehabilitation of teleology. They used the term "purposeful" to denote an act where "the act or behavior may be interpreted as directed to the attainment of a goal—namely, to a final condition in which the behaving object reaches a definite correlation in time or in space with respect to another object or event."[14] In this definition, the term "purpose" is co-extensive with a special understanding of the expression "final condition," which in this context means "end-state." In cybernetics, the terms "aim" or "final condition" mean "end-state," that is, the encounter between a behaving object (e.g., a missile) with a certain external object (e.g., a ship)—and this is merely a spatiotemporal event. On the basis of this concept of purpose that excludes every conceivable kind of first-person perspective—the behaving object does not have an *aim of its own*, as Jonas correctly states[15]—cyberneticists define teleologic behavior as the variety of purposeful behavior which reaches an end-state by means of a mechanism of *negative* feedback:

> We have restricted the connotation of teleological behavior by applying this designation only to purposeful reactions which are controlled by the error of the reaction—i.e., by the difference between the state of the behaving object at any time and the final state interpreted as the purpose. *Teleological behavior thus becomes synonymous with behavior controlled by negative feedback*, and gains therefore in precision by a sufficiently restricted connotation.[16]

There is an intrinsic relation between this understanding of teleology and the cybernetic concept of *information*. Wiener thinks of information as something that is used by a "behaving object" that is controlled by a negative feedback for steering it toward a predefined goal. Wiener developed his concept of information almost contemporaneously with Claude Shannon in the 1940s. Both authors employ the same formalism and connect the notion of information with the concept of statistical entropy which is a measure of a physical system's disorder.[17] By doing so, they identify information with the physical features of a material or energetic system (e.g., an electromagnetic signal), thus ignoring the meaning and reference of the message that this system potentially caries. The operations of cybernetic and information processing devices do not have meaning and value for the automata themselves, but only for human beings who determine the "goals" and "purposes" of the devices. Therefore, both renderings of the term "information" have only a syntactic aspect, and thus, are void of semantics (meaning and reference). Although a missile can process the electromagnetic signal of its radar in such a way that enables it to encounter a ship, the signal is void of meaning for the missile itself and so does the *telos* or purpose that it attains by processing the information of this signal. In cybernetics (Wiener) and information-theory (Shannon), teleology encompasses a concept of information that is not able to count for the semantic aspects which underlie the design of cybernetic and information-processing automata. These theories do not make any claims about the causality of the processes involved in creating, expecting, or evaluating the usefulness of information.

Second, some neo-Darwinists have welcomed the non-metaphysical conception of purpose and *telos* that was provided by the cyberneticists. They adopted and developed cybernetic teleology further. Ernst Mayr added that the mechanisms which orientate the negative feedbacks toward an end-state and activate them are programs. Mayr, Jakob, and Monod are the best-known proponents of the program metaphor in biology.[18] They consider programs as genetic or behavioral algorithms that were generated in evolution and brought *selective advantages* to the organisms carrying them out. Normally, neo-Darwinist theoreticians of teleology interpret purpose as *function*. They do not attempt to explain, for example, how the wing of a bird embryo develops step by step by molecular mechanisms, but rather content themselves with stating that wings develop in order to perform a function, leading to a positive selection of all its bearers which were progenitors of the bird embryo in question.[19] The "what for" questions and the "in order to" replies typical of teleological language were retained. They refer, however, only to natural selection:

> The sense in which what-for questions and their answers are teleological can now be clarified. Put cryptically, we explain A's existence in terms of A's function. More fully, A's existence is explained in terms of effects of past instances of A; but not just any effects: we cite only those effects relevant to the adaptedness of possessors of A.[20]

Griffiths described this kind of neo-Darwinist teleological reasoning that arose from the identification of purpose with function into perspective as follows: "Where there is [natural] selection there is teleology."[21] The neo-Darwinist idea of the genetic program or genetic information is based on the concept of information as it is introduced in cybernetics and informatics; in part, this makes it difficult to ascribe semantic aspects to this idea. Nevertheless, this should be possible, since survival, reproduction, or death cannot be conceived of without any reference to their meaning and value for the organisms, in question.

Third, despite the problematic conception of a "program," the lack of even a simple concept of organism remains a decisive weakness of neo-Darwinism. Neo-Darwinian teleologism only considers single functions. But, in evolution, a whole phenotype is selected, that is, a complex structure of mutually conditioning functions and elements. Two hundred years ago, Kant emphasized this most essential aspect of the organism with his concept of "self-organized things" (see below). Because of the inability of neo-Darwinism to consider whole organisms, an organismic turn is currently taking place in the philosophy of biology in which dynamic systems theory (or the theory of self-organization) plays an essential role. The directedness of embryonic and other processes toward a certain end-state or an "aim," as it is often called, is understood as the outcome of the *self-organized* complex molecular dynamics of organisms. The proponents of this conception integrate cybernetic and neo-Darwinian ideas into a more sophisticated model of organismic complexity: the organism would be a self-organized dynamical physico-chemical system, the dynamics of which results in virtue of an extremely complex structure of interdependent positive and negative feedbacks[22] that were successful in natural selection. Dynamic systems theory and theories of self-organization and complexity build the theoretical foundation of the third and most recent kind of biological neo-teleologism.

Philosophy of biology gives all three neo-teleological approaches great credit for providing interpretations of "purpose," "aim," and "telos," without any reference to mental factors. Philosophers of biology differentiate sharply between versions of special internal teleology divested of all psychological or mental connotations and others which assume mental factors. According to philosophers of biology, only *non-mental special internal teleology* comes into question for biology, thereby distancing themselves from many biophilosophers, including Aristotle, as the following diagram shows:

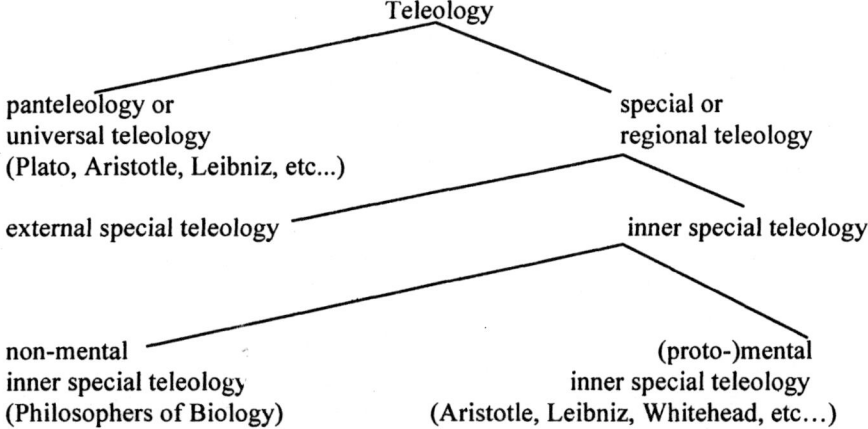

Figure 14.1: Different kinds of teleology. This diagram is based mainly on the ideas of Toepfer, and Mahner and Bunge.[23]

All three approaches serve the creation of a new form of teleologism, allowing biology to use teleological language without neglecting scientific metaphysics.

Mental Teleology in the Writings of Aristotle and Kant

The philosopher and scientist Mark Bedau rightly noted that "we are unsure whether teleological notions apply in roughly the same cases as those in which we are unsure whether value notions apply."[24] This statement is fully applicable to Aristotelian and Kantian considerations of organismic teleology. The second book of the *Physics* occupies a crucial position in Aristotle's theory of biological teleology that, from a present-day perspective, can be assigned to special inner teleology. In this text, Aristotle makes clear that the concepts: "end" and "purpose"—the Greek expression for "purpose" is "*ou heneka*" (οὗ ἕνεκα) which means "for the sake of which"—are mutually related and inseparable. The key query, however, is whether the Aristotelian term, "purpose" refers to an experiencing unity that strives to attain its aims. Neo-teleologically-minded biologists and philosophers tend strongly to interpret the Aristotelian concept of purpose in a functionalist manner[25] by overestimating the importance of certain passages in the second book of the *Physics* in which Aristotle explains the end-directedness of certain biologic processes by their functions in the organism, for example, "roots extend downwards . . . for the sake of nourishment"[26] and "sharp teeth are located in the front of the mouth for the sake of tearing."[27] Appropriately, Jonas criticized this myopic restriction of Aristotle's thinking to functionality when reminding us that his teleology is only "in the second place a fact of structure or physical organization, as exemplified in the relation of organic parts to the whole and in the functional fitness of organism generally."[28] Indeed, something beyond functionalism is much more important to Aristotle's teleology: his worldview simply *forbids* considering a natural process, controlled by blind, non-mental forces, as being able to achieve the kind of ordered result attained by an appropriately formed organic structure that serves the purpose of staying alive, like an organism or an organ, rather than degenerating into chaotic malformation.[29] Aristotle applies to blind mechanistic processes the term "automaton" (αὐτόματον), which may be translated as "senseless in itself," since "maten" (μάτην) means "in vain."[30] He refers to all processes that are not grounded in any kind of mental purpose as "automata." These may sometimes look *as if* there were a purpose behind their movement: the roof tile that falls on somebody's head could have been thrown at him purposefully by someone else. Only very rarely, however, do blind forces lead to an end-state that could be considered as an intentional one. Aristotle would subsume all processes which we today consider to be regulated by physico-chemical interactions under the category of "automaton." Accordingly, from the point of view of Aristotle, all phenomena of material self-organization constituting the third type of contemporary neo-teleologism would be cases of "automatic" becoming. He would never assume that these non-mental processes would be able to produce something as ordered as even a single cell. As stated above, this does not mean that Aristotle ascribes a human- or animal-like mentality to biological processes.

Over two thousand years later, Kant reflected on biological teleology against the background of Newtonian mechanics, which had clearly shown that non-mental forces like gravitation may produce well-ordered results like the solar system. The famous summary of his thinking about the ability of eighteenth century physics to explain the special inner teleology of organisms is contained in section 75 of his *Critique of Judgment*, which reads as follows:

> It is indeed quite certain that we cannot adequately cognize, much less explain, organized beings and their internal possibility according to *mere mechanical* principles of nature, and we can say boldly it is alike certain that it is *absurd* for men to make any such attempt or to hope that another *Newton* will arise in the future *who shall make comprehensible by us the production of a blade of grass according to natural laws* which no *intention* has ordered.[31]

Obviously, although Kant was very experienced in mechanistic physics—in 1755, he had published one of the very first cosmological studies about the generation of the planets of our solar system out of the primeval solar nebula through Newtonian gravitational pull—his skepticism concerning the applicability of physics to the study of organisms is very similar to Aristotle's refusal to consider living beings as automata. The concept of purpose is also of great importance to Kant's theory of the organism:

> For a body then which is to be judged in itself and its internal possibility as a natural purpose, it is requisite that its parts mutually depend upon each other both as to their form and their combination, and so produce a whole by their own causality. . . . In such a product of nature every part not only exists *by means of* the other parts, but is thought as existing *for the sake of* the others and the whole, that is as an (organic) instrument. Thus, however, it might be an artificial instrument, and so might be represented only as a purpose that is possible in general; but also its parts are all organs reciprocally *producing* each other. This can never be the case with artificial instruments, but only with nature which supplies all the material for instruments (even for those of art). Only a product of such a kind can be called a *natural purpose*, and this because it is an *organised* and *self-organising being*.[32]

Of course, most contemporary philosophers of biology try to interpret Kant's understanding of "purpose" and "for the sake of" in a non-mentalistic way, both in the passage quoted above and in his theory of teleology altogether. This would be misleading, since the Kantian concept of purpose implies the concepts of *will* and *reason*. As he states,

> In order to see that a thing is only possible as a purpose, that is, to be forced to seek the causality of its origin not in the mechanism of nature but in a cause whose faculty of action is determined through concepts, it is requisite that its form be not possible according to mere natural laws, i.e., laws which can be cognised by us through the Understanding alone when applied to objects of Sense; but that even the empirical knowledge of it as regards its cause and effect presupposes concepts of Reason. This *contingency* of its form in all empirical natural laws in reference to Reason affords a ground for regarding its causality as possible only through Reason. For Reason, which must cognise the necessity of every form of a natural product in order to comprehend even the conditions of its genesis, cannot assume such natural necessity in that particular given form. The causality of its origin is then referred to the faculty of acting in

accordance with purposes (a will); and the Object which can only thus be represented as possible is represented as a purpose.[33]

A "faculty of action" which is "determined through concepts" can only be a mental faculty since something non-mental could never capture concepts; a "causality" that is "possible only through Reason" can therefore only arise from a mental agent. It is not possible here to elaborate on the details of Kant's epistemological (and not ontological) treatment of teleology as it is developed in paragraph 68 of the third *Critique*. Let us just say that for Kant the idea of natural purposiveness does not mirror the essence of natural things but only the constitution of human reason.[34] However, what is much more important than talking about Kant's emphasis on epistemology is to ask why the very thinker who introduced the concept of self-organization developed it as something distinct from physics. This question makes sense, especially in view of the third neo-teleological approach that is grounded on a physicalistic conception of self-organization. Kant realized that the mechanistic thinking in the physics of his day was only able to explain processes controlled by strictly forward-directed causal connections.[35] In such connections, which nowadays are called *linear*, "things which as effects presuppose others as causes cannot be reciprocally at the same time causes of these."[36] But Kant also knew that "a causal combination according to a concept of Reason (of purposes) can also be thought, which regarded as a series would lead either forwards or backwards; in this the thing that has been called the effect may with equal propriety be termed *the cause of that of which it is the effect*."[37]

It is exactly this kind of causal connection that governs every organism or "natural purpose," since "its parts . . . are reciprocally cause and effect of each other's form."[38] The reason why Kant was as certain as to whether a *Newton* of even the simplest organism will never arise is the inability of the physics of his time to operate with reciprocal causal connections. Therefore, he introduced a species of causality into biophilosophy that is based on a mentalistic concept of purpose. For us today the main question is, of course, whether Kant's criticism is applicable only to eighteenth century physics. Would he agree that today's theory of self-organization, which operates with *non-linear* feedbacks and "downward causation" without employing a mentalistic concept of purpose, is able to adequately describe organisms? I believe that he would not.

On the Limitations of the Conception of Organisms as Self-Organized Dynamic Systems

Kant's use of the term "self-organizing," which he coined, is clearly far more generic than its use in current physics and biosciences. For Kant, "self-organization" refers to organic reciprocities. When contemporary theorists say that systems are self-organizing, they refer to energetically and/or materially open systems that operate far from thermodynamic equilibrium. These systems may be living or nonliving, but in all cases their organization emerges out of non-mental physicochemical interactions. Aided by the theory of dynamic systems, the modern paradigm of complexity or self-organization has become a main pillar of modern theoretical biology. However, there are good reasons to criticize theories and findings based on this development, which is the foundation of the third and newest neo-teleologic approach (outlined above), since it has at least one decisive weakness. In order to explain this, it is necessary to introduce some of the "technical details" of dynamical systems theory.

A system is defined as *dynamic system* if its state at any given moment can be described as a limited set of time-dependent, or state variables $x(t) = x_1(t), x_2(t), \ldots, x_n(t)$, for which a function F can be formulated by stating mathematically the connection between states at times t and $t + \delta t$. The properties of this function reflect the *causal* relationships at work within the system. The set of state variables $[x_1(t), x_2(t), \ldots, x_n(t)]$ spans an *abstract space*, the system's so-called "state-space."

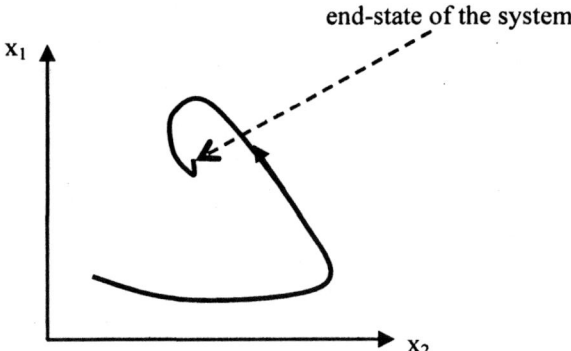

Figure 14.2: State-space and trajectory of a dynamic system with two state variables.

It is important to keep in mind that the development of a dynamic system is not merely the result of the function F, but depends also on a group of externally fixed parameters. The most abstract formula for a dynamic system must therefore be:[39]

$$x(t + \delta t) = F(x(t), p, \delta t); p = p_1, p_2, \ldots, p_m$$

The letter p represents a set of parameters. All parameters are *externally fixed constants*. Their role is to constrain the development of the state variables $x(t)$. Obviously, parameters play a very important role in dynamic systems theory. Dynamic systems can be subdivided into *conservative* and *dissipative* dynamic systems. The energy of the latter "dissipates"—that is, it disperses and must be replaced by the environment. *Dissipative systems produce entropy.* As we shall see, it is precisely because they produce entropy that, under certain conditions, allows dissipative systems the kind of spontaneous structuring of their behavior in space-time that is commonly labeled "self-organization."

The Paradox of Self-Organization: The System Organizes Itself in Order to Oppose the Causes of Its Self-Organization in a More Efficient Way

Statistical entropy is a concept applicable only to systems with a large number of particles, and it serves as a measure of disorder. Boltzmann and Planck define the statistical entropy of a system as the average value of its uncertainty.[40] A system is uncertain if there are many possible states in which it might be. Ideally, each of these states corresponds to a point in its state-space, which, with a particular degree of probability, can be the actual state of the system. The statistical entropy of a system is the average value of the probabilities of all possible states that the system might occupy. Accordingly, statistical entropy is related to the concept of *possibility*.

The order and the statistical entropy of a system that has n state variables can be depicted using an n-dimensional state-space. For each of the possible states there is a particular point in this abstract space to which it corresponds. Thus a limited area of the state-space represents all the possible states that the system can occupy at a given time.

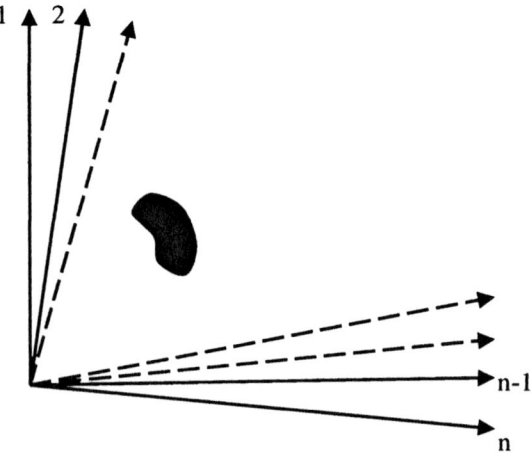

Figure 14.3: The high-dimensional state-space-volume of a system with n variables at a given time. Each one of its points represents a possible state of the whole system.

Considering living beings to be self-organized complex dynamical systems has become essential for theoretical biology in the last thirty years. "Self-organization" is a technical term. It means that the increase of a system's order—that is, the decrease of its entropy—is the result of interactions between its elements and not the outcome of the action of a single real or ideal entity, such as an acting person or a program. Self-organization does not mean elimination of entropy or uncertainty, only their diminution.

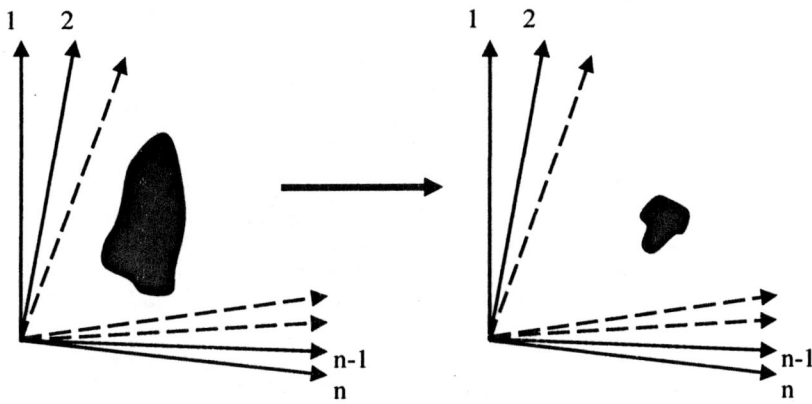

Figure 14.4: The diminution of a system's entropy equals a decrease in the number of possible states.

Systems that serve as models of self-organization require *gradients* of energy and/or material. A typical example of such a gradient is the difference of temperature in the so-called Bénard convection. This effect appears when the lower layer of a fluid is heated and the upper layer is kept at a cooler temperature. At a certain difference in temperature between the bottom and the top of the fluid, the heat flux reaches a critical value and convection arises. Coherent macroscopic movements emerge in the fluid and form a highly structured pattern of hexagonal cells. Of course, the examples of self-organization that are most interesting for biology are of chemical and biochemical nature.

One finding in thermodynamics has little-understood consequences for the applicability of the theory of complex dynamical systems to biology: every form of self-organization of a physico-chemical system amounts to a *decrease* in the gradients which are imposed on the system and which move it away from thermodynamic equilibrium, that is, from the state of a total lack of physical becoming. *Each self-organized system tends to return to equilibrium.* The hexagonal Bénard cells transport heat upwards faster than simple heat conduction, thus *increasing the rate of gradient destruction.* Two well-known physicists, Schneider and Kay, got to the heart of this finding:

> As systems are removed from equilibrium, they will utilize all avenues available to counter the applied gradients. As the applied gradients increase, so does the system's ability to oppose further movement from equilibrium. . . . No longer is the emergence of coherent self-organizing structures a surprise, but rather it is an expected response of a system as it attempts to resist and dissipate externally applied gradients which would move the system away from equilibrium.[41]

All self-organized structure-formation that is conceivable in physics assumes that the higher the order of the system's actions, the more efficient the degradation of the causes of this order (i.e., the degradation of the gradients).

This essential property of dissipative dynamic systems is also known as the principle of maximum entropy production, which can be easily explained. According to the second law of thermodynamics, all real (and not ideal) physical processes produce entropy. Production of entropy means dissipation or degradation of energy. Of course, energy cannot be created or destroyed in a system (as this is forbidden by the first law of thermodynamics or the law of conservation of energy), and the system cannot upgrade already degraded energy in order to degrade it again (since this is forbidden by the second law of thermodynamics). Thus the system can only produce entropy if it degrades the energy with which it is supplied from outside—namely, by means of the externally applied gradients. And the system can only be efficient in degrading gradients that distance it greatly from the state of thermodynamic equilibrium, if it maximizes the production of entropy in its own processes. The entropy inside the system decreases in order to enable it to produce entropy more quickly. The emergence of the higher-order macroscopic structure, which exerts downward causation on the lower-order events (on the molecular level), *only serves the degradation of gradients through the maximization of entropy production*. It is important to keep in mind that, in all mathematical models, gradients are represented by a number of parameters.

On Modeling Biomolecular Processes in Systems-Biology

Some bioscientists, primarily systems biologists, maintain that organisms *are nothing more* than physico-chemical dynamical systems, and they dream of future computer-simulations of whole organisms, possibly within the next fifty years.[42] This development is philosophically interesting because, in doing so, these bioscientists attribute ontological and not just heuristic relevance to the theory of self-organized dynamic systems. The solving of non-linear differential equations and the concomitant computer simulations are fundamentally important to formal reductions of organismic processes. Both operations require, among other things, a special condition: the sharp distinction between dynamic and static quantities—namely, between variables and parameters. This distinction is present in an enormous number of sources in systems-biological literature.[43] A very characteristic example can be found in Panning et al. (2007), who published a model of a very simple network of three interconnected biochemical reactions from the cell cycle of a frog's egg.[44] It is important to keep in mind that the computation of the self-organization of this system with only three dynamic quantities requires thirteen parameters—that is, thirteen quantities which take no part in the dynamics of the modeled self-organization. In modeling the cell cycle of yeast, the same authors use 36 coupled differential equations (i.e. 36 variables) to which they impose 143 parameters.[45]

Parameters do not necessarily represent specific chemical substances. Some of them symbolize *gradients,* while others are *abstractions,* summarizing relations between variables of the cell such as volume, temperature, pressure, pH-value, etc... Some parameters represent the degree of activity of specific molecules in systems-biological simulations, while others represent the rate coefficient of reactions in chemical kinetics. In simulations, parameters are determined by the systems-biologists themselves. They are either experimentally derived or estimated or simply taken from literature. There are many ways to manipulate the values of parameters in experiments. They are, however, kept constant in every experiment and corresponding computer simulation as well. The reason for this is simple. As has already been said, parameters constrain the self-organized processes studied by dynamic systems theory. They are one of the two factors that constrain the development of the variables. The other factor is the variables themselves, since the causal connections between them allow their perpetually changing quantities to actualize only a part of the possible values that would be actualized if the connections were not there.

Organisms Are More Than Complex Dynamic Systems

It is typical of *all* mathematical accounts that rely on differential equations—whether in physics, chemistry, or biology—to *depend essentially on a high number of externally set parameters.* Scientists set these quantities in their experiments and calculations. Within theories of self-organization and complexity there is a strict distinction between dynamic state variables and externally set parameters—in other words, between constrained and constraining quantities. It is obvious, however, that the quantities, whose causal influence on the system's dynamics in systems-biological models are summarized in parameters, will, in *real organisms,* vary constantly. For example, the variation of cell volume during a real cell cycle would imply that some parameters must vary. Therefore, with regard to the sufficiency of self-organized dynamic systems theory to deepen our theoretical understanding of the organism, the crucial question is whether the aforementioned distinction between dynamic variables and parameters could ever be overcome. While this division is not problematic within physics, even the most primitive organisms go beyond this dimension of self-organization. In stark contrast to formal models, almost all quantities in real organismic networks are highly dependent on the network's own inner dynamics. That which Kant claims for the parts of the organism—namely, "that its parts should so combine in the unity of a whole that they are reciprocally cause and effect of each other's form"[46] may, with regard to contemporary biomathematical modeling, be rephrased as follows: "The organism's quantities are reciprocally the cause and effect of each other."

In order to preserve their own adaptation, organisms trigger multiple changes within themselves.[47] In modeling organisms as dynamic systems, these changes ought to be described as *internally controlled* changes of many parameters, if the model makers are to make good on their claim to have created a model that does not just have heuristic value for biotechnology, but rather, gives an insight into a real organism's causality. A model which realistically mirrors the organism's autonomy must be able, at least in principle, *to calculate a significant part of its (the model's) parameters*—namely, to dynamize those quantities which in today's modeling are kept constant, or, in other words, to convert most of the parameters into variables and let the overall system's dynamics calculate their value. In addition, the model must be able to calculate and adjust independently, and not only those parameters describing the organism's self-supply with energy and materials from its environment, since this is something all organisms do constantly. Accordingly, the modeling of real biological self-organization would demand of systems that their dynamics calculate the self-imposition of energetic-material gradients. But within current physics this is *impossible*, primarily because all dissipative dynamic systems, which are governed strictly by non-mental physico-chemical forces, have an inherent *entropic tendency*. They organize themselves in order to reduce the gradients imposed on them and, therefore, cannot control these. The more the system would attain influence over its parameters—especially over those representing energetic-material openness (i.e., the existence of gradients)—the more it would become disorganized.

In contemporary systems-theory there is no clear evidence that a formal system is able to vary these quantities by itself. Of course, this does not exclude the possibility that, in the future, a new formalism based on a more advanced mathematics will appear, which will found a new kind of self-organized dynamics that will be able to convert a significant part of its parameters in variables. Within current systems-theoretical formalism, however, the entropic tendency of a dynamic system can only be limited if its parameters are *externally* controlled and kept constant. Often this criticism is rejected because, in real organisms, many processes take place under constant conditions as well. This is, of course, the case, but the holding constant of such conditions is something that cannot be taken for granted. On the contrary, it is an achievement of the organism itself: its overall dynamics hold certain quantities at least nearly constant. In terms of dynamic systems theory this could be expressed as follows: in certain processes the overall dynamics repeatedly calculate nearly the same value for particular variables.

So, within current systems-theoretical formalism, systems governed only by causes without any mental capacities, but able to influence most of their parameters would exhibit *an enormous number of causally indefinite states* in their state-spaces. This is the case because the considerable variation of parameters implies a change in the overall dynamics, and therefore, the generation of many new possible developments. Such systems would be much more unstable than some of the ordinary dynamic systems whose state-spaces have areas where closely adjacent trajectories tend to diverge strongly. Instability is a phenomenon often encountered in the theory of dynamic systems and in systems-biology, and even if all parameters are firmly set and not varied by the model makers. Instable states are causally indefinite (i.e., indeterminate). The *entropic tendency* inherent in every dissipative dynamic system means that the more the system is able to attain influence over its parameters—especially on those representing energetic-material openness—namely, the existence of gradients—the more it will become disorganized. The development of such a system would only lead to a continual increase of the system's own entropy (i.e., to a permanent increase of the number of its possible states).

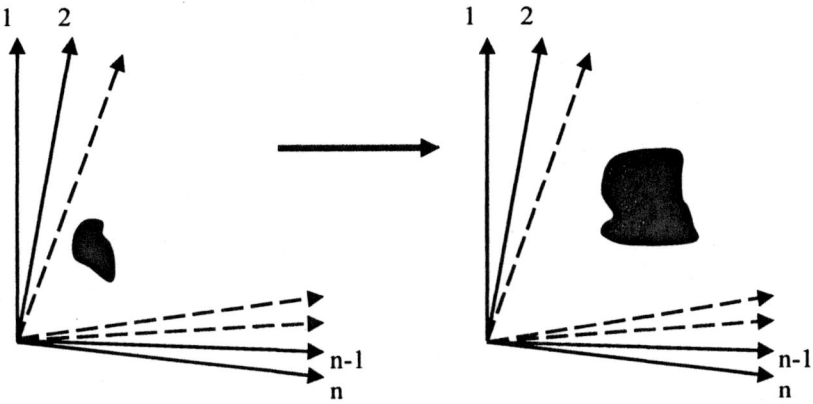

Figure 14.5: The increase of entropy within a system depicted as the increase in the number of possible states.

Figure 14.6 (below) shows another way of describing the increase of entropy—caused by the loss of constraints—as the increase of the number of possible trajectories.

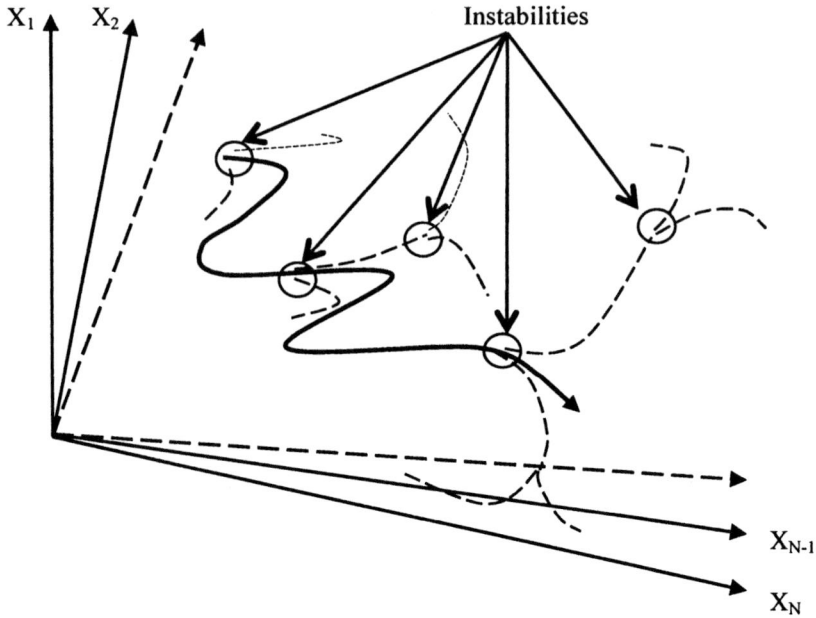

Figure 14.6: Permanently occurring instabilities in the development of a hypothetical dissipative dynamic system which influence the value of its parameters. If understood as displaying a real organism, the variables X_1 to X_n represent important dynamic quantities, like concentrations of proteins and signal substances, etc…, the coherent variation of which is characteristic of an organism. The curves do not symbolize single trajectories but rather bundles of these.

If the system were a model of a real organism, *only a very limited number of these possible trajectories would be biologically viable.* In other words, *very few would represent states of being alive.* This is so because biological structures constitute only a vanishingly small number of all possible physiochemical structures. In figure 14.6, the long curve represents a thin bundle of biologically viable trajectories, while the dotted lines stand for developments that are possible in terms of physics and chemistry, but fatal from a biological point of view. They show the derailment into areas of increasing entropy or uncertainty (i.e., of lethal deformation of the organism's structure). This leads necessarily to the following *aporia*: how does an organism succeed in avoiding *derailments* into areas of disorder, if it often faces possibilities equally valid from the point of view of physics allowing a biologically adequate choice between these possibilities?[48]

There is also tried and tested experimental evidence proving that the causal order of organisms stays beyond a physicalistic understanding of self-organization. For example, biologists have known for a long time that the exchange of energy and material between real (i.e., not virtual) organisms and their environment does *not* follow the principle of maximal production of entropy. This is logical, of course, since a high production of entropy means a high degradation, or waste, of energy. The year-long experimental and theoretical research of Gernot and Renate Falkner clearly shows that the metabolic exchange between cyanobacteria and their environment only exhibits a high production of entropy, if the physiological adaptation of the bacteria to their environment has been disturbed and is merely being readjusted.[49] The act of physiological re-adaptation, as they describe it, brings about the minimization of entropy production and not its maximization, which one would expect when starting out from the physico-chemical theory of self-organization. Only an organism that is no longer in a state of optimal adaptation seems to function in a way consistent with the physicalistic conception of self-organization: it is situated at a state of low entropy, but it produces a lot of entropy. But as soon as re-adaptation is reached, the organism is situated at a state of low entropy, and low entropy production as well. This fact contradicts physicalism. So, while the act of physiological adaptation begins as if the organism were following the laws of a merely physical dynamics, in the latter stages it proceeds in a manner that can only be explained in biological terms. Following from the Falkners' research with cyanobacteria, it seems plausible to assume that all organisms aim at the minimization of entropy production. Physiological re-adaptation is a biological act that requires the *internally conditioned* coordinated variation of many dynamic quantities or variables that, in the models of systems biologists, are described as parameters (i.e., as non-variables).

There is also experimental evidence suggesting that a process of minimization of entropy production also takes place during embryonic development. The results of numerous measurements in several studies led to the conclusion that the entropy production "indeed decreases at separate stages of the ontogenesis (if early stages of development are excluded)."[50] To sum up: both theoretical considerations and experimental evidence make clear that *the theory of self-organization of physics is too weak to account for real biological self-organization*.

Teleology beyond Dynamic Systems Theory

Every organism must act anti-entropically *without* requiring most conditions of its self-organization to be externally set. As stated above, parameters are quantities that constrain the development of dynamic systems. Hence, the simulation of a dynamic that could calculate a significant portion of its own parameters would be nothing less than the simulation of a *self-constraining dynamic*. But the latter is not conceivable within contemporary dynamic systems theory. Failing to see this (and thereby considering the models of systems biology as appropriate descriptions of organismic causality) would be a clear case of Whitehead's "fallacy of misplaced concreteness,"[51] that is, confusing an abstract model of the real with the concretely real.

This result invites us to go beyond the ontological limitations of current scientific teleology. Since it is not capable of accounting for anti-entropic behavior in real organisms, we should endeavor to think about concepts of teleology beyond contemporary mainstream physicalism. This brings the idea of genetic information back to the stage, which seems to play no role in physicalistic reductions of organismic causality. There is need of a new conception of genetic information that is not borrowed from informatics and is thus without any problematic semantic aspects, as is the case with neo-Darwinism. With that said, we leave behind dynamic systems theory, which is the most advanced scientific foundation of neo-teleologism today. In the next section, we will become acquainted with another way of thinking about organismic teleology and genetic information which goes beyond all kinds of neo-teleologism.

On the Autogen Model[52]

Terrence Deacon (2006) has introduced an abstract model system, which he calls an *autogen*, that exhibits a certain degree of self-constrainment. Deacon does not assume that mental factors operate in autogens; rather, he developed the autogen model in order to demonstrate that a special form of reciprocity between self-organizing molecular processes is sufficient to explain both self-replication and evolvability.[53]

Autogenic Dynamics: Teleodynamics

Deacon's autogen model offers a theoretical demonstration that some organismic properties can be exhibited by systems that are not alive in current biological terms. It also illuminates the fundamental difference between simple self-organizing processes and processes that exhibit the most basic features of organismic dynamics. The autogen model embodies a *partially* self-constraining self-constituting dynamic. The simplest type of autogen is an autonomous molecular system consisting of two coupled, non-linear, self-organizing processes: *autocatalysis* and *self-assembly*, which *mutually support and constrain each other*. The property of these structures is the result of the following reciprocity of molecular properties and interaction processes:

1. Autocatalysis (also known as "reciprocal collective catalysis") requires a local concentration of substrate molecules and a source of energy (e.g.,

embodied in chemical bonds) in order to occur. It depletes these supportive conditions at an ever-increasing rate as it produces more catalysts.
2. Autocatalysis depends on the collocation of reciprocal catalysts, and as long as it continues it produces more catalysts that replace those that have diffused away.
3. Autocatalysis thus produces a gradient of molecular concentration that temporarily generates a region of high concentration of a few molecular forms despite the tendency for them to diffuse away.
4. Autocatalysis rapidly depletes the immediate environment of substrate molecules and the energy gradient to drive this reaction.
5. Self-assembly is dependent on the structural symmetries of like molecules to spontaneously form regular molecular tessellations (producing crystals, sheets, polyhedrons, and tubular forms).
6. Self-assembly depends on a high local concentration of such molecules.
7. As a self-assembling structure grows, it depletes this local concentration.
8. The growth of self-assembled sheets and related structures tends to impede molecular diffusion, especially when it produces a closed or partially closed structure like a polyhedron or tube, respectively.

Notice that the persistent generation of a high local concentration of like molecules by autocatalysis (3) is a necessary boundary condition, or *constraint*, required for self-assembly (6), and that the prevention of the diffusion of reciprocal catalyst molecules (8) is the necessary boundary condition, or *constraint*, required for persistent autocatalysis (2). In rare cases, an autocatalytic process produces a molecular form susceptible to self-assembling into an enclosing structure, as a by-product. In these cases, enclosure is likely to encapsulate the very molecules that in proximity with one another tend to produce this containment. Both coupled processes require a gradient. Both deplete that gradient, and in the process, produce more regularity, more catalysts, and more enclosing structure. Moreover, if such an enclosed structure is broken up by external factors in an environment with supportive conditions, then the whole complex will tend to reform or even produce replicas from the partially diffused components.

The dynamics of an autogenic process does not merely use the constraints that enable the utilization of local gradients—*autogen closure halts both of its component self-organizing processes before supportive local energetic and substrate gradients are destroyed.* Cessation of these processes halts depletion of the conditions that make them possible. An autogen thereby preserves the potential for future autocatalysis and self-assembly when external factors break them up. Thus, an autogen acts like a ratchet: it produces something when active, and preserves it from getting lost when it is not active.

So, *autogens do not fully deplete essential environmental conditions (i.e., energetic and/or material gradients), in the way that ordinary self-organizing systems do*. This is the most essential feature of autogenic dynamics, since it goes beyond the principle of maximum entropy production that characterizes all kinds of self-organized dynamics. It results directly from the specific organization of autogenic dynamics which is an "additional emergent transition . . . dynamically supervenient"[54] on self-organized processes. Autogenic dynamics exerts an additional and highly constraining effect on the two simpler self-organized dynamics (autocatalysis and self-assembly) of which it consists. Dynamic systems that lack this reciprocity thus exist at the whim of extrinsic parameters, whereas autogenic systems actively generate parameters (constraints) critical to their existence. Deacon calls the specific organization of autogenic dynamics "teleodynamics"[55] because it exhibits "end-directedness and consequence-organized features."[56] Teleodynamics is always constituted by the mutual constraint and reciprocal synergy of two or more strongly coupled self-organized processes.[57]

How the Autogen Model Could Account for Organismic Dynamics

Deacon hypothesizes that autogenesis manifests the most basic form of self-production. But he admits that *autogens cannot be counted as living organisms*. Although they preserve environmental gradients and intrinsic constraints, in contrast to self-organized systems that destroy them, they merely prevent critical gradients in the environment from being depleted too rapidly and preserve critical constraints that enable these gradients to be utilized in order to maintain this higher-order property. *Autogens are thus passive entities*. They preserve themselves passively since they are passive to potential changes in their environment and they do not seek better environmental conditions. Furthermore, autogens do not maintain a persistent energetic and material openness (metabolism). They are in this sense similar to virus particles, except that they are not parasitic—namely, they do not rely on other organisms for their self-constituting and self-reproducing properties. Because of this passivity they cannot adapt if conditions should change. *They are unable to vary internally in terms of critical quantities (i.e., parameters) of their dynamics with respect to these conditions*. But although they therefore lack any internal representation of their environment, autogens meet some of the basic criteria of the organismic mode of being.

Teleodynamics is an emergent level beyond what can be understood through current dynamic systems-, self-organization-, and complexity theories. As such, autogenic dynamics exerts a constraining influence on the multitude of possible trajectories that would be allowed if organisms were nothing but self-organized dynamic systems (see figure 14.6). The highly constraining effect of the teleodynamical level on the lower-level self-organized processes reduces the number of possible physicochemical trajectories enormously. On the basis of this reduction, Deacon introduces a new concept of information.[58]

Obviously, what lies at the heart of the concept of autogenesis is the idea that autogenic dynamics are a unique form of self-constraint. Therefore, the degree to which autogenic dynamics impose such self-constraint should be examined. In order to do this, we have first to identify the factors that constrain each other. They are the concentration of the autocatalytically generated molecules and the concentration of the self-assembling molecules. In a formal description of the interconnection of autocatalysis and self-assembly as a system of differential equations, both mutually constraining quantities would be variables. There are, however, two other quantities that are very critical to the generation of both sorts of molecules: substrate-energy availability (i.e., the external gradient of matter) and diffusability. In the dynamic systems approach both of these quantities would be parameters that are set externally. In contrast, even the simplest teleodynamics is partially able to regulate these quantities autonomously, since the closure of the autogen changes both. But on the other hand, as in dynamic systems theory, the computation of the concentration of the autocatalytic and self-assembling molecules would require some other parameters or static quantities, for example, the rate coefficient of the reactions to be set from outside. Thus, at the present stage of its development, autogenic dynamics is self-constrained in a way that does not extend itself over all quantities, which in the dynamic systems approach, would be described as parameters of differential equations. But in any case, an autogen can be minimally said to change its dynamics in a way that controls at least some quantities that in dynamic systems theory are parameters, so that persistence is optimized. This manifests, even for the simplest autogen model, a degree of self-regulation unattainable in dynamic systems theory that clearly indicates a big step beyond neo-teleologism.

It should be noticed that as long as the autogen model accounts only for the formation of inanimate pre-biotic structures it does not need to display a dynamics able to regulate a significant part of its parameters autonomously. The reason for this is that the simulation of pre-biotic organization parameters (e.g., the rate coefficient of the reactions) would represent quantities the values of which depend on environmental conditions (e.g., temperature) that cannot be influenced by autogenic dynamics. At present, Deacon's model can show how pre-biotic entities produce their spatial border autonomously. If the autogen model is expected to provide a scenario for the transition from pre-biotic to proto-biotic organization, it will have to be demonstrated how autogens eventually overcome the distinction between parameters and variables to a higher degree than they already do. The future development of the autogen model should show that, under the primordial environmental conditions of the early planet Earth, autogens were able to evolve in such a way that would allow them to control more and more of the internal quantities which systems biologists describe as being parameters in their models. To show this would mean to demonstrate that evolution would progressively provide autogens with an increasingly self-constraining and end-directed dynamics typical of real cells and even of the first archaic cells. Hence, the further development of autogen theory would demand the introduction of models with a complex structure of interconnected causal loops in which more and more parameters become replaced by variables.

Although at present this problem has not been solved, Deacon's theory shows a way to deal with it. In *Incomplete Nature*, Deacon has introduced a new view on genetic information (and on information in general), which is based on dynamics. Therefore, it essentially deviates from the neo-Darwinistic understanding of information. Deacon describes how a lineage of teleodynamically-organized entities (i.e., autogens) could undergo variations which in evolutionary terms would be adaptations. If the surface of the autogen has molecular features to which environmental substrate molecules supportive to the autogen's autocatalytic process tend to bind, and in so doing weaken its structural stability, then the probability that the autogen breaks up in a supportive environment will increase. Natural selection would progressively endow autogens with the ability to break up in environments that contain substances supportive to their autocatalysis, so that the probability of autogenic replication will significantly rise.[59] It is possible that during this evolutionary process in environments with high concentrations of high-energy phosphate molecules some autogens produce free nucleotides as by-products. This could lead to polymerization of nucleotides inside the autogens. Polynucleotides would form and be passed to the next generation by the replication of the whole autogen of which they are a part. The sequence of the nucleotides of these macromolecules is of special importance because it will influence the tempo of autocatalysis when the autogen breaks up again:

> Although the order of nucleotide binding will be unbiased, the resulting sequence of nucleotides can serve as a substrate onto which various free molecules within the autogen (e.g. catalysts) will differentially bind due to sequence-specific stereochemical affinities. In this way, catalysts and other free molecules can become linearly ordered along a polynucleotide template, such that relative proximity determines reaction probability. Thus, for example, if this template molecule releases catalysts according to linear position (e.g. by depolymerization) they will become available to react in a fixed order.[60]

The breaking up of the autogen will expose the polynucleotide to external factors that will cause its progressive dissolution. Since the dissolution proceeds from one end to the other, the different nucleotides that occupy different positions on the macromolecule will be released at different times. This will cause a non-random order of different chemical reactions between the released catalysts and other molecules (bound on the released nucleotides) of the autogen and the substrates of the environment. Deacon writes, "To the extent that this order correlates with the order of reactions that is most efficient at reconstituting the autogenic structure there will be favored template sequences."[61] In other words, the time-asymmetry of chemical reactions matters, since every reaction will change the conditions under which the subsequent reactions will take place. The result will be that those autogens whose polynucleotides give rise to the most efficient order of reactions will replicate faster than others. In turn, this will lead to the replacement of some species of autogens by others in the evolutionary long run. Obviously, the polynucleotide sequences of the faster replicating autogens happen to fit better to their environment. One could say that they better *represent* the chemical constitution of the autogen's surroundings, and that the autogen in question *interpret* their environments more efficiently for the purpose of self-replication.

It is noteworthy that Deacon's speculative model does not assume that autogenic adaptation and natural selection require the ability of processing information. On the contrary, Deacon's hypothesis shows how the internal representation of the relationship between autogen dynamics and environmental conditions can evolve out of teleodynamically organized physicochemical processes, thereby demonstrating that reference and semiotic abilities could emerge from this specific form of dynamics. Instead of placing information at the very beginning of evolution, he shows how it could evolve out of physicochemical interactions of teleodynamically organized entities with their environments. As such, Deacon is providing insight into how genetic information, which represents external (i.e., environmental factors) appropriately, could emerge. In this way, he shows a way how evolution could enable autogens to control external quantities, e.g., the influx of energy and substances (which in terms of the dynamic systems theory would be parameters since those systems are not able to control this kind of quantities by themselves) to a progressively increasing degree. There is no reason, however, to *restrict* Deacon's intuition about the emergence of genetic information *only to the conversion of those parameters which represent environmental quantities to variables*. It is conceivable that his model could also show how evolution endows genetic information with the ability to represent, to a progressively increasing degree, the *internal* constitution of teleodynamically organized entities. Those entities would be able to significantly vary quantities representing the degree of activity of specific molecules inside the cell and relations between variables of the cell such as volume, pressure, free energy, etc... In other words, *they would be able to vary in terms of most of their internal quantities, which in dynamic systems theory are also described by parameters*. The emergence of such an autonomous teleodynamics would clearly mark the transition from non-living autogens to living archaic cells.

Conclusion: Toward a Biological Neo-Naturalism

While Deacon's theory is usually assigned to naturalism, I think, however, that this generic term is not appropriate, since it suggests that the autogen model operates within the same frame of basic assumptions as neo-teleological theories. If "naturalism" means "non-supernaturalism," "non-mentalism," or "non-intentionalism," then the autogen model may indeed be subsumed into this category. But contemporary biological naturalism cannot be captured by these very general ideas. It expresses rather the belief that all aspects of life are in principle explainable through the metaphysical categories that are currently employed by the natural sciences, such that they will not need to revisit their commitments to modern materialism which presupposes a concrete ontology and methodology. In contrast, Deacon questions, for example, explanatory attempts to reduce mental activity to localized patterns of material processes in the brain. He criticizes the typical figures of materialistic thought in today's neurobiological naturalism *without, however, introducing any immaterial factors*.

Starting from the assumption of the teleodynamic organization of the brain, he emphasizes the importance of the *absence* of certain material configurations and patterns in the brain for consciousness,[62] where "absence" does not mean "immateriality." Since the subject of this chapter is not the nature of mind, but rather the logic of organismic causality, these insights cannot be further expounded here. Nevertheless, many of these same issues concerning consciousness have their analogues in explanations of the nature of life. In both computational and dynamical models of the brain, consciousness is considered an epiphenomenon. Representation, anticipation, and intentionality are considered to be naïve metaphorical descriptions of neuronal computation or patterns of neuronal activity. This leads necessarily to the rejection of teleology in animal and human behavior. Analogously, their psychological epiphenomenalism, cybernetic, or information-theoretical, neo-Darwinistic, and dynamic systems approaches also introduced a biological epiphenomenalism. They consider any conception of biological *self* to be a vitalistic or substantialist remnant, since they reduce organismic autonomy to DNA or to complex patterns of molecular processes. In sharp contrast to those naturalistic theories of life, in the teleodynamic approach, the fundamental feature that discriminates life from non-life are never macromolecules or localized patterns of molecular processes, but rather, constraints. Analogous to his rejection of epiphenomenalism, Deacon criticizes the typical figures of both gene-centric and dynamic systems thought in today's biological naturalism, without introducing any immaterial vital factors. These fundamental differences should have made clear that it would be disadvantageous to a better understanding of the innovation capacity of teleodynamics, if it were classified as belonging to naturalism. For this reason, I suggest that Deacon's ideas as well as some other congenial ideas that are introduced by this volume be considered as heralds of an arising *biological neo-naturalism*. I turn now to a treatment of some of the important differences between these new approaches to organismic causality and the kinds of neo-teleologism presented above.

One of the most important innovations of the autogen model is that it introduces a new kind of teleological organization that *merges the logic of organismic causality with its physics*. This clearly goes beyond cybernetics and informatics. Norbert Wiener and Claude Shannon's concepts of information are based mainly on the implicit assumption that the logic of organization is independent of its physics, since it can be realized in many different physical ways.

Because of their multiple realizability, cybernetic and information processing automata do not display a necessary connection between their functionality (logic) and the maintenance of their material form (physics). Rather, they can be deactivated without becoming disintegrated since their material constitution follows a static physics. The sharp distinction between hardware and software, which is the most essential characteristic of all information processing machines to this day, necessarily includes a sharp distinction between physics and logic. Therefore, information processing (logic) never serves the maintenance of the material constitution (physics) of an operating machine, but serves the individuals who built or own this machine, namely—the operations have meaning and value for somebody outside the operating entity. For these reasons, there is only syntax but no semantics in the traditional concepts of information introduced by Wiener and Shannon.

Autogenic teleology also transcends one of the most important implicit assumptions of the second kind of neo-teleological theories, since neo-Darwinism presupposes a distinction between logic and physics as well. Although this distinction, which was implicitly introduced by Darwin himself, is a methodological one, it is as sharp as in cybernetics and informatics. The idea of natural selection is the logic of Darwinian evolution. Since "natural selection is a process defined by multiple realizability,"[63] it also does not presuppose a certain physical organization. One of the strengths of Darwin's theory of natural selection is its compatibility with significantly different physical mechanisms of organismic causality, which results in an agnosticism about physics. But the fact that natural selection "leaves out nearly all of the mechanistic detail of the processes involved in generating organisms, their parts, and their offspring" constitutes a serious shortcoming, since "it is precisely the process of generating *physical* bodies and maintaining metabolism that constitutes the coin of the natural selection economy. Variations do not exist in the abstract. Rather, they are always variations of some organism structure or process or their outcome."[64] Organisms must maintain their far-from-equilibrium dynamics, which is a physical process. But neo-Darwinists have not been able to propose a physical model of organismic causality up to now, since they have not gone substantially beyond the scope of Darwin. Therefore, neo-Darwinists did not challenge the strict separation between hardware and software typical to informatics. On the contrary, they have rendered it a main principle of their thinking, for they consider genetic information to be independent of the material processes of the cell. The idea that such information could be "saved" requires that the logic of genetic information be agnostic to physics. Thus the separation of logic and physics, essential to informatics (i.e., Shannon-information), underlies the neo-Darwinistic concept of a "genetic program."

The third kind of neo-teleological thinking, which is grounded on dynamic systems theory, is certainly the most sophisticated naturalistic approach of the three. The logic of causal relations within a dynamic system as well as between the system and its environment is dictated by pure physics. However, this physicalistic approach does not pay the appropriate attention to the essential differences between biology and physics. Its deep rootedness in metaphysics and methodology of physics does not allow it to question whether both the principle of maximum entropy production and the strict distinction between dynamic state variables and externally set parameters are appropriate for accounting for organismic causality. Failing to see the limits of physics explains why most exponents of this approach do not take offense at the lack of an adequate concept of information in their theory. Biological naturalism, which is grounded on dynamical systems theory, operates with a causality that is defined in terms of cause and effect, like all theories of physics. What clearly differentiates the theory of teleodynamics, along with other new biological ideas presented in this volume, from physicalistic approaches to organisms, such as self-organization and systems-biology, is that they include semiotic concepts in their biological explanations. Instead of introducing *ad hoc* semiotic terms, Deacon's thought experiment about the evolution of autogens provides a way to bridge the gap between the distinct concepts of causality that are held in physics and biosemiotics. It shows how a specific physical organization (i.e., teleodynamics) gives rise to the emergence of phenomena like representation, sentience, interpretation, anticipation, etc..., which Deacon summarizes under the term "ententional phenomena."[65]

With this in mind, "biological neo-naturalism" could be defined as a generic term referring to all kinds of biological thinking in which emphasis is placed on the explanatory power of ententional phenomena in the understanding of organismic causality. It would represent an attempt to re-invent semiotic biology a hundred years after its first introduction by Jacob von Uexküll. Biological neo-naturalism, as proposed here, is a scientific biophilosophy. Its metaphysical preassumptions do not leave the territory of what can be considered a possible scientific metaphysics today, as mentalistic teleology, for example, does. But, as stated at the beginning of this chapter, the borders between philosophy of biology and biophilosophy are fluid. It is hoped that biological neo-naturalism will become an essential part of philosophy of biology in the near future.

Notes

1. Among the best-known philosophers of biology are Michael Ruse, David Hull, Richard Lewontin, Stephen Jay Gould, Francisco Ayala, Michael Dupré, Sandra Mitchell, Peter Godfrey-Smith, Ernst Mayr, Paul Griffiths, Susan Oyama, Alexander Rosenberg, Robert Brandon, Elliott Sober, and Kim Sterelny.

2. Some pivotal occidental representatives of this tradition are Aristotle, Theophrastus, Harvey, Leibniz, Kant, Goethe, Carus, Fechner, Darwin, Haeckel, Nietzsche, Bergson, Driesch, Whitehead, Peirce, von Uexküll, Portmann, and Jonas.

3. Ruth Millikan, *Language, Thought and Other Biological Categories* (Cambridge, MA: MIT Press, 1984); Karen Neander, "The Teleological Notion of 'Function'," in *Australasian Journal of Philosophy* 69 (1991): 454-468; Paul Griffiths, "Functional

Analysis and Proper Functions," *British Journal for the Philosophy of Science* 44 (1993): 409-422; Phillip Kitcher, "Function and Design," *Midwest Studies in Philosophy* 18 (1993): 379-397; Peter Godfrey-Smith, "A Modern History Theory of Functions," in *Noûs* 28 (1994): 344-362; Collin Allen and Marc Beckoff, "Biological Function, Adaptation and Natural Design," *Philosophy of Science* 62 (1995) : 609-622.

4. Robert Cummins, "Neo-Teleology," in *Functions. New Essays in the Philosophy of Psychology and Biology*, ed. André Ariew, Robert Cummins, and Mark Perlman (New York: Oxford University Press, 2002), 162.

5. Paolo Costa, "Beyond Teleology?" in *Purposiveness: Teleology between Nature and Mind*, ed. Luca Illeterati and Francesca Michelini (Frankfurt: Ontos Verlag, 2008), 183-188.

6. Georg Toepfer, "Teleologie," in *Philosophie der Biologie*, ed. Ulrich Krohs and Georg Toepfer (Frankfurt: Suhrkamp, 2005), 36f; Martin Mahner and Mario Bunge, *Philosophische Grundlagen der Biologie* (New York: Springer, 2000), 348.

7. Toepfer, "Teleologie," 36.

8. Mahner and Bunge, *Philosophische Grundlagen der Biologie*, 348.

9. Toepfer, "Teleologie," 36.

10. Hans Jonas, *Das Prinzip Leben* (Frankfurt: Suhrkamp, 1997), 203, my translation.

11. Alfred North Whitehead, *Process and Reality*, eds. David Ray Griffin and Donald W. Sherburne (1929; corr. ed., New York: Free Press, 1978), 53.

12. Aristotle, *Physics*. The Internet Classics Archive, 199b26-30, http://classics.mit.edu/Aristotle/physics.2.ii.html. Date of access: 07/11/2012

13. Ernst Mayr, *Eine neue Philosophie der Biologie* (Munich, Germany: Piper, 1991), 61.

14. Arturo Rosenblueth, Norbert Wiener, and Julian Bigelow, "Behavior, Purpose and Teleology," *Philosophy of Science* 10, no. 1 (1943): 18.

15. Jonas, *Das Prinzip Leben*, 202, my translation.

16. Rosenblueth, Wiener, and Bigelow, "Behavior, Purpose and Teleology," 23-24, my italics.

17. Norbert Wiener, *Cybernetics* (New York: MIT Press, 1961), 62, 11.

18. Mayr, *Eine neue Philosophie der Biologie*, 61; François Jakob, *The Logic of Life* (Princeton, NJ: Princeton University Press, 1993), 1-17; Jacques Monod, *Chance and Necessity* (New York: Knopf, 1971).

19. André Ariew, "Teleology," in *The Cambridge Companion to the Philosophy of Biology*, ed. David L. Hull and Michael Ruse (New York: Cambridge University Press, 2007), 179; Mayr, *Eine neue Philosophie der Biologie*, 75, 61.

20. Robert Brandon, *Adaptation and Environment* (Princeton, NJ: Princeton University Press, 1990), 188.

21. Quoted by Toepfer, "Teleologie," 42, my translation.

22. Wayne Christensen, "A Complex Systems Theory of Teleology," *Biology and Philosophy* 11 (1996): 308f.; Robert Rosen, "Organisms as Causal Systems Which Are Not Mechanisms: An Essay into the Nature of Complexity," in *Theoretical Biology and Complexity: Three Essays on the Natural Philosophy of Complex Systems*, ed. Robert Rosen (New York: Academic Press, Inc., 1985), 173f.; Brian Goodwin, "A Structuralist Programme in Developmental Biology," in *Dynamic Structures in Biology*, ed. Brian Goodwin, Atuhiro Sibatani, and Gerry Webster (Edinburgh: Edinburgh University Press, 1989), 49-61.

23. Toepfer, "Teleologie," 36f; Mahner and Bunge, *Philosophische Grundlagen der Biologie*, 348.

24. Marc Bedau, "Where Is the Good in Teleology?" in *Nature's Purpose: Analysis of Function and Design in Biology* (Cambridge, MA: MIT Press, 1998), 272-273.

25. Ariew, "Teleology," 173.

26. Aristotle, *Physics*, 199a29.

27. Aristotle, *Physics*, 198b24.

28. Jonas, *Das Prinzip Leben*, 163, my translation.

29. Aristotle, *Physics* II, 198b33-199.

30. Aristotle, *Physics* II, 197b22-31.

31. Immanuel Kant, *Critique of Judgment* (London: Macmillan, 1914), §75, my italics. I made a minor change in the English translation: I replaced "design" with "intention," since the latter seems to me to be a more appropriate translation of the German word "Absicht."

32. Kant, *Critique of Judgment*, §65, my italics.

33. Kant, *Critique of Judgment*, §64.

34. In the *Critique of Judgment*, §68, Kant writes,

There are objects, alone explicable according to natural laws which we can only think by means of the Idea of purposes as principle . . . we speak in Teleology, indeed, of nature *as if* the purposiveness therein were intended, but in such a way that this intention is ascribed to nature, i.e. to matter. Now in this way there can be no misunderstanding, because no intention in the proper meaning of the word can possibly be ascribed to inanimate matter; we thus give notice that this word here only expresses a principle of the reflective not of the determinant Judgment, and so is to introduce no particular ground of causality; but only adds for the use of the Reason a different kind of investigation from that according to mechanical laws, in order to supplement the inadequacy of the latter even for empirical research into all particular laws of nature. . . . [T]here should be only signified thereby a kind of causality of nature *after the analogy of our own* in the *technical* use of Reason, in order to have before us the *rule* according to which certain products of nature must be investigated.

The words "designed" and "design" were replaced with "intended" and "intention." Kant's statement that the idea of purposiveness in nature "expresses a principle of the reflective not of the determinant Judgment" means that humans are not allowed to say that some natural things (organisms) are actually ruled by mental faculties, but merely that there are natural things in consideration of which we are compelled to derive the idea of purposiveness.

35. Kant, *Critique of Judgment*, §65.

36. Kant, *Critique of Judgment*, §65.

37. Kant, *Critique of Judgment*, §65.

38. Kant, *Critique of Judgment*, §65.

39. Werner Ebeling and Igor Sokolov, *Statistical Thermodynamics and Stochastic Theory of Nonequilibrium Systems* (New Jersey: World Scientific Publishing, 2005), 40.

40. Werner Ebeling, *Strukturbildung bei irreversiblen Prozessen* (Leipzig: Teubner, 1976), 13.

41. Eric Schneider and James Kay, "Order from Disorder: The Thermodynamics of Complexity in Biology," in *What Is Life? The Next Fifty Years*, ed. Michael Murphy and Luke O'Neill (Cambridge, MA: Cambridge Univ. Press, 1997), 165. Concerning the

connection of entropy and uncertainty see Ebeling and Sokolov, *Statistical Thermodynamics*, 85f.

42. Lewis Wolpert, "Development: Is the Egg Computable or Could We Generate an Angel or a Dinosaur?" in *What Is Life? The Next Fifty Years*, ed. Michael Murphy and Luke O'Neill (Cambridge, MA: Cambridge Univ. Press, 1997), 57-66; Masaru Tomita, "Whole Cell Simulation," *Trends in Biotechnology* 19, no. 6 (2001): 205-210; Dennis Normile, "Building Working Cells 'in Silico'," *Science* 284, no. 5411 (1999): 80-81; Wayt Gibbs, "Simulierte Zellen," *Spektrum der Wissenschaft* 11 (2001): 54-57.

43. James Murray, *Mathematical Biology* (New York: Springer, 1993); Alan Turing, "The Chemical Basis of Morphogenesis," *Philosophical Transactions of the Royal Society of London*, Series B, 237, no. 641 (1952): 37-72; Albert Goldbeter, *Biochemical Oscillations and Cellular Rhythms* (Cambridge: Cambridge University Press, 1997), 46; John Tyson, Katherine Chen, and Bela Novak, "Sniffer, Buzzers, Toggles and Blinkers: Dynamics of Regulatory and Signaling Pathways in the Cell," *Current Opinion in Cell Biology* 15 (2003): 222; Joanne Collier, Nicholas Monk, Philip Maini, and Julian Lewis, "Pattern Formation by Lateral Inhibition with Feedback: A Mathematical Model of Delta-Notch Intercellular Signaling," *Journal of Theoretical Biology* 183, no. 4 (1996): 429-446; Michael Elowitz and Stanislas Leibler, "A Synthetic Oscillatory Network of Transcriptional Regulators," *Nature* 403, no. 6767 (2000): 336-337; James Ferrel and Wen Xiong, "Bistability in Cell Signaling: How to Make Continuous Processes Discontinuous, and Reversible Processes Irreversible," *Chaos* 11, no. 1 (2001): 231-233; Timothy Gardner, Charles Cantor, and James Collins, "Construction of a Genetic Toggle Switch in *Escherichia coli*," *Nature* 403, no. 6767 (2000): 339-340.

44. Thomas Panning, Layne Watson, Nicholas Allen, Katherine Chen, Clifford Shaffer, and John Tyson, "A Mathematical Programming Formulation for the Budding Yeast Cell Cycle," *SIMULATION* 83 (2007): 498.

45. Panning et al., "A Mathematical Programming Formulation," 499.

46. Kant, *Critique of Judgment*, §65.

47. See in this volume, Gernot Falkner and Renate Falkner, "The Incompatibility of the Neo-Darwinian Hypothesis with Systems-Theoretical Explanations of Biological Development"; Kristjan Plaetzer, Randall Thomas, Renate Falkner, and Gernot Falkner, "The Microbial Experience of Environmental Phosphate Fluctuations," *Journal of Theoretical Biology* 235 (2005): 540-554.

48. In the past the answer might have been "the genes" or "genetic information." But both these notions are much less clear today than they were some decades ago. Now we understand that genes are massively co-determined by the organism's dynamics.

49. Falkner and Falkner, "The Incompatibility of the Neo-Darwinian Hypothesis."

50. L. Martyushev and V. Seleznev, "Maximum Entropy Production Principle in Physics, Chemistry and Biology," *Physics Reports* 426 (2006): 40.

51. Alfred North Whitehead, *Process and Reality*, 7-8.

52. I gratefully acknowledge Terrence Deacon's advice and great assistance in this short description of his autogen model.

53. Terrence Deacon, *Incomplete Nature* (New York: W. W. Norton & Company, 2012), 305-324; Terrence Deacon, "Reciprocal Linkage between Self-Organizing Processes Is Sufficient for Self-reproduction and Evolvability," *Biological Theory* 1(2) (2006): 141.

54. Deacon, *Incomplete Nature*, 265.

55. Deacon, *Incomplete Nature*, 265-287; Terrence Deacon, "Emergence: The Hole at the Wheel's Hub," in *The Re-Emergence of Emergence,* ed. Philip Clayton and Paul Davies (New York: Oxford University Press, 2006), 137-149.
56. Deacon, *Incomplete Nature*, 552.
57. Deacon, *Incomplete Nature*, 552.
58. Deacon, *Incomplete Nature*, 371-420.
59. Deacon, *Incomplete Nature*, 442.
60. Deacon, *Incomplete Nature*, 445.
61. Deacon, *Incomplete Nature*, 445.
62. Deacon, *Incomplete Nature*, 485-538.
63. Deacon, *Incomplete Nature*, 423.
64. Deacon, *Incomplete Nature*, 422, my italics.
65. Deacon, *Incomplete Nature*, 549.

Bibliography

Allen, Collin, and Marc Beckoff. "Biological Function, Adaptation and Natural Design." *Philosophy of Science* 62 (1995): 609-622.
Ariew, André. "Teleology." In *The Cambridge Companion to the Philosophy of Biology*, edited by David L. Hull and Michael Ruse, 160-181. New York: Cambridge University Press, 2007.
Aristotle. *Physics*. In *The Internet Classics Archive*, translated by R. P. Hardie and R. K. Gaye, http://classics.mit.edu/Aristotle/physics.2.ii.html.
Bedau, Mark. "Where Is the Good in Teleology?" In *Nature's Purposes. Analysis of Function and Design in Biology*, edited by Colin Allen, Marc Bekoff, and George Lauder, 261-291. Cambridge, MA: MIT Press, 1998.
Brandon, Robert. *Adaptation and Environment*. Princeton, NJ: Princeton University Press, 1990.
Christensen, Wayne. "A Complex Systems Theory of Teleology." *Biology and Philosophy* 11 (1996): 301-320.
Collier, Joanne, Nicholas Monk, Philip Maini, and Julian Lewis, "Pattern Formation by Lateral Inhibition with Feedback: A Mathematical Model of Delta-Notch Intercellular Signaling." *Journal of Theoretical Biology* 183, no. 4 (1996): 429-446.
Costa, Paolo. "Beyond Teleology?" In *Purposiveness: Teleology between Nature and Mind*, edited by Luca Illeterati and Francesca Michelini, 183-199. Frankfurt: Ontos Verlag, 2008.
Cummins, Robert. "Neo-Teleology." In *Functions. New Essays in the Philosophy of Psychology and Biology*, edited by André Ariew, Robert Cummins, and Mark Perlman, 157-172. New York: Oxford University Press, 2002.
Deacon, Terrence. "Reciprocal Linkage between Self-Organizing Processes Is Sufficient for Self-reproduction and Evolvability." *Biological Theory* 1, no. 2 (2006): 136-149.
———. "Emergence: The Hole at the Wheel's Hub." In *The Re-Emergence of Emergence*, edited by Philip Clayton and Paul Davies, 111-150. New York: Oxford University Press, 2006.
———. *Incomplete Nature*. New York: W. W. Norton & Company, 2012.
Ebeling, Werner. *Strukturbildung bei irreversiblen Prozessen*. Leipzig: Teubner, 1976.

Ebeling, Werner, and Igor Sokolov. *Statistical Thermodynamics and Stochastic Theory of Nonequilibrium Systems*. New Jersey: World Scientific Publishing, 2005.
Elowitz, Michael, and Stanislas Leibler. "A Synthetic Oscillatory Network of Transcriptional Regulators." *Nature* 403, no. 6767 (2000): 335-338.
Ferrel, James, and Wen Xiong. "Bistability in Cell Signaling: How to Make Continuous Processes Discontinuous, and Reversible Processes Irreversible." *Chaos* 11, no. 1 (2001): 227-236.
Gardner, Timothy, Charles Cantor, and James Collins. "Construction of a Genetic Toggle Switch in *Escherichia coli*." *Nature* 403 (6767) (2000): 339-342.
Gibbs, Wayt. "Simulierte Zellen." *Spektrum der Wissenschaft* 11 (2001): 54-57.
Godfrey-Smith, Peter. "A Modern History Theory of Functions." *Noûs* 28 (1994): 344-362.
Goldbeter, Albert. *Biochemical Oscillations and Cellular Rhythms*. Cambridge: Cambridge University Press, 1997.
Goodwin, Brian. "A Structuralist Programme in Developmental Biology." In *Dynamic Structures in Biology*, edited by Brian Goodwin, Atuhiro Sibatani, and Gerry Webster, 49-61. Edinburgh: Edinburgh University Press, 1989.
Griffiths, Paul. "Functional Analysis and Proper Functions." *British Journal for the Philosophy of Science* 44 (1993): 409-422.
Jakob, François. *The Logic of Life*, Princeton, NJ: Princeton University Press, 1993.
Jonas, Hans. *Das Prinzip Leben*. Frankfurt: Suhrkamp, 1997.
Kant, Immanuel. *Critique of Judgment, 2nd Edition Revised*. Translated by J. H. Bernard, London: Macmillan, 1914, http://ebooks.adelaide.edu.au/k/kant/immanuel/k16ju/.
Kitcher, Phillip. "Function and Design." *Midwest Studies in Philosophy* 18 (1993): 379-397.
Mahner, Martin, and Mario Bunge. *Philosophische Grundlagen der Biologie*. New York: Springer, 2000 (in German).
Martyushev, L. M., and V. D. Seleznev. "Maximum Entropy Production Principle in Physics, Chemistry and Biology." *Physics Reports* 426 (2006): 1-45.
Mayr, Ernst. *Eine neue Philosophie der Biologie*. Munich, Germany: Piper, 1991.
Millikan, Ruth. *Language, Thought and Other Biological Categories*. Cambridge, MA: MIT Press, 1984.
Monod, Jacques. *Chance and Necessity*. New York: Knopf, 1971.
Murray, James. *Mathematical Biology*. New York: Springer, 1993.
Neander, Karen. "The Teleological Notion of 'Function'." *Australasian Journal of Philosophy* 69 (1991): 454-468.
Normile, Dennis. "Building Working Cells 'in Silico'." *Science* 284, no. 5411 (1999): 80-81.
Panning, Thomas, Layne Watson, Nicholas Allen, Katherine Chen, Clifford Shaffer, and John Tyson. "A Mathematical Programming Formulation for the Budding Yeast Cell Cycle." *SIMULATION* 83 (2007): 497-514.
Plaetzer, Kristjan, Randall Thomas, Renate Falkner, and Gernot Falkner. "The Microbial Experience of Environmental Phosphate Fluctuations. An Essay on the Possibility of Putting Intentions into Cell Biochemistry." *Journal of Theoretical Biology* 235 (2005): 540-554.
Rosen, Robert. "Organisms as Causal Systems Which Are Not Mechanisms: An Essay into the Nature of Complexity." In *Theoretical Biology and Complexity: Three Essays on the Natural Philosophy of Complex Systems*, edited by Robert Rosen, 165-203. New York: Academic Press, Inc., 1985.

Rosenblueth, Arturo, Norbert Wiener, and Julian Bigelow. "Behavior, Purpose and Teleology." *Philosophy of Science* 10, no. 1 (1943): 18-24.

Schneider, Eric, and James Kay. "Order from Disorder: The Thermodynamics of Complexity in Biology." In *What Is Life? The Next Fifty Years*, edited by Michael Murphy and Luke O'Neill, 161-173. Cambridge, MA: Cambridge University Press, 1997.

Toepfer, Georg. "Teleologie." In *Philosophie der Biologie*. Edited by Ulrich Krohs and Georg Toepfer, 36-52. Frankfurt: Suhrkamp, 2005.

Tomita, Masaru. "Whole Cell Simulation." *Trends in Biotechnology* 19, no. 6 (2001): 205-210.

Turing, Alan. "The Chemical Basis of Morphogenesis." *Philosophical Transactions of the Royal Society of London*. Series B, 237, no. 641 (1952): 37-72.

Tyson, John, Katherine Chen, and Bela Novak. "Sniffer, Buzzers, Toggles and Blinkers: Dynamics of Regulatory and Signaling Pathways in the Cell." *Current Opinion in Cell Biology* 15 (2003): 221-231.

Whitehead, Alfred North. *Process and Reality: Corrected Edition*, edited by D. R. Griffin and D. W. Sherburne. 1929. Corrected ed., New York: The Free Press, 1978. Page references to corrected edition.

Wiener, Norbert. *Cybernetics*. New York: MIT Press, 1961.

Wolpert, Lewis. "Development: Is the Egg Computable or Could We Generate an Angel or a Dinosaur?" In *What Is Life? The Next Fifty Years*, edited by Michael Murphy and Luke O'Neill, 57-66. Cambridge, MA: Cambridge University Press, 1997.

Section 6:
Epigenetics

Chapter Fifteen
Epigenesis, Epigenetics, and the Epigenotype: Toward an Inclusive Concept of Development and Evolution

Brian K. Hall

> It is not my aim to present epigenetic or genetic models for biological phenomena, to describe new phenomena, to derive predictions from models, or to offer tests of predictions from models. Thus, I do not aim to make an empirical contribution to epigenetics.[1]

Introduction

Epigenetics is a term and concept that embraces the regulation of gene activity during embryonic development, animal and plant ontogeny, organismal evolution, and some animal (including human) diseases and cancers.[2] Coined in the 1940s to reflect control of gene action from outside the genome (Waddingtonian or environmental epigenetics), epigenetics now embraces molecular mechanisms that regulate gene activity but that are not encoded in the base pairs of DNA (molecular epigenetics). Although often cast as the genetic extension of *epigenesis* in discussions of *epigenesis* and *preformation*, epigenetics reflect a much deeper search in biology for the factor(s) that control the specification-determination of cells, tissues, organs, and individuals and for an inclusive theory of how information is activated within a generation (ontogeny, development) and transferred from generation to generation (phylogeny-evolution).

My aim is not to discuss epigenesis versus preformation. Rather, I trace the threads of thought used to explain embryonic development (and more recently evolution, disease, and cancer), not as the unfolding of a preformed substance or structure, but as the consequence of hierarchically organized and integrated *material structures* (genes, cells, tissues, and so forth), *partners* (genes, epigenetics, the environment, other species), and *processes* (gene regulation, inductive interactions within and between levels, including organism-environmental interactions). Until recently, emphasis was on mechanisms of change that are internal to organisms, internal to eggs or to fertilized eggs (zygotes) or internal to DNA. In the past seventy-five years and with increasing speed over the past fifteen years, genes as sequences of DNA have been joined on center stage by mechanisms of gene regulation (molecular epigenetics) and by the environment (Waddingtonian epigenetics, phenotypic plasticity). As summarized in a 2006 newsletter issued by the United States National Institutes of Health: "Which is more important in shaping who we are and what we will become—our genes or the environment around us? For centuries, people have debated whether nature or nurture decides how we look or act. Now, a field of research called epigenetics is showing that we can't really separate one from the other."[3]

Epigenesis: From Aristotle to Caspar Friedrich Wolff

How animals develop from egg to adult has fascinated humankind for millennia. Is the adult present fully formed in the egg (or in the sperm), as preformationism maintains, or does the adult emerge gradually through a series of ever more complex stages, as proponents of epigenesis maintain?[4] As with so much in biology, knowledge of embryonic development is traced through a written record back to the Greek philosopher Aristotle (384 BC–322 BC). Were he alive today, Aristotle would be recognized as a philosopher, ethicist, comparative anatomist, and comparative embryologist, but not as an evolutionary or evolutionary-developmental biologist.[5] As a comparative embryologist and anatomist, Aristotle examined developing fish, bird, and mammalian embryos seeking connections and similarities that would inform his even broader interest in the living world.

Aristotle developed the view that embryos developed by gradual transformation, what we now call *epigenesis*. Epigenetics (and so the modern foundations of the 'epi' part of epigenetics) is typically traced to Aristotle through epigenesis. The traditional reading of the history of the next two thousand years is that it was the sixteenth century before Aristotle's description of the embryonic development of the domestic chicken was added to by a Dutch physician and anatomist Volcher Coiter (1534–1576). Not until the seventeenth century did epigenesis made any real impact on prevailing preformationist views. Two individuals stand out: an English physician, anatomist, and physiologist, William Harvey (1578–1657), with his treatise on the generation of animals published in 1651, and a German physiologist and embryologist Friedrich Wolff (1733–1794), with his treatises on the same subject published in 1759 and 1764.[6]

William Harvey, who was physician to both James I and Charles I and an influential lecturer at the Royal College of Physicians in London, is best known for his discovery of the continuous circulation of the blood. In 1651, Harvey produced an important work in embryology, *Exercitationes de Generatione Animalium* (Disputations Touching the Generation of Animals). As Harvey concluded concerning animal embryology:

> The structure of these animals (developing chick embryos) commences from one part as its nucleus and origin, by the instrumentality of which the rest of the limbs are joined on, and this we say takes place by the method of epigenesis, namely by degrees, part after part, and this is, in preference to the other mode, generation [preformation] properly so called.[7]

His proposal of one part replacing another (budding out of another) during embryonic development placed Harvey at odds with the prevailing preformationist view of development.

Born in 1733, a century and a half after Harvey, Caspar Wolff used the opportunity of a position provided by Catherine the Great that left him ample time for his own investigations into developing embryos. Observant and insightful, Wolff demonstrated that embryos only developed from eggs that had been fertilized by sperm and so concluded that sperm must make a contribution to the embryo. In his doctoral dissertation, *Theoria Generationis*, published in 1759, Wolff used direct observation of embryonic development and developed a theoretical basis to reject preformation and revive epigenesis to explain embryonic development.[8] This publication is often used, and probably rightly so, to claim Wolff as the founder of modern embryology.

Much as Aristotle had done, Wolff presented each successive stage of development as a temporal sequence. As had Harvey before him, Wolff described embryonic development in animals and plants as processes based on groups and layers of cells. Wolff went beyond Harvey in grounding embryonic development in the *behaviour of cells* and in proposing the origination of developing animal organs from embryonic layers. For Wolff, successive stages were causally connected: "Each part is first of all an effect of the preceding part, and itself becomes the cause of the following part."[9] Wolff applied the same reasoning to plant development, recognizing that leaves, shoots, and roots grew from the tips of branches that at the outset bore no resemblance to leaves, shoots, or roots. Unsuccessfully, Wolff sought a theory of essential forces that governed the gradual development of plants and animals. With increasing levels of understanding we now appreciate, as did Wolff, that embryonic development, whether animal or plant, is successive, hierarchical, and progressive, moving from simple to ever more complex features of the phenotype.

Wolff had to battle for acceptance of his thesis of epigenesis. Theories of embryonic development based on preformation were entrenched firmly in the eighteenth century.[10] The Swiss botanist and physiologist Albrecht von Haller (1708–1777),[11] an influential proponent of preformation, was especially persistent in his attacks on Wolff and on epigenesis.[12] Wolff (1764) produced a second treatise, *Theorie von der Generation* in which he defended his dissertation and reinforced his evidence for epigenesis as the basis of animal development.

Epigenesis as the Basis of Comparative Embryology

These foundations laid by Harvey and Wolff, attacked as they were, stayed in place to form the foundation upon which Karl von Baer (1792-1876) established a comparative embryology based on germ layers shared between embryos of animals and demonstrated through direct observation that vertebrate embryos alter their form, size, and substance during their development.[13] Discoverer of the ovum and the notochord, author of the germ layer theory of animal development, and the first to place embryos into a classification based on shared features (homology), if anyone deserves the title, "Father of Comparative Embryology," it is Karl von Baer.

Cuvier has classified animals into four *embranchements* whose members shared fundamental features and which were separated from other groups by these same features. Cuvier's groups were: Vertebrates, Mollusks, Articulates (arthropods and segmented worms), and Radiates (cnidarians and echinoderms). Von Baer's genius was to show that the Cuvierian classification of animals into major groups was echoed in their embryology; organisms belonged to the same groups because of similar adult morphology and because of shared embryonic types; von Baer's embryonic types matched Cuvier's *embranchements*. They completed the circles of life history and of classification.

Following upon Wolff and the thesis that embryology is based on differentiating layers, von Baer divided animal embryonic development into three fundamental stages: (a) primary differentiation when the germ layers form; (b) histological differentiation when different cell types develop within the germ layers; and (c) morphological differentiation when organs were initiated. Embryos of one group recapitulated neither the embryonic development nor the adult structures of animals in other groups. Within a given group, earlier stages were more alike than were later stages because development progressed from the general or primitive, to the specific or advanced. During embryonic development, first the features of the phylum appeared, then those of the class, then of the order, family, genus, and species.

On the basis of a lifetime spent thinking about animal embryos (especially in the fifteen-year period from 1819 to 1834 while at Königsberg and then as Professor at St. Petersburg from 1834 until his death forty-two years later), von Baer formulated four general statements about animal development. Usually now referred to as "von Baer's laws," they demonstrate his unwavering conclusion that animal development occurs through epigenesis.

- The first law states that the general features of a large group of animals (the hair of mammals, for example, appears earlier in embryos than do the more

special features associated with subgroups of the larger group, for example, tooth form in cats or rodents).
- The second law states that less general characters of a subgroup (cats) develop from the more general characters of the group (mammals), that less general characters (say of a species of cat) develop from more general features of cats, and so forth, until the most specialized characters appear at the end of development.
- The third law states that rather than passing through the stages of other animals, embryos of a given species depart more and more from the embryos of other animals during development.
- The fourth law follows from the third: embryos of a "higher" animal (say a mammal) are never like the adult of a lower animal (say a fish), but only like its embryos.

Pangenesis, Germ Plasm, and the Foundations of Epigenetics

Darwin took his embryology from von Baer and used embryological evidence to demonstrate that similarities between animals reflected their classification because they all were members of a tree of life that reflected their shared evolutionary history.[14] Darwin (1859) provided two of the three essential components required for a theory of organismal evolution. One was to document the widespread existence of *variation*. The second was to propose a mechanism—*natural selection*—to explain the presence of such variation. What Darwin lacked was the third essential component, a theory of *inheritance*. Lack of a mechanism for inheritance led to the strongest criticism of Darwin's theory of evolution by natural selection. Darwin saw the lack, was sensitive to the criticisms, and spent the rest of his life seeking a mechanism of inheritance.

Knowing the importance of eggs and sperm in the reproduction of animals, Darwin went back to the theory of *pangenesis*. A Lamarckian concept, in pangenesis the essence of each tissue and organs is sent to the gonads in the form of gemmules that accumulate in eggs and sperm. Inheritance of these gemmules (and by inference, the use to which organs had been put during the lifetime of the individual) ensures that the features of the organism are passed from generation to generation.[15] Darwin's cousin, Francis Galton used blood transfusions between rabbits to test the theory of pangenesis but could find no evidence that gemmules were passed from animal to animal.[16] Darwin's response was that he had never proposed that gemmules were transferred in the blood. Nevertheless and despite Darwin's advocacy, pangenesis soon died for lack of evidence.[17]

Indeed, by 1885, pangenesis was ruled out as even a theoretical possibility when a German biologist, August Weismann (1834–1914) developed the theory of the germ plasm and soma in which only the germ plasm contributes to the next generation. Weismann proposed that protoplasm of the germ cells (germ plasm) was separate from the protoplasm of body cells (soma). Germ plasm was not involved in producing the tissues and organs of the body, but was maintained intact as the source of the germ cells for the next generation. Hence, the theory of the *continuity of the germ plasm*, and germ plasm as the physical embodiment of past evolutionary history and of future evolutionary potential.[18]

Pangenesis was not the only theory of development and evolution to be eliminated by Weismann. Lamarck's inheritance of acquired characters also met its demise. With respect to the passing on of hereditary information, there could be no feedback from soma to germ plasm, no matter what changes were wrought by the circumstances of the individual's life. Weismann's germ plasm is a linear precursor of the central dogma of molecular biology, which is that information flow is from germ plasm (i.e., DNA) to protein (i.e., organism).[19] Weismann's theory also laid the foundations for much of our current thinking about the significance of sex, embryology, and aging: sexual reproduction is possible because of the separation of germ plasm and soma; embryos are the vehicles that carry germ plasm from generation to generation; aging occurs after organisms have passed on their germ plasm. Even death found its evolutionary explanation in the continuity of the germ plasm.

Weismann proposed the existence of *determinants* that resided in the germ plasm and constituted the material and preformed basis for inheritance, thus reducing organisms (and inheritance) to their material parts. But neither determinants nor genes are preformed adults, so this is not the old version of preformation that was espoused before Harvey and Wolff. Preformation became more subtle, and increasing knowledge of interactions between nucleus and cytoplasm rendered epigenesis much more complex (see below).

The fundamental distinction between the continuity of the germ plasm and the discontinuity of the soma is reflected in the traditional separation of genotype from phenotype, heredity from development, and ontogeny from phylogeny. Because Weismann's germ plasm theory leaves no place for environmental influences, Weismannism cannot embrace epigenetics or epigenetic inheritance.[20] Maynard Smith (1989) as always, ahead of the rest of us, recognized that epigenetics provided an additional mode of inheritance: "There is a second inheritance system—an epigenetic inheritance system—in addition to the system based on DNA sequence that links sexual generations."[21]

Nuclear and/or Cytoplasmic Control

Weismann's recognition of the separation of germ plasm and soma, coupled with the rediscovery of Mendelian genetics, the development of the concept of the gene, and studies on embryos of marine invertebrates that demonstrated cytoplasmic control of cell fate, resulted in the development and centrality of cell biology in the first decades of the twentieth century. Is development controlled by genes in the nucleus, by factors in the cytoplasm (presumably inherited from the mother), or by interactions between nucleus and cytoplasm? Would the explanation found for one species apply to another or to all species? How simple or complex is control of embryonic development and how did increased understanding at the cellular level relate to the nineteenth century theories of epigenesis and preformation?

As laid out in detail in the third edition of his treatise on the cell (and in anticipation of the central role of genes in development), the influential cell biologist, E. B. Wilson (1856–1939)[22] maintained a prominent role for preformation in his evaluation of the relevant roles of nuclear and cytoplasmic control over development.[23] As summarized in 1925,

> Fundamentally, however, we reach the conclusion that in respect to a great number of characters heredity is effected by the transmission of a nuclear preformation which in the course of development finds expression in a process of cytoplasmic epigenesis.[24]

It appears on first reading that Wilson anticipated the roles later given to nuclear genes and cytoplasmic determinants. Indeed, in many respects, the decades after the discovery of the central dogma of DNA→ RNA→ protein and the realization that gene activity was regulated, could be categorized as "nuclear preformation" and "cytoplasmic epigenesis." Wilson, however, only allowed the most limited epigenetic influences in embryos and then only for some of their "external" features and as an expression of a nuclear preformation. By external features, Wilson meant trivial features.

Genes and Environment: Waddingtonian Epigenetics

Interactions between nuclear and cytoplasmic components remain central to our understanding of embryonic development but now in the context of genes (genetics) regulated from above ("epi")—epigenetics—a term coined by British embryologist, geneticist and theoretical biologist, Conrad Hal Waddington (1905–1975). Struck by the limited number of developmental pathways that seemed available to fruit flies after mutagenesis—some phenotypes were more sensitive and some less sensitive to mutation—Waddington developed the concept of the *canalisation* of embryonic development. Convinced of the important role of genes in development, but aware that gene action must be regulated, Waddington (1942) coined epigenetics as "the branch of biology which studies the causal interactions between genes and their products which bring the phenotype into being."[25] Bear in mind that, at the time the biochemical basis of genes was unknown. For Waddington, epigenetics was a conceptual model of genes interacting with their products.

This definition limits epigenetic signals to "information" derived from genes (à la Wilson) and limits epigenetic interactions to those taking place within the nucleus or between nucleus and cytoplasm. Signals that do not have a genetic basis (temperature, pH, for example) are excluded. With his discovery of genetic accommodation Waddington extended control to environmental influences—non-genetic signals that could trigger gene action and be subject to natural selection.[26]

This wider concept of, and more inclusive basis for, epigenetics are reflected in subsequent definition: "the study of the mechanisms of temporal and spatial control of gene activity during the development of complex organisms. Thus epigenetic can be used to describe anything other than DNA sequence that influences the development of an organism."[27] Hall defined epigenetics as "the sum of the genetic and non-genetic factors acting upon cells to control selectively the gene expression that produces increasing phenotypic complexity during development."[28] To this, I would only add "and evolution" at the end. While emphasizing that signals could have a genetic or a non-genetic basis, Hall maintained that: "it is a mistake to speak of epigenetics as nongenetic or of genetic versus epigenetic factors as if one is always in the ascendancy or acting to the exclusion of the other. . . . Epigenetic control is control of gene expression. . . . The genotype is the starting point and the phenotype is the endpoint of epigenetic control."[29] Réne Thom (1923–2002), a French mathematician and topologist responsible for the introduction and development in biology of catastrophe theory in dynamical systems theory, captured the essence of the transition from Wilson to Waddington and the linkage from Aristotle to the present day:

> If you were to follow Aristotle's theory of causality . . . you would say that from the point of view of material causality in embryology, everything is genetic as any protein is synthesized from reading a genomic molecular pattern. From the point of view of efficient causality, everything is also 'epigenetic.' as even the local triggering of a gene's activity requires, in general, an extra genomic factor.[30]

Wolf (1995) is typical of those who expanded Waddington's definition to include what are often referred to as environmental interactions: "interactions between genes and their products, and the various other conditions composing the milieu required for developmental processes to take place. Epigenetic changes are the result of these interactions, and may contribute significantly to the phenotype."[31] Genes interact with gene products but with "various other conditions." Reflecting a growing trend, Wolf emphasized a lack of a hierarchical arrangement between genes (genotype) and structures (phenotype); assigning genes a privileged position is an arbitrary decision made by men not the reality of development. The phenotype is, thus, the product of ontogenetic development rather than the mere consequence of the genetic constitution of the zygote, with the consequence that "the phenotype is not deducible from the genotype."[32]

A broader account of epigenetics began to emerge, one capable of including genetic interactions and non-genic factors, and both genetic and non-genetic interactions between these components. Epigenetics "is an ensemble of processes that propagate phenotypic characteristics throughout development. These processes derive from either indirect effects of gene action (emergent properties) or from non-genetic phenomena (e.g., cell-cell or hormone-target communications)."[33] Epigenetics writ large encompasses interactions within genes (DNA sequences, mRNA, histone- and non-histone proteins), cells (enzymes, cell-to-cell interactions) or organisms (hormones, metabolites), and interactions between genes, cells and organisms and their developmental-environmental contexts, including parental effects and temperature.[34] Epigenetic interactions at one level lead to emergent properties at a higher level. Indeed, one of those emergent properties is the gene itself. The knowledge that genes are regulated brings us to a new and recently emerged concept of epigenetics.

Molecular Epigenetics

Simultaneous with the use of epigenetics in the sense coined by Waddington, recent years have seen epigenetics used by molecular geneticists to refer to a quite specific class of interactions occurring within nuclei and fundamental to gene regulation:

> The term—"epigenetics" was introduced by Conrad Waddington to describe changes in gene expression during development. Nowadays, epigenetics in the Waddington sense refers to alterations in gene expression without a change in nucleotide sequence. However, this definition is so broad that an issue in Trends in Genetics devoted to epigenetics would read more like a modern biology textbook than a series of critical reviews. A more focused description of epigenetics refers to modifications in gene expression that are brought about by heritable, but potentially reversible, changes in chromatin structure and/or DNA methylation.[35]

I use the term *molecular epigenetics* to distinguish this more recent usage from Waddington's usage. Molecular epigenetics refers to inherited mechanisms of gene expression that are independent of the DNA sequences they regulate. Research into molecular epigenetics has revealed heritable codes in addition to the genetic code. Two of a number of mechanisms of epigenetics and two epigenetic codes are introduced briefly.

Methylation and Genetic Imprinting

Chromatin proteins associated with DNA are activated or silenced in what is often termed chromatin remodeling, either by post-translational modification or by *methylation* of cytosine to 5-methylcytosine. Methylation is not uniform throughout the genome. More highly methylated regions of DNA are less transcriptionally active then are less highly methylated regions.

Progress on methylation has been rapid. The first organismal methylation map came from the study by Zhang et al. (2006)[36] on the watercress, *Arabidopsis thaliana*. A DNA methylation profile of human chromosomes 6, 20, and 22 was published in the same year (Eckhardt et al. 2006)[37] and methylation shown to vary between individuals (Lister et al. 2009).[38] The pattern of methylation can persist into the germ cells of the next generation; profiles of sperm methylation were used by Molaro et al. (2011)[39] to show epigenetic inheritance in primate sperm and a role for epigenetics in primate evolution. This process of *genetic imprinting* forms part of the *epigenetic code* passed by *epigenetic inheritance* from generation to generation.[40]

Unraveling the nature of methylation and genetic imprinting have both played major roles in the development of evolutionary medicine, as it has become apparent that epigenetic inheritance plays a previously unrecognized role in the acquisition and transmission of disease states.[41]

Acetylation and the Histone Code

Acetylation of histones is another molecular epigenetic mechanism. Because DNA chains are coiled around histones—eight histones wrapped around 220 base pairs of DNA forming a nucleosome—methylation and histone acetylation influence the compactness of the DNA, and therefore, its availability for transcriptional regulation. An important consequence of acetylation is that a histone at one position may have a different function from the same histone at another position in the genome. Histone acetylation provides a *histone code* that is in addition to the code based on genetic imprinting.

Epigenotype

Both Waddingtonian and molecular epigenetics have made it abundantly clear that in addition to the genotype and phenotype, organisms possess an *epigenotype*. Indeed, "the epigenotype" was the title of Waddington's 1942 paper in which epigenetics and the epigenotype were proposed and defined: "Between genotype and phenotype, and connecting them to each other, there lies a whole complex of developmental processes. It is convenient to have a name for this complex; 'epigenotype' seems suitable."[42] For Waddington, the epigenotype represented the inherited potential to develop along a particular path or in a particular way. The epigenotype is then the sum of the agents and the interactions between agents linking genotype with phenotype, phenotype with environment, and genotype-phenotype with natural selection and evolutionary change.[43]

As with epigenetics, the definition of the epigenotype has changed as molecular studies of gene regulation have accelerated. In alignment with molecular epigenetics, the epigenotype is the stable and heritable pattern of gene expression found in individual cells or organisms.[44] No matter how narrow or wide the definition, it is clear that we can no longer think of genotype and phenotype as the only two representations of where heritable information is contained or how it is expressed.[45]

Given the presence of epigenetic codes, epigenetics reaches far beyond the control of embryonic development. A recent volume dealing with epigenetics as the link between the genotype and phenotype (the epigenotype) in development and evolution it was concluded that: "Epigenetic interactions link ecology to development and to evolution, and therefore, provide mechanisms underpinning phenotypic plasticity and life-history strategies."[46] In its broadest sense, epigenetics integrates processes at molecular, developmental, organismal, and environmental levels of biological organization In atomizing epigenetics at different levels of the biological hierarchy, Hallgrímsson and Hall identified the roles of (a) interactions *within individuals* in the origination of cell types, tissues, and organs; (b) interactions *between individuals of the same species* in polymorphisms; and (c) interactions *between individuals of different species* in polymorphisms seen especially in the coupling of predator and prey, as representing "a single class of mechanisms linking embryonic inductions, life-history morphs and ecological adaptation, and evolutionary plasticity."[47]

Systems Theory

Waddington's development of epigenetics and the epigenotype is a reflection of his philosophical approach and his conception of organisms as complex systems, amenable to analysis only by integrated investigation across the various levels of the biological hierarchy carried out with the framework of a theoretical biology.[48] Waddington's philosophy was grounded in the philosophy of English mathematician, logician, and philosopher, Alfred North Whitehead (1861–1947), coauthor with Bertrand Russell (1872–1970) of *Principia Mathematica*.[49] Whitehead developed what came to be known as "process philosophy": process, not substance or structure or pattern, is the fundamental level at which life can be understood. In Waddington's hands process was epigenetic as reflected in the epigenotype (or epigenotypes) that link genotype and phenotype.

Knowledge of many levels of biological organization and integration across those levels will be required to document and then decode the epigenetic code and the epigenotype. By comparison, revealing the nature of the hereditary material and the genetic code required knowledge of the nature and organization of a single molecule, DNA. Uncovering the central dogma of molecular biology required knowledge of only three classes of molecules, DNA, RNA, and proteins. While these are monumental discoveries, they required only a tiny fraction of the knowledge required to unmask the epigenetic code and epigenotype. To these ends, contemporary systems theory and systems biology recognize that understanding the nature of development and evolution, indeed of life itself, will entail a comprehension of far greater degrees of complexity than previously thought.

Waddington's insight into the approaches required to understand organismal development and evolution lives on. As Terrence Deacon indicated in the abstract for chapter 13 of this volume (that was provided in the proposal to the publisher): "Evidence is mounting that epigenesis depends on diverse self-organizational, self-assembly, and constraint-dependent processes." Because the epigenotype is an emergent property of interactions at multiple levels of biological organization, the consequences of epigenetic processes cannot be predicted from the properties of the components that interact.[50] Epigenetics, therefore, is as an important component of *systems biology*, which is a subclass of systems theory.

Systems theory was promulgated by the biologist Ludwig von Bertalanffy (1901–1972)[51] and developed by several others, notably British psychiatrist and pioneer in cybernetics W. Ross Ashby (1903–1972) with his concept of the organism as a homeostatic machine.[52] The principle of self-organization and regulatory models developed last century are reflected today in theories of complex adaptive systems, of community-level interactions in ecology, of interactions between the hundreds of bacterial species in the human gut; of gene regulatory networks in development and evolution, and of the organization of the genomes.[53]

Publication of the methylation map for the watercress, *Arabidopsis thaliana* by Zhang et al. (2006) provides a window into one way in which epigenetics will advance our understanding, but will require a systems biology approach involving the detection of epigenetic variation in populations. Epigenetic regulation varies between regions of the genome, between individuals and between populations, and in response to biotic and abiotic components of environments.[54] Consequently, "as more and longer genomes become complemented by epigenomes and population epigenomes, we may very shortly find we have more data than sense."[55] The website of the Department of Systems Biology at the Harvard Medical School captures the approaches that will be required:

> Living systems are dynamic and complex, and their behavior may be hard to predict from the properties of individual parts. To study them, we use quantita-

tive measurements of the behavior of groups of interacting components, systematic measurement technologies such as genomics, bioinformatics and proteomics, and mathematical and computational models to describe and predict dynamical behavior. Systems problems are emerging as central to all areas of biology and medicine.[56]

Conclusion

Epigenetics is a term and concept that embraces the regulation of gene activity during embryonic development, animal and plant ontogeny, organismal evolution, and some animal diseases and cancers. Often cast as the genetic extension of epigenesis in discussions of epigenesis and preformation, epigenetics reflects a much deeper search in biology for the factor(s) that control the specification-determination of germ layers, cells, tissues, organs, and individuals, and for an inclusive theory of how information is activated within a generation (ontogeny, development) and transferred from generation to generation (phylogeny-evolution).

Epigenesis (and so the modern foundations of the "epi" part of epigenetics) is typically traced to Aristotle and from the sixteenth to the eighteenth century comparative anatomists and comparative embryologists William Harvey, Caspar Friedrich Wolf and Karl von Baer, who demonstrated through direct observation that vertebrate embryos alter their form, size, and substance during their development. Beginning in the late nineteenth century with the recognition of the cellular basis of life, three major types of interactions were proposed as mechanistic bases for epigenesis. The first was proposed before the discovery of genes but after the rediscovery of Mendelian genetics in 1900. Embryologists, especially those analyzing cell lineages in the embryos of marine invertebrates, asked whether the nucleus or the cytoplasm "controlled" cell fate, with the implicit and sometimes explicit reductionist and preformationist understanding that control would reside in one or the other. With the discovery in the first decades of the twentieth century that cytoplasmic factors regulated nuclear activity, interactions between nucleus (genes) and cytoplasm were proposed as the basis for the specification of cell fate using experimental approaches that focused on cytoplasmic regulation of nuclear factors. Thirdly were proposals for the central role for genes, either acting alone (a reductionist genes only approach), acting in response to cytoplasmic signals (an internalist approach), and/or acting in response to environmental signals (an approach incorporating interaction between organism and environment). Depending on the approach and underlying theory brought to the embryos being studied, the environment was either the internal environment of the embryo or mother or the external environments in which organisms developed. In any case, the genetic code was identified as the only informational code available to individuals and the only code that could be passed on from generation to generation.

The joining of genes and control above the gene—epigenetics—is traced to concepts developed by Conrad Hal Waddington, whose philosophical background positioned him to develop an epigenetic approach at a time (the 1940s-1960s) when almost all others were either seeking reductionist gene-centered mechanisms for the origin and development of features in embryonic development, or were propounding natural selection and changes in gene pools as the origination of population-level variation and the primary or only mechanism of evolution. Increasing understanding of gene structure and function from the 1970s on, establishment of molecular genetics, and the entrenchment of a reductionist concept of the gene as *the* informational unit of development and inheritance, the term epigenetics came to be used for molecular information that affected the function of DNA sequences within the nucleus; the ultimate primacy of nucleus over cytoplasm.

Those who sought a greater understanding of the way in which development and environment influence evolutionary change *above* the level of the gene through such mechanisms as canalization and phenotypic plasticity, continued to use epigenetics in its original Waddingtonian sense.[57] The concept of the epigenotype outlined by Waddington in 1942 and developed more recently, coupled with the recognition that genes do not occupy a privileged position as the sole regulators of cellular processes, have brought epigenetics to the forefront of approaches to understanding development and evolution. Integration between levels and the application of systems biology approaches show that epigenetics will form a major integrating component of the inclusive theory of biology we expect the twenty-first century to bring.

Notes

1. James Griesemer, "What Is 'Epi' about Epigenetics?" *Annals of the New York Academy of Science* 981 (2002): 97.

2. For overviews of epigenetics and its historical roots in epigenesis, see the following sources: Linda Van Speybroeck, "From Epigenesis to Epigenetics: The Case of C. H. Waddington," *Annals of the New York Academy of Science* 981 (2002): 61-81; Gerd B. Müller and Olsson Leonard, "Epigenesis and Epigenetics," in *Keywords and Concepts in Evolutionary Developmental Biology*, ed. Brian K. Hall (Cambridge MA: Harvard University Press, 2003), 114-123; Brian K. Hall, "Genetic and Epigenetic Control of Vertebrate Development," *Netherlands Journal of Zoology* 40 (1990): 352-361; Brian K. Hall, "Epigenetics: Regulation Not Replication," *Journal of Evolutionary Biology* 11 (1998): 201-205; Brian K. Hall, "A Brief History of the Term and Concept Epigenetics," in *Epigenetics: Linking Genotype and Phenotype in Development and Evolution*, ed. Benedikt Hallgrímsson and Brian K. Hall (Berkeley, CA: University of California Press, 2011), 9-13; Benedikt Hallgrímsson and Brian K. Hall, eds., *Epigenetics: Linking Genotype and Phenotype in Development and Evolution* (Berkeley CA: University of California Press, 2011a).

3. Keller, Evelyn Fox, *The Mirage of a Space between Nature and Nurture* (Durham & London: Duke University Press, 2010), 4.

4. For evaluations of epigenesis and preformation, see the following sources: Charles Otis Whitman, "Evolution and Epigenesis," *Biological Lectures of the Marine Biology Laboratory Woods Hole* (1894), 205-224; Edward Stuart Russell, *Form and Function: A Contribution to the History of Animal Morphology* (London: John Murray, 1916; repr. Chicago: The University of Chicago Press, 1982); Francis Joseph Cole, *Early Theories of Sexual Generation* (Oxford: Oxford University Press, 1980); Donna Haraway, *Crystals, Fabrics and Fields* (New Haven and London: Yale University Press, 1976); John Farley, *Gametes & Spores: Ideas about Sexual Reproduction 1750–1914* (Baltimore: The Johns Hopkins University Press, 1982); John A. Moore, *Science as a Way of Knowing: The Foundations of Modern Biology* (Cambridge, MA: Harvard University Press, 1993); Clara Pinto-Correia, *The Ovary of Eve: Egg and Sperm and Preformation* (Chicago: The University of Chicago Press, 1997).

5. For Aristotle, see the following sources: Harry B. Torrey and F. Felin, "Was Aristotle an Evolutionist?" *Quarterly Review of Biology* 12 (1937): 1-18; Ernst Mayr, *The Growth of Biological Thought. Diversity, Evolution, and Inheritance* (Cambridge MA: The Belknap Press of Harvard University Press, 1982); María Claudia Cecchi, Carlos Guerrero-Bosagna, and Jorge Mdodozis, "Aristotle's Crime?" *Revista Chilena de Historia Natural* 74 (2001): 507-514.

6. For the work of Harvey and Wolff, see the following sources: A. E. Gaissinovitch, "Wolff, Caspar Friedrich," in *Dictionary of Scientific Biography, Volume 15*, ed. C. Gillespie, Supplement I), S 524-526 (New York: Charles Scribners' Sons, 1978); Shirley A. Roe, "Rationalism and Embryology: Caspar Friedrich Wolff's Theory of Epigenesis," *Journal of the History of Biology* 12 (1979): 1-43; Shirley A. Roe, *Matter, Life and Generation: 18th Century Embryology and the Haller-Wolff Debate* (Cambridge: Cambridge University Press, 1981); Charles E. Dinsmore, "Conceptual Foundations of Metamorphosis and Regeneration: From Historical Links to Common Mechanisms," *Wound Repair and Regeneration* 6 (1998): 291-301.

7. William Harvey, *Exercitationes de Generatione Animalium* (*Disputations Touching the Generation of Animals*), trans. G. Whitteridge (1651; repr. London: Blackwells, 1981), 148.

8. Caspar Friedrich Wolff, *Theoria Generationis* (Halae ad Salam, 1759).

9. Caspar Friedrich Wolff, *Theorie von der Generation in zwo Abhandlungen* (Berlin: Friedrich Wilhelm Birnstiel, 1764), 211.

10. See Pinto-Correia. *The Ovary of Eve*.

11. Haller is regarded as the first to use the term *evolution* (L. *evolutio*, unrolling) for the gradual unfolding of preformed embryos:

But the theory of evolution proposed by Swammerdam and Malpighi prevails almost everywhere [*Sed evolutionem theoria fere ubique obtinet a Swammerdamio et Malphighio proposita*]. . . . Most of these men teach that there is in fact included in the egg a germ or perfect little human machine. . . . And not a few of them say that all human bodies were created fully formed and folded up in the ovary of Eve and that these bodies are gradually distended by alimentary humor until they grow to the form and size of animals (Haller 1774, cited from Howard B. Adelmann, *Marcello Malpighi and the Evolution of Embryology, Five Volumes* (Ithaca NY: Cornell University Press, 1966), 893-894.

12. Shirley A. Roe, "The Development of Albrecht von Haller's Views on Embryology," *Journal of the History of Biology* 8 (1975): 167-190; Roe, *Matter, Life and Generation*.

13. Karl Ernst von Baer, *Über Entwickelungsgeschichte der Thiere: Beobachtung und Reflexion* (Königsberg: Gebrüder Borntråger, 1828; Culture et Civilisation, Bruxelles, 1967); Jane Oppenheimer, "von Baer, Karl Ernst," *Dictionary of Scientific Biography, Volume 1*, ed. C. Gillespie (New York: Charles Scribners' Sons, 1971), 385-389.

14. Robert J. Richards, *The Meaning of Evolution: The Morphological Construction and Ideological Reconstruction of Darwin's Theory* (Chicago: The University of Chicago Press, 1992).

15. Charles Darwin, *The Variation of Animals and Plants under Domestication* (London: John Murray, 1868).

16. See Michael Bullmer, "The Development of Francis Galton's Ideas on the Mechanism of Heredity," *Journal of the History of Biology* 32 (1999): 263-292.

17. Youngsheng Liu, "A New Perspective on Darwin's Pangenesis," *Biological Reviews of the Cambridge Philosophical Society* 83 (2008): 141-149.

18. August Weismann, *Über die Vererbung* (Jena: Gustav Fischer, 1883); August Weismann, *Die Kontinuitt des Keimplasmas als Grundlage einer Theorie der Vererbung* (Jena: Gustav Fischer, 1885); August Weismann, *The Germ Plasm: A Theory of Heredity*, trans. W. Newton Parker and H. Ronnfeld (London: Walter Scott Ltd., 1893).

For August Weismann and the continuity of the germ plasm, see the following sources: Edward Stuart Russell, *Form and Function: A Contribution to the History of Animal Morphology* (London: John Murray, 1916; repr. Chicago: The University of Chicago Press, Introduction by George V. Lauder, 1982); Frederick B. Churchill, "August Weismann and a Break From Tradition," *Journal of the History of Biology* 1 (1968): 91-112; Rasmus G. Winther, "August Weismann on Germ-Plasm Variation," *Journal of the History of Biology* 34 (2001): 517-555.

19. Griesemer, "What Is "Epi" about Epigenetics?" 97-110.

20. See Griesemer, "What Is 'Epi' about Epigenetics?" for a different perspective on this issue.

21. John Maynard Smith, "Weismann and Modern Biology," in *Oxford Surveys in Evolutionary Biology*, ed. P. H. Harvey and L. Partridge (Oxford: Oxford University Press, 1989), 11.

22. For evaluations of Wilson's contributions as a founder of *Cell Biology*, see the following sources: Thomas Hunt Morgan, "Edmund Beecher Wilson," *Biographical Memoirs Series, National Academy of Sciences* 21 (1940): 315-342; Hermann Joseph Muller, "Edmund B. Wilson—An Appreciation," *American Naturalist* 77 (1943): 5-37, 142-172; Jane Maienschein, *Transforming Traditions in American Biology, 1880–1915* (Baltimore: Johns Hopkins University Press, 1991).

23. Jane Maienschein, "Cell Lineage, Ancestral Reminiscence, and the Biogenetic Law," *Journal of the History of Biology* 11 (1978): 129-158; Brian K. Hall, *Evolutionary Developmental Biology* (London: Chapman and Hall, 1992).

24. Edward B. Wilson, *The Cell in Development and Heredity*, 3rd Edition (New York: Macmillan, 1925), 1112.

25. Quoted from Conrad Hal Waddington, *The Evolution of an Evolutionist* (Ithaca, NY: Cornell University Press, 1975), 218. Waddington reprinted many of his most significant publications in his autobiography, *The Evolution of an Evolutionist*.

26. For recent analyses of Waddingtonian epigenetics, see: Brian K. Hall, "Waddington, Conrad H.," in *New Dictionary of Scientific Biography, Volume 7*, ed. N. Koert-

ge (New York: Charles Scribners' Sons, 2008), 201-207; Van Speybroeck, "From Epigenesis to Epigenetics"; Brian K. Hall and Manfred Laubichler, "Conrad Hal Waddington, Theoretical Biology, and Evo-Devo," *Biological Theory* 3, no. 3 (2009): 185-289; Heather A. Jamniczky, Julia C. Boughner, Campbell Rolian, Paula N. Gonzalez, et al., "Rediscovering Waddington in the Post-Genomic Age: Operationalising Waddington's Epigenetics Reveals New Ways to Investigate the Generation and Modulation of Phenotypic Variation," *BioEssays* 32 (2010): 553-558. Also see the full thematic issue of *Biological Theory* 3, no. 3, Summer 2008 on Waddingtonian epigenetics.

27. Robin Holliday, "Mechanisms for the Control of Gene Activity during Development," *Biological Reviews of the Cambridge Philosophical Society* 65 (1990): 431-471; Robin Holliday, "Epigenetics: An Overview," *Developmental Genetics* 15 (1994): 453-457; Robin Holliday, "DNA Methylation and Epigenotypes," *Biochemistry (Moscow)* 70 (2005): 500–504.

28. Brian K. Hall, *Evolutionary Developmental Biology* (London: Chapman and Hall, 1992), 89.

29. Brian K. Hall, *Evolutionary Developmental Biology* (Boston: Kluwer Academic Publishers, 1999), 114, 399.

30. René Thom, "An Inventory of Waddingtonian Concepts" in *Theoretical Biology*, ed. Brian Goodwin and P. Saunders (Edinburgh: Edinburgh University Press 1989), 3.

31. U. Wolf, "The Genetic Contribution to the Phenotype," *Human Genetics* 95 (1995): 127-148.

32. Wolf, "The Genetic Contribution to the Phenotype," 128, 144, 127, 144.

33. Carl D. Schlichting and Massimo Pigliucci, *Phenotypic Evolution: A Reaction Norm Perspective* (Sunderland, MA: Sinauer Associates, 1998), 232.

34. For the breadth of epigenetics today, see the following sources: V. E. A. Russo, Robert A. Martienssen, and Arthur D. Riggs, eds. *Epigenetic Mechanisms of Gene Regulation* (Plainview, NY: Cold Spring Harbor Laboratory Press, 1996); Bryan M. Turner, "Epigenetic Responses to Environmental Change and Their Evolutionary Implications," *Philosophical Transactions of the Royal Society of London (B) Biological Sciences* 364 (2009): 3403–3418; Christian Biémont, "From Genotype to Phenotype. What Do Epigenetics and Epigenomics Tell Us?" *Heredity* 105 (2010): 1-3; Jamniczky, et al. "Rediscovering Waddington in the Post-Genomic Age"; Hallgrímsson and Hall, eds. *Epigenetics: Linking Genotype and Phenotype in Development and Evolution*.

35. Steven Henikoff and M. A. Matzke, "Exploring and Explaining Epigenetic Effects," *Trends in Genetics* 13 (1997): 293.

36. Xiaoyu Zhang, Junshi Yazaki, Ambika Sundaresan, Shawn Cokus, et al. "Genome-Wide High-Resolution Mapping and Functional Analysis of DNA Methylation in *Arabidopsis*," *Cell* 126 (2006): 1189-1201.

37. Florian Eckhardt, Joern Lewin, Rene Cortese, Vardhman K. Rakyan Vardhman, et al., "DNA Methylation Profiling of Human Chromosomes 6, 20 and 22," *Nature Genetics* 38 (2006): 1378-1383.

38. Ryan Lister, Mattia Pelizzola, R. H. Dowen, R. David Hawkins, et al. "Human DNA Methylomes at Base Resolution Show Widespread Epigenomic Differences," *Nature* 462 (2009): 315-322.

39. Antoine Molaro, Emily Hodges, Fang Fang, Qiang Song, W. Richard McCombie, et al. "Sperm Methylation Profiles Reveal Features of Epigenetic Inheritance and Evolution in Primates," *Cell* 146 (2011): 1029-1041.

40. Danesh Moazed, "Mechanisms for the Inheritance of Chromatin States," *Cell* 146 (2011): 510-517; Snait B. Gissis and Eva Jablonka, eds. *Transformations of Lamarckism: From Subtle Fluids to Molecular Biology* (Cambridge: The MIT Press, 2011). According to Griesemer in "What Is "Epi" about Epigenetics?" *Annals of the New York Academy of Science* 981 (2002), 108, "Inheritance is a special case of reproduction processes in which the developmental mechanisms have evolved to serve developmental functions. Genetic inheritance is a special case of inheritance in which the developmental mechanisms have evolved to the 'coding' grade of organization."

41. Discussion of evolutionary medicine is beyond the scope of this chapter. For recent analyses see the following sources: Peter D. Gluckman, Mark H. Hanson, and Alan S. Beedle, "Non-Genomic Transgenerational Inheritance of Disease Risk," *BioEssays* 29 (2007): 145-154; Peter D. Gluckman, Alan S. Beedle, and Mark H. Hanson, *Principles of Evolutionary Medicine* (Oxford: Oxford University Press, 2009); Felicia M. Low, Peter D. Gluckman, Mark A. Hanson, "Developmental Plasticity, Epigenetics and Human Health," *Evolutionary Biology* (January 11, 2012): 1-16.

42. Conrad Hal Waddington, "The Epigenotype," *Endeavour* 1 (1942): 19.

43. For the epigenotype, see the following sources: Hall, "Genetic and Epigenetic Control of Vertebrate Development"; Brian K. Hall, Roy D. Pearson, and Gerd Müller, eds., *Environment, Development, and Evolution: Toward a Synthesis* (MIT Press, Cambridge MA, 2004); Isaac Salazar-Ciudad and Jukka Jernvall, "Graduality and Innovation in the Evolution of Complex Phenotypes: Insights From Development," *Journal of Experimental Biology (Molecular Development and Evolution)* 6 (2005): 619-631; Hallgrímsson and Hall, eds. *Epigenetics: Linking Genotype and Phenotype in Development and Evolution*; Benedikt Hallgrímsson and Brian K. Hall, "Epigenetics: The Context of Development," in *Epigenetics: Linking Genotype and Phenotype in Development and Evolution*, ed. Benedikt Hallgrímsson and Brian K. Hall (Berkeley, CA: University of California Press, 2011b), 424-438; Benedikt Hallgrímsson and Brian K. Hall, "Introduction," in *Epigenetics: Linking Genotype and Phenotype in Development and Evolution*, ed. Benedikt Hallgrímsson and Brian K. Hall (Berkeley, CA: University of California Press, 2011), 1-5.

44. Robin Holliday, "DNA Methylation and Epigenotypes," 500–504.

45. Mechanisms of transmission of heritable information include genetic, epigenetic, cultural and linguistic. See Evelyn Fox Keller, *The Mirage of a Space between Nature and Nurture*; Gissis and Jablonka, eds. *Transformations of Lamarckism*.

46. Hallgrímsson and Hall, "Epigenetics: The Context of Development," 429.

47. Hallgrímsson and Hall, "Epigenetics: The Context of Development," 431.

48. See Waddington (1968–1972) for a four-volume treatise on theoretical biology, Waddington (1975) for his own take on his life's work, and see Thom (1989) and the papers in Hall and Laubichler (2009) for analyses of Waddington's conceptual impact (i.e.,): Conrad Hal Waddington, *Towards a Theoretical Biology, Volumes 1-4* (Edinburgh: Edinburgh University Press, 1968–1972); Conrad Hal Waddington, *The Evolution of an Evolutionist*; René Thom, "An Inventory of Waddingtonian Concepts," 1-7; Hall and Laubichler, "Conrad Hal Waddington, Theoretical Biology, and Evo-Devo."

49. Alfred North Whitehead and Bertrand Russell, *Principia Mathematica, Volumes 1-3* (Cambridge, MA: Cambridge University Press, 1910, 1912, 1913).

50. James Collins, Scott Gilbert, Manfred Laubichler and Gerd Müller, "Modeling in EvoDevo: How to Integrate Development, Evolution, and Ecology," in *Roots of Theoretical Biology: The Prater Vivarium Centenary*, ed. by Manfred Laubichler (Cambridge:

MIT Press, 2007), 355-378; Hallgrímsson and Hall, eds. *Epigenetics: Linking Genotype and Phenotype in Development and Evolution.*

51. Ludwig von Bertalanffy, "Untersuchungen über die Gesetzlichkeit des Wachstums. I. Allgemeine Grundlagen der Theorie; mathematische und physiologische Gesetzlichkeiten des Wachstums bei Wassertieren," *Archives Entwicklungsmechanic* 131 (1934): 613-652; Ludwig von Bertalanffy, "General System Theory: A Critical Review," *General Systems* 7 (1962): 1–20; Ludwig von Bertalanffy, *General System Theory: Foundations, Development, Applications* (New York: George Braziller, 1968).

52. W. Ross Ashby, *An Introduction to Cybernetics* (London: Chapman & Hall, 1956).

53. H. Kitano, "Systems Biology: A Brief Overview," *Science* 295 (2002): 1662-1664; Eric H. Davidson, "Developmental Biology at the Systems Level," *Biochimica et Biophysica Acta* 1789 (2009), 248-249; James A. Shapiro, *Evolution. A View from the 21st Century* (Upper Saddle River, NJ: FT Press Science, 2011); Robert Feil and Mario Fraga, "Epigenetics and the Environment: Emerging Patterns and Implications," *Nature Reviews Genetics* 13 (2012): 97-109.

54. Ryan Lister, et al., "Human DNA Methylomes at Base Resolution Show Widespread Epigenomic Differences"; Jamniczky et al. "Rediscovering Waddington in the Post-Genomic Age."

55. L. J. Johnson and P. J. Tricker, "Epigenomic Plasticity Within Populations: Its Evolutionary Significance and Potential," *Heredity* 105 (2010): 113.

56. Harvard Medical School Department of Systems Biology website: https://sysbio.med.harvard.edu/. Accessed May 24th, 2012.

57. Benedikt Hallgrímsson, Katherine Willmore, and Brian K. Hall, "Canalization, Developmental Stability, and Morphological Integration in Primate Limbs," *American Journal of Physical Anthropology Yearbook* S 45 (2002): 131–158.

Bibliography

Adelmann, Howard B. *Marcello Malpighi and the Evolution of Embryology, Five Volumes.* Ithaca NY: Cornell University Press, 1966.

Aristotle. *The Works of Aristotle, Volume IV. Historia Animalium.* Translated by D'Arcy Thompson. Oxford. Clarendon Press, 1910.

———. *The Works of Aristotle, Volume V. De Partibus animalium; De motu. De incessu animalium; De generatione animalium.* Translated by William Ogle, Arthur Spencer, Loat Farquharson, and Arthur Platt. Clarendon Press: Oxford, 1912.

Ashby, W. Ross. *An Introduction to Cybernetics.* London: Chapman & Hall, 1956.

Biémont, Christian. "From Genotype to Phenotype. What Do Epigenetics and Epigenomics Tell Us?" *Heredity* 105 (2010): 1-3.

Bullmer, Michael. "The Development of Francis Galton's Ideas on the Mechanism of Heredity." *Journal of the History of Biology* 32 (1999): 263-292.

Cecchi, María Claudia, Carlos Guerrero-Bosagna, and Jorge Mdodozis. "Aristotle's Crime?" *Revista Chilena de Historia Natural* 74 (2001): 507-514.

Churchill, Frederick B. "August Weismann and a Break from Tradition." *Journal of the History of Biology* 1 (1968): 91-112.

Cole, Francis Joseph. *Early Theories of Sexual Generation*. Oxford: Oxford University Press, 1980.

Collins, James, Scott Gilbert, Manfred Laubichler and Gerd Müller. "Modeling in Evo-Devo: How to Integrate Development, Evolution, and Ecology." In *Roots of Theoretical Biology: The Prater Vivarium Centenary*, edited by Manfred Laubichler, 355-378. Cambridge: MIT Press, 2007.

Darwin, Charles. *The Origin of Species by Means of Natural Selection*. London: John Murray, 1859.

———. *The Variation of Animals and Plants under Domestication*. London: John Murray, 1868.

Davidson, Eric H. "Developmental Biology at the Systems Level." *Biochimica et Biophysica Acta* 1789 (2009): 248-249.

Dinsmore, Charles E. "Conceptual Foundations of Metamorphosis and Regeneration: From Historical Links to Common Mechanisms." *Wound Repair and Regeneration* 6 (1998): 291-301.

Eckhardt, Florian, Joern Lewin, Rene Cortese, Vardhman K. Rakyan, et al. "DNA Methylation Profiling of Human Chromosomes 6, 20 and 22." *Nature Genetics* 38 (2006): 1378-1383.

Farley, John. *Gametes & Spores. Ideas about Sexual Reproduction 1750-1914*. Baltimore and London: The Johns Hopkins University Press, 1982.

Feil, Robert, and Mario F. Fraga, "Epigenetics and the Environment: Emerging Patterns and Implications." *Nature Reviews Genetics* 13 (2012): 97-109.

Gaissinovitch A. E. "Wolff, Caspar Friedrich." In *Dictionary of Scientific Biography, Volume 15*, edited by C. Gillespie, Supplement I, S 524-526. New York: Charles Scribners' Sons, 1978.

Gissis, Snait B., and Eva Jablonka, eds. *Transformations of Lamarckism: From Subtle Fluids to Molecular Biology*. Cambridge: The MIT Press, 2011.

Gluckman, Peter D., Alan S. Beedle, and Mark H. Hanson. *Principles of Evolutionary Medicine*. Oxford: Oxford University Press, 2009.

Gluckman, Peter D., Mark H. Hanson, and Alan S. Beedle. "Non-Genomic Transgenerational Inheritance of Disease Risk." *BioEssays* 29 (2007): 145-154.

Griesemer, James. "What Is "Epi" about Epigenetics?" *Annals of the New York Academy of Science* 981 (2002): 97-110.

Guerrero-Bosagna, Carlos. "Finalism in Darwinian and Lamarckian Evolution: Lessons from Epigenetics and Developmental Biology." *Evolutionary Biology* (in press, 2012, DOI 10.1007/s11692-012-9163-x)

Hall, Brian K. "Genetic and Epigenetic Control of Vertebrate Development." *Netherlands Journal of Zoology* 40 (1990): 352-361.

———. *Evolutionary Developmental Biology*. London: Chapman and Hall, 1992.

———. "Epigenetics: Regulation Not Replication." *Journal of Evolutionary Biology* 11 (1998): 201-205.

———. *Evolutionary Developmental Biology*. Boston: Kluwer Academic Publishers, 1999.

———. "Waddington, Conrad H." In *New Dictionary of Scientific Biography, Volume 7*, edited by N. Koertge, 201-207, New York: Charles Scribners' Sons, 2008.

———. "A Brief History of the Term and Concept Epigenetics." In *Epigenetics: Linking Genotype and Phenotype in Development and Evolution*, edited by Benedikt Hallgrímsson and Brian K. Hall, 9-13. Berkeley, CA: University of California Press, 2011.

———. "Lamarck, Lamarckism, Epigenetics and Epigenetic Inheritance." *Metascience* 21 (March 2012): 375-378.
Hall, Brian K., and Manfred Laubichler. "Conrad Hal Waddington, Theoretical Biology, and Evo-Devo." *Thematic issue of Biological Theory* 3, no. 3 (2009): 185-289.
Hall, Brian K., Roy D. Pearson, and Gerd Müller, eds. *Environment, Development, and Evolution: Toward a Synthesis.* MIT Press, Cambridge MA, 2004.
Hallgrímsson, Benedikt, and Brian K. Hall, eds. *Epigenetics: Linking Genotype and Phenotype in Development and Evolution.* Berkeley CA: University of California Press, 2011a.
———. "Epigenetics: The Context of Development." In *Epigenetics: Linking Genotype and Phenotype in Development and Evolution*, edited by Benedikt Hallgrímsson and Brian K. Hall, 424-438. Berkeley, CA: University of California Press, 2011b.
———. "Introduction." In *Epigenetics: Linking Genotype and Phenotype in Development and Evolution*, edited by Benedikt Hallgrímsson and Brian K. Hall, 1-5. Berkeley, CA: University of California Press, 2011c.
Hallgrímsson, Benedikt, Katherine Willmore, and Brian K. Hall. "Canalization, Developmental Stability, and Morphological Integration in Primate Limbs." *American Journal of Physical Anthropology Yearbook* S 45 (2002): 131–158.
Haraway, Donna. *Crystals, Fabrics and Fields.* New Haven and London: Yale University Press, 1976.
Harvard Medical School Department of Systems Biology website: https://sysbio.med.harvard.edu/. Accessed May 24th, 2012.
Harvey, William, *Exercitationes de Generatione Animalium (Disputations Touching the Generation of Animals),* translated with Introductions and notes by G. Whitteridge. London: Blackwells, 1651, reprinted 1981.
Henikoff, Steven, and M. A. Matzke. "Exploring and Explaining Epigenetic Effects." *Trends in Genetics* 13 (1997): 293-295.
Holliday, Robin. "Mechanisms for the Control of Gene Activity during Development." *Biological Reviews of the Cambridge Philosophical Society* 65 (1990): 431-471.
———. "Epigenetics: An Overview." *Developmental Genetics* 15 (1994): 453-457.
———. "DNA Methylation and Epigenotypes." *Biochemistry (Moscow)* 70 (2005): 500–504.
Jamniczky, Heather A., Julia C. Boughner, Campbell Rolian, Paula N. Gonzalez, et al. "Rediscovering Waddington in the Post-Genomic Age: Operationalising Waddington's Epigenetics Reveals New Ways to Investigate the Generation and Modulation of Phenotypic Variation." *BioEssays* 32 (2010): 553-558.
Johnson, L. J., and P. J. Tricker. "Epigenomic Plasticity within Populations: Its Evolutionary Significance and Potential." *Heredity* 105 (2010): 113-121.
Keller, Evelyn Fox. *The Mirage of a Space between Nature and Nurture.* Durham & London: Duke University Press, 2010.
Kitano, H. "Systems Biology: A Brief Overview." *Science* 295 (2002): 1662-1664.
Lewontin, Richard C. "The Units of Selection." *Annual Review of Ecology and Systematics* 1 (1970): 1-18.
———. *The Triple Helix: Gene, Organism, and Environment.* Cambridge: Harvard University Press, 2000.
Lister, Ryan, Mattia Pelizzola, R. H. Dowen, R. David Hawkins, et al. "Human DNA Methylomes at Base Resolution Show Widespread Epigenomic Differences." *Nature* 462 (2009): 315-322.

Liu, Youngsheng. "A New Perspective on Darwin's Pangenesis." *Biological Reviews of the Cambridge Philosophical Society* 83 (2008): 141-149.
Low, Felicia M., Peter D. Gluckman, and Mark A. Hanson. "Developmental Plasticity, Epigenetics and Human Health." *Evolutionary Biology* (January 11, 2012): 1-16.
Maienschein, Jane. "Cell Lineage, Ancestral Reminiscence, and the Biogenetic Law." *Journal of the History of Biology* 11 (1978): 129-158.
———. *Transforming Traditions in American Biology, 1880–1915*. Baltimore: Johns Hopkins University Press, 1991.
Marsh, David E. "The Origins of Diversity: Darwin's Conditions and Epigenetic Variations." *Nutrition and Health* 19 (2007): 103-132.
Maynard Smith, John. "Weismann and Modern Biology." In *Oxford Surveys in Evolutionary Biology*, edited by P. H. Harvey and L. Partridge, 1-12. Oxford: Oxford University Press, 5 (1989).
Mayr, Ernst. *The Growth of Biological Thought. Diversity, Evolution, and Inheritance*. Cambridge MA: The Belknap Press of Harvard University Press, 1982.
Moazed, Danesh. "Mechanisms for the Inheritance of Chromatin States." *Cell* 146 (2011): 510-517.
Molaro Antoine, Emily Hodges, Fang Fang, Qiang Song, W. Richard McCombie, et al. "Sperm Methylation Profiles Reveal Features of Epigenetic Inheritance and Evolution in Primates." *Cell* 146 (2011): 1029-1041.
Moore, John A. *Science as a Way of Knowing: The Foundations of Modern Biology*. Cambridge, MA: Harvard University Press, 1993.
Morgan, Thomas Hunt. "Edmund Beecher Wilson." *Biographical Memoirs Series, National Academy of Sciences* 21 (1940): 315-342.
Müller, Gerd B, and Olsson Leonard. "Epigenesis and Epigenetics." In *Keywords and Concepts in Evolutionary Developmental Biology*, edited by Brian K. Hall, 114-123. Cambridge MA: Harvard University Press, 2003.
Muller, Hermann Joseph. "Edmund B. Wilson—An Appreciation." *American Naturalist* 77 (1943): 5-37, 142-172.
Oppenheimer, Jane. "von Baer, Karl Ernst." In *Dictionary of Scientific Biography, Volume 1*, edited by C. Gillespie, 385-389. New York: Charles Scribners' Sons, 1971.
Pinto-Correia, Clara. *The Ovary of Eve: Egg and Sperm and Preformation*. Chicago: The University of Chicago Press, 1997.
Richards, Robert J. *The Meaning of Evolution: The Morphological Construction and Ideological Reconstruction of Darwin's Theory*. Chicago: The University of Chicago Press, 1992.
Roe, Shirley A. "The Development of Albrecht von Haller's Views on Embryology." *Journal of the History of Biology* 8 (1975): 167-190.
———. "Rationalism and Embryology: Caspar Friedrich Wolff's Theory of Epigenesis." *Journal of the History of Biology* 12 (1979): 1-43.
———. *Matter, Life and Generation. 18th Century Embryology and the Haller-Wolff Debate*. Cambridge: Cambridge University Press, 1981.
Russell, Edward Stuart. *Form and Function: A Contribution to the History of Animal Morphology*. London: John Murray, 1916. Reprinted, Chicago: The University of Chicago Press, Introduction by George V. Lauder, 1982.
Russo, V. E. A., Robert A. Martienssen, and Arthur D. Riggs, eds. *Epigenetic Mechanisms of Gene Regulation*. Plainview, NY: Cold Spring Harbor Laboratory Press, 1996.

Salazar-Ciudad, Isaac, and Jukka Jernvall. "Graduality and Innovation in the Evolution of Complex Phenotypes: Insights From Development." *Journal of Experimental Biology (Molecular Development and Evolution)* 6 (2005): 619-631.

Schlichting, Carl D., and Pigliucci, Massimo. *Phenotypic Evolution: A Reaction Norm Perspective*. Sunderland, MA: Sinauer Associates, 1998.

Shapiro, James A. *Evolution. A View from the 21st Century*. Upper Saddle River, NJ: FT Press Science, 2011.

Thom, René. "An Inventory of Waddingtonian Concepts." In *Theoretical Biology*, edited by Brian Goodwin and P. Saunders, 1-7. Edinburgh: Edinburgh University Press 1989.

Torrey, Harry B., and F. Felin. "Was Aristotle an Evolutionist?" *Quarterly Review of Biology* 12 (1937): 1-18.

Turner, Bryan M. "Epigenetic Responses to Environmental Change and Their Evolutionary Implications." *Philosophical Transactions of the Royal Society of London (B): Biological Sciences* 364 (2009): 3403-3418.

Van Speybroeck, Linda. "From Epigenesis to Epigenetics: The Case of C. H. Waddington." *Annals of the New York Academy of Science* 981 (2002): 61-81.

von Baer, Karl Ernst. *Über Entwickelungsgeschichte der Thiere: Beobachtung und Reflexion*. Königsberg: Gebrüder Bornträger, 1828 / Culture et Civilisation, Bruxelles, 1967.

von Bertalanffy, Ludwig. "Untersuchungen über die Gesetzlichkeit des Wachstums. I. Allgemeine Grundlagen der Theorie; mathematische und physiologische Gesetzlichkeiten des Wachstums bei Wassertieren." *Archives Entwicklungsmechanic* 131 (1934): 613-652.

———. "General System Theory: A Critical Review." *General Systems* 7 (1962): 1–20.

———. *General System Theory: Foundations, Development, Applications*. New York: George Braziller, 1968.

Waddington, Conrad Hal. "The Epigenotype." *Endeavour* 1 (1942): 18–20.

———. ed. *Towards a Theoretical Biology, Volumes 1-4*. Edinburgh: Edinburgh University Press, 1968-1972.

———. *The Evolution of an Evolutionist*. Ithaca, NY: Cornell University Press, 1975.

Weismann, August. *Über die Vererbung*. Jena: Gustav Fischer, 1883.

———. *Die Kontinuitt des Keimplasmas als Grundlage einer Theorie der Vererbung*. Jena: Gustav Fischer, 1885.

———. *The Germ Plasm: A Theory of Heredity*. Translated by W. Newton Parker and H. Ronnfeld. London: Walter Scott Ltd., 1893.

Whitehead, Alfred North, and Bertrand Russell. *Principia Mathematica, Volumes 1-3*. Cambridge, MA: Cambridge University Press, 1910, 1912, 1913.

Whitman, Charles Otis. "Evolution and Epigenesis." *Biological Lectures of the Marine Biology Laboratory Woods Hole* (1894): 205-224.

Wilson, Edward B. *The Cell in Development and Heredity, 3rd Edition*. New York: Macmillan, 1925.

Winther, Rasmus G. "August Weismann on Germ-Plasm Variation." *Journal of the History of Biology* 34 (2001): 517-555.

Wolf, U. "The Genetic Contribution to the Phenotype." *Human Genetics* 95 (1995): 127-148.

Wolff, Caspar Friedrich. *Theoria Generationis*. Halae ad Salam, 1759.

———. *Theorie von der Generation in zwo Abhandlungen*. Berlin: Friedrich Wilhelm Birnstiel, 1764.

Zhang, Xiaoyu, Junshi Yazaki, Ambika Sundaresan, Shawn Cokus, et al. "Genome-Wide High-Resolution Mapping and Functional Analysis of DNA Methylation in *Arabidopsis*." *Cell* 126 (2006): 1189-1201.

Chapter Sixteen
Epigenetics, Soft Inheritance, Mechanistic Metaphysics, and Bioethics

Adam C. Scarfe

Epigenetics is an area of biological study that has existed for decades, but it is currently considered to be revolutionizing our understanding of life.[1] Quite generally, epigenetics ("epi": "on top of," "external to," "in addition to," "beyond") points to a "layer" of biological connections that sheathes, and/or to some degree, interpenetrates, an organism's genetic material, and which can regulate the expression of genes. Under the neo-Darwinian orthodoxy that stemmed from the hardening of the modern synthesis, it was postulated that phenotypes are fully under genotypic control, that the only important source of inheritable biological information that is transmitted from generation to generation is to be found in the DNA code, and that offspring inherit a biologically "clean slate," regardless of what their parents had experienced in their lives. However, epigenetics is a branch of biology that investigates how molecular, biochemical, hormonal, physiological, behavioral, experiential, and/or environmental factors can modify patterns of gene expression that, in turn, impact on development, and can have potentially inheritable effects. As such, to some degree, epigenetics would seem to implicate a "Lamarckian element" in evolution.

Epigenetics serves, in some ways, to undo the traditional "nature versus nurture" dichotomy. It helps to provide explanations, for example, why a caterpillar can turn into a butterfly, without changes occurring to its DNA, and why monozygotic twins with identical genotypes have phenotypic differences, one genotype being expressed in different ways. The importance of epigenetic processes can also be seen in recent studies that have shown that environmental influences, such as maternal diet or exposure to carcinogens, not only can have lifelong effects, but they can also have an affect on several generations of offspring.[2]

Returning to some of the themes that were presented in the introduction to this volume, this chapter offers a philosophical discussion concerning the putative need to reframe the neo-Darwinian emphasis on "hard inheritance," as opposed to "soft inheritance." And in outlining some of the ethical ramifications that emerge from the basic insights that epigenetics research has afforded to us, here I take issue with the reduction, without remainder, of epigenetic processes (such as DNA methylation, histone modification, chromatin remodeling, and genomic imprinting) to *mechanisms*. The machine metaphor currently holds wide sway, among researchers, in the quest to fully understand the functioning of the epigenome, and has entailed a rush on the part of biotechnology and pharmaceutical corporations to exploit these "epigenetic *switches*" in developing applications for agriculture and biomedicine. It is argued that a phase of holistic reflection in research is needed in order to offset the risk of scattering untargeted effects by way of such manipulations, and that Waddingtonian "biological wisdom" needs to be cultivated in relation to the ways that we approach the genome and epigenome, which are among the most sensitive of organismic structures.

Defining Epigenetics

While the meaning of the term *epigenetics* has evolved since the 1950s, and will continue to do so as more and more about this subject is discovered, today, a vast majority of literature on the subject deems it to refer to the study of modifications in gene expression that are not deemed to alter the genetic "code" or "sequence" itself, yet which may produce heritable effects.[3] The term largely owes its origin to Conrad Hal Waddington (1905-1975), who, in *Organizers and Genes* (1940), used the word to designate "a marriage synthesis between 'Epigenesis' ([i.e.,] the classical older theory opposed to 'preformation' in embryological theory) and [modern] 'genetics'."[4] Historically-speaking, in contrast to *preformationism*, the theory that the embryo resembles an adult of its species and grows, such that all the necessary elements comprising its potential future being are present in it, *epigenesis* involved the notion that the organism undergoes successive differentiation in developmental stages toward maturity, through which features emerge that were not represented in the embryo.[5] As such, connoting the merger of modern genetics and the theory of epigenesis, for Waddington, epigenetics involved the study of the reciprocal "causal interactions between genes and their products which bring the phenotype into being,"[6] although even this definition is perhaps still too gene-centered.

Influenced by the holistic process metaphysics of Alfred North Whitehead (1861-1947)[7] and especially his notion of "concrescence,"[8] Waddington's genetic assimilation experiments[9] were premised on the notion that not only does evolution involve changes in relation to phenotypes (e.g., morphological, physiological, behavioral) and gene frequencies in populations, but that the living organism is itself constituted by a process of development. For Waddington, this reflects "the Whiteheadian point . . . that the organisms undergoing the process of evolution are themselves processes."[10] Epigenetics, for Waddington, studies the "whole complex of developmental processes"[11] that are attributable to gene-phene and gene-environment interactions in development. He employed the term *epigenotype* as a non-dualistic analogue to the terms "genotype" and "phenotype," in order to designate this complex of creative processes. For him, the epigenotype emerges out of the reciprocal interactions of genotype and phenotype in development. In this Waddingtonian light, a good stipulative definition of the term epigenetics is that it is "the study of the epigenotype"—namely, of "the properties of the pathways and processes that link the genotype and phenotype."[12]

Molecular biologists have referred to epigenetics as "the study of mitotically and/or of meiotically heritable changes in gene function that cannot be explained by changes in DNA sequence"[13] (i.e., transmitted from one cell lineage to another and/or from parent to offspring through the germ line). In so doing, they have drawn a further distinction between *mutations*, or changes in the DNA sequence, and *epimutations*, meaning heritable changes that are not "coded" in the DNA sequence[14] in order to denote a separation between genetic and epigenetic inheritance systems.

A Brief Reference to Some Epigenetic Processes

In general, the field of molecular epigenetics aims to discover or to study epigenetic processes—generally termed "mechanisms"—in the context of development, which contribute to the regulation of gene expression. Such processes include: (1) DNA methylation, (2) histone modifications, (3) chromatin remodeling, (4) genomic imprinting, and a host of others. First, *DNA methylation* is an epigenetic process that was discovered in the mid-1970s.[15] It occurs in most lifeforms.[16] Cellular differentiation and heredity work, in part, due to the addition of chemicals to, and/or their subtraction from, DNA bases. In respect to DNA methylation, whereas the addition of a methyl group to cytosine bases, forming 5-methyl-cytosine, tends to make it less probable that genes are expressed, its removal (i.e., *demethylation*) is likely to heighten gene activity.[17] As a normal part of the processes, for example, by which cellular differentiation occurs, diverse patterns of DNA methylation help to account for how the specialized cells making up the diverse tissues and organs in the body are replaced. Patterns of DNA methylation are mitotically heritable, but in some cases they can be meiotically heritable as well, in ways that may bypass the reprogramming that occurs in embryogenesis.[18] Abnormal methylation patterns are associated with diseases such as Alzheimer's disease, Rett syndrome, and cancer.[19]

Second, a large number of epigenetic processes related to *histone modification* have also been identified. These processes involve biochemical alterations to histone proteins, comprising the part of chromosomes around which a DNA strand is "wrapped" and which help to hold it in tight configuration. One such process is *histone acetylation* (and its converse, *de-acetylation*), in which acetyl groups are added to, or subtracted from, residues of the amino acid lysine at the N-terminal tails of the histone, a reaction that is catalyzed by acetyltransferase enzymes and which affects histone binding, DNA repair, and replication. While acetylation upstream of a gene tends to heighten its expression, de-acetylation is said to diminish it. Other histone modifications may have the reverse effect. Abnormal patterns of histone acetylation have been linked to diseases such as leukemia and epithelial cancers. Some other processes of histone modification are: *histone methylation, ubiquitylation, phosphorylation*, and *sumoylation*.[20]

Third, *chromatin remodeling* in animals involves the constant modification of the dynamic structure of the chromatin that comprises the nucleus of a eukaryotic cell. Chromatin remodeling events occur by way of multiprotein complexes "alter[ing] histone—DNA interactions to form, disrupt, or reposition nucleosomes."[21] Chromatin remodeling is said to regulate DNA replication, repair, and recombination. In the process, alterations in terms of gene expression occur as a result of "the exposure of different regions of DNA to transcription factors and RNA polymerases."[22] Chromatin remodeling also occurs in plants and plays an important role in their developmental processes, from seed maturation to flowering.[23]

Fourth, *genomic imprinting*, which, in humans, normally occurs in relation to less than 1 percent of genes, involves the biochemical marking of genes that suppresses their expression depending on the parent of origin.[24] Genomic imprinting indicates a kind of "epigenetic memory" in relation to parental origin. While it is not clear exactly why imprinting occurs, and why certain genes are targeted and others not, it has been suggested that it is the evolutionary result of a conflict in terms of the differing reproductive and fitness interests of males and females, evolutionarily construed.[25] Imprinting takes place during egg and sperm production, and imprinted genes are silenced, bypassing the normal epigenetic reprogramming that occurs in the newly formed embryo. While having copies of each parent's genes gives mammals the evolutionary benefit of the potential to offset a defective gene, in the relatively few cases where imprinting occurs the default copy of the gene, or allele, has been silenced, rendering a greater susceptibility to the deleterious effects of mutation. The effects of genomic imprinting have been linked to Angelmann's syndrome, Prader-Willi syndrome, Beckwith-Wiedemann syndrome, cancer, and schizophrenia.[26] The increased susceptibility to various diseases that is associated with both *in vitro* fertilization and cloning (i.e., Somatic Cell Nuclear Transfer) is thought to be the product of the fact that, in these procedures, imprinting processes do not take place in the normal way.

What is also being studied are the complex interactions between the various epigenetic processes, which are all tied together. For example, both DNA methylation and histone modification are key actors that figure prominently in chromatin remodeling and in genomic imprinting, and DNA methylation and histone modification pathways intersect.[27] Other subjects of study are how factors like experience, behavior, and environment (e.g., high glycemic diet, work stress, lack of exercise, prolonged exposure to air pollution) can effect aberrancies in such epigenetic functioning on the molecular level, leading to alterations in gene expression,[28] and whether such aberrancies, stemming from experiential, environmental, and behavioral factors, are inducers of genetic mutations and misspellings, polymorphisms, and/or congenital diseases in humans.

From an evolutionary point of view, it has been hypothesized that epigenetic systems of inheritance allow for a balance between the need to consistently generate cells of a specialized sort pertaining to the various tissues and organs, and of their diverse functions, and the plasticity in relation to gene expression that enables the organism to respond to environmental changes.[29] However, in recent literature, the distinction between the genetic and the epigenetic systems of inheritance has been somewhat blurred, leading to a renewed debate over "hard" and "soft" inheritance. I turn now to providing a synopsis of the contrast between the neo-Darwinian and the epigenetics perspectives on heredity, respectively, as well as of the present controversy over "hard" and "soft" inheritance.

Epigenetics, Neo-Darwinism, and the Extended Contention over "Hard" and "Soft" Inheritance

The central insights that the contemporary field of epigenetics has generated, which have been alluded to above, seem to call for a need to "reframe" the traditional neo-Darwinian emphasis on "hard inheritance."[30] Hard inheritance is a notion that was coined by Ernst Mayr (1904-2005). Hard inheritance stood for the neo-Darwinian view of heredity (as will be outlined in what follows), which he contrasted with the term "soft inheritance." This latter term was synonymous with the Lamarckian and/or neo-Lamarckian theory of the inheritance of acquired characters, which was discredited in the modern synthesis.

As a cornerstone of the consensus that comprised the modern synthesis, the notion of "hard inheritance" followed from an interpretation of August Weismann's (1834-1914) putative thesis that the germplasm is sequestered off from somatoplasm, and is unaffected by it, and that nothing an organism does or experiences during its lifetime gets transmitted to its offspring.[31] Weismann understood that heredity, implying "the continuity of the substance of the germ plasm," logically depended on the assumption of a separation between the somatoplasm and the germ plasm, as belonging to distinct biological compartments.[32] He wrote that "all permanent—i.e. hereditary—variations of the body proceed from primary modifications of the primary constituents of the germ. . . . Neither injuries, functional hypertrophy and atrophy, structural variations due to the effect of temperature or nutrition, nor any other influence of the environment on the body, can be communicated to the germ-cells, and so become transmissible."[33] But Weismann's position here requires the qualification that

> The germ cells are contained in the organism, and the external influences which affect them are intimately connected with the state of the organism in which they lie hid. If it be well nourished, the germ-cells will have abundant nutriment; and, conversely, if it be weak and sickly, the germ-cells will be arrested in their growth. It is even possible that the effects of these influences may be more specialized; that is to say, they may act only upon certain parts of the germ-cells. But this is indeed very different from believing that the changes of the organism which result from external stimuli can be transmitted to the germ-cells and will re-develop in the next generation at the same time as that at which they arose in the parent, and in the same part of the organism.[34]

What Weismann appears to be saying in this quotation is that while the germ cells can be altered by somatic factors and by external stimuli, the underlying *substance* of heredity—the germplasm (i.e., what we would today call the genetic "code")—cannot be, and remains self-sufficient and continuous.[35]

While Julian Huxley (1887-1975) admitted that "the distinction between soma and germplasm [was] not always so sharp as Weismann supposed," he commended him for this differentiation, since it had contributed "a great clarification" in preparing the way for the modern synthesis.[36] In *Genetics and the Origin of Species* (1937), a central work laying the ground for the modern synthesis, and exemplifying Mayr's notion of "hard inheritance," Theodosius Dobzhansky (1900-1975) remarked that "the genotype possesses tremendous self-regulatory powers, and can withstand unchanged the impact of most environmental agencies," evolution being possible only because the "conservative force" that is heredity is "counteracted by another force opposite in effect, namely, mutation."[37] In outlining the *mechanisms* of evolution, Dobzhansky identified: (1) gene changes (i.e., mutations which he defined as "any change in the genotype that is not due to recombination of Mendelian factors," or as a usually random "change in the structure of a gene,") which were followed in terms of importance by (2) "rearrangements of the genic materials within the chromosomes," and (3) "reduplications and losses of whole chromosome sets."[38] Heritable variations, for Dobzhansky, were due mainly to chemical changes in genes and mechanical changes in chromosomes. They were to be considered *random* because they occurred both unpredictably, in terms of timing and the genes that they affected, and without reference to adaptive needs—namely, whether they were beneficial or harmful in adaptation.[39]

Dobzhansky speculated about environmental causes of gene mutation, citing, for example, shortwave radiation in nature, X-rays, and sublethal high temperatures, yet he admitted that much remained unknown in relation to potential external causes of mutation. Nevertheless, Dobzhansky proclaimed that "the production of mutations by external agents has nothing to do with the theories of the so-called direct adaptation and the inheritance of acquired characteristics," since mutation-producing agents only caused "an increase of the spontaneous mutation rate and not a genetic transformation of masses of individuals" and that there was no experimental data to support the Lamarckian hypothesis.[40] Having failed the test of empirical verification, he claimed that the Lamarckian theory of the inheritance of acquired characters was obsolete, and he refrained from further discussion of it.[41] Later in the book, Dobzhansky relegated most instances of maternal cytoplasmic inheritance to the control of genes, and suggested that the few cases in which it may occur non-genically are minor and subordinate when compared with "genic inheritance."[42] Consistent with (2) and (3) above, Dobzhansky did make the case that chromosomal changes had hereditary effects by stating that "gene mutations and chromosomal changes are not necessarily as fundamentally distinct as they first appear."[43] In sum, for Dobzhansky, the major sources of the phenotypic variations on which Darwinian natural selection acted were random gene mutations and chromosomal changes, although he admitted that the mechanisms underlying these processes had not yet been fully understood and brought under human control.

It was in "Thoughts on the History of the Evolutionary Synthesis" (1980) that Ernst Mayr coined and distinguished between "hard" and "soft" inheritance, championing the former over the latter. He related that the main target of criticism on the part of the architects of the modern synthesis was the widespread belief in soft inheritance, which he claimed was assumed in Lamarckist, neo-Lamarckist, Geoffroyist, and orthogenetic theories. As Mayr defined it, the term "soft inheritance" indicated the belief that

> the genetic basis of [phenotypic] characters could be modified either by direct induction of the environment, or by use and disuse, or by an intrinsic failure of constancy [of the genetic material], and that this modified genotype was then transmitted to the next generation. Soft inheritance is usually referred to as a belief in an inheritance of acquired characters, but [it] includes a broader range of phenomena. It is also sometimes called Lamarckism or neo-Lamarckism, even though Lamarck's own theory was only one subdivision in this group of theories.[44]

Later in the article, he adds,

> The most far-reaching consequence of the failure to distinguish between genotype and phenotype was that it fostered a belief in soft inheritance. I use this term to designate belief in a gradual change of the genetic material itself, either by use or disuse, or by some internal progressive tendencies, or through the direct effect of the environment [and generally holding that] as the genetic material itself is affected, these changes are (by definition) inherited. . . . The concept of soft inheritance includes the inheritance of acquired characters, but it is more comprehensive because it also includes various theories of a direct influ-

ence on the germ plasm. . . . All theories of soft inheritance deny the complete constancy of the genetic material.[45]

Elsewhere, Mayr describes soft inheritance as involving the belief that "phenotypic adaptation could be converted into a genotypically based adaptation."[46] According to Mayr, the belief in soft inheritance had been widespread since Darwin, both in the public's perception of evolution, as well as in terms of the scientific community, and it was "a major stumbling block in the path of a neo-Darwinian interpretation" of life.[47] But in the context of the modern synthesis, it had been successfully eclipsed by his generation of biologists. While even Darwin had been open to a variety of factors contributing to variation, including "external conditions" and use and disuse,[48] and while in the 1930s, Thomas Hunt Morgan (1866-1945) had pointed to a two-way, reciprocal interaction between the chromatin of the cells and the rest of the protoplasm in development,[49] Mayr went so far as to define the brand of neo-Darwinism reigning in the modern synthesis as the view, "established by Weismann," that stood opposed to the original Darwinism "primarily by excluding all possibility of an inheritance of acquired characteristics."[50] Later, Mayr claimed that "the pathway of the DNA of the genome to the proteins of the cytoplasm (transcription and translation) is strictly a one way track. The proteins of the body cannot induce any changes in the DNA."[51] This stance was consistent with Crick's "central dogma of molecular biology" that information is transmitted to proteins through RNA, but that there is no reciprocal transfer by proteins on other proteins, on RNA, or on DNA.[52]

The neo-Darwinian paradigm that became entrenched through the hardening of the modern synthesis continues to dominate mainstream biology in the present day. In *Developmental Plasticity and Evolution* (2003), Mary Jane West-Eberhard, suggests that "if pressed to name the mechanism behind the origin of novel traits . . . most biologists would argue that [genetic] mutation is ultimately the *only* legitimate source of evolutionary novelty, simply because evolution is defined as involving genetic change."[53] However, recent findings in epigenetic research as well as theoretical work seem to challenge the assumption that genetic mutation is the only efficient cause of phenotypic novelty, and/or the "prime motor"[54] of evolutionary change alongside natural selection. Several scholars are looking to epigenetics, and other new research fields, to supplement the synthesis of Darwinian natural selection and Mendelian and population genetics and/or to erect a "third pillar" of biology.

As alluded to above, in the vast majority of scholarly and scientific literature on the subject, the term epigenetics is defined as the study of modifications in gene expression *that are not deemed to alter the genetic "code" or "sequence" itself*, and which may produce heritable effects. Defined in this light, it can be said that epigenetics neither refutes hard inheritance nor legitimates soft inheritance, although it does suggest that the "gene-centered" picture regarding inheritance that was painted by the architects is not representative of the whole story. Additionally, by pointing to a degree of reciprocal causal interaction between genotype and phenotype and between genotype and environment, this interpretation of the meaning of the term epigenetics serves to undo the rigidity of the dualism that was postulated by Mayr, and others, between hard and soft inheritance. However, more recently, some scholars and scientists, such as the avant-garde Eva Jablonka and Marion Lamb, in their book *Evolution in Four Dimensions* (2005); in their article "Soft Inheritance: Challenging the Modern Synthesis"; and in their anthology chapter, "Transgenerational Epigenetic Inheritance" (2010)—have sought to go further. And many others have joined into a movement to go further, as can be seen, for example, in Snait B. Gissis and Eva Jablonka (eds.), *Transformations of Lamarckism: From Subtle Fluids to Molecular Biology* (2011).

In addition to claiming that research in epigenetics points to the existence of a distinct epigenetic inheritance system interacting with three other inheritance systems (genetic, behavioral, and symbolic), Jablonka and Lamb assert that the ongoing uncovering of the *mechanisms* of epigenetic inheritance is legitimating soft inheritance. They clarify that "for most biologists, . . . when inheritance is what Ernst Mayr called 'soft inheritance'—that is, when the hereditary material (or process) is not constant from generation to generation but can be modified by the effects of the environment or the organism's activities—it qualifies as Lamarckian inheritance."[55] Appearing to go by this criterion, with reference to recent epigenetics research, Jablonka and Lamb suggest that soft inheritance—namely, "the inheritance of developmentally induced and regulated variations, exists, and is likely to be important. . . . It involves both non-DNA variations and developmentally induced variations in DNA sequences."[56] Therefore, according to them, soft inheritance needs to be included in the emerging extended synthesis that they anticipate and work toward.[57]

Subsequent to emphasizing the importance of a multiple-dimensional understanding of variation and of a multiplicity of interacting inheritance systems, Jablonka and Lamb are also confident that research will uncover exactly how "epigenetic control systems are involved in the generation of systematic mutations."[58] They hypothesize that genetic mutation is not entirely a random affair, emphasizing that some changes may be induced and adaptive, potentially serving to calibrate the organism in response to environmental stimuli, and that some mutations and restructurings may be directed or guided by cellular systems as a normal part of development. Jablonka and Lamb also attempt to overcome the gene-centrism of the neo-Darwinist orthodoxy by pointing to epigenetic processes. They assert that, in light of recent findings in epigenetics, "the genome should be seen not just as a repository of genes that are inputs into development, but as a developmental system."[59] Finally, Jablonka and Lamb argue toward a more holistic, inclusive biology, one in which reciprocal interactions and internal relations are taken more seriously, and the abstractions created as a result of the sharp conceptual distinctions made, for example, between somatoplasm and germplasm, and genotype and phenotype, are laid bare. That said, like most of their peers, they employ mechanistic language in unpacking epigenetic processes, as can be seen in their several references to epigenetic "switches," "machines," and "control mechanisms."

Numerous responses to Jablonka and Lamb have been waged on the part of defenders of the neo-Darwinian orthodoxy. In "Weismann Rules! Epigenetics and the Lamarckian Temptation" (2007), David Haig, who identifies himself as a "rather orthodox neo-Darwinist,"[60] concedes that the discovery of epigenetic processes does allow for a Lamarckian element in evolutionary theory, but he claims that there is nothing in Jablonka and Lamb's championing of epigenetics that challenges the main tenets of the modern synthesis.[61] One of Haig's chief arguments is that epigenetic switches, the reality of which he very readily admits, have evolved by natural selection, much like any other trait, and that epigenetic variation is most likely "subservient to the evolving DNA sequence."[62] In other words, for him, "the ability to adaptively switch epigenetic state is a property of the DNA sequence . . . and that any increase of adaptedness has come about by a process of natural selection."[63] Haig maintains the primacy of DNA by suggesting that the "*machinery* of epigenetic modification ([e.g.,] DNA methyltransferases, histone deacetylases, etc...) is 'encoded' by the DNA sequence and that whether or not a particular sequence is subject to a particular epigenetic modification will be partly dependent on properties of the sequence itself."[64] For him, Jablonka and Lamb need to qualify and to provide more evidence for their admission of "non-random" mutations, for that would imply that mutations are issued by their causes in a specific, intentional, and goal directed way—namely, "producing a predetermined adaptive end."[65] While adaptive directive systems do exist, for him, they are explainable as by-products of evolution. Haig ends his article by suggesting that one reason for the contention is that the terms such as "soft inheritance" and the "inheritance of acquired characteristics" are defined subjectively and generally so as to be potentially meaningless. The confusion, he thinks, is heightened by the notion that gene mutations issuing from epigenetic factors could be said to constitute acquired characters.[66]

Similar arguments against Jablonka and Lamb's claims that research in epigenetics legitimates soft inheritance, and that it does so in a way that challenges the main tenets of the extended evolutionary synthesis, are presented by Thomas Dickens and Qazi Rahman, in a May 2012 article entitled, "The Extended Evolutionary Synthesis and the Role of Soft Inheritance in Evolution." They defend the notion that epigenetic processes are dependent on mechanisms (e.g., DNA methylation, histone modification) that are designed by natural selection and are genetically inherited—namely, they are part of the phenotype that is under genetic control and subject to natural selection. Furthermore, they claim that while Jablonka and Lamb present much proximate detail regarding epigenetic processes and mechanisms, they have not presented adequate evidence outlining how such processes enable organisms a degree of phenotypic plasticity in the context of development and that, therefore, they have not provided an adequate treatment of the ultimate causes of epigenetic processes which would warrant their claims. As such, epigenetic inheritance, for Dickens and Rahman, does not eclipse the main "explanatory and conceptual resources of the modern synthesis, which are sufficient."[67] They add that a serious investigation of the ultimate causes of epigenetic processes will require the employment of the conceptual tools of the modern synthesis which remain in the hands of more "instrumentally-minded biologists."[68]

Even more recently, in a June 2012 article, "Rethinking Heredity, Again," Russell Bonduriansky states, "We are witnessing a re-emergence of the debate on the role of soft inheritance in heredity and evolution."[69] He claims that the meaning of the term "soft inheritance" has shifted, and that the version that is defended by Jablonka and Lamb is not the same as that which was purged from the modern synthesis by its architects. Specifically, he makes the case that the version of soft inheritance that was rejected as a "chemical impossibility"[70] in the modern synthesis was that of "genetic encoding," meaning that the environment or somatoplasm can "modify the germ line DNA sequence so as to produce an inheritance of acquired traits."[71] For Bonduriansky, this contrasts with a second, more contemporary, version of soft inheritance, involving non-genetic inheritance systems such as epigenetic inheritance, which are parallel to the genetic system, and can be mediated by it, but transmit their elements alongside alleles. Bonduriansky seems to implicate Jablonka and Lamb in emphasizing the latter rather than the former. However, I would suggest that Jablonka and Lamb, and others, are emphasizing both forms of soft inheritance: the second (non-genetic processes altering patterns of gene expression) as an indication that more about the parameters of the first will be brought to light by research, as can be seen by several of the quotations above. This is the case, given several of the quotations provided above[72] and the emphasis, for example, on the possibilities of "genetic assimilation," "genetic accommodation," and/or "mutational assimilation" that are discussed in the closing chapters of *Transformations of Lamarckism*.[73] As such, we might here ask: are Jablonka and Lamb, and other epigeneticists, suggesting that the standard definition of epigenetic processes as involving *modifications to gene expression that do NOT alter the genetic "code" or "sequence"* is itself another "dogma" in biology that will be eclipsed? I believe the answer is "Yes!"

The Repercussions of the Dominance of the Mechanistic Metaphysical Framework in Neo-Darwinian Biology and Epigenetics Research

The extended contention between hard and soft inheritance, and especially the key question of whether epigenetics legitimates the form of soft inheritance that was "purged" by the architects of the modern synthesis, is likely to be one of the exciting main themes in biology over the course of the twenty-first century. Terms such as "epigenetics" and "soft inheritance" are evolving, and will continue to do so, as more and more is found out about epigenetic processes and inheritance. With reference to the findings of epigenetics research, Jablonka and Lamb, and others, have gone a long way to present a more inclusive, holistic, and multi-dimensional picture of evolution that can inform an extended modern synthesis. This is a welcome development. Even if it was ultimately found out that somatoplasm or environment do not modify the genetic sequence, but can only affect gene expression, or that epigenetic processes are under genetic control, no longer is the evolutionary spotlight dogmatically centered exclusively on genes as "unmoved movers,"[74] blind to their environment and only externally related to somatoplasm, yet unilaterally responsible for inheritance.

This development would seem to require that we see phenotypes as something more than just "survival machines—robot vehicles blindly programmed to preserve the selfish molecules known as genes,"[75] as was postulated by Richard Dawkins in *The Selfish Gene* (1976). And in light of the recent findings of epigenetics research, genes are now to be seen as efficient causes that are also responsive subjects in developmental processes, interacting reciprocally with other genes as well as with their somatic counterparts. To be sure, not only does less than 2 percent of the genome "code" for proteins, but "'decisions' as to which genes will engage in protein synthesis at any point in time is a function of the cell, not the genes themselves."[76] At the same time, the decentering of the genome will undoubtedly have consequences for ethical deliberation, as will be discussed later in this chapter.

Recent developments in epigenetics would seem to signal a thorough illumination of, and liberation from, the abstractions created by the mechanistic understanding of life that has been emphasized by the neo-Darwinian orthodoxy. Unfortunately, what remains common to both sides in the debate over the legitimacy of soft inheritance is an entrenched employment of mechanistic metaphysics of the reductionist variety. On the one side, we find the claim that the self-sufficient, efficient causal substances it champions have sovereignty, power, and primacy over the other, subservient slave machinery. As one commentator describes, the gene-centered view is a "hegemonic 'genetic fundamentalism' paradigm"[77] that is "dominated by an informational myth and a mechanistic Cartesian body / mind and form / substance dualism [as can be seen in the dichotomy between genetic code and information and phenotype], . . . [which] considers the genome as an ensemble of discrete units of information governing human body and behavior, and remains hegemonic in life sciences and in the public imagination."[78] As for the other side, there is an aim to undo the previous side's claims as to the independent primacy of the master material, and arguments are presented in support of the notion that the alternative efficient causes it champions are just as important in evolution and development. Genes are here considered "more as followers than leaders in evolutionary change."[79] In this way, in their dispute over the locus of regulatory control, both sides lose focus on the fact that living organisms are not machines that are comprised of isolated and externally related parts, each making separate contributions to developmental outcomes and which can be judged in terms of the degree of causal power they have in evolution and development. Rather, organisms are complex self-organizing systems that function as seamless irreducible wholes. Genome and epigenome, germplasm and somatoplasm, genotype and phenotype, phenotype and environment, genotype and environment, are *internally related*.[80] Thus, if a novel evolutionary synthesis is to be reached, both sides will have to build a consensus, and this consensus will perhaps be able to occur only if both sides are willing to look to the organic whole beyond the abstractions of their machine metaphors, at least in a phase of their deliberations.

The widespread employment of mechanistic metaphysics in epigenetics research has, thus far, not been subjected to much in the way of critical questioning, and yet there are very important implications to consider. Any survey of studies in epigenetics will reveal that the ultimate concern of the researchers is to identify the *epigenetic mechanisms* governing gene expression (e.g., DNA methylation, histone modification, chromatin remodeling, genomic imprinting, etc...) and to understand their functioning, presumably so as to bring them under human control. The epigenome is presented by the vast majority of researchers, be they neo-Darwinian or neo-Lamarckian, as essentially "the switchboard of the genome,"[81] and epigenetic processes are understood as causal *mechanisms*—namely, as "switches" turning genes "on" and "off," "dimmer switches," "volume dials,"[82] "control mechanisms," and "bits of machinery," entailing that they are at-the-ready for manipulation via biotechnological and pharmaceutical applications.

In *Evolution in Four Dimensions*, Jablonka and Lamb do seem genuinely concerned about the ecological impact of biotechnological manipulations, the morality of non-human cloning and our treatment of non-human animals, as well as other ethical issues. However, in earlier works, certain passages, where they are obviously trying to underscore the great importance of epigenetics research for applications in a variety of different areas, including agriculture and biomedicine, give a bit of a different impression. The emphasis here is on employing our newfound knowledge of epigenetic "mechanisms" to manipulate organisms for our own benefit. For example, Jablonka and Lamb write:

> It is quite clear that for normal development the somatic cell or nucleus that is used for cloning needs to be epigenetically reprogrammed. The frequency of successful animal clones is still low, and many of the animals that manage to reach adulthood have abnormalities that can be attributed to aberrant reprogramming of the original somatic nucleus. Knowing which cell types to choose for cloning, how to treat them before their fusion with the enucleated egg, whether or not to do several serial transfers and so on *is going to be crucial for the success of this important technique*. Obviously, a good understanding of epigenetics is required.[83]

> One field in which it has been impossible to ignore epigenetic inheritance is genetic engineering. Transgenerational epigenetic effects have often been found (mainly in plants, but also in animals) when inserted transgenes became stably inactive after a few generations of activity (Matzke and Matzke, 1995). A good understanding of [Epigenetic Inheritance Systems] and how the cell responds to various insults will be necessary for effective genetic engineering. Since carry-over effects in plants and animals can last for several generations, understanding the basis of these might lead to *epigenetic engineering through manipulation of the environmental conditions.*[84]
>
> In agriculture, the importance of epigenetic inheritance is already widely acknowledged, because it has caused many problems in genetic engineering aimed at crop improvement. Commonly, newly inserted foreign genes are heritably silenced through extensive DNA methylation, so ways of circumventing this problem have had to be developed. *On the positive side, since some epigenetic variations can be induced by environmental changes, it may be possible to develop agricultural practices that exploit these inducing effects and thus develop improved, "epigenetically engineered" crops.*[85]

What is problematic here, at least from a multi-perspectival bio- and environmental ethics standpoint, is that there is a lack of concern for the long-term evolutionary consequences that such epigenetic manipulations have for the organisms involved. There is merely a focus on the instrumental value of the "living machines" to be manipulated, as opposed to on the intrinsic and aesthetic value of such creatures. Furthermore, here, Jablonka and Lamb do not articulate a concern to cultivate "biological wisdom" in regards to the potential of scattering unintended epigenetic and genetic effects in the course of such manipulations, which would be revealed by a phase of holistic reflection in their discussions (see the introduction to this volume). Rather, they indicate that the problems that have resulted from previous manipulations (e.g., somatic cell nuclear transfer and genetic manipulation) can be fixed through additional manipulations of the epigenetic "switchboard."

To be fair, although Waddington was inspired by Whitehead's holistic process-relational metaphysics, he himself employed the language of the mechanistic framework in some of his research. And he once wrote, "We can scarcely claim to have anything like a satisfying scientific understanding of any natural process until we can feel that we are in a position to control it, or at least to see how it might be controlled, even if we are not actually able to carry out the necessary measures."[86] That said, he qualified that there is a "considerable difference" between the "effort to control" and to "understand the world," and that today, we tend "somewhat to neglect the natural-philosophical aspect of science."[87] Furthermore, in his description of the ideal path carried out by scientific research, which is quite reminiscent of Whitehead's rhythm of learning and/or research (discussed in the introduction to this volume), he included phases of: (1) curiosity and imagination, (2) ratiocination, (3) manipulative experimentation, and (4) humble generalization, whereby the scientist engages in holistic reflection on his or her attempt to "force [nature] into his categories of thought" or "trick her into doing what he [or she] wants."[88] Nevertheless, when considering epigenetic and genetic manipulation for agricultural, biotechnological, and biomedical purposes, we must remember that we are dealing with some of the most sensitive of organismic structures belonging to living creatures. In the next sections, I discuss some of the ramifications of epigenetics for biomedical ethics.

Epigenetics and Bioethics

The decentering of the genome by epigenetics research will undoubtedly have an impact on ethical debates, for example, over the genetic manipulation of organisms. It is safe to say that bio- and environmental ethicists in North America have not had much luck, thus far, in arresting biotechnology corporations from genetically manipulating organisms for agricultural purposes on any widespread scale. Traditionally, the genetic code has been associated with an organism's fundamental biological "essence" or "identity." But if, as a result of epigenetics research, the genome is seen to be less associable with an organism's "core biological identity" or "essence," then it may be deemed "less biologically valuable," and hence, some people may be more inclined to simply accept genetic manipulation wholeheartedly. Common gene-centered conceptions of what a "species" is, may also have to be modified in light of the findings of epigenetics research. Additionally, the minimization of the genome may impact on ethical deliberation, as organisms may be deemed lacking in terms of "having an essential selfhood or nature" and hence, they may be viewed as void of intrinsic value. However, it may be replied that it still remains the case that the genome is a vital organismic structure, for example, in respect to cell differentiation and maintenance.[89] Furthermore, just because epigenetics reveals that the genome does not as adequately satisfy the metaphysical requirements of Cartesian substance-hood (i.e., a sovereign, detached, self-sufficient, self-encapsulated entity, standing apart from any reference to other entities), and instead ought to be considered more as "an embedded and plastic entity"[90]—a process-relational event of the Whiteheadian sort—does not make it somehow devoid of value. As Whitehead pointed out, it is the scourge of the modern age to place value only on that which appears to be a Cartesian substance.[91] As such, it is hoped that a more biologically wise position would issue from the notion that the genome is not safely sequestered from the rest of its protoplasmic counterparts by the Weismannian barrier, but is permeable and more sensitive to external factors than previously thought.

In the ground-breaking Överkalix study that provided evidence of a transgenerational epigenetic response in humans, Marcus Pembrey, a pioneer in the field of epigenetics who, working with epidemiological data from the town, discovered that "well fed women whose mothers had starved during mid-pregnancy were producing smaller babies than women whose mothers had not starved at that point,"[92] derives a key ethical insight from his findings. According to Pembrey, in that lifestyle, behavior, and environment can have transgenerational effects—namely, they "can change the biological inheritance of [our] children and grandchildren,"[93] especially in relation to health and disease, and that "you can't, in life, in ordinary development and living, separate out the gene from the environmental effect, [since in reality] . . . [t]hey're so interwined," it follows that we are all "guardians of [our] genome."[94] Epigenetics entails that we are agents who are responsible for securing our own health and well-being over the long term, as well as that of our children, grandchildren, great grandchildren, and of our species.

In this light, epigenetics has profound implications for biomedical and environmental ethics. Factors like maternal environment and nutrition; malnutrition; prolonged drug use; alcoholism; smoking;[95] stress; anxiety; trauma; obesity; famine; dehydration; sociality; lack of exercise; consumption of genetically modified foods; exposure to cold, heat, (second hand) smoke; pollution; pesticides; herbicides (e.g., atrazine); fungicides (e.g., vinclozolin); asbestos; pharmaceuticals; chemicals (e.g., bisphenol-A, radon gas); environmental toxins; ultraviolet radiation; depleted uranium; and nuclear radiation, etc... all have the potential to produce heritable epigenetic effects, and thus our immediate actions take on great moral significance. We have a direct Jonasian imperative of responsibility[96] to preserve the soundness of the human genome and epigenome, to secure the health, well-being, and the wealth of opportunity not only of those presently existing, but also of future generations of people.

Epigeneticists point out that we may be more susceptible to specific epigenetic "triggers" at various sensitive periods in our lives (e.g., in infancy, adolescence). And much more work still needs to be done to pinpoint exactly how factors like these "trigger" epigenetic effects at the molecular level, how they may alter gene expression, genetic material in sperm and eggs, and/or cellular material connected with the genome. Nevertheless, epigenetics points to our duty to future generations to engage in practices, and to adopt lifestyles, that take epigenetic inheritance into account—for example, acting to ensure that air pollution is curtailed and that water quality standards are upheld in our society. Passing down a legacy of environmental toxicity to future generations may present an epigenetics-related obstacle that threatens "the health of future generations in a self-perpetuating cycle of poor health and reduced quality of life."[97]

Epigenetics should force us to see medicine more holistically and more preventatively. Disease is no longer to be viewed as something whose causes and effects are merely localized within the body. Rather, promoting health and curing disease will also involve looking to our environment, our experiences, our behaviors, our habits, and our lifestyles in order to find the ultimate causes of disease, and taking concrete action to change them if necessary (e.g., doing what we can to alleviate stress, getting regular exercise, supporting action to regulate industrial pollution, etc...). Emphasis ought to be placed on educational and social initiatives for securing long-term health and well-being, rather than on the mere prescription of drugs to alleviate whatever ails us.

Physicians will have to do more to cultivate an awareness of the potential epigenetic effects of the pharmaceuticals they prescribe, dosage, as well as of the interactions of such drugs. To be sure, in "Epigenetic Side-Effects of Common Pharmaceuticals: A Potential New Field in Medicine and Pharmacology" (2008) and in "Epigenetics, DNA Methylation, and Chromatin Modifying Drugs" (2009), Moshe Szyf and Antonei Csoka warn of the fact that many commonly prescribed drugs may promote epigenetic effects such as DNA demethylation, which "may be involved in the etiology of heart disease, cancer, neurological and cognitive disorders, obesity, diabetes, infertility, and sexual dysfunction."[98] They advise that the level of a drug's ability to effect DNA demethylation is something that should be measured in safety tests, so that biomedical professionals and patients can become aware of the potential for these drugs to amplify the expression of disease-promoting genes or to produce aberrant methylation patterns.[99] Szyf and Csoka also suggest that a new field of "pharmaepigenomics" be inaugurated, which would focus on the "epigenomic screening"[100] of pharmaceuticals. In this endeavor, they recommend that a "systems biology approach employing microarray analyses of gene expression and methylation patterns can lead to a better understanding of long-term side-effects of drugs."[101]

Epigenetic Pharmaceuticals and the Question of Untargeted Effects

The advances made in molecular epigenetics research have opened up opportunities to develop therapies and pharmaceuticals that, for example, reverse abnormal, and potentially inheritable, disease-promoting DNA methylation and histone acetylation patterns. Currently, pharmaceutical companies are racing to develop the next generation of epigenetic drugs for the treatment of cancer, leukemia, Alzheimer's disease, AIDS, muscular dystrophy, atherosclerosis, and numerous other diseases. It is currently estimated that twelve to fifteen epigenetic drugs are in clinical trials, that several more are in pre-clinical development, and that the market value of sales of epigenetic drugs will grow to $18.2 billion by 2015 from $519 million in 2009.[102] Many such diseases are held to involve and accentuate aberrancies in relation to "signaling pathways"[103] in cells that lead to abnormal gene expression. In treating cancer, epigenetic drugs may target and re-amplify underactive tumor suppressor gene pathways and/or de-amplify overactive oncogene pathways. The first FDA-approved epigenetic drugs allow physicians some control over the processes of DNA methylation and histone acetylation in their patient's bodies and cells, and/or effect chromatin modification, so that aberrant patterns of gene expression can be normalized, thereby "reprogramming" cancer cells.

To get a sense of how epigenetic pharmaceuticals work, there is a great deal of literature to suggest that the employment of histone deacetylases (HDACs) can suppress the expression of genes involved in cancer promotion, and that chromatin modifying histone deacetylase inhibitors (HDACis) can assist in treatments for mental pathologies.[104] Successful tests and clinical trials have led to the approval of several pharmaceuticals by the FDA, such as vorinostat[105] and romidepsin, for the treatment of cutaneous T-cell lymphoma. Vorinostat (suberoylanilide hydroxamic acid) was the first approved HDACi in 2006, and is sold by Merck as "Zolinza," while romidepsin is Celgene's "Istodax."[106] Moshe Szyf (2009) describes the functioning of HDACs and HDACis in the treatment of cancer somewhat as a means to rebalance histone acetylation states between hyper- and hypo-acetylation. Szyf writes, the "blockage of HDACs [by HDACis] tilts the balance of acetylation-deacetylation reactions towards acetylation . . . [resulting] in hyper-acetylation of histone tails and induction of genes that suppress the cancer phenotype such as tumor suppressor genes and metastasis- and invasion-inhibitory genes."[107] Szyf slso reports that recent studies suggest that hypo-methylation is systemic in certain cancers, and it can cause genomic instability. Furthermore, demethylation can activate metastatic genes and plays a role in metastasis, such that drugs and other therapies inhibiting demethylation may be a key to beating many types of cancer.[108] Two of the first four FDA-approved epigenetic drugs: azacitidine (Celgene's "Vidaza") and decitabine (Eisai's "Dacogen"), which are used for the treatment of myelodysplastic syndrome (MDS), are hypo-methylating agents.[109]

The intention behind the development of drugs to manipulate epigenetic processes is to affect the way that genes are expressed, and, crudely, to do so "at the flick of a switch."[110] In the pharmaceutical industry and in biomedical fields, it is generally held that they change gene expression, but do not permanently modify genes or cause heritable change. However, some of the perspectives that have been articulated and examined both in this chapter (e.g., in relation to whether epigenetic processes imply soft inheritance) and volume force us to ask whether we can, in fact, draw such an unbroken, impermeable line between the effects of epigenetic therapy and alterations of DNA. Epigenetic drugs target specific signaling pathways and genes in order to modify patterns of gene expression in order to treat disease. Furthermore, epigenetic processes are complex and interact—for example, histone modification enzymes interact with DNA methylating enzymes. Given these facts, we must ask: (1) How confident are we in the adequacy of the mechanistic metaphysic that has been employed in the quest to understand the epigenome?; (2) To what extent is there a potential for epigenetic therapies to scatter harmful untargeted effects through the epigenome and genome (e.g., unintentionally altering gene expression in healthy cells)?; and (3) To what extent can they do so in ways that are inheritable?

While it is probable that a majority of biomedical professionals are satisfied with the adequacy of the experimentation involved, the clinical trials, the safety tests, and the federal regulatory oversight, several reporters, biologists, and scholars have raised questions. A first concern is that the "causal mechanism" which explains why the demethylation of DNA and/or the histone deacetylation induced by the first round of epigenetic drugs produces the response has not yet been "identified."[111] In this way, to some extent, doctors may be "operating blindly" when prescribing them—namely, without a comprehensive understanding of the real reasons why the epigenetic pharmaceuticals produce the effects they do.

Second, reflecting on the potential impacts of the first generation of epigenetic drugs, one science and technology reporter writes,

> There is a perilous side to leveraging epigenetic mechanisms in pursuit of therapeutic payoff. The wrong kind of epigenetic interventions can silence a gene or turn it on when it shouldn't be. Either way, the biological responses can be indistinguishable from those due to deleterious genetic mutations.
>
> This is what makes Vidaza, vorinostat . . . [and] other clinically approved epigenetics drugs ([e.g.,] valproic acid, a long-prescribed anticonvulsant that is also an HDAC inhibitor, and decitabine, a demethylating agent . . .) so harrowing to drug developers. They all work in a genomically global fashion, not merely on particular genetic switches that might be implicated in a disorder. . . .
>
> This worries Melanie Ehrlich, a molecular biologist at Tulane University, in New Orleans, and founder of the DNA Methylation Society. Analyses of methylation patterns have revealed that cancerous cells and tissues often are associated with genome-wide reductions of DNA methylation. "Moreover, tumor progression is frequently marked by these decreases in DNA methylation," she says. "So a sledge-hammer attempt to decrease DNA methylation in cancer patients may lead to more metastases later on." A demethylating agent that saves you today could end up killing you later, she says.
>
> But Ehrlich . . . and others . . . predict [that] pharmaceuticals will [eventually] develop more sophisticated agents that target genomic locations, perhaps using recognition-components such as anti-bodies or snippets of RNA.[112]

Additionally, there are concerns about the potential side effects generated by the currently approved epigenetic drugs,[113] as well as about their interaction with other courses of treatment like chemotherapy.

Third, there has been some concern articulated in relation to targeting the germline through the use of epigenetic and genetic therapies in order to treat diseases related to genomic imprinting, such as cancer, Prader-Willi Syndrome, Beckwith-Wiedemann, Angelman Syndrome, uniparental disomies, autism, schizophrenia, as well as growth-related defects, metabolic abnormalities, and hyperinsulism. It has been speculated that pharmaceuticals could be used, for example, to alter methylation patterns so as to "restor[e] the original epigenetic state of . . . modified genes."[114] However, two evolutionary biologists suggest:

> Diseases caused by mutations or epimutations at imprinted loci make intriguing candidates for gene therapy. In particular, clinical useable [epigenetic] tools for activating or inactivating alleles could provide treatment for many . . . disorders . . . . However, given the complex patterns of regulation and expression at many imprinted loci, this approach may prove technically challenging, [and] is not without potential dangers. *Unintended consequences such as a predisposition for tumor formation, will be a danger of any therapy that attempts to quantitatively modify the expression level of imprinted genes. Furthermore, while an intervention of this sort might be beneficial for the patient, the possibility exists that induced epigenetic changes could be passed on to offspring.*[115]

Given these assessments of the potential for significant health risks associated with pharmaceuticals targeting the epigenome, including the possibility of transmitting harmful effects onto offpring, it would be biologically wise for pharmaceutical corporations and biomedical professionals to proceed with extreme caution; to reflect on the limits of clinical trials conducted as well as on the abstractions created as a result of the mechanistic metaphysic that may be assumed therein, rather than rushing to exploit the findings of epigenetics research in a manner driven by the prospect of reaping multi-billion dollar profits.

Epigenetic Pharmaceuticals and Canada's *Assisted Human Reproduction Act* (2004)

In the United States, genetic therapies and epigenetic drugs are regulated by the *Food and Drug Administration* (FDA), for the most part under the authority of the *Federal Food, Drug, and Cosmetic Act* (FFDCA), in conjunction with *The Center for Biologics Evaluation and Research* (CBER). The FDA website reports that it has "not yet approved any human gene therapy product for sale,"[116] although it states that it has received many requests from medical researchers and recognizes the potential medical importance of such research. Given that it has approved several epigenetic drugs, this is an indication of the maintenance of a strict conceptual separation between epigenetic and genetic therapies, which, as has been discussed in this chapter, may be in the process of being blurred by some recent theoretical and scientific research.

In Canada, gene therapies are regulated by Health Canada's *Biologics and Genetic Therapies Directorate*, under the *Food and Drugs Act*, in conjunction with the *Canadian Institutes of Health Research* (CIHR). On its website, the CIHR defines epigenetics as "the study of changes in the regulation of gene activity and expression that are not dependent on alterations in gene sequence,"[117] which, as has been suggested in this chapter, is a definition that is currently being placed in question. In addition, gene therapy, stem cell research, cloning (i.e., Somatic Cell Nuclear Transfer), and reproductive technologies and procedures, such as *in vitro* fertilization and embryo twinning, have been subject to regulation, since 2004, under the *Assisted Human Reproduction Act* (Bill C-6, AHRA). Under this Act, it is stated that breaches of the legislation are punishable by up to a $500,000 fine and ten years in prison. However, in 2010, a Supreme Court ruling struck down several of the provisions of the Act, prompting the Conservative Federal government to begin "winding down" the regulatory board that was set up to oversee governance of the controlled technologies and procedures (e.g., *in vitro* fertilization and stem cell research), thereby weakening the prospect of enforcing many of the bill's provisions even further.

In any event, for our present purposes, one of the fundamental purposes of the AHRA is to uphold the principle that "human individuality and diversity, and the integrity of the human genome must be protected and preserved."[118] The key section is 5f, which states, "No person shall knowingly alter the genome of a cell of a human being or *in vitro* embryo *such that the alteration is capable of being transmitted to descendants.*"[119] This clause effectively bans *positive gene therapy* (i.e., the manipulation of the genome in ways ranging from attempts to arrest inheritable genetic diseases like cancer, to outright selective genetic enhancement) in Canada. Here, the word "gene" is defined in the Act as including "a nucleotide sequence" whereas "genome" means "the totality of the deoxyribonucleic acid sequence of a particular cell."[120]

Biotechnologies and pharmaceuticals that target the epigenome so as to alter gene expression for therapeutic purposes, such as for arresting cancer, potentially run up against the limits of these provisions, given, for example, some of the recent theoretical and scientific research that has been discussed in this chapter, as well as the closeness of association of histone proteins and DNA in chromosomes. As such, we must ask: is the line so clear as to suggest that therapeutic pharmaceuticals and biotechnologies that target the epigenome do *not* alter the germ line of a human being in such a way that is capable of being transmitted to descendants? Do epigenetic pharmaceuticals qualify for being scrutinized under the Act? The Act's wording of "*knowingly alter* the genome," i.e., a nucleotide sequence, may also have profound legal importance. Does *knowingly* mean "in known ways," or "known to alter the sequence of the genome"?

From these considerations, it is evident that epigenetics "raises [extremely] difficult questions about the obligations of society to preserve the soundness of the human genome and epigenome for the benefit of future generations,"[121] especially given the fact that it is currently an evolving discipline. The arrival of epigenome-targeting biotechnological applications and pharmaceuticals adds but one more layer of complexity to social, ethical, and legal issues in biomedicine, as society flounders to keep up with, and on top of, rapid scientific and technological progress.

Notes

1. See Nessa Carey, *The Epigenetics Revolution: How Modern Biology Is Rewriting Our Understanding of Genetics, Disease, and Inheritance* (London, UK: Icon Books, 2011), 6.

2. Examples of data to support these claims are now well worn. For example, as reported by Carey, *The Epigenetics Revolution*, 3-4, from research on the long-term effects of the Dutch Hunger Winter (1944-1945) on pregnant women and the sons and daughters they gave birth to, it was found that

> If a mother was well-fed around the time of conception and malnourished only for the last few months of the pregnancy, her baby was likely to be born small. If, on the other hand, the mother suffered malnutrition for the first three months of the pregnancy only (because the baby was conceived toward the end of this terrible episode), but then was well-fed, she was likely to have a baby with a normal body weight. . . . The babies who were born small stayed small all their lives, with lower obesity rates than the general population. For forty or more years, these people had access to as much food as they wanted, and yet their bodies never got over the early period of malnutrition. . . . Even more unexpectedly, the children whose mothers had been malnourished only early in pregnancy, had higher obesity rates than normal. Recent reports have shown a greater incidence of other health problems as well, including certain tests of mental activity. . . . Even more extraordinarily, some of these effects seem to be present in the children of this group, i.e. in the grandchildren of the women who were malnourished during the first three months of their pregnancy. So something that happened in one pregnant population affected their children's children.

As pointed out by Eva Jablonka and Marion Lamb in "The Changing Concept of Epigenetics," *Annals of the New York Academy of Science* 981 (2002): 90, "we know that the offspring of mice treated with carcinogens are predisposed to tumors and other abnormalities. Moreover, some environmental effects go beyond the first generation: drug-induced abnormalities in endocrine function, as well as starvation-induced physiological and behavioral abnormalities, are heritable for at least three generations.

3. As defined by Daniel E. Gottschling in "Epigenetics: From Phenomenon to Field," in *Epigenetics*, eds. C. David Allis, Thomas Jenuwein, Danny Reinberg, Marie-Laure Caparros (Cold Spring Harbor, NY: Cold Spring Harbor Laboratory Press, 2007), 2, "An epigenetic phenomenon [is] a change in phenotype that is heritable but does not involve DNA mutation. Furthermore, the change in phenotype must be switch-like, 'ON' or 'OFF,' rather than a graded response, and it must be heritable even if the initial conditions that caused the switch disappear."

4. René Thom, "An Inventory of Waddingtonian Concepts," in *Theoretical Biology: Epigenetic and Evolutionary Order from Complex Systems*, ed. Brian Goodwin and Peter Saunders (Edinburgh: Edinburgh University Press, 1989), 3, my additions.

5. It is interesting to note that in the second edition of the *Critique of Pure Reason* (1787), §27: B166-169, Immanuel Kant (1724-1804) explained the origin of the *a priori* concepts of the understanding (e.g., substance, self, causality, etc...) by way of an analogy to a "weak" form of epigenesis. Kant took the term to mean an individualized process of development within the context of generic preformation. In describing epigenesis later in his *Critique of Judgment* (1790), Kant wrote,

an organized being is thus not a mere machine, for that has only a motive power, while the organized being possesses in itself a formative power, and indeed one that it communicates to the matter, which does not have it (it organizes the latter): thus it has a self-propagating formative power, which cannot be explained through the capacity for movement alone (that is, mechanism) (§5:374).

Kant's analogy to epigenesis pointed suggestively to the notion that biology could hold answers to the question of the origin of the *a priori* categories, such that more recent evolutionary neo-Kantians like Konrad Lorenz (1903-1989) have claimed that they are ideas that been selected over eons of evolutionary time, which are now hardwired in us, having had their mettle tested in the struggle for existence.

See Adam C. Scarfe, "Skepticism Concerning Causality: An Evolutionary Epistemological Perspective," *Cosmos and History* 8, no. 1 (2012): 227-288.

6. Conrad Hal Waddington, *The Evolution of an Evolutionist* (Ithaca, NY: Cornell University Press, 1975), 218.

7. Given the epistemological and metaphysical commitments expressed in his philosophy of organism, Whitehead would have been quite interested in epigenetics, which gets away from the mechanistic view of genes as self-sufficient substances that was advanced by neo-Darwinian biology. In *Modes of Thought* (1938; repr. New York: The Free Press, 1966), 138-139, Whitehead notes that:

In truth, the notion of the self-contained particle of matter, self-sufficient within its local habitation, is an abstraction. Now an abstraction is nothing else than the omission of part of the truth. The abstraction is well-founded when the conclusions drawn from it are not vitiated by the omitted truth.

This general deduction from the modern doctrine of physics vitiates many conclusions drawn from the application of physics to other sciences, such as physiology. . . . For example, when geneticists conceive genes as the determinants of heredity. The analogy of the old concept of matter sometimes leads them to ignore the influence of the particular animal body in which they are functioning. They presuppose that a pellet of matter remains in all respect self-identical whatever its changes in environment. So far as modern physics is concerned, any characteristics may, or may not, effect changes in the genes, changes which are as important in certain respects, though not in others. In fact recently physiologists have found that genes are modified in some respects by their environment.

Whitehead may perhaps be referring to Thomas Hunt Morgan's *Embryology and Genetics* (1934) here.

Elsewhere, in Lucien Price's obscure and perhaps inaccurate *Dialogues of Alfred North Whitehead* (Boston: Little, Brown, and Company), 1953, 279-281, Whitehead is quoted as supporting soft inheritance.

According to process philosopher, John Cobb in *Back to Darwin: A Richer Account of Evolution* (Grand Rapids, MI: William B. Eerdmans Publishing Company, 2008), 240-241, "For a Whiteheadian, the possibility of some entities (like genes) being well insulated against the influence of others is entirely acceptable. The idea of total insulation is not." Cobb then goes on to quote Tim Ingold to the effect that we should see the gene "not as a discrete, pre-specified entity but as a particular locus of growth and development within a continuous field of relationships." In light of the findings of epigenetics, Cobb goes on to conclude that "the activity of organisms affects both the environment that selects and the genetic constitution of future organisms."

8. The term "concrescence" in Whitehead's process metaphysics indicates the growing together of actual occasions, and is a central part of the theory of prehensions, as articulated in *Process and Reality*. In *Evolution of an Evolutionist*, 9-10, Waddington relates that

> in the late thirties [he] began developing the Whiteheadian notion that the process of becoming (say) a nerve cell should be regarded as the result of the activities of large numbers of genes, which interact together to form a unified 'concrescence'.... The 'gene concrescence' itself undergoes processes of change; at one embryonic period a given concrescence is in a phase of 'competence' and may be switched into one or other of a small number of alternative pathways of further change—but the competence later disappears and if you've missed the bus the switch won't work. ... If I had been more consistently Whiteheadian, I would probably have realized that the 'specificity' involved does not need to lie in the switch at all, but may be a property of the 'concrescence' and the ways in which it can change. Because of course what I have been calling by the Whiteheadian term 'concrescence' is what I have later called a *chreod*.

9. Waddington, *Evolution of an Evolutionist*, 5.

10. As described by Bruce Ellis and David Bjorklund in *Origins of the Social Mind: Evolutionary Psychology and Child Development* (New York: Guilford Press, 2005), 52, in one experiment, Waddington

> subjected pupal fruit flies (*Drosophila melanogaster*) to heat shock. In response to this treatment, some of the surviving flies developed wings that contained few or no cross-veins. Waddington subsequently bred the no-cross-vein flies and exposed the pupal flies of that second generation to heat shock as well. This produced a second generation of fruit flies that also had few or no cross-veins in their wings. After fourteen generations of selective breeding, some fruit flies developed the no-cross-wing phenotype *without* the preexposure to heat shock; that is, Waddington showed that a new phenotype was eventually seen in the developing offspring, without exposure to the original activating environmental event. Waddington referred to this phenomenon as genetic assimilation, which he defined as "the conversion of an acquired character into an inherited one...."

11. Conrad Hal Waddington, "The Epigenotype," *Endeavour* (1942): 18, repr. *International Journal of Epidemiology* 41 (2012): 10.

12. Benedikt Hallgrímsson and Brian K. Hall, eds. *Epigenetics: Linking Genotype and Phenotype in Development and Evolution* (Berkeley CA: University of California Press, 2011), 2, citing Brian K. Hall, "Genetic and Epigenetic Control of Vertebrate Development," *Netherlands Journal of Zoology* 40 (1990): 352-361, and Isaac Salazar-Ciudad and Jukka Jernvall, "Graduality and Innovation in the Evolution of Complex Phenotypes: Insights from Development," *Journal of Experimental Biology (Molecular Development and Evolution)* 6 (2005): 619-631.

13. Arthur D. Riggs, Robert A. Martiensen, and Vincenzo E. A. Russo, eds., *Epigenetic Mechanisms of Gene Regulation* (Cold Spring Harbor, New York: Cold Spring Harbor Laboratory Press, 1996), 1.

14. See Francisco Úbeda and Jon F. Wilkins, "Imprinted Genes and Human Disease: An Evolutionary Perspective," in *Genomic Imprinting*, ed. Jon F. Wilkins (New York: Landes Bioscience and Springer, 2008), 109.

15. See Arthur D. Riggs, "X Inactivation, Differentiation, and DNA Methylation," *Cytogenetics and Cell Genetics* 14 (1975): 9-25; Robin Holliday and John Pugh, "DNA Modification Mechanisms and Gene Activity during Development," *Science* 187 (1975): 226-232; Eva Jablonka and Marion Lamb, "Transgenerational Epigenetic Inheritance," in *Evolution: The Extended Synthesis*, ed. Massimo Pigliucci, and Gerd B. Müller (Cambridge, MA: MIT Press, 2010), 142-143.

16. See Kovalchuk and Kovalchuk, *Epigenetics in Health and Disease* (Upper Saddle River, NJ: Pearson Education, Inc., 2012), 75.

17. See Theresa Phillips,"The Role of Methylation in Gene Expression," *Nature Education* 1, no. 1 (2008), http://www.nature.com/scitable/topicpage/the-role-of-methylation-in-gene-expression-1070.

18. See Kovalchuk and Kovalchuk, *Epigenetics in Health and Disease*, 11.

19. See Carey, *The Epigenetics Revolution*, 64.

20. See Kovalchuk and Kovalchuk, *Epigenetics in Health and Disease*, 119-145.

21. Kovalchuk and Kovalchuk, *Epigenetics in Health and Disease*, 20.

22. Theresa Phillips and K. Shaw, "Chromatin Remodeling in Eukaryotes," *Nature Education* 1, no. 1 (2008), http://www.nature.com/scitable/topicpage/chromatin-remodeling-in-eukaryotes-1082.

23. Kovalchuk and Kovalchuk, *Epigenetics in Health and Disease*, 49.

24. See Jablonka and Lamb, "Transgenerational Epigenetic Inheritance," 143.

In "Imprinted Genes and Human Disease: An Evolutionary Perspective," 101, Úbeda and Wilkins describe genomic imprinting as follows:

In the case of imprinting, the maternally and paternally inherited genes within a single cell have epigenetic differences that result in divergent patterns of gene expression. In the simplest scenario, only one of the two alleles at an imprinted locus is expressed. In other cases, an imprinted locus can include a variety of maternally expressed, paternally expressed and biallelically expressed transcripts.

25. See Robert L. Trivers, "Parent-Offspring Conflict," *American Zoologist* 14 (1974): 249-264; David Haig, *Genomic Imprinting and Kinship* (New Brunswick, NJ: Rutgers University Press, 2002).

26. See Jill U. Adams, "Imprinting and Genetic Diseases: Angelman, Prader-Willi and Beckwith-Weidemann Syndromes," *Nature Education* 1, no. 1 (2008), http://www.nature.com/scitable/topicpage/imprinting-and-genetic-disease-angelman-prader-willi-923.

27. Kovalchuk and Kovalchuk, *Epigenetics in Health and Disease*, 142-143.

28. For instance, the link between behavior and epigenetic processes is studied by the discipline of behavioral epigenetics. See Greg Miller, "The Seductive Allure of Behavioral Epigenetics," *Science* 329 (July 2, 2010): 24-27.

29. See Carey, *The Epigenetics Revolution*, 90.

30. Ernst Mayr, "Prologue: Some Thoughts on the History of the Evolutionary Synthesis," in *The Evolutionary Synthesis: Perspectives on the Unification of Biology*, ed. Ernst Mayr and William B. Provine (Cambridge, MA: Harvard University Press, 1980), 4.

31. See Kovalchuk and Kovalchuk, *Epigenetics in Health and Disease*, 4.

32. August Weismann, *The Germ-Plasm: A Theory of Heredity*, trans. W. Newton Parker and Harriet Rönnfeldt (New York: Charles Scribner's Sons, 1893), 183.

33. Weismann, *The Germ-Plasm*, 395.

34. August Weismann, *Essays upon Heredity and Kindred Biological Problems*, ed. Edward Poulton, Selmar Schönland, and Arthur E. Shipley (Oxford: Clarendon Press, 1889), 103-104.

35. See Kovalchuk and Kovalchuk, *Epigenetics in Health and Disease*, 4, who describe the Cartesian metaphysic underlying the Weismannian and/or neo-Darwinian view of the germplasm. According to them, it is considered to be: "immortal," a "self-sufficient substance that is not influenced by the environment."

Contemporary scholarship on Weismann has provided an insightful view of his main theses, which deviates from the standard view. For example, in "August Weismann on Germ-Plasm Variation," *Journal of the History of Biology* 34 (2001): 518-520, Rasmus G. Winther (2001), following from Griesemer and Wimsatt (1989), argues that there is a discrepancy between the position Weismann is generally known for, which he calls "Weismannism," and the many positions that he actually held throughout his life. Winther claims that Weismann himself did not "conceptually cleave heredity and development, which Weismannism does," and that for much of his career Weismann championed the notion that "*changes in external conditions, acting during development*, ultimately caused all novel *variation* in the hereditary material," a notion that Winther calls "the inheritance of acquired germ-plasm *variations*." To be sure, Weismann in *The Germ Plasm*, 463, wrote that "the ultimate source of *variation* [on which natural selection will act] is always the effect of external influences." So, for Weismann, external influences are the chief source of variation in the germplasm, but this is very different than saying that they are the chief source of mutation of the germplasm, or that they can be the basis of an inheritance of acquired characteristic that is directed by the environment or that necessarily provides the organism with progressive, useful characters. In any case, the architects of the modern synthesis largely followed the received Weismannismic distinctions between germplasm and soma, heredity and development, in defining their positions, whereas Winther claims that recent findings, for example, in epigenetics are forcing us to move more toward the original Weismann and away from the received Weismannism.

36. Julian Huxley, *Evolution: The Modern Synthesis* (Cambridge, MA: MIT Press, 2010), 17-18. Preceding this statement he says,

Weismann drew a sharp distinction between soma and germplasm, between the individual body which was not concerned in reproduction, and the hereditary constitution contained in the germ-cells, which alone was transmitted in heredity. Purely somatic effects, according to him, could not be passed on: the sole inheritable variations were variations in the hereditary constitution.

37. Theodosius Dobzhansky, *Genetics and the Origin of Species* (1937; repr., New York: Columbia University Press, 1982), 16.

38. Dobzhanksy, *Genetics and the Origin of Species*, 12-13; 16-17.

39. See Dobzhansky, *Genetics and the Origin of Species*, 45; Theodosius Dobzhansky, *Genetics of the Evolutionary Process* (New York: Columbia University Press, 1970), 65, 92; Theodosius Dobzhansky, Francisco J. Ayala, Ledyard G. Stebbins, and James W. Valentine, *Evolution* (San Francisco: W. H. Freeman and Company, 1977), 66.

40. Dobzhansky, *Genetics and the Origin of Species*, 31.

41. In *Genetics and the Origin of Species*, 31, Dobzhansky states that the Lamarckian theory of the inheritance of acquired characteristics had "been discussed almost *ad nauseum* in the old biological literature, and any text book of genetics [could] give the reader a review of the present status of this problem, so that we may refrain from the discussion of it altogether."

Also, see Stephen Jay Gould, *The Structure of Evolutionary Theory* (Cambridge, MA: Belknap Press of Harvard University Press, 2002), 454.

42. Dobzhansky, *Genetics and the Origin of Species*, 72.

43. Dobzhansky, *Genetics and the Origin of Species*, 117.

44. Mayr, "Prologue: Some Thoughts on the History of the Evolutionary Synthesis," 4.

45. Mayr, "Prologue: Some Thoughts on the History of the Evolutionary Synthesis," 15.

46. Ernst Mayr, "The Role of Systematics in the Evolutionary Synthesis," in *The Evolutionary Synthesis: Perspectives on the Unification of Biology*, ed. Ernst Mayr and William B. Provine (Cambridge, MA: Harvard University Press, 1980), 130.

47. Mayr, "Prologue: Some Thoughts on the History of the Evolutionary Synthesis," 15.

48. In the last paragraph of the *The Origin of Species* (New York: Penguin Books, 1968), 459, Darwin remarked that he was open to the notion of "variability from the indirect and direct action of the external conditions of life, and from use and disuse." He also wrote on page 225 that "we may sometimes attribute importance to characters which are really of little importance, and which have originated from quite secondary causes, independently of natural selection."

49. In *Embryology and Genetics* (New York: Columbia University Press, 1934), 234, Thomas Hunt Morgan inquired about the "possible interaction between the chromatin of the cells and the protoplasm during development." He continues, "the most common genetic assumption is that the genes remain the same throughout this time. It is, however, conceivable that the genes also are . . . changing in some way, as development proceeds . . . and that these changes have a reciprocal influence on the protoplasm."

50. Mayr, "Prologue," 5.

51. Ernst Mayr, *The Growth of Biological Thought* (Cambridge, MA: Belknap Press of Harvard University Press, 1982), 828.

52. See Francis Crick, "Central Dogma of Molecular Biology," *Nature* 227 (August 8, 1970): 561-563.

53. Mary Jane West-Eberhard, *Developmental Plasticity and Evolution* (New York: Oxford University Press, 2003), 143.

54. West-Eberhard, *Developmental Plasticity and Evolution*, 143.

55. Eva Jablonka and Marion Lamb, *Evolution in Four Dimensions: Genetic, Epigenetic, Behavioral, and Symbolic Variation in the History of Life* (Cambridge, MA: MIT Press, 2005), 229.

56. Eva Jablonka and Marion Lamb, "Trangenerational Epigenetic Inheritance," 139. Elsewhere, in *Evolution in Four Dimensions*, 7, they define soft inheritance as "the inheritance of genomic changes induced by environmental factors."

57. Jablonka and Lamb, "Trangenerational Epigenetic Inheritance," 168.

58. Eva Jablonka and Marion Lamb, "Soft Inheritance: Challenging the Modern Synthesis," *Genetics and Molecular Biology* 31, no. 2 (2008): 393, my addition.

59. Jablonka and Lamb, "Trangenerational Epigenetic Inheritance," 169.

60. David Haig, "Weismann Rules! OK? Epigenetics and the Lamarckian Temptation." *Biology and Philosophy* 22 (2007): 427.

61. Later in his "Lamarck Ascending! A Review of *Transformations of Lamarckism: From Subtle Fluids to Molecular Biology*, ed. Snait B. Gissis and Eva Jablonski, *Philosophy and Theory in Biology* 3 (November 2011): 5, Haig states that the neo-Darwinian response to such neo-Lamarckian challenges to the modern synthesis general-

ly employ the "rhetorical maneuver . . . of deny[ing] a connection to Lamarck and argu[ing] that phenotypic plasticity, epigenetic inheritance, and the like, can be accommodated easily within existing theory and do not challenge reigning paradigms."

62. Haig, "Weismann Rules!," 422.
63. Haig, "Weismann Rules!," 423.
64. Haig, "Weismann Rules!," 422.
65. Haig, "Weismann Rules!," 426. A similar concern is raised against Jablonka and Lamb by Francesca Merlin in "Evolutionary Chance Mutation: A Defense of the Modern Synthesis' Consensus View," *Philosophy and Theory in Biology* 2 (September 2010): 1-22.
66. However, here, one might interject that rather than criticize Jablonka and Lamb for the apparent terminological confusion, we might do better to look to discrepancies in Mayr's formulation of the terms.
67. Thomas E. Dickens and Qazi Rahman, "The Extended Evolutionary Synthesis and the Role of Soft Inheritance in Evolution," *Proceedings of the Royal Society: Biological Sciences* 279, no. 1740 (May 16, 2012): 2913.
68. Dickens and Rahman, "The Extended Evolutionary Synthesis and the Role of Soft Inheritance in Evolution," 2917.
69. Russell Bonduriansky, "Rethinking Heredity, Again," *Trends in Ecology and Evolution* 27, no. 6 (2012): 330.
70. Bonduriansky, "Rethinking Heredity, Again," 330.
71. Bonduriansky, "Rethinking Heredity, Again," 331.
72. See notes 55-59 above. Also, for example, in their 2002 article, "The Changing Concept of Epigenetics," *Annals of the New York Academy of Science* 981 (2002): 88, 93, Jablonka and Lamb argue that "although one can usefully distinguish between DNA and non-DNA inheritance, there are no simple criteria for distinguishing between genetic and epigenetic phenomena."
73. See Adam Wilkins, "Epigenetic Inheritance: Where Does the Field Stand Today? What Do We Still Need To Know," in *Transformations of Lamarckism: From Subtle Fluids to Molecular Biology*, ed. Snait B. Gissis and Eva Jablonka, 389-393 (Cambridge, MA: MIT Press, 2011), who speaks of "epimutations [that] have been converted to a DNA sequence mutation" and calls for more research in the same vein as Waddington's genetic assimilation experiments (392). Also see the "Final Discussion," in *Transformations of Lamarckism*, 395-409, where the participants discuss "genetic accommodation" designating a "process through which, following developmental response to environmental change or to a new mutation, selection leads to genotypic change. . . . Genetic accommodation may lead to canalization (which decreases the range of phenotypic responses, making them more buffered), to enhanced plasticity (which increases the range of responses, making them more condition-dependent), or to the amelioration of the deleterious side effects of a new environmental or mutational change" (397). And see Evelyn Fox Keller's and Peter Gluckman's respective suggestions of "mutational assimilation"—namely, the "possibility that epigenetic variation can have a direct effect on mutability" (398).
74. Mae Wan Ho, "Environment and Heredity in Development and Evolution," in *Beyond Neo-Darwinism: An Introduction to the New Evolutionary Paradigm*, ed. Mae Wan Ho and Peter T. Saunders (London: Academic Press, 1984), 285, quoted by Gilbert Gottlieb, *Individual Development and Evolution: The Genesis of Novel Behavior* (New York: Oxford University Press, 1991), 146.

75. Richard Dawkins, *The Selfish Gene (30th Anniversary Edition)* (1976; repr., Oxford: Oxford University Press, 2006), xxi.

76. Richard C. Francis, *Epigenetics: The Ultimate Mystery of Inheritance* (New York: W. W. Norton & Company, 2011), 19.

77. Guido Nicolosi and Guido Ruivenkamp, "The Epigenetic Turn: Some Notes about the Epistemological Change of Perspective in the Biosciences," *Medicine, Health Care, and Philosophy* 15, no. 3 (2012), 312, citing David Le Breton, "Genetic Fundamentalism or the Cult of the Gene," *Body and Society* 10, no. 4 (2004): 1-20.

78. Nicolosi and Ruivenkamp, "The Epigenetic Turn," 309, my addition.

79. West-Eberhard, *Developmental Plasticity and Evolution*, 614.

80. See my introduction to this volume for a synopsis of the notion of "internal relations."

81. In Aaron D. Goldberg, C. David Allis, and Emily Bertstein, "Epigenetics: A Landscape Takes Shape," *Cell* 128 (February 23, 2007): 637, there is an illustration of the "epigenetic molecular machinery" likening it to a pinball machine.

82. Admittedly, I have used a music mixing board analogy in a second year biomedical ethics course in order to provide a brief overview of epigenetics to philosophy students. I tell them that the post-rock band, "Gene Therapy" has four players (a drummer, a guitarist, a bassist, and a vocalist, giving them names whose first letter corresponds to the four nucleotide bases: A, T, C, G) playing a gig. Some of the players (e.g., the guitarist) get out of control with their solos in trying to overemphasize themselves, while others sometimes lag back in their playing, and that it is the duty of the mix-master at the back of the pub (whose name corresponds to the epigenetic "mechanism" in question) to dial and tune them up or down in the mix, so that the music does not sound terrible. Amusing as these sorts of mechanistic analogies may be, and regardless of how useful they are for explanatory purposes, it is important to lay the abstractions bare via a phase of holistic reflection.

83. Eva Jablonka and Marion Lamb, "The Changing Concept of Epigenetics," 91, my emphasis.

84. Eva Jablonka and Marion Lamb, "Genic Neo-Darwinism—Is It the Whole Story?," *Journal of Evolutionary Biology* 11 (1998): 258, my emphasis.

85. Jablonka and Lamb, "The Changing Concept of Epigenetics," 91-92, my emphasis.

86. Conrad Hal Waddington, *The Nature of Life* (New York: Athenaum, 1961), 11.

87. Waddington, *The Nature of Life*, 13.

88. See Waddington, *The Nature of Life*, 16.

89. As Carey states in *The Epigenetics Revolution*, 42, my emphasis,

There was a period when the prevailing orthodoxy seemed to be that the only thing that mattered was our DNA script, our genetic inheritance. . . . This can't be the case, as the same script is used differently depending on its cellular context. The field is now possibly at risk of swinging a bit too far in the opposite direction, with hard-line epigeneticists almost minimizing the significance of the DNA code. The truth is, of course, [probably] somewhere in between.

90. See Alfred North Whitehead, *Science and the Modern World* (1925; repr., New York: The Free Press, 1967), 194-196.

91. Nicolosi and Ruivenkamp, "The Epigenetic Turn," 310.

92. Howard Wolinsky, "Paths to Acceptance: The Advancement of Scientific Knowledge Is an Uphill Struggle against 'Accepted Wisdom'," *EMBO Reports* 9, no. 5 (2008), 416.

93. Wolinsky, "Paths to Acceptance," 217.
94. NOVA. "Ghost in Your Genes," PBS, Airdate: October 16, 2007, transcript: http://www.pbs.org/wgbh/nova/transcripts/3413_genes.html.
95. For example, see Yuk Ting Ma's research with the UK Institute of Cancer Research, as reported in European Society for Medical Oncology, "Cancer-Linked Epigenetic Effects of Smoking *Found*," *ScienceDaily* (October 11, 2010), http://www.sciencedaily.com/releases/2010/10/101009082825.htm, and elsewhere.
96. See Hans Jonas, *The Imperative of Responsibility: In Search of an Ethics for the Technological Age* (Chicago: University of Chicago Press, 1984).
97. Mark A. Rothstein, Yu Cai, and Gary E. Marchant. "The Ghost in Our Genes: Legal and Ethical Implications of Epigenetics," *Health Matrix* 19, no. 1 (Winter 2009): 50, my addition.
98. Antonei B. Csoka and Moshe Szyf, "Epigenetic Side-Effects of Common Pharmaceuticals: A Potential New Field in Medicine and Pharmacology," *Medical Hypotheses* 73 (2009): 770.
99. In "Epigenetics, DNA Methylation, and Chromatin Modifying Drugs," *Annual Review of Phamacology and Toxicology* 49 (2009): 252, Szyf says that in addition to 5-Azacytidine and Zebularine, "other commonly used drugs [have been] shown to bring about demethylation," citing procainamide, Hydralazine, and Valproic acid. Szyf continues, "These data raise the concern that other heavily used drugs affect the DNA methylation pattern and thus can promote the expression of disease-promoting genes. Future drug safety tests should include measures of DNA demethylation."
100. Csoka and Szyf, "Epigenetic Side-Effects of Common Pharmaceuticals," 776.
101. Csoka and Szyf, "Epigenetic Side-Effects of Common Pharmaceuticals," 770.
102. See Marks, Paul A. "Epigenetic Anti-Cancer Drugs: An Unfolding Story," *Oncology* 25, no. 3 (March 16, 2011), http://www.cancernetwork.com/cancer-genetics/content/article/10165/1822124.
See also "The World's Largest Epigenetics Meeting," webpage for the *3rd World Epigenetics Summit: Unlocking the Commercial and Therapeutic Potential of Epigenetics Research*, http://epigenetics-summit.com.
103. A cell functions by way of a complex set of signaling pathways, which convey messages between different parts of the cell, such as from the extracellular domain to the nucleus to effect changes in gene expression. Such pathways often involve the phosphorylation of proteins, which are normally only transiently activated or deactivated. However, in cancer, some of these enzymes may be overactive, causing the pathways to also be overactive, which, in turn, overstimulates the nucleus to promote proliferation and metastasis. Epigenetic drugs may target such pathways and may re-amplify an underactive tumor suppressor pathway, and/or de-amplify an overactive oncogenic pathway. (Personal communication with Dr. Andrew Scarfe of the Cross-Cancer Institute at the University of Alberta, Canada, July 29th, 2012).
104. Szyf, "Epigenetics, DNA Methylation, and Chromatin Modifying Drugs," 244.
105. Interestingly, in a 2010 study, Heidrun Karlic, Julia Varga, Roman Thaler, Cornelia Berger, Silvia Spitzer, Michael Pfeilstöcker, Klaus Klaushofer, and Franz Varga, "Effects of Epigenetic Drugs (Vorinostat, Decitabine) on Metabolism-Related Pathway Factors in Leukemic Cells," *The Open Leukemia Journal* 3 (2010): 34-42, data was given to support the notion that vorinostat also affected "epigenetic processes which are tightly associated with malignancy-associated features of energy metabolism" and raised the question as to "whether [it] could be pressed into service against the rising tide of metabolic diseases and the associated risks for malignancies."

106. See Al Doig, "Epigenetics and the Role of Phenotypic Changes to DNA," *Bio-ITWorld* (November 16, 2010),
http://www.bio-itworld.com/BioIT_Article.aspx?id=102922, 2.

107. Szyf, "Epigenetics, DNA Methylation, and Chromatin Modifying Drugs," 245.

108. Szyf, "Epigenetics, DNA Methylation, and Chromatin Modifying Drugs," 253.

109. See Doig, "Epigenetics and the Role of Phenotypic Changes to DNA," 2.

110. Leslie Pray, "At the Flick of a Switch: Epigenetic Drugs," *Chemistry and Biology* 15 (July 21, 2008): 640.

111. See Rabiya S. Tuma, "Epigenetic Therapies Move into New Territory, but How Exactly Do They Work?," *Journal of the National Cancer Institute News* 101, no. 19 (October 7, 2009): 1300; Guillermo Garcia-Manero, "Epigenetic Therapy: Current Status and Future Potential," *Horizon Symposia* website, 2005, http://www.nature.com/horizon/epigenetics/kq/8_Garcia-Manero.html.

In response to the questions, "What are the consequences of epigenetic drugs? Will there be secondary malignancies / drastic side effects associated with the effects of epigenetic therapy on normal gene silencing?," Dr. Garcia-Manero says,

Because DNA methylation is a normal characteristic of the genome, and a neoplastic "histone code" has not been identified, the use of these drugs may induce epigenetic changes that could result in toxicities and the development of second malignancies. Indeed, animal models of severe global hypomethylation have been associated with the development of tumors and genomic instability. Despite these observations, there is no clinical evidence to support the idea that exposure to these drugs could increase the incidence of second malignancies or any other severe, unanticipated toxicity.

This could be explained by the fact that the levels of hypomethylation achieved in patients are lower than those observed in animal models. However, it is also possible that the period of time required for these second malignancies to become detectable is longer than the median follow-up or survival of these patients. Therefore, despite the clinical evidence to the contrary, the simple answer to the above question is "potentially yes," especially if the drugs are used in the long term and in patients with non-neoplastic diseases. However, the answer is obviously far more complicated than this.

112. Ivan Amato, "The First Four: The Founding Generation of Epigenetics-Based Drugs Combine Promise and Peril," *Science and Technology* 87, no. 14 (April 6, 2009), http://pubs.acs.org/cen/science/87/8714sci1a.html, 2.

113. For example, there are considerable side effects associated with the therapeutic use of vorinostat, see:
http://www.merck.com/product/usa/pi_circulars/z/zolinza/zolinza_ppi.pdf.

In "The First Four," 2, Amato reports that "from the metric of drug specificity, 'these are awful compounds', says epigenetic researcher Norbert O. Reich of the University of California, Santa Barbara. They only work at all, because, like conventional chemotherapies, they can do more harm to a patient's disease-associated cells than to the healthy cells, he notes."

As such, pharmaceutical companies are moving to develop new HDACis, such as CJR-3996, which has been shown to be safe in a phase 1 clinical trial. See Institute of Cancer Research, "Next Generation 'Epigenetic' Cancer Pill Shown to Be Safe in Phase I Trial," *ScienceDaily* (April 30, 2012),
http://www.sciencedaily.com/releases/2012/05/120501085508.

114. Francisco Úbeda and Jon F. Wilkins, "Imprinted Genes and Human Disease: An Evolutionary Perspective," *Genomic Imprinting*, ed. Jon F. Wilkins (New York: Landes Bioscience and Springer, 2008), 113.
115. Úbeda and Wilkins, "Imprinted Genes and Human Disease," 113, my emphasis.
116. U.S. Food and Drug Administration website, "Vaccines, Blood & Biologics," http://www.fda.gov/BiologicsBloodVaccines/CellularGeneTherapyProducts/default.htm. Accessed August 11, 2012.
117. Canadian Institutes of Health Research (CIHR) website: http://www.cihr-irsc.gc.ca/e/43734.html.
118. Minister of Justice, Government of Canada, *Assisted Human Reproduction Act* (2004), 1-40, http://laws-lois.justice.gc.ca/PDF/A-13.4.pdf, Principles, §2g, 2.
119. *Assisted Human Reproduction Act* (2004), §5f, 5.
120. *Assisted Human Reproduction Act* (2004), §3, 3.
121. Rothstein, Cai, and Marchant, "The Ghost in Our Genes: Legal and Ethical Implications of Epigenetics," 58.

Bibliography

Adams, Jill U. "Imprinting and Genetic Diseases: Angelman, Prader-Willi and Beckwith-Weidemann Syndromes." *Nature Education* 1, no. 1 (2008), http://www.nature.com/scitable/topicpage/imprinting-and-genetic-disease-angelman-prader-willi-923.
Allis, C. David, Thomas Jenuwein, Danny Reinberg, and Marie-Laure Caparros, eds. *Epigenetics*. New York: Cold Spring Harbor Laboradory Press, 2007.
Amato, Ivan. "The First Four: The Founding Generation of Epigenetics-Based Drugs Combine Promise and Peril." *Science and Technology* 87, no. 14 (April 6, 2009), http://pubs.acs.org/cen/science/87/8714sci1a.html.
Bateson, Patrick, and Peter Gluckman. *Plasticity, Robustness, Development, and Evolution*. Cambridge, UK: Cambridge University Press, 2011.
Bondurianksy, Russell. "Rethinking Heredity, Again." *Trends in Ecology and Evolution* 27, no. 6 (2012): 330-336.
Canadian Institutes of Health Research (CIHR) website: http://www.cihr-irsc.gc.ca/e/43734.html.
Carey, Nessa. *The Epigenetics Revolution: How Modern Biology Is Rewriting Our Understanding of Genetics, Disease, and Inheritance*. London, UK: Icon Books, 2011.
Cobb, John, ed. *Back to Darwin: A Richer Account of Evolution*. Grand Rapids, MI: William B. Eerdmans Publishing Company, 2008.
Crews, David. "Epigenetics, Brain, Behavior, and the Environment." *Hormones* 9, no. 1 (2010): 41-50.
Crick, Francis. "Central Dogma of Molecular Biology." *Nature* 227 (August 8, 1970): 561-563.
Csoka, Antonei, and Moshe Szyf, "Epigenetic Side-Effects of Common Pharmaceuticals: A Potential New Field in Medicine and Pharmacology," *Medical Hypotheses* 73 (2009): 770-780.
Darwin, Charles. *The Origin of Species by Means of Natural Selection*. New York: Penguin Books, 1968.

Dawkins, Richard. *The Selfish Gene (30th Anniversary Edition)*. 1976. Reprinted, Oxford: Oxford University Press, 2006. Page references are to the 2006 edition.
Dennett, Daniel. *Darwin's Dangerous Idea: Evolution and the Meanings of Life*. New York: Simon & Schuster, 1995.
Dickens, Thomas E., and Qazi Rahman. "The Extended Evolutionary Synthesis and the Role of Soft Inheritance in Evolution." *Proceedings of the Royal Society: Biological Sciences* 279, no. 1740 (May 16, 2012): 2913-2921.
Dobzhansky, Theodosius. *Genetics and the Origin of Species*. 1937. Reprinted, New York: Columbia University Press, 1982.
———. *Genetics of the Evolutionary Process*. New York: Columbia University Press, 1970.
Dobzhansky, Theodosius, Francisco J. Ayala, Ledyard G. Stebbins, and James W. Valentine. *Evolution*. San Francisco: W. H. Freeman and Company, 1977.
Doig, Al. "Epigenetics and the Role of Phenotypic Changes to DNA." *Bio-ITWorld* (November 16, 2010), http://www.bio-itworld.com/BioIT_Article.aspx?id=102922.
Ebrahim, Shah. "Epigenetics: The Next Big Thing." *International Journal of Epidemiology* 41 (2012): 1-3.
Ellis, Bruce, and David Bjorklund. *Origins of the Social Mind: Evolutionary Psychology and Child Development*. New York: Guilford Press, 2005.
Esteller, Manel, ed. *Epigenetics in Biology and Medicine*. Boca Raton, FL: Taylor & Francis Group, 2009.
European Society for Medical Oncology, "Cancer-Linked Epigenetic Effects of Smoking Found," *ScienceDaily* (October 11, 2010), http://www.sciencedaily.com/releases/2010/10/101009082825.htm.
Francis, Richard C. *Epigenetics: The Ultimate Mystery of Inheritance*. New York: W. W. Norton & Company, 2011.
Fritah, Sabrina, Edwige Col, Cyril Boyault, Jérôme Govin, Karin Sadoul, Susanna Chiocca, Elisabeth Christians, Saadi Khochbin, Caroline Jolly, and Claire Vourc'h. "Heat-Shock Factor 1 Controls Genome-Wide Acetylation in Heat-Shocked Cells." *Molecular Biology of the Cell* 20 (December 1, 2009): 4976-4984.
Garcia-Manero, Guillermo. "Epigenetic Therapy: Current Status and Future Potential." *Horizon Symposia* website, 2005, http://www.nature.com/horizon/epigenetics/kq/8_Garcia-Manero.html.
Gilbert, Scott F. "Diachronic Biology Meets Evo-Devo: C. H. Waddington's Approach to Evolutionary Developmental Biology." *American Zoologist* 40 (2000): 729-737.
Gissis, Snait B., and Eva Jablonka, eds. *Transformations of Lamarckism: From Subtle Fluids to Molecular Biology*. Cambridge, MA: MIT Press, 2011.
Goldberg, Aaron D., C. David Allis, and Emily Bertstein, "Epigenetics: A Landscape Takes Shape." *Cell* 128 (February 23rd, 2007): 635-638.
Goodwin, Brian, and Peter Saunders, eds. *Theoretical Biology: Epigenetic and Evolutionary Order from Complex Systems*. Edinburgh: Edinburgh University Press, 1989.
Gottlieb, Gilbert. *Individual Development and Evolution: The Genesis of Novel Behavior*. New York: Oxford University Press, 1991.
Gould, Stephen Jay. *The Structure of Evolutionary Theory*. Cambridge, MA: Belknap Press of Harvard University Press, 2002.
Griesemer, James R., and William C. Wimsatt, "Picturing Weismannism: A Case Study of Conceptual Evolution." In *What Philosophy of Biology Is: Essays Dedicated to*

David Hull, edited by Michael Ruse, 75-137. Dordrecht, Netherlands: Kluwer Academic Publishers, 1989.
Haig, David. *Genomic Imprinting and Kinship*. New Brunswick, NJ: Rutgers University Press, 2002.
———. "The (Dual) Origin of Epigenetics." *Cold Spring Harbor Symposia on Quantitative Biology* 69 (2004): 67-70.
———. "Weismann Rules! OK? Epigenetics and the Lamarckian Temptation." *Biology and Philosophy* 22 (2007): 415-428.
———. "Lamarck Ascending! A Review of (Snait B. Gissis and Eva Jablonski, eds.) *Transformations of Lamarckism: From Subtle Fluids to Molecular Biology*." *Philosophy and Theory in Biology* 3 (November 2011): 1-9.
Hall, Brian K., and Manfred D. Laubichler. "Conrad Hal Waddington, Theoretical Biology, and EvoDevo (Thematic Issue)." *Biological Theory: Integrating Development, Evolution, and Cognition* 3, no. 3, 2008, 185-287.
Hallgrímsson, Benedikt, and Brian K. Hall, eds. *Epigenetics: Linking Genotype and Phenotype in Development and Evolution*. Berkeley CA: University of California Press, 2011.
Hanson, Mark A., Felicia M. Low, Peter D. Gluckman. "Epigenetic Epidemiology: The Rebirth of Soft Inheritance." *Annals of Nutrition and Metabolism* 58, no. 2 (August 12, 2011): 8-15.
Ho, Mae Wan. "Environment and Heredity in Development and Evolution." In *Beyond Neo-Darwinism: An Introduction to the New Evolutionary Paradigm*, edited by Mae Wan Ho and Peter T. Saunders, 267-287. London: Academic Press, 1984.
Holliday, Robin, and John Pugh. "DNA Modification Mechanisms and Gene Activity during Development." *Science* 187 (1975): 226-232.
Huxley, Julian. *Evolution: The Modern Synthesis*. 1942. Reprinted, Cambridge, MA: MIT Press, 2010.
Institute of Cancer Research, "Next Generation 'Epigenetic' Cancer Pill Shown to Be Safe in Phase I Trial," *ScienceDaily* (April 30, 2012), http://www.sciencedaily.com/releases/2012/05/120501085508.
Jablonka, Eva, and Marion J. Lamb. "Genic Neo-Darwinism—Is It The Whole Story?" *Journal of Evolutionary Biology* 11 (1998): 243-260.
———. "The Changing Concept of Epigenetics." *Annals of the New York Academy of Science* 981 (2002): 82-96.
———. *Evolution in Four Dimensions: Genetic, Epigenetic, Behavioral, and Symbolic Variation in the History of Life*. Cambridge, MA: MIT Press, 2005.
———. "Soft Inheritance: Challenging the Modern Synthesis." *Genetics and Molecular Biology* 31, no. 2 (2008): 389-395.
———. "Trangenerational Epigenetic Inheritance." In *Evolution: The Extended Synthesis*, edited by Pigliucci, Massimo and Gerd B. Müller, 137-174. Cambridge, MA: MIT Press, 2010.
———. "The Pillars of Darwinism." *Project Syndicate*, http://www.project-syndicate.org/commentary/the-pillars-of-darwinism.
Jonas, Hans. *The Imperative of Responsibility: In Search of an Ethics for the Technological Age*. Chicago: University of Chicago Press, 1984.
Kant, Immanuel. *Critique of Pure Reason*. Translated and edited by Paul Guyer and Allen W. Wood. New York: Cambridge University Press, 1998.
———. *Critique of the Power of Judgment*. Translated by Paul Guyer and Eric Mathews; edited by Paul Guyer. New York: Cambridge University Press, 2000.

Karlic, Heidrun, Julia Varga, Roman Thaler, Cornelia Berger, Silvia Spitzer, Michael Pfeilstöcker, Klaus Klaushofer, and Franz Varga. "Effects of Epigenetic Drugs (Vorinostat, Decitabine) on Metabolism-Related Pathway Factors in Leukemic Cells." *The Open Leukemia Journal* 3 (2010): 34-42.

Kovalchuk, Igor, and Olga Kovalchuk. *Epigenetics in Health and Disease*. Upper Saddle River, NJ: Pearson Education, Inc., 2012.

Le Breton, David. "Genetic Fundamentalism or the Cult of the Gene." *Body and Society* 10, no. 4 (2004): 1-20.

Marks, Paul A. "Epigenetic Anti-Cancer Drugs: An Unfolding Story," *Oncology* 25, no. 3 (March 16, 2011), http://www.cancernetwork.com/cancer-genetics/content/article/10165/1822124.

Mayr, Ernst. "Prologue: Some Thoughts on the History of the Evolutionary Synthesis." In *The Evolutionary Synthesis: Perspectives on the Unification of Biology*, edited by Ernst Mayr and William B. Provine, 1-48. Cambridge, MA: Harvard University Press, 1980.

———. "The Role of Systematics in the Evolutionary Synthesis." In *The Evolutionary Synthesis: Perspectives on the Unification of Biology*, edited by Ernst Mayr and William B. Provine, 123-136. Cambridge, MA: Harvard University Press, 1980.

———. *The Growth of Biological Thought*. Cambridge, MA: Belknap Press of Harvard University Press, 1982.

McGowan, Patrick O., Michael J. Meaney, and Moshe Szyf. "Epigenetics, Phenotype, Diet, and Behavior." In *Handbook of Behavior, Food and Nutrition*, edited by Victor R. Preedy, Colin R. Martin, and Ronald R. Watson, 17-31. New York: Springer, 2011.

Merlin, Francesca. "Evolutionary Chance Mutation: A Defense of the Modern Synthesis' Consensus View." *Philosophy and Theory in Biology* 2 (September 2010): 1-22.

Miller, Greg. "The Seductive Allure of Behavioral Epigenetics." *Science* 329 (July 2, 2010): 24-27.

Minister of Justice, Government of Canada. *Assisted Human Reproduction Act* (2004): 1-40, http://laws-lois.justice.gc.ca/PDF/A-13.4.pdf.

Morgan, Thomas Hunt. *Embryology and Genetics*. New York: Columbia University Press, 1934.

Nicolosi, Guido, and Guido Ruivenkamp. "The Epigenetic Turn: Some Notes About the Epistemological Change of Perspective in the Biosciences." *Medicine, Health Care, and Philosophy* 15, no. 3 (2012): 309-319.

NOVA. "Ghost in Your Genes," PBS, Airdate: October 16, 2007, transcript: http://www.pbs.org/wgbh/nova/transcripts/3413_genes.html.

Packard, Alpheus S. *Lamarck: The Founder of Evolution: His Life and Work With Translations of His Writings on Organic Evolution*. New York: Longmans, Green, and Co., 1901.

Phillips, Theresa. "The Role of Methylation in Gene Expression." *Nature Education* 1, no. 1 (2008), http://www.nature.com/scitable/topicpage/the-role-of-methylation-in-gene-expression-1070.

Phillips, Theresa, and K. Shaw. "Chromatin Remodeling in Eukaryotes." *Nature Education* 1, no. 1 (2008), http://www.nature.com/scitable/topicpage/chromatin-remodeling-in-eukaryotes-1082.

Pigliucci, Massimo, and Gerd B. Müller, eds. *Evolution: The Extended Synthesis.* Cambridge, MA: MIT Press, 2010.
Pray, Leslie. "At the Flick of a Switch: Epigenetic Drugs." *Chemistry and Biology* 15 (July 21, 2008): 640-641.
Price, Lucien. *Dialogues of Alfred North Whitehead.* Boston: Little, Brown, and Company, 1953.
Richards, Eric J. "Inherited Epigenetic Variation—Revisiting Soft Inheritance." *Nature Reviews: Genetics* 7 (May 2006): 395-401.
Riggs, Arthur D. "X Inactivation, Differentiation, and DNA Methylation." *Cytogenetics and Cell Genetics* 14 (1975): 9-25.
Riggs, Arthur D., Robert A. Martiensen, and Vincenzo E. A. Russo, eds. *Epigenetic Mechanisms of Gene Regulation.* Cold Spring Harbor, New York: Cold Spring Harbor Laboratory Press, 1996.
Rothstein, Mark A., Yu Cai, and Gary E. Marchant. "The Ghost in Our Genes: Legal and Ethical Implications of Epigenetics." *Health Matrix* 19, no. 1 (Winter 2009): 1-62.
Scarfe, Adam C. "Skepticism Concerning Causality: An Evolutionary Epistemological Perspective." *Cosmos and History* 8, no. 1 (2012): 227-288.
Sippl, Wolfgang, and Manfred Jung, eds. *Epigenetic Targets in Drug Discovery.* Weinheim, Germany: Wiley-VCH Verlag GmbH & Co., 2009.
Slack, Jonathan M. W. "Conrad Hal Waddington: The Last Renaissance Biologist?" *Nature Reviews—Genetics* 3 (November 2002): 889-895.
Szyf, Moshe. "Epigenetics, DNA Methylation, and Chromatin Modifying Drugs." *Annual Review of Phamacology and Toxicology* 49 (2009): 243-263.
Third World Epigenetics Summit: Unlocking the Commercial and Therapeutic Potential of Epigenetics Research, "The World's Largest Epigenetics Meeting," website: http://epigenetics-summit.com. Accessed August 5th, 2012.
Trivers, Robert L. "Parent-Offspring Conflict." *American Zoologist* 14 (1974): 249-264.
Tuma, Rabiya S. "Epigenetic Therapies Move into New Territory, but How Exactly Do They Work?" *Journal of the National Cancer Institute News* 101, no. 19 (October 7, 2009): 1300-1301.
Úbeda, Francisco, and Jon F. Wilkins. "Imprinted Genes and Human Disease: An Evolutionary Perspective." In *Genomic Imprinting*, edited by Jon F. Wilkins, 101-115. New York: Landes Bioscience and Springer, 2008.
U.S. Food and Drug Administration website, "Vaccines, Blood & Biologics," http://www.fda.gov/BiologicsBloodVaccines/CellularGeneTherapyProducts/default.htm. Accessed August 11, 2012.
Waddington, Conrad Hal. "The Epigenotype," *Endeavour* (1942): 18-20, reprinted in *International Journal of Epidemiology* 41 (2012): 10-13.
———. "Genetic Assimilation of an Acquired Character." *Evolution* 7, no. 2 (June 1953): 118-126.
———. "Genetic Assimilation of the Bithorax Phenotype." *Evolution* 10, no. 1 (March 1956): 1-13.
———. *The Strategy of the Genes: A Discussion of Some Aspects of Theoretical Biology.* London, UK: George Allen & Unwin, Ltd., 1957.
———. *The Nature of Life.* New York: Athenaum, 1961.
———. *The Ethical Animal.* New York: Athenaum, 1961.
———. *Biology, Purpose, and Ethics.* Clark University Press with Barre Publishers, 1971.
———. *The Evolution of an Evolutionist.* Ithaca, NY: Cornell University Press, 1975.

Weismann, August. *Studies in the Theory of Descent, Volumes 1-2*, translated by Raphael Meldola. London, UK: Sampson Low, Marston, Searle & Rivington, 1882.
——. *Essays upon Heredity and Kindred Biological Problems*. Edited by Edward Poulton, Selmar Schönland, and Arthur E. Shipley. Oxford: Clarendon Press, 1889.
——. *The Germ-Plasm: A Theory of Heredity*. Translated by W. Newton Parker and Harriet Rönnfeldt. New York: Charles Scribner's Sons, 1893.
West-Eberhard, Mary Jane. *Developmental Plasticity and Evolution*. New York: Oxford University Press, 2003.
Whitehead, Alfred North. *Science and the Modern World*. 1925. Reprinted, New York: The Free Press, 1967.
——. *Process and Reality*. 1929. Corrected edition edited by D. R. Griffin and D. W. Sherburne. New York: The Free Press, 1978. Page references to corrected edition.
——. *Modes of Thought*. 1938. Reprinted, New York: The Free Press, 1966. Page references are to the 1966 edition.
Winther, Rasmus G. "August Weismann on Germ-Plasm Variation." *Journal of the History of Biology* 34 (2001): 517-555.
Wolinski, Howard. "Paths to Acceptance: The Advancement of Scientific Knowledge Is an Uphill Struggle against 'Accepted Wisdom'." *European Molecular Biology Organization Reports* 9, no. 5 (2008): 416-418.

Section 7:
Organism and Mechanism

Chapter Seventeen
From Organicism to Mechanism—and Halfway Back?

Michael Ruse

The most significant event in the history of Western science was the so-called Scientific Revolution that occurred in the sixteenth and seventeenth centuries. It involved many major changes, including Copernicus's putting the Sun at the center of the universe, Kepler's work on the planets, Galileo's and Descartes' articulation of mechanics, and finally Newton's great synthesis. This is not to mention related events such as Harvey's work on the heart and Gilbert's work on magnetism. Underlying these great scientific advances was a fundamental change in worldview, more specifically a fundamental change in the root metaphor which informs everything else.[1] The Greeks, notably Plato and Aristotle, subscribed to an *organic* view of the world. They saw things in terms of living beings, and for this reason thought it made sense to ask about purposes or ends.[2] Famous was Aristotle's insistence that for full understanding one must ask about final causes. After the Scientific Revolution, people subscribed to a *mechanical* view of the world.[3] He who was (in my opinion) the greatest philosophical commentator on the Revolution, the English chemist Robert Boyle, saw things clearly. From now on, we work in terms of machines, of artifacts.[4] Organicists like Aristotle who see living forces directing nature are just plain wrong:

> And those things which the school philosophers ascribe to the agency of nature interposing according to emergencies, I ascribe to the wisdom of God in the first fabric of the universe; which he so admirably contrived that, if he but continue his ordinary and general concourse, there will be no necessity of extraordinary interpositions, which may reduce him to seem as if it were to play aftergames — all those exigencies, upon whose account philosophers and physicians seem to have devised what they call nature, being foreseen and provided for in the first fabric of the world; so that mere matter, so ordered, shall in such

and such conjunctures of circumstances, do all that philosophers ascribe on such occasions to their omniscient nature, without any knowledge of what it does, or acting otherwise than according to the catholic laws of motion. And methinks the difference between their opinion of God's agency in the world, and that which I would propose, may be somewhat adumbrated by saying that they seem to imagine the world to be after the nature of a puppet, whose contrivance indeed may be very artificial, but yet is such that almost every particular motion the artificer is fain (by drawing sometimes one wire or string, sometimes another) to guide, and oftentimes overrule, the actions of the engine, whereas, according to us, it is like a rare clock, such as may be that at Strasbourg, where all things are so skillfully contrived that the engine being once set a-moving, all things proceed according to the artificer's first design, and the motions of the little statues that as such hours perform these or those motions do not require (like those of puppets) the peculiar interposing of the artificer or any intelligent agent employed by him, but perform their functions on particular occasions by virtue of the general and primitive contrivance of the whole engine.[5]

In Boyle's opinion therefore, what you have is just blind law, functioning endlessly. For this reason, he thought it made no sense to ask about purposes or ends. One might object of course that machines do indeed have ends, but historically people quickly dropped this implication of the metaphor. The focus was much more on blind laws working without stop. In the words of one of the great historians of the Scientific Revolution, purposes dropped out and God became a "retired engineer."[6]

The Challenge of Biology

There was nevertheless a major fly in the ointment of the new science. Organisms did not seem to fit comfortably into this world picture. Whilst it was agreed that it made no sense to talk about purposes when dealing with inanimate objects—no one asks about the purpose of the moon—it surely did make very good sense to ask about purposes when dealing with inanimate objects. The eye and hand are surely for seeing and grasping, respectively. Organisms, or at least parts of organisms, seemed to have purposes. This was a point fully recognized by Boyle himself. He wrote in *A Disquisition about the Final Causes of Natural Things*:

> For there are some things in nature so curiously contrived, and so exquisitely fitted for certain operations and uses, that it seems little less than blindness in him, that acknowledges, with the Cartesians, a most wise Author of things, not to conclude, that, though they may have been designed for other (and perhaps higher) uses, yet they were designed for this use. As he, that sees the admirable fabric of the coats, humours, and muscles of the eyes, and how excellently all the parts are adapted to the making up of an organ of vision, can scarce forbear

to believe, that the Author of nature intended it should serve the animal to which it belongs, to see with.[7]

Boyle continued that supposing that "a man's eyes were made by chance, argues, that they need have no relation to a designing agent; and the use, that a man makes of them, may be either casual too, or at least may be an effect of his knowledge, not of nature's." But not only does this then take us away from the urge to dissect and to understand—how the eye "is as exquisitely fitted to be an organ of sight, as the best artificer in the world could have framed a little engine, purposely and mainly designed for the use of seeing"—but it takes us away from the designing intelligence behind it.

Note that while Boyle did not see this position of his as something threatening to the mechanical position, he certainly did see it as something taking us beyond mechanism. Moreover, although he saw the role of final causes in theology—"it is rational, from the manifest fitness of some things to cosmical or animal ends or uses, to infer, that they were framed or ordained in reference thereunto by an intelligent and designing agent"[8]—he still saw a role for final causes in science: "In the bodies of animals it is oftentimes allowable for a naturalist, from the manifest and apposite uses of the parts, to collect some of the particular ends, to which nature destinated them. And in some cases we may, from the known natures, as well as from the structure, of the parts, ground probable conjectures (both affirmative and negative) about the particular offices of the parts."[9]

In the study of organisms, final causes—a legacy (some might say hangover) from the old organic metaphor—were still around. The machine metaphor could not move right in. It was also a belief that persisted right through the eighteenth century. For instance, the great German philosopher Immanuel Kant (1790) argued explicitly that in order to understand organisms, we must ask questions in terms of ends. Final causes may have been banished from physics and chemistry, but they were very much alive in botany and zoology. Unhappily, Kant admitted that final causes seem to point to the designing efforts of a deity; but, realizing that such talk is inadmissible in modern science, the best he could do was to say biologists have to act as if there was a deity, without necessarily committing themselves to the real existence of such a being. Somewhat petulantly, having failed to find a fully adequate solution to the problem of final causes, Kant turned round and blamed the science itself! Famously, he said: "There will never be a Newton of the blade of grass." Biology is forever barred from being as good a science as physics.

Charles Darwin

It was this challenge that Charles Darwin set out to meet and vanquish in his work on evolutionary theory, discovered and formulated mainly in the late 1830s but published only twenty years later in 1859 in his great work *On the Origin of Species.* Darwin accepted fully the existence of final causes, in the sense that he thought the distinguishing aspect of organisms is that their parts are complex, exquisitely design-like, and thus in need of explanation in terms of ends or purposes. Darwin accepted that the eye is for seeing and that the hand is for grasping. This way of thinking was inculcated when Darwin was an undergraduate at the University of Cambridge, where he was required to read the *Natural Theology* (1802) of Archdeacon William Paley. He accepted that viewpoint then and forevermore. Final causes were a given for Charles Darwin. However, he saw that the Kantian solution is really no solution at all. It is really not good enough to say that everything looks as though it were designed and created by God, but that we must as scientists treat this as a pretend story. It was not that Darwin had anything against God. In fact, until almost the end of his life he believed in a deity of some sort or another. But he thought that it is impermissible to bring even a pretend God into science. One must provide a fully mechanical solution of an order such as is to be found in physics and chemistry.

The solution Darwin offered to this problem was his theory of evolution through natural selection. He argued that all organisms are the products of a long, slow, natural process of change, meaning that they are the products of an entirely law-bound process of development without any special guidance or design. To speak specifically to the design-like nature of organisms, Darwin invoked his special mechanism of natural selection. He argued first that there will always be an ongoing process of struggle brought on by the pressures of population growth.

> A struggle for existence inevitably follows from the high rate at which all organic beings tend to increase. Every being, which during its natural lifetime produces several eggs or seeds, must suffer destruction during some period of its life, and during some season or occasional year, otherwise, on the principle of geometrical increase, its numbers would quickly become so inordinately great that no country could support the product. Hence, as more individuals are produced than can possibly survive, there must in every case be a struggle for existence, either one individual with another of the same species, or with the individuals of distinct species, or with the physical conditions of life. It is the doctrine of Malthus applied with manifold force to the whole animal and vegetable kingdoms; for in this case there can be no artificial increase of food, and no prudential restraint from marriage. Although some species may be now increasing, more or less rapidly, in numbers, all cannot do so, for the world would not hold them.[10]

He then combined this struggle with the undoubted existence of variation within populations, arguing that not all organisms will survive and reproduce and that those that are successful will on balance be different from those that are not.

> Can the principle of selection, which we have seen is so potent in the hands of man, apply in nature? I think we shall see that it can act most effectually. Let it be borne in mind in what an endless number of strange peculiarities our domestic productions, and, in a lesser degree, those under nature, vary; and how strong the hereditary tendency is. Under domestication, it may be truly said that the whole organisation becomes in some degree plastic. Let it be borne in mind how infinitely complex and close-fitting are the mutual relations of all organic beings to each other and to their physical conditions of life. Can it, then, be thought improbable, seeing that variations useful to man have undoubtedly occurred, that other variations useful in some way to each being in the great and complex battle of life, should sometimes occur in the course of thousands of generations? If such do occur, can we doubt (remembering that many more individuals are born than can possibly survive) that individuals having any advantage, however slight, over others, would have the best chance of surviving and of procreating their kind? On the other hand, we may feel sure that any variation in the least degree injurious would be rigidly destroyed. This preservation of favourable variations and the rejection of injurious variations, I call Natural Selection.[11]

The important thing to note about natural selection is that it does not just produce change, but it produces what modern biologists call "adaptations." Things like the hand and the eye come about because of those would-be ancestors who had proto-eyes and hands survived and reproduced, and those that did not have such features vanished without trace.[12]

Mendel and Beyond

As is well known, after *The Origin*, although people very rapidly accepted Darwin's message of evolution, few indeed were convinced of the power of natural selection.[13] Various other mechanisms were proposed, from random jumps—what were often known as "saltations"—to the inheritance of acquired characteristics, so-called Lamarckism. This was a state of affairs that persisted for over half a century. However, early in the twentieth century people rediscovered the work of the Austrian monk Gregor Mendel, and before long a powerful theory of heredity, or what came to be known as "genetics," was developed.[14] Particular credit must go to the American biologist Thomas Hunt Morgan, working with students at Columbia University in New York in the second decade of the new century. Then around 1930 a number of mathematically gifted biologists—notably, Ronald Fisher and J. B. S. Haldane in England and Sewall Wright in

America—brought together what was now called Mendelian genetics with the Darwinian mechanism of natural selection. Very quickly, a new theory was developed and before long empirical scientists were working hard both in the field and in the laboratory to find evidence of what was known as "neo-Darwinism" or "the synthetic theory of evolution."[15]

To use a hackneyed phrase, evolutionary theory had gotten its paradigm, and for some eighty or so years now biologists have been extending and developing this paradigm theory. Today's evolutionary biologists are simply of the opinion that Kant was wrong. There was a Newton of the blade of grass and his name was Charles Darwin—aided a little by Gregor Mendel! Mechanism has conquered in the biological sciences, no less than it conquered in the physical sciences. As always, Richard Dawkins puts the matter forthrightly. Organisms are, in his felicitous phrase, no more and no less than "survival machines."

> We are survival machines, but 'we' does not mean just people. It embraces all animals, plants, bacteria, and viruses. The total number of survival machines on earth is very difficult to count and even the total number of species is unknown. Taking just insects alone, the number of living species has been estimated at around three million, and the number of living insects may be a million, million, million.
>
> Different sorts of survival machines appear very varied on the outside and in their internal organs. An octopus is nothing like a mouse, and both are quite different from an oak tree. Yet in their fundamental chemistry they are rather uniform, and, in particular, the replicators which they bear, the genes, are basically the same kind of molecule in all of us—from bacteria to elephants. We are all survival machines for the same kind of replicator—molecules called DNA—but there are many different ways of making a living in the world, and the replicators have built a vast range of machines to exploit them. A monkey is a machine which preserves genes up trees, a fish is a machine which preserves genes in the water; there is even a small worm which preserves genes in German beer mats. DNA works in mysterious ways.[16]

Noting that there have been great explanatory successes right across the spectrum from sociobiology to development, it should be emphasized how one important factor has been the way in which the molecular advances in biology have been brought into the evolutionary picture, and how today's neo-Darwinian biology relies very heavily on our understanding of micro-phenomena and processes. This illustrates strongly a point to which we shall be coming in a moment—namely, that together with mechanism today's biology is firmly committed to a related notion, that of "reductionism." There is a strong belief that in order to understand machines, you must break them down into smaller components, before building them up again into working entities. In biology, as in physics and chemistry, small really is beautiful.

Objectors

Is this all that there is to be said? Is Richard Dawkins right? Many think not. From the time of *The Origin* on—from well before the Origin, in fact—there have always been those who challenge the presumptions of mechanism in biology. There are those who argue that the full picture can only be achieved by making reference to the deity—if not to the Christian deity, at least to some form of Intelligence. Darwin's great American supporter, Asa Gray,[17] always wanted to supplement the evolutionary process with (what we would call) guided mutations, meaning new variations which, in some very strong sense, are caused and directed by the divine being (which for Gray, an evangelical Presbyterian, was certainly the God of the Bible). Today we have the so-called Intelligent Design Theorists who argue that the living world manifests instances of supposed "irreducible complexity."[18] They argue that these call for direct interventions. I shall not tarry over these and like thinkers. Gray was certainly an important scientist and a loyal friend of Darwin. I shall simply say, as did Darwin against Gray, that to make appeals to divine interventions is take the discussion out of science. I shall agree with Darwin and the majority of evolutionary biologists since his day that the cure is worse than the disease. Not, I would hasten to add, that Darwin or today's evolutionary biologists will talk of "disease" in the first place. They think that natural selection is entirely adequate. In speaking of selection as "entirely adequate" this is not to claim that mechanists think that natural selection is the only mechanism of change. Darwin, for one, relied on many subsidiary mechanisms, including his own secondary form of selection that operates purely at the sexual level. The point of importance here is that Darwin and his followers think that we can offer entirely natural mechanisms and a turn to God is simply not needed.

About one hundred years ago, there was another movement that tried to break with mechanism in a dramatic way. I refer to the so-called "vitalists," notably in Germany, Hans Driesch with his concept of an "entelechy"[19] and in France, Henri Bergson with his concept of an "*élan vital*."[20] These figures argued in an overt neo-Aristotelian fashion that we must suppose some vital forces which are involved in the life processes—forces governing living organisms as well as directing the overall course of life's history. However, again I think we can dismiss this challenge to mechanism with little or no discussion. Even at the time of their introduction, critics noted that it really is not acceptable in the light of modern science to posit such forces as entelechies or *élans vitaux*, nor (as with overt appeals to God) does the introduction prove particularly helpful. Postulating vital causes leads to no new insights or to a stronger theory methodologically. No new prediction ever came from an appeal to entelechies. No new unification ever came from an appeal to *élans vitaux*. This is not to deny that some very influential thinkers, including scientific thinkers, were inspired by vitalism. Notably, Julian Huxley—the grandson of Thomas Henry Huxley and the author of the seminal work *Evolution: The Modern Synthesis*—was very much taken with Bergson's ideas.[21] One suspects strongly that Huxley's lifelong commitment to the progressive nature of the evolutionary process was thoroughly Bergsonian. Yet Huxley himself was the first to say that vitalism is a philosophy and not science. It may inspire, but it does not explain. Anyone who is to be a serious scientist simply cannot go that way.

So with these two contenders out of the way, the remaining really crucial question is whether or not there is a third way between the hard-line mechanists of the Richard Dawkins variety and a religious or vitalistic approach that truly takes one outside the confines of genuine science. There are certainly those who think that Darwinian natural selection is totally inadequate as a force for evolutionary change, and who hark back to the earlier viewpoint of people like Plato and Aristotle. However, whether they are actually breaking with or supplementing mechanism is another matter. One such person was the early twentieth century Scottish morphologist D'Arcy Wentworth Thompson.[22] He had little time for natural selection. He preferred rather to stress the formal aspects of organisms, arguing that there is some kind of self-organization that comes about through the normal processes of physics and chemistry. Much influenced by Aristotle, he instanced such phenomena as the morphology of jellyfish, which he thought could be explained entirely in terms of the mechanics of heavy liquids falling through lighter liquids, and which thus had no explanatory need of natural selection. However, whether he thought that he was actually breaking with mechanism is perhaps another matter. He can certainly be read as saying that we can have completely mechanical explanations of the living world. It is just that there is no need for natural selection. I would add that people like Darwin and Dawkins undoubtedly would disagree and would argue that, apart from a few, isolated, unimportant instances, the adaptive complexity that we see in the living world simply cannot be explained by physics and chemistry. If D'Arcy Thompson thought otherwise, it can only be because in some way he was putting special direction into his physical models. He may not have been an explicit vitalist, but there is certainly the odor of spirit forces about what he claims.

Organicism Redux

More interesting and pertinent here are those who are explicit in their desire, if not to break with mechanism, then at least to supplement it. Unlike the Intelligent Design Theorists, past or present, unlike the vitalists, these are people who are firmly in the camp of science. That is a given. These people go under a variety of names including "emergentist" and "holist"—and, perhaps most revealingly, as "organicist," the term I shall favor here. These are people who agree that the machine metaphor is powerful and important but who do hark back in a very distinctive way to the organic metaphor also. They think that the adoption of the new need not and should not mean total repudiation of the old. There is some confusion about what is meant by these various terms, but one can see a general pattern. Aristotle again is often quoted, particularly his claim that the whole is more than the sum of the parts. The essential idea is that if one looks at the total integrated entity, then there's something more than just the individual components. A new form, perhaps a new existent, emerges. A triangle is made of three lines, but as a triangle is it something more than the lines. In a way, its triangle-ness could not have been predicted from its line-ness. An organism is a collection of molecules, but the organism as organism could not be deduced or predicted from those molecules. The organism is a new level of being.

One has to be careful here. Suppose one thinks of the clock—the paradigmatic example of a machine, in the eyes of the mechanists—it is surely more than merely the individual wheels, cogs, and springs. It is a complex mechanism telling the time. So obviously holists think that there is something more than just this. They argued that at some level one has a more powerful kind of emergence, where new properties or entities appear. John Stuart Mill writes thus about the idea:

> All organised bodies are composed of parts, similar to those composing inorganic nature, and which have even themselves existed in an inorganic state; but the phenomena of life, which result from the juxtaposition of those parts in a certain manner, bear no analogy to any of the effects which would be produced by the action of the component substances considered as mere physical agents. To whatever degree we might imagine our knowledge of the properties of the several ingredients of a living body to be extended and perfected, it is certain that no mere summing up of the separate actions of those elements will ever amount to the action of the living body itself.[23]

I should say that I think that one of the most important sources of organicism was the Romantic movement at the beginning of the nineteenth century, especially as represented by the poet Johann Wolfgang von Goethe, and (above all) the philosopher Friedrich Schelling.[24] It was a movement that spread from Germany to other countries, and notably was a major element of the American Transcendentalist movement as represented by such major thinkers as Ralph Waldo Emerson and Henry David Thoreau. Staying with science, an important emergentist was the physiologist J. S. Haldane, the father of J. B. S. Haldane. Writing in the journal *Mind* in the 1880s, Haldane argued that the machine metaphor (his example was a steam engine) simply cannot work or apply in the case of organisms. He stated, "The energy which leaves the machine is dissipated on the surroundings in a manner which, so far as our conception of the machine is concerned, has no reference to the future maintenance of its mechanism in action. When the fuel is finished, the machine cannot replace it, but simply runs down." Continuing: "If, now, we turn to the organism, we find that the energy which leaves it is by no means dissipated at random on the surroundings. On the contrary it is a characteristic fact that a part at least of that energy is so expended as directly or indirectly to bring about the maintenance of the organism in activity."[25] Note the Platonic-Aristotelian concern here with the end-directed nature of the process. Organisms are as if designed to promote their own well-being and continued existence. They do not just work. They function. Haldane was keen to stress that organisms can respond to changed circumstances. They have within them the ability to regroup and try again. He saw in the processes of life something more than just the sum of the parts.

Holism-emergentism was a very important motivating force for many biologists, especially German biologists, in the early decades of the twentieth century. In this particular case, it clearly resonated with broader socio-political concerns about the need of a unified society, first reflecting earlier yearnings for a Germany that united politically as it had earlier been only by language and culture, and then (particularly after the First World War) by desires to have a state that was more than (what many perceived to be) something cobbled together in the aftermath of defeat. Writing at the end of the War, one prominent biologist lamented the situation, but in a rather chilling premonition of what was to come cheered with this reflection:

> A healthy basic energy exists in the Germans, and there will also never be a dearth of spiritual leaders who can cause a state organism to develop out of this material. When the sickness has worn itself out, and the stuff responsible for infection has been neutralized, the new process of growth can begin. For me, the cry for Order that is now ringing loose from hearts is the first sign of a beginning resistance.[26]

Whether from common cause or cause and effect, the sad fact is that this kind of thinking persisted in Germany. A good many National Socialists were deeply committed to holism, and with good reason saw the philosophy as exemplified by their State. "All in all, the National Socialistic conception of state and culture is that of an organic whole." What are the implications? "As an organic whole, the *völkisch* state is more than the sum of its parts, and indeed because these parts, called individuals, are fitted together to make a higher unity, within which they in turn become capable of a higher level of life achievement, while also enjoying an enhanced sense of security. The individual is bound to this sort of freedom through the fulfillment of his duty in the service of the whole."[27] Naturally enough, Jews were identified as ultra-reductionistic in their thinking: "The Jew is always attempting to split all things, to break them down to their atoms and in this way can make them complicated and so incomprehensible that a healthy person can no longer find his way in the jumble of contradictory theories."[28]

Harvard Holists

Crossing the Atlantic, holism took a very different turn. Particularly influential were a number of scientists at Harvard University. These included the biochemist L. J. Henderson,[29] the physiologist W. B. Cannon,[30] and above all the ant specialist William Morton Wheeler.[31] They all saw a need to go beyond the workings of blind mechanism to a form of holism, as they viewed nature. In Cannon's case this led to his most famous work on homoeostasis, where he saw the body—particularly the human body—as a kind of self-regulating machine, but one which taken as a whole is more than just a machine. In the same mode, Wheeler thought of the ant colony as a kind of superorganism.

> An ant society, therefore, may be regarded as little more than an expanded family, the members of which cooperate for the purpose of still further expanding the family and detaching portions of itself found other families of the same kind. There is thus a striking analogy, which is not escaped the philosophical biologist, between the ant colony and the cell colony which constitutes the body of a Metazoan animal....

Wheeler went on to say that the

> queen mother of the ant colony displays the generalized potentialities of all the individuals, just as a Metazoan egg contains *in potentia* all the other cells of the body. And, continuing the analogy, we may say that since the different castes of the ant colony are morphologically specialized for the performance of different functions, they are truly comparable with the differentiated tissues of the Metazoan body.[32]

This superorganismic idea was picked up by a number of thinkers, including the Chicago termite specialist Alfred Emerson, and, more recently, Wheeler's own intellectual grandson, Edward O. Wilson, also of Harvard.[33] Wheeler was the supervisor of Frank Carpenter, who was in turn the supervisor of Wilson.

For all of these thinkers, as you pull things together, you get something more, a kind of new biological entity emerging. Moreover, thinking within this perspective really has bite when it comes to thinking about mechanisms. Most particularly, although probably none of these holists wants to deny the significance of Darwin's mechanism of natural selection, they would all push it in ways unacceptable to Darwin, and to his successors (especially to Richard Dawkins). Mention has already been made of reductionism, another philosophy which often goes hand-in-hand with mechanism. In Darwin's case, this link came out very strongly in the way that he regarded the working of selection. Specifically, Darwin thought that large-scale biological phenomena are to be explained in terms of natural selection working not at the group level but rather at that of the individual.[34] It is individual organisms that struggle, and groups do so only inasmuch as their members struggle. It is individuals that get variations not groups. Hence, it is individuals that get selected and adapted, not groups. The human species is not intelligent—it is individual humans. And when we have group features, for instance cooperation among members, these features exist only inasmuch as they benefit the individuals. No one cooperates for the good of the group; one cooperates only because it is in one's own interests. Mentioning Richard Dawkins yet again, since the days of Darwin this reductionism has been taken to an even-further extreme because now we have natural selection working at the level of the gene—the so-called "selfish gene." Obviously this is a metaphor, but the point is clear. Genes get passed on if and only if they have features that favor their own success. Being nice for its own sake is a lousy biological strategy.

Holists like Wheeler and Wilson deny absolutely that such an approach can give a full explanation. They argue that only by allowing that natural selection can operate at the level of the group can one explain many intra-group qualities, particularly those involving reciprocation ("altruism") between group members. For these thinkers, mechanism is an inadequate perspective. To capture the richness of biological phenomena, one must in some sense revert to the organicist model. Think of the group as if it were an individual.

> Consider genetic variation of traits such as nest construction, nest defence, provisioning the colony for food, or raiding other colonies. All of these activities provide public goods at private expense. All entail emergent properties based on cooperation among the colony members. Slackers are more fit than solid citizens within a single colony, but colonies with more solid citizens have the advantage of the group level.[35]

Meaningless or Moral?

We have now a sense both of the coming of mechanism and of those who resist its calls. I do not pretend to have given a full picture. For instance, one thing I have not talked about is consciousness. As it happens, it too in recent years has felt the blast of mechanism, particularly with the sub-metaphor (of the overall machine metaphor) of the brain as a computer.[36] Cognitive scientists have made great inroads on the problem with their assumption that our thinking organs are indeed machines, although machines of a particular kind and sophistication. How far cognitive science has been or ever can be successful in making mechanism triumphant in this area is a matter of some considerable debate. Some think it has succeeded already.[37] Some do not.[38] And there are some who now argue that we will never explain consciousness simply in terms of machines.[39] Interesting and important though these issues are, however, I will pass over them here. They do not alter the overall point of this discussion. The machine metaphor has been incredibly successful, and there are those—I suspect most scientists (including biologists)—who see no reason to doubt its continuing success, and adequacy. There is, however, a group, including very good scientists, who continue to look back to the organicist metaphor and who would (at the least) supplement mechanism by using (updated versions of) this older perspective.

Let me turn now from the historical to the more philosophical. Let us grant that the mechanical model or metaphor is very powerful and has led to stunning scientific advances. Let us grant, at least for the sake of discussion, that this has occurred not just in the physical sciences but (thanks to Darwin) in the biological sciences also and that today cognitive science is showing the power of the metaphor when applied to the brain and to thinking and so forth. The question then is why people are dissatisfied with the metaphor, thinking that it must be supplemented in some way, and as a related question why people would want to return to the organic metaphor. I am not such a cynic as to pretend or claim that the facts of the matter are irrelevant. They are. If we think of Darwin as a mechanist, then it is equally correct to think of the co-discoverer of the mechanism of natural selection, Alfred Russel Wallace, as a holist, as an organicist. I am not sure he ever used these words of himself, but his thinking as a scientist is deeply committed to such a perspective. And a good mark is that whereas Darwin was ever an individual selectionist, Wallace from the first thought that group selection was a powerful force in nature. And he had facts, and problems to back up this commitment. For instance, when it came to the sterility of hybrids like the mule, Darwin could see no way of explaining this from an individual selection perspective. It could never be in the interests of the parental horses and donkeys to have sterile offspring. He had to argue that the sterility is just a by-product of having two different reproductive systems combined. Wallace, to the contrary, argued strongly that it is in the interests of the parental species not to have hybrid offspring competing with the good true offspring of the species and so sterility occurred, for the good of the species not the individual.[40]

In like manner, in the twentieth century, we find people arguing the same way. For instance, the late Stephen Jay Gould argued that we should (in a fashion started by the Romantics) look at organisms as wholes, bound by underlying groundplans or *Baupläne*, rather than fragmented in a reductionistic fashion into separate characteristics, each part produced independently by natural selection.[41] By thus freeing biology from the tyranny of individualistic selection, Gould argued that we can see that evolution does not have always to be smooth and in adaptive focus. It can move or break sharply at times, going from one form to another. This insight became the foundation of his celebrated (some would say, notorious) theory of punctuated equilibrium, where he (and his colleague Niles Eldredge) saw periods of non-change, stasis, followed by periods of rapid change, going from one form or *Bauplan* to another.[42] Although admittedly all of this was contested strongly by the Darwinians, the point is that the discussion was at the level of fact and theory. What is the fossil record like, does it show gaps, and how do we explain them? Is it all a matter of incomplete fossilization or is there something more profound (and biological) at work?

One could keep this argument going for some time. The brain and consciousness are obviously things that organicists are going to say are inexplicable under the mechanistic viewpoint. But let us now simply grant that facts and theory are important—I mean let us grant this fully, willingly, without reservation. Is there anything more? I confess that I am struck with the bitterness that is often expressed across the divide between hard-line mechanists and those with organismic inclinations. Listen, for example, to John Maynard Smith, the doyen of Darwinian evolutionists in the second half of the twentieth century, when he gets going on the subject of the science of Stephen Jay Gould.

> Gould occupies a rather curious position, particularly on his side of the Atlantic. Because of the excellence of his essays, he has come to be seen by non-biologists as the preeminent evolutionary theorist. In contrast, the evolutionary biologists with whom I have discussed his work tend to see him as a man whose ideas are so confused as to be hardly worth bothering with, but as one who should not be publicly criticized because he is at least on our side against the creationists. All this would not matter, were it not that he is giving non-biologists a largely false picture of the state of evolutionary theory.[43]

When I come across comments like this, I start to think of the points made by Thomas Kuhn, when writing about what he called "paradigms," in his great *The Structure of Scientific Revolutions*.[44] He stressed that we have different world pictures, different perspectives, that the divides are not simple facts or reason, but different commitments, as we find in religion or politics. It is not just a question of right or wrong, or brighter or more stupid, but really something more a-rational. I should say that there is some encouragement in Kuhn's later writings for my suspicion, because he stressed the significance of metaphorical thinking in science, even going so far as to identify paradigms with metaphors. Yet, although I think this may be an encouraging line of thought, I would stress that I don't think one can simply apply Kuhn's thinking to the present situation. Most importantly, he seems to have thought of paradigms as existing sequentially. Now you have one paradigm. Now you change to another. It seems to me that in the organicist-mechanist case, apart from some overlap (I do not see why a person strongly inclined one way should not also accept some aspects of the alternative view), we have ongoing perspectives. Call them paradigms if you will, but recognize that they are not about to be defeated or rejected. In some (often uneasy) way, they co-exist.[45]

But agree there is something in Kuhn's analysis. What then? Although Kuhn does not stress this, I (with many who have written on the history of science in the years since Kuhn) start to wonder if moral commitments are also playing a role. After all, if people do care strongly about things, then often values are not far from the surface. Let me say straight out that in the case of organicism, while I do not want to say that it is impossible that someone has no value motives at all, I think for many organicists such a case can be made fairly readily. Take just the case of Edward O. Wilson. It is clear that he is deeply motivated by moral issues. Understanding the biological world leads straight into understanding the social world. Most importantly, as the novelist E. M. Forster says in *Howards End*: "Only connect!" Extending the group perspective out from the species to the rest of life, Wilson sees humans as having evolved in symbiotic relationship with the entire living world. We need animals and plants, not just for physical sustenance—although that is very true—but also for spiritual sustenance.[46] A world of plastic would be, quite literally, deadly. We have an urge to "biophilia." This means that our highest ethical imperative must be to save the planet, the living things upon it. Preserving the Brazilian rainforests, preserving biodiversity generally, is not just a whim of middle-class people in the First World. It is the ultimate ethical imperative. "To summarize: a sense of genetic unity, kinship, and deep history are among the values that bond us to the living environment. They are survival mechanisms for ourselves and our species. To conserve biological diversity is an investment in immortality."[47] Note that, very much in the tradition of emergentism-holism, Wilson happily derives moral dictates from the state of nature. "The biologist, who is concerned with questions of physiology and evolutionary history, realizes that self-knowledge is constrained and shaped by the emotional control centers in the hypothalamus and limbic systems of the brain." We must ask, what "made the hypothalamus and limbic system?" The answer and its consequences follow at once. "They evolved by natural selection. That simple biological statement must be pursued to explain ethics and ethical philosophers, if not epistemology and epistemologists, at all depths."[48] For a man with this philosophy, it follows naturally that nature is no dead, unfeeling thing. It is in a very real sense living and morally good.

And this is a point reiterated recently by Wilson in a magazine article about his most recent work.

> Wilson posits that two rival forces drive human behavior: group selection and what he calls "individual selection"—competition at the level of the individual to pass along one's genes—with both operating simultaneously. "Group selection," he said, "brings about virtue, and—this is an oversimplification, but—individual selection, which is competing with it, creates sin." That, in a nutshell, is an explanation of the human condition.

> "Our quarrelsomeness, our intense concentration on groups and on rivalries, down to the last junior-soccer-league game, the whole thing falls into place, in my opinion. Theories of kin selection didn't do the job at all, but now I think we are close to making sense out of what human beings do and why they can't settle down."

> By settling down, Wilson said, he meant establishing a lasting peace with each other and learning to live in a sustainable balance with the environment.[49]

Conclusion

Famously, the great Pragmatist William James divided thinkers into two groups: the tender-minded and the tough-minded. Is this what I am suggesting now? The organicists are tender-minded folk, who put values into nature, who see it as essentially good because ultimately it is all organic or living and life is a good thing? If you felt like pushing this, you could drive it all the way back to Plato and the *Timaeus*, where he argues that the world is an organism. Since the world spirit is in some sense a manifestation of the Form of the Good, everything is in some sense good. I have to say that if you look at some of the writings of the mechanists on their perspective, they certainly portray themselves as being tough-minded. Richard Dawkins, unexpectedly, has thoughts on this issue. For his reading of mechanism, all is blind law without meaning. In a universe of blind physical forces and genetic replication, some people are going to get hurt, other people are going to get lucky, and you will not find any rhyme or reason in it, nor any justice. The universe we observe has precisely the properties we should expect if there is, at bottom, no design, no purpose, no evil and no good, nothing but blind, pitiless indifference. As that unhappy poet A. E. Houseman put it:

> For Nature, heartless, witless Nature
> Will neither know nor care.
> DNA neither knows nor cares. DNA just is. And we dance to its music.[50]

Is this all there is to be said? It makes sense. Mechanism came along, having expelled meaning and morality from the world. David Hume is the patron saint here with his famous distinction between matters of fact and matters of morality or obligation, and the barrier he puts up to those who would derive the latter from the former. "For as this *ought*, or *ought not*, expresses some new relation or affirmation, 'tis necessary that it should be observed and explained; and at the same time that a reason should be given; for what seems altogether inconceivable, how this new relation can be a deduction from others, which are entirely different from it."[51] Some people did not like this. They wanted to stay within the bounds of science, and they did not want to invoke God or unverifiable forces. Nevertheless, they reverted to the older organic model, keeping some form of meaning and morality right in the world.

But I suspect that it is a little more complicated than this. I cannot believe that in the case of someone like Richard Dawkins, of all people, values of some kind are not involved. I suspect at least at one level, there is the desire to keep values out of the science because of the worry that the wrong values may get into the world. Remember the Nazis. Better to add the values to the world, rather than find them in the world. I suspect at another level the very belief that science is value free is itself a value commitment! There is felt to be something clean and decent, invigorating, about a value-free world. If it is stern and harsh, then so be it. The grown-up does not flinch from this but accepts it and makes his or her own meaning nevertheless. I certainly get the sense that when Darwin opts for the struggle for existence between individuals and not groups, this is a motivating factor. And everything that Dawkins writes drips this sentiment. It's a tough world out there, and we had better start to realize this.

So I conclude. Mechanists argue one way. Organicists argue another way. After many years of fighting in the trenches on these issues, my suspicion is that ultimately the mechanist-organicist debate is not going to be solved by facts or even by reasoning. It is much more a matter of commitments, almost in the way that Thomas Kuhn speaks of commitments that come through paradigms.[52] Some of us see things one way, and some of us see things another way. Some of us are happy with mechanism. Some of us are not. I stress again that it is not a matter of being in science or not—although certainly there will be partisans who make this kind of charge—nor is it really a matter of being tough-minded or tender-minded—although again there will be partisans who make this kind of charge—and it is really not something where there are no pertinent facts and theories. It is a matter of different metaphors and, as we philosophers know only too well, that is as deep and profound a difference as one can get.

Notes

1. Michael Ruse, *Science and Spirituality: Making Room for Faith in the Age of Science* (Cambridge: Cambridge University Press, 2010).
2. David Sedley, *Creationism and Its Critics in Antiquity* (Berkeley: University of California Press, 2008).
3. A. Rupert Hall, *The Revolution in Science, 1500–1750* (London: Longman, 1983).
4. Michael Ruse, "Robert Boyle and the Machine Metaphor," *Zygon* 37 (2002): 581-596.
5. Robert Boyle, *A Free Enquiry into the Vulgarly Received Notion of Nature*, eds. Edward B. Davis and Michael Hunter (Cambridge: Cambridge University Press, 1996), 12-13.
6. Eduard J Dijksterhuis, *The Mechanization of the World Picture* (Oxford: Oxford University Press, 1961), 491.
7. Robert Boyle, "A Disquisition about the Final Causes of Natural Things," in *The Works of Robert Boyle*, ed. Thomas Birch (Hildesheim, Germany: Georg Olms, 1966), 397-398.
8. Boyle, "Final Causes," 428.
9. Boyle, "Final Causes," 424.
10. Charles Darwin, *On the Origin of Species by Means of Natural Selection, or the Preservation of Favoured Races in the Struggle for Life* (London: John Murray, 1859), 63-64.
11. Darwin, *Origin of Species*, 80-81.
12. Michael Ruse, *Darwin and Design: Does Evolution Have a Purpose?* (Cambridge, MA: Harvard University Press, 2003).

13. Michael Ruse, *The Darwinian Revolution: Science Red in Tooth and Claw* (Chicago: University of Chicago Press, 1979).
14. William B. Provine, *The Origins of Theoretical Population Genetics* (Chicago: University of Chicago Press, 1971).
15. Michael Ruse, *Monad to Man: The Concept of Progress in Evolutionary Biology* (Cambridge, MA: Harvard University Press, 1996).
16. Richard Dawkins, *The Selfish Gene* (Oxford: Oxford University Press, 1976), 22.
17. Asa Gray, *Darwiniana* (New York: D. Appleton, 1876).
18. William A. Dembski and Michael Ruse, eds., *Debating Design: Darwin to DNA* (Cambridge: Cambridge University Press, 2004).
19. Hans Driesch, *The Science and Philosophy of the Organism* (London: Black, 1908).
20. Henri Bergson, *L'évolution Créatrice* (Paris: Alcan, 1907).
21. Julian Huxley, *Evolution: The Modern Synthesis* (London: Allen and Unwin, 1942).
22. D'Arcy W. Thompson, *On Growth and Form* (Cambridge: Cambridge University Press, 1917).
23. John Stuart Mill, *A System of Logic Ratiocinative and Inductive,* ed. J.M. Robson (Toronto: University of Toronto Press, 1974), III.6. §1.
24. Robert J. Richards, *The Romantic Conception of Life: Science and Philosophy in the Age of Goethe* (Chicago: University of Chicago Press, 2003).
25. John S. Haldane, "Life and Mechanism," *Mind* 9 (1884): 27-47.
26. Anne Harrington, *Reenchanted Science: Holism in German Culture from Wilheim II to Hitler* (Princeton: Princeton University Press, 1999), 58.
27. Harrington, *Reenchanted Science,* 176.
28. Harrington, *Reenchanted Science,* 182-184.
29. Lawrence J. Henderson, *The Order of Nature* (Cambridge, MA: Harvard University Press, 1917).
30. Walter B. Cannon, *The Wisdom of the Body* (Cambridge, MA: Harvard University Press, 1931).
31. William M. Wheeler, *Ants: Their Structure, Development and Behavior* (New York: Columbia University Press, 1910).
32. Wheeler, *Ants,* 7.
33. Gregg Mitman, *The State of Nature: Ecology, Community, and American Social Thought, 1900-1950* (Chicago: University of Chicago Press, 1992).
34. Michael Ruse, "Charles Darwin and Group Selection," *Annals of Science* 37 (1980): 615-630.
35. David S. Wilson and Edward O. Wilson, "Rethinking the Theoretical Foundation of Sociobiology," *The Quarterly Review of Biology* 82 (2007): 341.
36. Paul Thagard, *Mind: Introduction to Cognitive Science* (Cambridge, MA: MIT Press, 2005).
37. Paul M. Churchland, *Matter and Consciousness* (Cambridge, MA: MIT Press, 1984).
38. John Searle, *The Mystery of Consciousness* (New York: New York Review of Books, 1997).
39. Colin McGinn, *The Mysterious Flame: Conscious Minds in a Material World* (New York: Basic Books, 2000).
40. Ruse, "Charles Darwin."

41. Stephen J. Gould and Richard C. Lewontin, "The Spandrels of San Marco and the Panglossian Paradigm: A Critique of the Adaptationist Programme," *Proceedings of the Royal Society of London, Series B: Biological Sciences* 205, no. 1161 (1979): 581-598.

42. Niles Eldredge and Stephen J. Gould, "Punctuated Equilibria: An Alternative to Phyletic Gradualism," in *Models in Paleobiology*, ed. T. J. M. Schopf (San Francisco: Freeman Cooper, 1972); Stephen J. Gould and Niles Eldredge, "Punctuated Equilibria: The Tempo and Mode of Evolution Reconsidered," *Paleobiology* 3 (1977): 115-151.

43. John Maynard Smith, "Genes, Memes, and Minds," *New York Review of Books* 42, no. 19 (1995): 46.

44. Thomas Kuhn, *The Structure of Scientific Revolutions* (Chicago: University of Chicago Press, 1962).

45. Michael Ruse, "The Darwinian Revolution: Rethinking Its Meaning and Significance," *Proceedings of the National Academy of Science* 106, supplement 1 (2009): 10040-10047.

46. Edward O. Wilson, *Biophilia* (Cambridge, MA: Harvard University Press, 1984).

47. Edward O. Wilson, *The Future of Life* (New York: Vintage Books, 2002), 133.

48. Edward O. Wilson, *Sociobiology: The New Synthesis* (Cambridge, MA: Harvard University Press, 1975), 3.

49. Howard French, "E. O. Wilson's Theory of Everything," *The Atlantic*, November 2011.

50. Richard Dawkins, *A River out of Eden* (New York: Basic Books, 1995), 133.

51. David Hume, *A Treatise of Human Nature* (Oxford: Oxford University Press, 1940), 3.1.1.

52. Kuhn, *Scientific Revolutions*.

Bibliography

Bergson, Henri. *L'évolution Créatrice*. Paris: Alcan, 1907.
Boyle, Robert. "A Disquisition about the Final Causes of Natural Things." In *The Works of Robert Boyle, Vol. 5*, edited by Thomas Birch, 392-444. Hildesheim, Germany: Georg Olms, 1966.
Boyle, Robert. *A Free Enquiry into the Vulgarly Received Notion of Nature*. Edited by Edward B. Davis and Michael Hunter. Cambridge: Cambridge University Press, 1996.
Cannon, Walter B. *The Wisdom of the Body*. Cambridge, MA: Harvard University Press, 1931.
Churchland, Paul M. *Matter and Consciousness*. Cambridge, MA: MIT Press, 1984.
Darwin, Charles. *On the Origin of Species by Means of Natural Selection, or the Preservation of Favoured Races in the Struggle for Life*. London: John Murray, 1859.
Dawkins, Richard. *A River out of Eden*. New York: Basic Books, 1995.
———. *The Selfish Gene*. Oxford: Oxford University Press, 1976.
Dembski, William, and Michael Ruse, eds. *Debating Design: Darwin to DNA*. Cambridge: Cambridge University Press, 2004.
Dijksterhuis, Eduard J. *The Mechanization of the World Picture*. Oxford: Oxford University Press, 1961.

Driesch, Hans. *The Science and Philosophy of the Organism*. London: Black, 1908.
Eldredge, Niles, and Stephen J. Gould. "Punctuated Equilibria: An Alternative to Phyletic Gradualism." In *Models in Paleobiology*, edited by T. J. M. Schopf, 82-115. San Francisco: Freeman Cooper, 1972.
French, Howard. "E. O. Wilson's Theory of Everything." *The Atlantic*, November 2011.
Gould, Stephen J., and Niles Eldredge. "Punctuated Equilibria: The Tempo and Mode of Evolution Reconsidered." *Paleobiology*, no. 3 (1977): 115-151.
Gould, Stephen J., and Richard C. Lewontin. "The Spandrels of San Marco and the Panglossian Paradigm: A Critique of the Adaptationist Programme." *Proceedings of the Royal Society of London, Series B: Biological Sciences* 205, no. 1161 (1979): 581-598.
Gray, Asa. *Darwiniana*. New York: D. Appleton, 1876.
Haldane, John S. "Life and Mechanism." *Mind* 9 (1884): 27-47.
Hall, A. Rupert. *The Revolution in Science, 1500-1750*. London: Longman, 1983.
Harrington, Anne. *Reenchanted Science: Holism in German Culture from Wilheim II to Hitler*. Princeton: Princeton University Press, 1999.
Henderson, Lawrence J. *The Order of Nature*. Cambridge, MA: Harvard University Press, 1917.
Hume, David. *A Treatise of Human Nature*. Oxford: Oxford University Press, 1940.
Huxley, Julian S. *Evolution: The Modern Synthesis*. London: Allen and Unwin, 1942.
Kant, Immanuel. *Critique of Judgment*. New York: Haffner Press, 1951.
Kuhn, Thomas. *The Structure of Scientific Revolutions*. Chicago: University of Chicago Press, 1962.
McGinn, Colin. *The Mysterious Flame: Conscious Minds in a Material World*. New York: Basic Books, 2000.
Mill, John Stuart. *A System of Logic Ratiocinative and Inductive*. Edited by J.M. Robson. Toronto: University of Toronto Press, 1974.
Mitman, Gregg. *The State of Nature: Ecology, Community, and American Social Thought, 1900-1950*. Chicago: University of Chicago Press, 1992.
Paley, William. "Natural Theology." In *Collected Works IV*. London: Rivington, 1819.
Provine, William B. *The Origins of Theoretical Population Genetics*. Chicago: University of Chicago Press, 1971.
Richards, Robert J. *The Romantic Conception of Life: Science and Philosophy in the Age of Goethe*. Chicago: University of Chicago Press, 2003.
Ruse, Michael. "Charles Darwin and Group Selection." *Annals of Science* 37 (1980): 615-630.
———. *Darwin and Design: Does Evolution Have a Purpose?* Cambridge, MA: Harvard University Press, 2003.
———. "The Darwinian Revolution: Rethinking Its Meaning and Significance." *Proceedings of the National Academy of Science* 106, supplement 1 (2009): 10040-10047.
———. *The Darwinian Revolution: Science Red in Tooth and Claw*. Chicago: University of Chicago Press, 1979.
———. *Monad to Man: The Concept of Progress in Evolutionary Biology*. Cambridge, MA: Harvard University Press, 1996.
———. "Robert Boyle and the Machine Metaphor." *Zygon* 37 (2002): 581-596.
———. *Science and Spirituality: Making Room for Faith in the Age of Science*. Cambridge: Cambridge University Press, 2010.

Searle, John. *The Mystery of Consciousness*. New York: New York Review of Books, 1997.
Sedley, David. *Creationism and Its Critics in Antiquity*. Berkeley: University of California Press, 2008.
Smith, John Maynard. "Genes, Memes, and Minds." *New York Review of Books* 42, no. 19 (1995): 46-48.
Thagard, Paul. *Mind: Introduction to Cognitive Science*. Cambridge, MA: MIT Press, 2005.
Thompson, D'Arcy W. *On Growth and Form*. Cambridge: Cambridge University Press, 1917.
Wheeler, William M. *Ants: Their Structure, Development and Behavior*. New York: Columbia University Press, 1910.
———. *Emergent Evolution and the Social*. London: Kegan Paul, Trench, Trübner, and Co., 1927.
Wilson, David S., and Edward O. Wilson. "Rethinking the Theoretical Foundation of Sociobiology." *The Quarterly Review of Biology* 82 (2007): 327-348.
Wilson, Edward O. *Biophilia*. Cambridge, MA: Harvard University Press, 1984.
———. *The Future of Life*. New York: Vintage Books, 2002.
———. *Sociobiology: The New Synthesis*. Cambridge, MA: Harvard University Press, 1975.

Chapter Eighteen
Machines and Organisms:
The Rise and Fall of a Conflict

Philip Clayton

In these pages, I challenge the view that over the course of modern history the successes of the natural sciences lend ever-increasing support to a mechanistic worldview. What *is* true is that the scientific method continued to produce better and better knowledge of the biosphere; bit by bit we have come to understand more and more pieces of the puzzle. But that is not the same as a victory for mechanism. For at the same time we have also gained amazing knowledge of what organisms are—of their unique properties and of how they function.

Commentators up to, and including, the present have found it difficult to sort out the data and concepts that pertain to this question. After all, if the quality of scientific explanations grows stronger over time, does that not mean that the mechanistic worldview has triumphed? And what does "mechanism" mean, anyway? It is worth devoting a few pages to the attempt to sort out some of the widespread confusion on this topic.

There is another reason why this topic is important. The debate about the scientific study of organisms lies at the heart of the present book. Many of the authors will defend the claim that the more accurate understanding of the biological sciences—and of the philosophy of biology—comes when one does not identify science with mechanism. The scientific method, these authors argue, is about more than "mechanistic" interactions; science is competent to provide reliable knowledge for a vast variety of systems. These include classically Newtonian systems, of course, but also a vast variety of other kinds of systems. The goal of the present chapter is to clarify the underlying concepts and to review the history that one will need in order to make sense of the claims that follow.

Foundations of the Mechanistic Worldview

The discussion of biological systems is dominated by a confusion about what makes something a machine and what constitutes an organism. Several of the reasons for the confusion have historical roots. Only when we have exposed those roots will we be able to untangle the confusions that still dominate discussions of biological systems today.

The scientific method was invented for classical systems. Historians often cite Galileo's early experiments with rolling objects down inclined planes. Detailed observations by the Danish astronomer Tycho Brahe played a crucial role in achieving an adequate theory of celestial dynamics. Yet nothing seemed more to support a mechanistic worldview than Newton's three laws: the Law of Inertia, $F = ma$, and Newton's third law of motion, "for every action there is an equal and opposite reaction."

"There is also a rhythm and a pattern between the phenomena of nature," writes Richard Feynman, "which is not apparent to the eye, but only to the eye of analysis; and it is these rhythms and patterns which we call Physical Laws."[1] Newton's Law of Gravitation, for example, recognizes that two bodies exert a force between each other, which varies inversely as the square of the distance between them and varies directly as the product of their masses. Feynman calls this "the greatest generalization achieved by the human mind."[2] With it in hand, one recognizes that a vast range of distinct motions—the cannonball fired from a cannon; the orbits of the earth, moon, and all other planets; the movement of galaxies; and the apple falling from the tree—all are manifestations of a single regularity, tied together by a single law linking every object in the universe. The brilliance of Newton's laws was that they proposed a single quantitative dynamical system that could in principle account for the motion of all physical bodies whatsoever. It was this assumption, among others, that made possible the birth of the scientific method.

The fascination with machines, with automata, also played a central role in the rise of mechanistic philosophies. As Derek Price writes,

> By the time of Shakespeare, man's ancient dream of simulating the cosmos, celestial and mundane, had been vividly recaptured and realized through the fruition of many technological crafts. . . . The new automata were to capture the imagination of the next generation, including Boyle and Digby and Descartes himself. Their very perfection would lead to the next phases: automation of rational thought—a stream that leads from Pascal and Leibnitz through Babbage to the electronic computer, of memory by means of the punched tape first used in sixteenth century Augsburg hodometers, and of the cybernetic stuff of responsive action perceived dimly in the Chinese south-pointing chariot, decisively in the thermostatic furnace of Cornelius Drebbel, and more usefully than either in the steam-engine governor of James Watt.[3]

Newton's *Principia Mathematica* (1687) and the quest for automation are linked. Newton's three laws were valuable not only for providing knowledge of the natural world. They also held out the promise of manipulating that world. The law-bound world of Newton only became the foundation for a *mechanistic* worldview through the idea of, and fascination with, building machines. Mathematical physics met *Homo faber*, the builder, and with their marriage modernity as we know it was born.

But Newton is critical for a second and more complex reason. His inverse square law recognized an extremely general gravitational attraction. Applying the law required one to treat each object as a point-mass, its actual dimensions being irrelevant to the computation. It also allowed one to do the computations without having the foggiest idea what kind of force gravity is. Remember, in the era of mechanism there should be no "spooky action at the distance." Work is supposed to be done by things bumping into things (to put it crassly) or, if one is building a machine, through the "lawful ordering of the parts." Newton's three laws are much more concrete, easily conceivable in terms of objects colliding. But what kind of force can it be that draws distant planets together? How does this gravitational force function, and through what medium is it conveyed?

Historically, probably no other single development in modern physics more encouraged the development of a mechanistic philosophy of science than did the inverse square law. Ironically, it played this role by calculating without explaining. Gravity was able to support a mechanistic worldview only because one acted *as if* gravity were a mechanistic force, the kind of force we utilize when (say) we throw a rock. In the first place, metaphysical questions were suspended so that scientists could concentrate on making and testing predictions based on the inverse square law. Then, after a while, the mechanists forgot the "as if" status of this assumption and began to take the resulting physics as the triumph of mechanism. Obviously, however, not raising a question is not the same as answering it. Scientists today are beginning to make sophisticated conjectures of how the four forces of fundamental physics might be related. But one cannot say that the natural philosophers of the seventeenth century made great progress in closing in on an answer to this question.

Some Misguided Foundations for Organicism

I have introduced the early history of mechanism because it sheds some significant light on the "mechanism *versus* organicism" debate that dominated the history of biology through much of the modern period.[4] To be a "mechanist" about biology meant to insist that explanations of living systems could ultimately be given in terms of the interactions of the forces and particles of physics. These forces were understood to work "mechanically," that is, in the way that man-made machines and automata work, or the way one billiard ball strikes and moves another, or (to jump forward to the nineteenth century) in the way that a steam piston operates. To be an "organicist" then meant, by implication, to deny that premise.

Unfortunately, the proponents of organicism did not have a very impressive arsenal of theories, or even decent metaphors, at their disposal—it would take some 150 years after Darwin's *Origin of Species* before that situation could be rectified. The proposals that the organicists brought forward to defend themselves against the mechanists were drawn from outside of science, and frequently from pre-scientific philosophical and religious systems. As the scientific weaknesses of these various proposals came to light, intelligent people interpreted the results as falsifying organicism in all its forms (present and future), and therefore as demonstrating the truth of mechanism. The failed suggestions included divine sources of movement, "occult" sources of movement, conscious sources of movement, vitalism (whether based on Aristotle's entelechy or on the *vis vivendi* or "living force"), and agent-based explanations. Let us consider each of these in turn.

(1) If you want to understand a philosophical position, sometimes the most efficient way is to observe its most extreme manifestation. Why was it plausible for Descartes to think that a spiritual or intellectual substance (*res cogitans*) could directly influence physical objects (*res extensa*)? Surely it was because *God*, as a purely spiritual substance, could and did regularly influence any and all objects in the natural world. One can see the logic of this position with remarkable clarity in the twentieth century defense of miracles by C. S. Lewis. He writes,

> Nature (at any rate the surface of our own planet) is perforated or pock-marked all over by little orifices at each of which something of a different kind from herself—namely reason—can do things to her.... If God annihilates or creates or deflects a unit of matter He has created a new situation.... Immediately all Nature domiciles this new situation, makes it at home in her realm, adapts all other events to it.[5]

On this view, God the Creator is Lord over natural law; He thus can and does set aside natural laws to bring about His purposes. Nature's task is to "domesticate" the new situation and to adapt all subsequent events to it. Descartes may have been particularly foolish to locate a specific *physical* point of contact, the pineal gland in the brain. But the logic of his argument had impeccable theological credentials at the time.

The most extreme form of this theological form of explanation came in the doctrine of occasionalism. We associate occasionalism with the seventeenth century philosopher Malebranche, although it actually has its roots in one of the most influential medieval Arabic philosophers, al-Ghazali. For the occasionalists, *every* action in the physical world is directly caused by God. We may postulate that the physical event that preceded that action caused it, but we are confusing their conjunction with a real causal relationship. (Notice the overtones of the "constant conjunction" argument that David Hume would develop a century later). The better argument, Malebranche believed, is that the omnipotent God directly causes each result for His greater glory. The doctrine of occasional divine miracles tells a similar causal story but decreases the frequency of the interventions.

(2) The strong influence of alchemy, hermeticism, and the magical arts, paired with the growth of secularity after the Thirty Years War (1618-1648), inspired another of the other major alternatives to mechanism. When later scientists and naturalist philosophers spoke derisively of explanations based on "occult forces," they had these three groups in mind.[6] Alchemists claimed to offer better explanations of physical events than the mechanists could provide, and for a while it seemed that they might be right. They worked madly in their labs to convert base metals into gold and to demonstrate empirically the existence of dark and not-so-dark occult forces. Had they been right, they would have demonstrated the existence of a wide variety of natural forces that would have supplemented what is today the standard repertoire of energies and particles. Over time, however, the alleged demonstrations failed to materialize, while the experiments in what would eventually become modern chemistry were producing more and more reliable results.

(3) Leibniz's influential philosophy was not alchemy. His theory of monads was, however, the paradigmatic example of an alternative to mechanism that relied on an appeal to conscious forces. According to the *Monadology*, all reality consists of individual mental units or monads. Each monad acts out its own individual essence. The monads are "windowless," but God coordinates their actions through a "pre-established harmony," so that the net result is a perfectly aligned world, indeed, the best of all possible worlds. Leibniz's philosophy is not occasionalism. Instead of one divine agent being responsible for all the events that occur in the world, all the individual conscious agents together are responsible for the patterns that we observe and that we attribute to natural laws. This move makes consciousness—whether that of the individual monads or of God, the monad of monads (*monadus monadum*), as the Creator and Organizer of the whole system—the ultimate cause for all patterns in the biosphere.

Leibniz was a major architect of a second central feature of pre-Darwinian biology, the *plenum* or "fullness" of eternal and unchanging species. God, being perfectly good, would have to create a natural order with no holes, no missing spots.[7] If between two species that there could possibly be a third, then that third must exist, lest the creation be less than complete, i.e., perfect. Once the "great chain of being" was created, it could not change, or the perfect order would be disturbed. Hence all species must have had their origin at the same moment of creation, and all must be preserved by divine providence, *without change*, until the end of history. The influential work of the Comte de Buffon, Georges-Louis Leclerc, in France in the late eighteenth century, especially in his massive *Histoire naturelle*, still assumed something like this Leibnizian framework as its basis. It was the collapse of this worldview that Alfred Lord Tennyson was alluding to in his famous poem, to which we also owe the phrase "red in tooth and claw":

> Are God and Nature then at strife
> That Nature lends such evil dreams?
> So careful of the type she seems
> So careless of the single life, . . .
>
> "So careful of the type?" but no.
> From scarped cliff and quarried stone
> She cries a thousand types are gone;
> I care for nothing, all shall go.[8]

For in the "scarped cliff and quarried stone" biologists found fossils of species that clearly no longer existed—"a thousand types [species] are gone." With this discovery the firm foundation for a theologically oriented Leibnizian biology was cast into question. Once again, by default, it seemed the victory was being handed to the mechanists.

(4) Vitalism has a long and noble history; one finds versions of it in most of the indigenous cultures and in most ancient civilizations. The great physician Galen was a vitalist, and Aristotle has been called one as well. Early modern vitalism owes much to Leibniz's successor, Caspar Friedrich Wolff (1733-1794). Wolff explained biological development by appealing to an "essential force" or *vis essentialis*. By 1781 Johann Friedrich Blumenbach began to speak of epigenesis—a separate kind of becoming—and attributed it to the "formative drive." The debate between epigenesis and "preformationism" would rage for most of the following century. Famous nineteenth century vitalists included the physicians Marie François Xavier Bichat and John Hunter, Jons Jakob Berzelius, Johannes Peter Muller, Carl Reichenbach, and Hans Driesch. All advocated for the existence of a "living principle," a life-energy distinct from physical laws and forces, in addition to mechanics. The connection between organicism and holism has its source here as well, for it was the living principle that made of each organism an indivisible whole.

By the 1930s, neo-Darwinism was on the rise, and vitalism fell out of favor.[9] Unfortunately, the organismic perspective in biology had been closely tied with vitalism, and through it with holism, for over a hundred years. When empirical science broke with vitalist assumptions, organicism went out with the bath water as well.

Agent-Based Explanations

(5) The result was unfortunate, however. By this point, organism-level studies in biology had some more impressive empirical data to draw on. If these data and studies had been separated from the flawed philosophical models, the history of biology over the last fifty years might have looked rather different. Whatever may be the strengths and weaknesses of British Empiricism, it does tend to make one look more closely at the natural world. One effect of the establishment of the Royal Society in 1660 was to foster careful observation of species. The activity of carefully observing led to better and better classification systems, and those in turn gradually produced sciences such as botany, ornithology, and ethology as we know them today.

The strength of observation-based science is that it is non- or even anti-ideological. The great tradition of fieldwork in Britain and under British influence in the eighteenth and nineteenth centuries had as its primary goal to find descriptive categories that were adequate to the plant and animal species and behaviors being observed. These early scientists were not trying to make their mark by defending any "-ism," whether mechanism or organicism. They approached the natural world with a certain natural piety, seeking categories suggested by the structures and functions that they observed in the organisms before them.

Among the categories that turned out to be crucial for this purpose was, not surprisingly, the category of organism. When one observes plants and animals in the natural world, it certainly appears that organisms do things. The lion pursues and eats its prey, the bird feathers its nest, and the mother protects her young. Organisms also clearly engage in purposive actions, although it is a philosophical question whether or not they do so consciously. For observers of animals, the category of purposive action is a metaphysically minimalist one. The beaver gnaws through the tree in order to fell it, and she fells it in order to build her dam. To observe a bird, fox, or squirrel carefully is to recognize a myriad of means-ends behaviors. It is difficult to spend much time watching animals at work or at play without concluding that they are engaging in these various actions in order to produce specific outcomes. The best natural observers are those who are able to identify which actions individuals and groups are carrying out and what results they are seeking to achieve by their actions.

This natural attitude of the field empiricists came under severe attack, however. Scientists, motivated by the long-term goal of giving the most rigorous possible explanation of natural phenomena, wanted to know *why* plants and animals respond as they do. It turns out that explanations at smaller scales are generally more rigorous. There is also a progression: research in genetics and microbiology establishes the mechanisms that underlie behaviors; biochemistry and eventually physical chemistry identify the substratum of genetics; and fundamental physics describes the forces and particles that underlie chemistry.

One would think that there would be a strong cooperative relationship between both types of knowledge. Fieldwork explains living systems at the level of individual organisms as they interact with other individuals and with their environment, while lower-level explanations describe some of the mechanisms that help to bring about these actions and reactions. In principle there is no inconsistency between these two approaches. Observational knowledge of animal behaviors on the one hand, and the knowledge of structures that microbiology affords on the other, could be complementary aspects of the one scientific quest.

But the history of microbiology did not in fact unfold with this sort of complementarity. Instead, the lower-level explanations tended to *replace* explanations at the level of organisms. The implication was clearly that when one understands the biochemistry or the mechanisms of gene expression, one does not really need the organism-level explanations any longer. The large tended to be superseded by the small. What lay behind this "replacement thesis," presumably, was this same centuries-long debate between the mechanists and the organicists that we have been tracing. Thanks to that debate, the "both-and" became "either-or."

Beyond the Mechanism-Organicism Dichotomy

The various chapters in this book have shown convincingly that this dichotomy is unnecessary and damaging to the progress of science. Some hold that more recent developments in science *de facto* are moving decisively between the horns of the old dilemma. Systems biology, evolutionary developmental biology, complex systems dynamics, and the dynamics of self-organizing systems—just to mention a few of the examples cited in these pages—are not saddled with the sharp contrasts of the early modern battle between mechanism and organicism.

Another group argues that the mistake lies in the theory of knowledge upon which the old "either-or" was grounded. Others defend an understanding of evolutionary biology after Darwin that leaves place for distinct levels of evolutionary explanations. Group selection theory is one example; another involves theories of the evolving systems that are based less on competing genes and more on symbiosis and cooperation between organizations and groups.

Still other authors (the present author included) update what "mechanism" and "organicism" mean, so that the two are not sworn enemies but complementary dimensions of biological explanation. We believe that scientific developments have either falsified mechanism or demand a radically different definition. For example, the constant velocity of light (186,282 miles per second) does not obey the Galilean velocity addition formula, $v'=v1+v2$. The laws of classical mechanics could not be reconciled with the laws of electromagnetic fields.

Space and time are no longer understood in classical terms. In short, contemporary physics has superseded classical mechanics on a wide variety of fronts.

Others, finally, propose philosophical frameworks other than the early modern opposition between machines and organisms, in order to develop a philosophy of biology that allows for a more integrated partnership between microbiology and the study of organisms and ecosystems.

Three of these integrative philosophical models in particular—panpsychism, process philosophies, and philosophies of emergence—played a significant role in the modern period, and each deserves a quick word before we close. I do not here include theological models and models based on human consciousness, since those theories have tended more to support the competition between mechanistic and organism-based approaches to biology. But one should acknowledge for the record that some of the proposals in theology and consciousness studies during the modern period were in fact aimed at integrating rather than dividing.

Pan-psychism is the view that psyche or mind or experience permeates all that exists.[10] Proponents of pan-psychism can be found from the Presocratics through the twentieth century, notably in the sixteenth century Italian Renaissance (G. Cardano, G. Bruno, T. Campanella), in seventeenth century Continental Rationalism (Spinoza, Leibniz), and at the close of the nineteenth century (W. Wundt, R. H. Lotze, William James, Josiah Royce). The attractiveness of the position lies in the fact that, whereas a life principle or consciousness found only in humans and other organisms would separate biology from "mechanistic" physics, an experiential dimension that pervades all of reality might serve as a common ground between them. In the end, however, pan-psychism did not serve as a significant bridge. It was widely seen—also by many of its adherents!—as opposed to the "mechanistic" explanations of early modern science, which made it *de facto* an ally of organicism and a contributor to the early modern split.

Process philosophy[11] has an equally ancient pedigree. It is traceable back to the Presocratic philosopher Heraclitus; it also continually reemerges as the dark heresy alongside the time-transcending categories of Greek substance metaphysics and static medieval worldviews that were based on the immutability of God. But whereas mental units that permeate all reality are a hard pill for empirical scientists to swallow, the pervasive nature of change or motion is a well-established empirical datum and a continuing invitation to scientific study. Natural scientists might not accept the process philosopher's ultimate explanation for change—for example, Hegel's theory of the necessary unfolding of Absolute Spirit. Yet they are very interested in describing discernible patterns in the midst of pervasive change, just as process philosophers also seek to identify and to explain these patterns.

A second connection is just as significant. Process philosophers in the modern period were drawn to the pervasiveness of process itself, which they interpreted as a unifying feature of reality.[12] This focus put them in strong opposition to dualist philosophers such as Descartes and to all forms of theological dualism.

They looked instead for features shared in common by all the various forms of process in the natural world. Is it that reality consists ultimately of events or "occasions"? Is it that systems—webs of connections—are primary, and what we call individuals or things are actually just nodes in the web? Could it be that the *organization* of matter-energy into structures is actually what scientists should be explaining? If so, studying the continual evolution of these structures, with their concomitant properties and functions, becomes the real task of science, rather than the study of "things" in which qualities somehow inhere.

Because process philosophers, like natural scientists, were working to specify commonalities in natural processes, they were not inclined to set mechanistic and organism-based explanations against each other. In the philosophy of biology, for example, they sought for ways to understand changes in objects and changes in organisms as ultimately related. This fact brought process philosophers into close proximity with a third philosophical school, the philosophy of emergence. A famous passage from Samuel Alexander in the early twentieth century nicely expresses some central features of emergence-based approaches:

> The higher quality emerges from the lower level of existence and has its roots therein, but it emerges therefrom, and it does not belong to that level, but constitutes its possessor a new order of existent with its special laws of behaviour. The existence of emergent qualities thus described is something to be noted, as some would say, under the compulsion of brute empirical fact, or, as I should prefer to say in less harsh terms, to be accepted with the "natural piety" of the investigator.[13]

More detailed presentations of the history of emergence philosophies in modern thought are available.[14] Without a doubt, the emergentists gave competing accounts of what constituted emergence, appealing respectively to novelty, unpredictability, new structures, new functions, new properties, and new types of explanations. They differed on the necessary and sufficient conditions for identifying genuine emergence. Only some defended the "strong" emergentist thesis that new objects with new types of causal powers come into being over evolutionary history.[15] The relationship with science was also differently construed: do the phenomena of emergence support scientific research or undercut it? Some identified emergence with what science has not yet explained (but someday will), whereas others equated emergence with what science could never explain, even in principle.

Despite these differences, emergentist philosophies consistently and often powerfully challenged the opposition between mechanism and organicism that dominated much of the modern period. It is just obvious, they claimed, that chemical systems gave rise to the biochemistry of living cells and that without biochemistry there would be no life. We have known this since at least 1828, when Wöhler produced organic compounds from inorganic reactants, effectively giving birth to synthetic organic chemistry.[16] Contrary to vitalism, which maintained that life could only come from the vital force of organic matter or living things, suddenly the boundary between organic and inorganic compounds was breached. Few textbooks today would call chemistry "mechanistic"—that's a philosophical term that just does not do any work in biochemical research. But if one really must use the term, then the right answer is that "mechanical" processes have given rise to living organisms and continue to undergird them moment by moment.

Yet it is patently false to say that organisms are "nothing but" machines. You can know all there is to know about a steam engine, a clock, or a computer, and still fail your first freshman examination in biology. Biological research is not only about qualities that organisms share with non-living objects; it is also, and crucially, about the distinctive features of living systems at all levels, from cells to ecosystems. In many cases, emergentists do in philosophical language what biologists do in scientific language: they describe the relationships between lower-level systems and the higher-level systems—cells, groups of cells, organs, organisms, and ecosystems—to which they give rise.[17] The lower-level phenomena are the necessary components of the systems in which they play a role. Yet these emergent systems *also* manifest properties, functions, and dynamics that are not reducible to the lower-level systems. The contrasts between mechanistic systems and organisms are not grounds for battles or philosophical posturing; they are the stuff of which good biological science is made.

Conclusion

These pages have not been about an easy reconciliation. I have not claimed to offer a final synthesis of microbiology and the studies of organisms in their environments. Biologists are still in the early stages of work on this set of problems. It remains one of the most challenging (and fascinating) tasks of the biological sciences today.

But one thing should be clear: the state of the question has moved far beyond the centuries-old debate between mechanism and organicism. We have found that the classical understanding of both terms falls far below the current level of biological knowledge. Contemporary physics is not mechanistic in anything like the sense that the followers of Newton thought it should be. And the sort of vitalism that dominated the study of organisms for most of the nineteenth century finds few adherents today.

The failure of the "machines versus organisms" debate leaves us right where we should be today. Microbiology—biochemistry, genetics, proteomics, systems biology, and related fields—offers indispensable resources for understanding living systems. Yet, as the various authors in this volume have convincingly shown, a variety of new forms of study above and beyond microbiology are being brought to bear on the questions as well. If we can set aside the old philosophical battles and ideologies, we will presumably be able to understand the unique features of living systems more fully than ever before. Biology has become a far more interesting field of inquiry after the demise of the "machines versus organisms" dichotomy. It would be not just ironic but downright tragic if the understanding of living systems remains hampered by outworn philosophical categories from the past, when an amazing world of interacting organisms lies spread us before us for our observation and study.

Acknowledgments

I am grateful to soon-to-be Dr. Brianne Donaldson for her scholarly help in the background research for this chapter and in compiling the bibliography and references. This chapter is adapted with permission from Philip Clayton, "Why Emergence Matters: A New Paradigm for Relating the Sciences," in *Adventures in the Spirit*, ed. Zachary Simpson (Minneapolis: Fortress, 2008), 64-74.

Notes

1. Richard Feynman, *The Character of Physical Law* (1965; repr. Cambridge, MA: MIT Press, 1992), 13.
2. Feynman, *The Character of Physical Law*, 14.
3. Derek J. de Solla Price, "Automata and the Origins of Mechanism and Mechanistic Philosophy," in *Technology and Culture* 5 (1964): 9-23, quote 22f.
4. See Margaret Dauler Wilson, *Ideas and Mechanism: Essays on Early Modern Philosophy* (Princeton: Princeton University Press, 1999); Robert N. Brandon, "Reductionism versus Wholism versus Mechanism," in *Concepts and Methods in Evolutionary Biology*, ed. Robert N. Brandon (Cambridge: Cambridge Univ. Press, 1996), 179-204.
5. C. S. Lewis, *Miracles: A Preliminary Study* (1947; repr. New York: Macmillan, 1974).
6. Stephen Clucas, *Magic, Memory and Natural Philosophy in the Sixteenth and Seventeenth Centuries* (Farnham, UK: Ashgate Publishing Company, 2011); Roberto Poma, *Magie et guérison: la rationalité de la médecine magique, XVIe-XVIIe* (Paris: Orizons, 2009); Ryan J. Stark, *Rhetoric, Science, and Magic in Seventeenth-Century England* (Washington, DC: Catholic University of America Press, 2009); Penelope Gouk, *Music, Science, and Natural Magic in Seventeenth-Century England* (New Haven: Yale University Press, 1999). See also the classic work of Frances Yates on the hermetic tradition, e.g., *Giordano Bruno and the Hermetic Tradition* (London: Routledge, 2002).
7. Arthur O. Lovejoy, *The Great Chain of Being: A Study of the History of an Idea* (New Brunswick, NJ: Transaction Publishers, 2009).
8. Alfred, Lord Tennyson, *In Memoriam A.H.H.*, sections LV-LVI, http://www.online-literature.com/donne/718/. Accessed July 21, 2012.

9. Mark A. Bedau and Carol E. Cleland, *The Nature of Life: Classical and Contemporary Perspectives from Philosophy and Science* (Cambridge: Cambridge University Press, 2010), 95, quoting John Scott Haldane.

10. Whiteheadians have preferred the term "panexperientialism" since David Ray Griffin first introduced it several decades ago. I have chosen not to use that term here because the connotations of the traditional term are important in a historical exposition such as the present one.

11. Nicholas Rescher, *Process Philosophy: A Survey of Basic Issues* (Pittsburgh: University of Pittsburgh Press, 2000).

12. Douglas Browning and William T. Myers, eds. *Philosophers of Process* (New York: Fordham University Press, 1998); James R. Gray, *Modern Process Thought: A Brief Ideological History* (Washington, DC: University Press of America, 1982); George R. Lucas, *The Genesis of Modern Process Thought* (Metuchen, NJ: Scarecrow Press, 1983).

13. Samuel Alexander, *Space, Time, and Deity*, The Gifford Lectures for 1916-18, 2 vols. (London: Macmillan, 1920), II: 260.

14. See Clayton, "Conceptual Foundations of Emergence Theory," chap. 1 of Clayton and Paul Davies., eds., *The Reemergence of Emergence* (Oxford: Oxford Univ. Press, 2006), 1-31. The work of the British Emergentists in the late nineteenth and early twentieth centuries is nicely summarized in Rudolf Metz, *A Hundred Years of British Philosophy* (London: G. Allen and Unwin, 1938).

15. Clayton, *Mind and Emergence* (Oxford: Oxford University Press, 2004).

16. Robert H. Carlson, *Biology Is Technology: The Promise, Peril, and New Business of Engineering Life* (Cambridge: Harvard University Press, 2010), 33.

17. "Lower" and "higher" are used here without any connotations of higher or lower value. Lower-level systems are usually (relatively speaking) simpler and come earlier in evolutionary time.

Bibliography

Alexander, Samuel. *Space, Time, and Deity*. The Gifford Lectures for 1916-18, 2 vols. London: MacMillan, 1920.

Bedau, Mark A., and Carol E. Cleland, *The Nature of Life: Classical and Contemporary Perspectives from Philosophy and Science*. Cambridge: Cambridge University Press, 2010.

Brandon, Robert N. "Reductionism versus Wholism versus Mechanism." In *Concepts and Methods in Evolutionary Biology*, edited by Robert N. Brandon, 179-204. Cambridge: Cambridge Univ. Press, 1996.

Browning, Douglas, and William T. Myers, eds. *Philosophers of Process*. New York: Fordham University Press, 1998.

Carlson, Robert H. *Biology is Technology: The Promise, Peril, and New Business of Engineering Life*. Cambridge: Harvard University Press, 2010.

Clayton, Philip. *Mind and Emergence*. Oxford: Oxford University Press, 2004.

———. "Conceptual Foundations of Emergence Theory." In *The Reemergence of Emergence*, edited by Philip Clayton and Paul Davies, 1-31. Oxford: Oxford University Press, 2006.

Clucas, Stephen. *Magic, Memory and Natural Philosophy in the Sixteenth and Seventeenth Centuries.* Farnham, UK: Ashgate Publishing Company, 2011.

Feynman, Richard. *The Character of Physical Law.* Cambridge, MA: MIT Press, 1965, reprint 1992.

Gouk, Penelope. *Music, Science, and Natural Magic in Seventeenth-Century England.* New Haven: Yale University Press, 1999.

Gray, James R. *Modern Process Thought: A Brief Ideological History.* Washington, DC: University Press of America, 1982.

Lewis, C. S. *Miracles: A Preliminary Study.* New York: Macmillan, 1947, reprint 1974.

Lovejoy, Arthur O. *The Great Chain of Being: A Study of the History of an Idea.* New Brunswick, NJ: Transaction Publishers, 2009.

Lucas, George R. *The Genesis of Modern Process Thought.* Metuchen, NJ: Scarecrow Press, 1983.

Metz, Rudolf. *A Hundred Years of British Philosophy.* London: G. Allen and Unwin, 1938.

Poma, Roberto. *Magie et guérison: la rationalité de la médecine magique, XVIe-XVIIe.* Paris: Orizons, 2009.

Price, Derek J. de Solla. "Automata and the Origins of Mechanism and Mechanistic Philosophy." *Technology and Culture* 5 (1964): 9-23.

Rescher, Nicholas. *Process Philosophy: A Survey of Basic Issues.* Pittsburgh: University of Pittsburgh Press, 2000.

Stark, Ryan J. *Rhetoric, Science, and Magic in Seventeenth-Century England.* Washington, DC: Catholic University of America Press, 2009.

Wilson, Margaret Dauler. *Ideas and Mechanism: Essays on Early Modern Philosophy.* Princeton: Princeton University Press, 1999.

Yates, Frances. *Giordano Bruno and the Hermetic Tradition.* London: Routledge, 2002.

Index

Abrahamsen, Adele, 28, 50n4, 51n24
abstraction, 32, 36, 39, 42-44, 46-48, 99, 205, 208, 216, 323, 377, 380-81, 388, 391n7, 397n82
acquired characteristics: inheritance of (Lamarck), 95, 188, 259, 350, 375-77, 378, 394n41, 413; transmission of, 95; *See also* Lamarck, Jean-Baptiste; Lamarckism; soft inheritance
actual entities (Whitehead), 100; society of, 44, 238-39
actual occasions (Whitehead), 42, 44, 100, 172, 238, 392n8
adaptability driver (Bateson), 272n12
adaptation, 3, 9-11, 15, 31-32, 52n26, 55n44, 67-68, 70, 77, 84, 94, 96-97, 99-101, 103, 105, 107-11, 148, 155-56, 161, 170, 172, 184, 186-88, 190, 195-96, 214, 244n8, 252, 260-61, 263-64, 274n23, 277n49, 287-89, 310, 314, 324, 327, 330, 332-33, 355-56, 374-78, 410, 413, 416, 420-21, 434; Darwinian pre-, 12-13
adapted states, 98-103, 107, 195
adaptive behavior, 103-5, 107
adaptive boundary, 192-93

adaptive energy converter, 99, 101
adaptive events. *See* events
adaptive landscape, 20, 70, 195-96
adaptive operation modes, 100-1, 103, 107-8
adaptive processes, 99, 107
adaptive radiation, 69
adaptive representational network. *See* network
adaptive response, 99, 186
adaptive state space, 195-96
adenosine triphosphate (ATP), 98, 106-08
adjacent possible, 13-18, 22, 119
Adler, Stephen, 78
agency, 4-6, 70, 76, 85, 125-26, 147-49, 151-52, 155-61, 172-73, 176-77, 218-20, 251, 264-65, 271, 289, 303, 409-10; autobiographical, 176; autonomous, 298-99, 301-02; biotic, 176; causal, 79, 252; mental, 176; proto-experiencing, 311; selective, 265-270, 276n44; subjective, 175
agent, 6, 28, 52n27, 82, 85, 147-49, 155, 160, 172, 173, 177, 193, 218-19, 224, 240, 260, 264, 267n44, 312, 355, 374, 384, 410, 417, 434-35, 437; autonomous, 6, 126, 160; collective, 177; designing, 411; of homeostasis, 240; mental, 318; min-

imal molecular autonomous, 160; mutation-producing, 375; of selection, 46, 264, 265; selective, 266
agriculture, 31, 39, 266, 370, 381-83
alchemy, 435
alcoholism, 384
Alexander, Samuel, 37, 42, 440
al-Ghazali, 434
algorithm, 9-11, 13, 17-18, 21, 54n38-39, 135, 223, 235-37, 252, 278n68, 289, 314
anatomy, 76, 186
anthropocentrism, 152, 222, 224, 265
anthropomorphism, 156-57, 175, 222, 225
antibiotics, 214-15
anticipation, 99-101, 105, 110-11, 152, 158, 170, 172, 334, 336
a priori categories, 29, 31, 53n34, 54n41, 127, 141, 390-91n5
Arab Spring, 20, 22
Aristotle, 2, 7, 19, 34, 44, 136, 156-58, 165n29, 216, 220-23, 267, 290, 295, 309, 311-12, 315-17, 336n2, 346-47, 352, 357, 409, 415-18, 434, 436
artificial intelligence, 28
artificial life, 191, 291
asbestos, 384
Ashby, W. Ross, 292, 356
Ashkenasy, 6-8
Assisted Human Reproduction Act (Canada), 388-389
Augustine, 152
autocatalysis, 16, 121-26, 160, 288, 294-97, 299-300, 302, 328-32; *See also* catalysis; catalyst
autocatalytic chemical reactions, 223
autocatalytic configuration, 122, 124

autocatalytic cycle, 122, 220
autocatalytic molecules, 331
autocatalytic persistence, 296
autocatalytic process, 122, 295-99, 329, 332
autocatalytic set, 6-8, 15-19, 160, 295
autocatalytic systems, 70, 296-97, 299, 301
autocell, 89n40, 160, 166n45, 170, 299, 302
autogen, 160, 299-303, 328-334, 336, 339n52
autogenesis, 298-302, 305n24, 306n31, 328-331, 335
automaton, 34, 52, 236-37, 313, 316-17, 335, 432-33
autonomy, 27, 138, 175-76, 293, 297-98, 300, 303, 324, 334
autopoiesis, 27, 52n27, 54n39, 83-84, 176-77, 210, 222, 295-303, 306n31
Avagadro's Number, 119

bacteria, 5-6, 98, 102-106, 150, 188-89, 207-17, 220, 223, 225, 240, 327, 356, 414; X-, 215
bacteriophages, 208
Baer, Karl von, 348-49
Baer's laws, 328
Baetan, Elizabeth, 176
Baldwin, James Mark, 251, 259-60, 262, 264-65, 272-73n12, 274n29, 274n32, 278n64
Baldwin Effect, 14, 37, 48, 251-52, 259, 261-63, 270-71, 272-273n12, 273n23. *See also* Baldwin, James Mark; Organic Selection
Bary, Heinrich Anton de, 211
Bastian, Henry Charles, 189
Bateson, Gregory, 116-17, 121, 272n12, 274n30

Bechtel, William, 28, 50n4-5, 51n24
behavior, 3-5, 8, 12, 14, 20, 25, 34, 36-37, 39, 45-46, 51n24, 54n41, 60n115, 77, 80-81, 83, 85, 89n45, 94, 100-01, 103, 105, 107, 116-117, 120-21, 125, 139-40, 148, 150, 161-62, 172-73, 175-76, 186, 208-13, 216-19, 221-22, 224, 234-37, 244n8-9, 246n20, 251-52, 259-64, 266, 269-70, 272n12, 274n29, 290, 312-14, 320, 328, 334, 356-57, 369-70, 373, 377, 380, 384, 390n2, 393n28, 423, 437-38
Bénard cells, 183, 210, 321
Bergson, Henri, 237, 289, 311, 415-416
Bernard, Claude, 184-87, 191-93
Bertalanffy, Ludwig von, 310, 356
Berzelius, Jons Jakob, 436
Bichat, Marie François Xavier, 436
Big Bang, 127
biochemical reactions, 98, 101, 154, 174
biochemistry, 38, 76, 80, 82, 85, 54, 289, 369, 371-72, 323, 351, 419, 437-38, 440-41
biological wisdom (Waddington), 41, 267, 370, 382
biology, 4, 6, 10, 16, 25, 27-29, 31-33, 37-38, 41-43, 48-49, 50n4, 54n41, 71, 75, 78-79, 81, 86, 93-94, 96, 98, 115-20, 133-34, 140-41, 147-48, 155-56, 169, 173-76, 183-88, 190-91, 193, 195-96, 208, 210-11, 220, 237, 241, 243, 259, 271, 287, 289, 290, 304n12, 305n24, 309-11, 314-15, 336, 345-46, 351-52, 357, 369, 376, 379-80, 391n5, 410-11, 414-15, 421, 433, 435-37, 439, 442; cell, 50n4, 76, 79, 148, 151, 350-51; developmental, 67-68, 70, 147-48, 151, 193, 273n23, 310, 438; evolutionary, 29, 31-32, 49, 148, 185, 188, 240, 262, 268-69, 271, 289, 438; evolutionary developmental, 346, 438; genocentric, 210, 212; mechanistic, 97; micro-, 190, 437-39, 441; modern, 26, 32, 155, 187, 190-91 197, 236, 353; molecular, 67-69, 95, 149, 161, 185, 188, 190, 208, 236-37, 240, 243, 278n64, 288-89, 350, 356, 376; neo-Darwinian, 45, 278n68, 379, 391n7, 414; neuro-, 301; New Frontiers of, 27, 48, 271; philosophers of, 27, 41, 304n12, 309-311, 315, 317, 336n1; philosophy of, 41, 115, 310-12, 314-15, 336, 431, 439-40; socio-, 234, 244n8, 414; systems-, 3, 27, 48, 75, 78-79, 84, 86, 93, 95, 98, 100, 102, 110, 327, 356, 358, 385, 438, 441
Biology's First Law, 184, 191
Biology's Second Law, 183, 187, 191, 193-94, 197, 241
biomedicine. *See* medicine
biophilosophy, 309-11, 318, 336
biosemiosis, 8, 152, 163
biosemiotics, 8, 27, 48, 133, 152, 155-56, 159-63, 169-73, 265, 336
biosphere, 1, 6, 8-16, 18, 20, 22, 39, 40, 42, 75-76, 141, 162, 172, 223-24, 242, 267, 431, 435
biotechnology, 38-41, 48, 324, 370, 381, 383, 389
bisphenol-A, 384
Bohr, Niels, 158

Boltzmann, Ludwig, 68, 95, 205, 226n3, 320
Bonduriansky, Russell, 379
Boolean Networks. *See* network
boundary conditions, 2, 7, 10, 19, 98, 117, 294, 300, 306n31
Boyle, Robert, 33, 409-11
Brahe, Tycho, 432
Brenner, Sidney, 115-16
Brewster, David, 68
Brown, Gordon, 21
brown anole lizard, 263
Bruno, Giordano, 439
Buchler, Justus, 41, 134-36, 139
Bunge, Mario, 315

Cambrian period, 111; pre-, 37
canalization, 32, 358, 396n73
cancer, 75, 345, 357, 371-72, 385-87, 389, 398n95, 398n103, 399n113
Cannon, W. B., 419
Cardano, G., 439
Carson, Rachel, 163
catalysis, 6, 8, 16-19, 123, 189-90, 295, 300, 328; *See also* autocatalysis
catalyst, 18, 123-24, 295-97, 299-301, 329, 332
categoreal obligation (Whitehead), 108, 113n40
causality, 1, 5, 7, 10-12, 15, 19, 21, 26, 29-30, 41, 43-44, 47, 53n34, 59n100, 59n 103, 67-68, 76, 79-81, 83, 85-86, 94-95, 99, 103, 116, 133, 138, 141, 147, 149-151, 156-57, 159, 165n29, 170-71, 174-75, 189, 209, 213, 215-18, 221-22, 252, 259-60, 264-66, 272n12, 293, 297, 304n12, 331-32, 334-336, 338n34, 347, 351, 370, 374-75, 377-79, 381-82, 384, 386-388, 390n3, 390n5, 394n35, 395n48, 398, 415, 418, 434-35, 440; circular, 79; downward, 37, 78-80, 88n17, 88n21, 139, 169, 171, 295, 318, 322; efficient, 2, 7, 19, 26, 28, 31, 33, 41-42, 51n25, 94, 96, 101, 157-58, 165n32, 264, 295, 352, 376, 380; final, 26, 31, 35, 42, 94-97, 101, 156-58, 165n32, 221, 246n22, 290, 295, 301-03, 304n13, 409-12; formal, 7, 18-19, 94; material, 127, 352; organismic, 328, 334-36; mental, 79; reciprocal, 43, 59n100, 318, 377; semiotic, 156, 158; ultimate, 221, 378-79, 384, 435; *See also* mechanism
central dogma of molecular biology. *See* dogma
centripetality, 123-24
Chardin, Teilhard de, 237, 289
chemical processes, 154-55, 291
chemical reactions, 7, 16-17, 19, 78, 159, 223, 296, 297, 332; *See also* biochemical reactions
chemicals, 7-8m 16-19, 77-78, 82, 84, 85, 106, 123, 138, 141, 159, 169-172, 174, 176-77, 183, 189, 191, 208, 210, 216, 221-22, 224, 235, 242, 244n3, 289, 295, 314, 316, 319, 321-24, 326-27, 329-33, 371, 374, 379, 384, 440
chemistry, 19, 21, 26, 37-38, 59, 76, 78, 80-81, 141, 170, 173, 175, 187, 189-91, 226n2, 236, 243, 290, 294, 297, 298, 323, 326, 409, 411-12, 414, 416, 435, 437, 440-41; *See also* biochemistry
Chetverikov, Sergi, 68
chordate, 37
chromatin, 353-54, 372, 376, 395n49; modification, 39, 385-

86, 398n99; remodeling, 354, 370-73, 381
chromosomal changes, 375
chromosomal markings, 3
chromosomal rearrangements, 70
chromosomes, 95, 153-54, 214, 219, 354, 372, 374, 389
Clayton, Philip, 6, 49, 159-61, 296, 298, 442
climate change, 40, 267; *See also* global warming
cloning, 39, 268, 372, 381-82, 388
Cobb, John, 49, 252, 391n7
cognition, 53n34, 77, 84, 94, 100, 110, 149, 152, 184, 193, 237, 245n12, 304n13, 385; *See also* consciousness
cognitive science, 80, 141, 420-21
Coiter, Volcher, 346
collective individuals, 233-41, 243n1
collective instabilities, 78
collective properties, 57
collectives, 170, 177, 225
colony, 27, 40, 186, 192, 211, 234-37, 239-41, 243n1, 244n3, 244n6, 244n9-9, 245n20, 266, 419, 420
communication, 52, 82, 106-07, 110, 141, 149, 152, 161, 170, 172, 175, 217, 234-36, 244n3, 288, 292, 353, 374, 391n5
competition, 3, 45, 103, 109, 123-24, 195, 268-69, 421, 423, 438
complexes: multiprotein, 372; natural, 41, 134, 136-37, 140-41
complexity theory, 70
complex system, 9, 16, 57n74, 67, 69-70, 75, 78-79, 86, 96, 120, 124, 169, 174, 209-10, 222, 224-25, 251, 253, 287, 355-56, 438
compositionism, 289
compositionist explanation, 289-90

computer, 16, 20, 22, 28, 69, 107, 126, 134, 155, 171, 193, 236, 252, 266, 288, 290, 294, 310, 322-23, 420, 432, 441
concrescence (Whitehead), 42, 58n90, 109, 370, 392n8
connectedness, 45
consciousness, 6, 8, 34, 36, 56n64, 59n94, 77-78, 80, 93, 141, 152, 156, 169, 171-72, 175, 184, 209-10, 218-24, 236-37, 265, 276n49, 277n51, 312, 334, 420, 422, 434-35, 437, 439
constraint, 15, 81, 86, 101-02, 117, 120-21, 125-26, 139, 141, 160, 162, 292-96, 299-303, 306n31, 319, 323, 325, 327-31, 334, 356
coordination, 95, 99, 109, 234-35, 238
Corliss, John B., 128
crane (Dennett), 58, 252, 262
creationism, 224-25, 422, 435
creativity, 5, 14-15, 18, 21-22, 31, 47, 52n26, 59, 70, 80, 96, 100, 111, 151, 276n44, 279n74, 293-94, 297, 300, 371
Crick, Francis, 7, 149-50, 190, 376
Csoka, Antonei, 385
culture, 1, 14, 18, 85, 156, 162, 268, 272n12, 278n60, 418, 436
Cuvier, Georges, 186, 348
cybernetics, 28, 169, 174, 290, 312-14, 334-35, 356, 432
cyclols, 190
cytoplasm, 106-07, 212, 215, 298, 300, 350-52, 357-58, 375-76

Dagotto, Elbio, 78
Daisyworld, 210
Damasio, Antonio, 171-72
Darwin, Charles, 3, 9, 12, 16, 26, 56n64, 57n78, 67-68, 86, 94-

95, 120, 147-48, 151, 184-89, 214, 221, 251, 263, 265, 268-71, 276n45, 279n70, 279n74, 287, 290, 304n1, 306n32, 335, 349, 376, 395n48, 412-13, 415-16, 419-21, 425, 434, 438
Darwin, Erasmus, 186
Darwinian Research Tradition, 67, 69-70
Darwinism, 12-13, 26, 50n7, 67-70, 94, 117, 120, 123, 157, 184-85, 188-91, 196, 212, 214, 241, 251-52, 259, 261, 268-69, 274n32, 279n76, 335, 375-76, 414, 416, 421-22, 435; *See also* neo-Darwinism
Dawkins, Richard, 35, 45, 56n64, 85, 208, 212, 215, 221-22, 236, 240, 242-43, 279n76, 289, 380, 414-16, 419-20, 424-25
Deacon, Terrence, 49, 83-85, 155, 160, 252, 293, 298-300, 302, 305n24, 328, 330-34, 336, 339n52, 356
decision, 22, 101, 158, 208, 235-37, 244n9, 245n8, 245n12, 265-66, 277n51, 352, 380
decomposition, 36, 67, 69, 124, 150, 153, 173-75
deism, 21, 117
Delbruck, Max, 208
demands, 55, 97
Dennett, Daniel, 29, 38, 54n38, 57n78, 225, 252, 270, 278n68, 280n78
deoxyribonucleic acid (DNA), 6, 7, 18, 95, 108-09, 125, 160, 190, 208, 212, 215-16, 219, 222, 288, 294, 334, 345-46, 350, 351, 353-54, 356, 369, 372, 376-78, 386, 389, 396n72, 397n89, 414, 424; code, 68, 80, 369, 397n89; demethylation, 385, 387, 398n99; methylation, 353, 370, 371, 372-73, 378, 381-82, 385-87, 398n99, 399n111; molecule, 149, 169, 209, 356, 414; mutation, 390n3; replication, 8, 212, 279n76, 272; sequence, 3, 95, 110, 194, 289, 352-53, 358, 371, 377-79, 396n73, 389; survival, 279n76;
Depew, David, 67-69, 71
Derrida, Jacques, 220
Descartes, René, 1-2, 6, 9, 11, 19, 33-36, 43-44, 52n26-27, 79, 85, 93-96, 111, 149, 156, 185, 221-23, 238, 265, 380, 383, 394n35, 409-10, 432, 434, 439
descent, 157; with modification, 214
Descent of Man (Darwin), 269, 279n70
desire, 30, 157, 171, 183, 185, 196-97, 209, 221, 224, 261, 266-67, 271, 275n43, 312, 416, 418, 424
determinism, 2-3, 15, 19, 21, 44, 59n100, 67-69, 98, 100, 102, 109-10, 116-21, 124, 134, 138, 140, 148, 151, 156-57, 162, 169-70, 174, 195, 218, 237-38, 241, 245n20, 260, 265, 275n44, 297, 300-01, 312-13, 317-18, 323, 325, 332, 338n34, 339n48, 345, 350-351, 357, 378, 391n7
development, 3-4, 27, 31, 37, 39-40, 44, 53n28, 54n38, 68-70, 82, 84, 93, 95, 100, 102, 110-111, 111n5, 115, 127, 147-48, 150, 154, 172, 174, 188, 193, 211, 235, 236, 246n20, 252-53, 260, 264, 267-68, 274n32, 274n43, 291, 298, 300, 319, 323, 325-27, 331, 345-358, 362n40, 369-72, 376-78, 380-

81, 384, 390n5, 391n7, 395n35, 395n49, 396n73, 399n111, 412, 414, 436; *See also* biology
Dewey, John, 96-100, 237
Dickens, Thomas, 378-79
dissipative structures, 27, 291, 304n12, 306n31
dissipative systems, 27, 70, 287, 292, 303, 319-20, 322, 324-26
dissociation, 302
DNA. *See* deoxyribonucleic acid
Dobzhansky, Theodosius, 26, 68, 214, 216, 374-75, 394n41
dogma, 32, 37, 78, 184; central dogma of molecular biology, 95, 149, 150, 188, 190, 194, 288, 350-51, 356, 376, 379-80
Driesch, Hans, 165n32, 289, 336n2, 415, 436
dualism, 34, 76, 79, 85, 126, 133, 139, 158, 224, 289, 371, 377, 380, 439
Dutch Hunger Winter, 390n2

Earth System, 206, 224, 242-243
eco-evo-devo, 70
ecology, 4, 17, 36, 38-41, 45, 68, 70-71, 97, 102, 116, 119-20, 122-26, 140, 310, 355-56; evolutionary, 273n23; physiological, 273n23; population, 242; process 71, 126; systems, 67, 69, 123
econosphere, 14-15, 18, 20
eco-semiotic interaction structure, 162
ecosystem, 17, 39, 84-85, 116, 119-20, 126-27, 140-41, 162-63, 177, 195, 206, 209, 220, 223, 266, 275n43, 278n61, 439, 441; Clementsian, 192; engineering, 150
Edelman, Gerald, 77

efforts, 97, 222, 234
ego, 218; *See also* ethical egoism
eidos, 311
Einstein, Albert, 10, 14
Eldredge, Niles, 69, 214, 421
electromagnetic field, 83, 140, 313, 438
Elsasser, Walter, 81, 117-19
embodiment, 43, 52n26, 78, 135, 163, 191, 196, 241, 299, 301, 303, 329
embranchements, 348
embryogenesis, 193, 312, 371
embryology, 346-52, 357, 359n11, 370, 391n7, 395n49
emergence, 7, 12, 15, 19, 22, 27, 37, 48, 52n26, 59n94, 69, 75-86, 87n9, 89n45, 94-95, 108, 111, 125, 133, 137, 139-41, 156, 160, 169, 171, 173, 177, 184-85, 188, 236, 238, 242-43, 253, 287, 299, 305n24, 322, 330, 333, 336, 417, 440, 442; of chemical properties, 77; of classical physical systems, 77; of collectively autocatalytic sets, 18-19; of a distributed colony intelligence, 236; of forms of social order, 236; of genetic information, 33; of genetics, 190; of life, 160, 177; of living systems, 69; of mind, 69; of a network perspective, 78; of novelty, 70; radical, 12, 14-15, 22; of self-organizing structures, 322; of semiotic processes, 70; spontaneous, 7
emergence theory, 27, 42, 76-77, 80, 83, 85, 439, 440
emergent complexity, 75, 85
emergentism, 35, 37, 76, 81, 85-86, 137, 141, 416-18, 423, 441, 443n14

emergent processes, 67, 69, 71, 84, 223
emergent properties, 36-38, 42, 44, 54n39, 57n74, 78, 80, 84, 87n6, 133, 139, 170, 173, 175, 330, 353, 356, 420, 440
emergent system, 441
enablement, 8, 20-21
enabling constraint, 15
energy, 98-102, 106, 108, 123, 127, 140, 192, 205-10, 216, 219-20, 222-25, 226n3, 241, 260, 279n68, 293, 295, 297, 306n31, 321-22, 324, 327-29, 331-33, 398n105, 417-18, 436, 440; conversion, 98-101, 107; dissipation, 99, 107-08, 319; flow, 69, 97-99, 208, 210, 225, 306n31; gradient, 216, 293, 329
Enlightenment, 2, 14, 21-22, 116, 124
entailing law, 1, 4, 8-11, 16
entailment, 2, 18, 86, 165n29
ententional phenomena, 326
entropy, 11, 99, 107, 127, 157, 192, 205-06, 209, 221, 223, 225, 226n2-3, 292-93, 295, 298, 300, 302, 313, 320-22, 324-29, 336, 339n41
enzyme, 7-8, 78, 155, 190, 194, 217, 353, 372, 386, 398n103
epigenesis, 7, 70, 345-48, 350-51, 356-57, 358n2, 359n4, 359n6, 361n26, 370, 390n5, 436
epigenetic code, 353-56
epigenetic inheritance, 149, 350, 354, 373, 377, 379, 382, 384, 396n61, 399n113
epigenetic mechanism 354, 378, 381, 387
epigenetic pharmaceuticals, 40, 385-88, 398n103, 398n105, 399n111; *See also* pharmaepigenomics
epigenetic process, 370-72, 377-81, 386, 398n105
epigenetics, 3, 27, 39, 48, 57n65, 58n90, 79, 95, 151, 193, 266, 345-46, 349-58, 358n2, 360n26, 361n34, 362n40, 369-73, 376-89, 390n2-3, 391n7, 393n24, 394n35, 396n72-73, 397n81-82, 397n89, 398n95, 398n99, 398n111, 398n1133; environmental, 345; molecular, 345-46, 353-54, 371, 385
epigenetic switches, 79, 155, 370, 378, 381-82
epigenome, 40, 356, 370, 381, 384, 386, 388-89; *See also* pharmaepigenomics
epigenotype, 345, 354-56, 358, 362n43, 371
epimutation, 371, 388, 396n73
epistemology, 29, 35, 41, 54n44, 59n94, 95, 119, 158, 211, 216, 269, 294, 304n13, 318, 391n5, 391n7, 423
equilibrium, 5, 69, 97, 99, 107, 138, 191-92, 206, 209, 222, 224, 306n31, 319, 321-22, 421; far-from-, 87n9, 222, 287, 291-93, 335
ethical egoism, 270
ethical *praxis*, 268
ethics, 38, 40, 48, 163, 268-71, 279n76, 346, 370, 380-81, 383-84, 389, 423; bioethics, 39, 369, 382-84, 397n82; environmental, 39, 259, 265-66, 270-71, 278n68, 282, 382-84, 384; evolutionary, 259, 265-68, 270, 279n75, 280n79
ethology, 141, 259, 437
eukaryote, 9, 17, 37, 69, 99, 108-11, 305n24, 372

eutrophication, 136, 140
evapotranspiration, 206
events, 30, 44, 94, 116, 134, 136-37, 152-53, 156-57, 161, 170, 191, 207, 221, 237-38, 261-62, 290, 313, 322, 409, 434-35, 440; adaptive, 98-103, 107-08, 110-11, 113n37; biochemical, 155; causal, 44; chromatin remodeling, 372; developmental, 193; environmental, 79, 392n10, extinction, 16, 37; intentional, 14; internally related, 238; macromolecular, 155; mutational, 15, 148; non-gradual, 148; process-relational, 383; quantum 15; random, 15, 95, 119-20, 127; releasing, 31, 54n41; singular, 119, 125; spatiotemporal, 313; speciation 213, 215; stochastic, 119; world, 31, 271
evolution, 1, 3-4, 6, 8-18, 22, 25-26, 31-32, 37, 39, 42-43, 46, 48, 52n26, 54, 58n90, 60n115, 67, 69-71, 76-78, 84-86, 93-96, 102, 108-111, 120, 123, 138-39, 147-48, 150-51, 156-57, 161, 163, 170, 173, 175, 184, 186-88, 191-97, 206-07, 209-16, 220, 223-25, 236, 240-41, 251-52, 259-67, 270-71, 272n12, 274n29, 275n44, 276n44, 278n68, 289, 291, 293, 302-03, 310-11, 314, 331-33, 335-36, 345, 349-50, 352, 354-58, 359n11, 369-70, 372-76, 378-82, 412-14, 421, 423, 440; of the biosphere, 6, 8, 13-14, 16; cosmic, 78; of life, 11, 156, 184, 191; mechanism of, 148, 252, 269, 358, 374; of morality, 270; morphological,

37; by natural selection, 17, 412; psychosocial, 276n44
evolutionary change, 26, 68, 95, 148, 207, 210, 212, 216, 225, 261-62, 266, 271, 355, 358, 376, 380, 416
evolutionary developmental biology (evo-devo). *See* biology
evolutionary epoch, 26, 265, 271
evolutionary explanation, 221, 350, 438
evolutionary neo-Kantianism, 29, 31, 391n5; *See also* Konrad Lorenz
evolutionary process, 17, 52n26, 84-85, 110, 120, 162, 170, 191, 193-94, 211, 251, 264, 269, 271, 272n12, 276n44, 278n68, 289, 291, 332, 371, 415-16
evolutionary psychology, 27, 80, 259
Evolutionary Synthesis. *See* German Synthesis.
evolutionary theory, 10, 67-68, 96, 120, 251, 259, 266, 273n12, 279n79, 289, 291, 310, 349, 378, 412, 414, 422
evolutionary time, 29, 210, 269, 391n5, 443n17
experience, 27, 29-31, 44, 53n34, 54n41, 96-103, 105-06, 108-11, 134, 148-49, 153, 156-57, 162, 171-72, 233, 237-39, 312, 316, 365, 373, 384, 439; conscious, 93, 312; intensity of, 109-10; organismic, 59n94, 111, 265; proto-, 311; second order, 109; subjective, 312; unity of, 239, 316; *See also* pan-experientialism; pan-psychism
exploratory activity, 260, 262

Extended Evolutionary Synthesis, 27, 69-70, 251-52, 259, 262, 271, 377-78; *See also* German Synthesis

externalism, 141, 148, 151, 156, 157

Falkner, Gernot, 49, 104-05, 327
Falkner, Renate, 49, 104-05, 327
fallacy of misplaced concreteness (Whitehead), 47, 328
feelings, 34, 94, 97, 109, 149, 162, 171, 265, 423; conceptual (Whitehead), 108; mental, 265; physical (Whitehead), 59n94, 109, 265; *See also* prehension
Fisher, Ronald, 3, 26, 68, 147, 195, 251, 413
Fleck, Ludwik, 212
flexibility, 116
flux, 100, 110, 127, 138, 192, 216, 321
Food and Drug Administration (FDA), 385-86, 388
function, 4-5, 7, 9, 11-14, 17, 19, 25, 28, 30-31, 33, 35-36, 38-39, 77, 81, 84, 86, 94, 99, 101-02, 108-09, 134-35, 138, 140, 150, 153-55, 160, 165n29, 169, 171, 172-73, 175, 184-86, 193-96, 209-10, 217, 219-20, 234-35, 239-43, 246n26, 260, 265, 270, 276n49, 279n75-76, 288-91, 294, 296, 303, 310-11, 314, 316, 319, 327, 335, 354, 358, 362n40, 370-71, 373, 380-81, 385-86, 390n2, 391n7, 398n103, 410, 418-19, 431, 433, 437, 440, 441; developmental, 69; imprinting, 354; molecular, 148, 151, 289, 290, 358

Gaia, 210, 223-24

Galapagos, 262-63
Galen, 436
Galileo, 33, 409, 432, 438
gemmules, 188, 349
gene, 45, 60n115, 70, 79, 85, 115, 147-52, 155, 184, 187, 191, 193-94, 196, 208-09, 214-16, 218-19, 224, 236-37, 240-42, 245n20, 259, 275n44, 259, 275n44, 289, 310, 339n48, 346, 350-53, 357, 358, 370-72, 374-75, 377, 380-81, 384-87, 391n7, 392n8, 393n24, 395n49, 398n99, 399n111, 414, 420, 423, 438; activity, 345, 351-53, 357, 371, 388; expression, 79, 95, 154, 190, 352-53, 355, 369-72, 376, 379-81, 385-86, 389, 393n24, 398n103; frequency, 46, 68, 212, 289, 370; mechanistic view of, 391n7; mitochondrial, 219; pool, 358; regulation, 353, 355-56; selection, 193, 195-96; selfish (Dawkins), 155, 219, 289, 380, 420; therapy, 387-89; tumor suppressor, 385
genetic accommodation, 352, 398n73
genetic assimilation (Waddington), 370, 396n73
genetic change, 26, 251, 371, 374, 376, 391n7
genetic code, 169, 172, 177, 208, 290, 353, 356-57, 370, 374, 376, 379, 380, 383
genetic copying, 222
genetic determinism, 169, 195
genetic diseases, 389
genetic engineering, 382
genetic enhancement, 389
genetic essentialism, 212

genetic fitness, 241
genetic information, 190, 289, 291, 301, 314, 332-33, 335, 339n48
genetic inheritance, 149, 362n40, 371, 378, 397n89
genetic instruction, 150
genetic/genomic imprinting, 370-73, 381, 387, 393n24
genetic manipulation, 40, 266, 268
genetic material, 369, 375, 384
genetic mechanism, 26, 41
genetic mutation, 148, 212, 215, 373-75, 377-78, 387
genetic program, 314
genetics, 3, 26, 39-40, 68, 70, 76, 82, 85-86, 95, 147-51, 154-55, 169-70, 172, 176-77, 185, 188, 190-91, 193-96, 207-09, 211-12, 214, 219-20, 222, 237, 241, 245n20, 251, 260, 262, 266, 274n29, 276n44, 288-90, 297, 300, 314, 328, 335, 345, 350-54, 357-58, 362n45, 370, 372-73, 375-76, 379, 382-83, 387-89, 391n7, 394n41, 396n72, 423-24, 437, 441; population, 26, 70, 147-48, 376; quantitative, 70 genetic sequence, 212, 380, 388 genetic support, 251-52; *See also* Mendel
genetic variation. *See* variation
genocentrism, 18, 45, 85, 148, 185, 210, 212, 219, 240-41, 252, 262, 334, 358, 370, 377, 380, 383
genotype, 45, 60n115, 95, 151-52, 173, 195, 260-61, 274n29, 350, 352-55, 369, 371, 374-77, 381, 396n73
geo-engineering, 40, 58n87
Geoffroyism, 375
Gerhart, John, 151

German Synthesis, 70.
germ cells, 350, 354, 357, 374, 394n36
germ layer, 348, 357
germ line, 45, 187-88, 190, 251, 371, 379, 387, 389
germ plasm, 95, 244n8, 349-50, 360n18, 373-74, 376-77, 381, 394n35-36
Gilbert, Scott, 70
global warming, 40, 266
Gödel, Kurt, 117
golden mean, 267
Goldschmidt, Richard, 147
good trick (Dennett), 260, 270
Goodwin, Brian, 82, 304n3, 304n12
Gould, Stephen Jay, 12, 37, 50n7, 68-70, 184, 221, 278n64, 395n41, 421-22
gradient reduction, 210, 221, 225
gradualism, 68, 148, 207
Grant, Robert, 186
gravitation, 2, 19, 140, 147, 316-17, 432-33
Gray, Asa, 415
Gray, Russell, 252
Griffiths, Paul, 252, 314

Haig, David, 378, 395n61
Haldane, J. B. S., 3, 26, 68, 147, 220-21, 228n53, 413, 417-18
Hall, Brian K., 49, 352, 355
Haller, Albrecht von, 347, 359n11
Hallgrímsson, Benedikt, 355
hard inheritance, 370, 373, 374, 376-77, 379
Harvard, 155, 419; holists, 419; medical school, 356
Hazen, Robert, 128

health, 21, 33, 35, 39-40, 126, 149, 177, 185, 215-16, 271, 346, 384-86, 388, 390n2, 399n113, 400n117, 418
Heidegger, Martin, 25, 53n31, 58n81
Heisenberg uncertainties, 119
helix: double, 7, 136; triple, 194
Henderson, L. J., 419
Heraclitus, 126, 439
heredity, 26, 39, 41, 159, 186-89, 190, 193, 196, 241, 245n20, 260, 350-51, 356, 371, 373-79, 391n7, 394n35-36, 413; *See also* inheritance, memory
hermeticism, 435
Herschel, John, 68
Hinton and Nowlan, 252
histone: acetylation, 3, 354, 372, 385-87; code, 354, 399n111; deacetylase inhibitors (HDACis), 386; deacetylases (HDACs), 378, 386; deacetylation, 386-87; methylation, 372; modification, 370-72, 378, 381, 386; proteins, 353, 389
HMS Beagle, 186, 263
Hobbes, Thomas, 265
Hoffmeyer, Jesper, 6, 49, 71, 169-71, 173
holism, 35-42, 44, 46-48, 80-81, 185, 242, 267, 370, 377, 380, 382, 384, 416-21, 423, 436
holistic reflection, 38-39, 42, 46-47, 268, 370, 382-83, 397n82
Holland, John, 81
Hölldobler, Bert, 233, 235-37, 244n3, 244n8, 245n9, 245n20, 246n22
Holmes III, Rolston, 176
homeostasis, 48, 176-77, 183-87, 191-94, 196-97, 240-42; *See also* machine
Hordijk, Wim, 16

hormones, 80, 264, 353, 369
house finch, 263-65
human exceptionalism, 223-24
Hume, David, 6, 30-31, 43-44, 53n34, 59n103, 424, 435
Hunter, John, 436
Hutchison, G. Eveleyn, 195
Huxley, Julian, 26, 32, 35, 37, 68, 252, 261, 272n11, 275n44, 278n65, 374, 394n36, 415-16
Huxley, Thomas H., 189, 268, 415
Hypercycles, 299

identity, 98, 100, 102, 110, 134, 175-76, 185, 213, 222-23, 236, 238, 243, 383
IMF, 20
indispensability, 29-32, 35, 39, 42, 53n34, 54n44, 55n46, 135, 441
information, 11, 57n74, 67, 79, 81, 83-84, 95, 99, 101, 103, 105, 107, 135, 139, 149-50, 155, 159, 161-62, 172, 176, 190, 193, 213, 217, 225, 233, 236-37, 244n3, 244n9, 245n12, 287-91, 300-01, 303, 313-14, 330, 332-34, 336, 345, 350, 352, 355, 357-58, 362n45, 369, 376, 380; genetic, 150, 171, 190, 194, 289, 301, 314, 328, 332-33, 339n48, 350, 355, 362n45, 369, 376, 380; molecular, 358; processing, 99-103, 105-06, 108, 150, 313, 333, 335; theory, 11-12, 313, 334
inheritance; *See* epigenetic inheritance, hard inheritance, heredity, soft inheritance, somatic inheritance
Intelligent Design, 415-16
intentionality, 116, 133, 135, 169-70, 172-73, 175, 177, 193-94, 334

interactionism, 79
interpretant, 153-55, 164n18, 171
inventory control system, 155, 159
is-ought fallacy. *See* naturalistic fallacy

Jablonka, Eva, 266, 377-82, 390n2, 395n56, 396n65-66, 396n72
James, William, 54n44, 237, 311, 424, 439
Jeon, Kwang, 215-16
Jonas, Hans, 36, 133, 175-76, 269, 280n79, 311, 313, 316, 336n2, 284
Jukes, Thomas, 69

Kant, Immanuel, 5-6, 16, 25, 28-30, 44, 51n26, 53n28, 53n34, 54n41, 59n100, 59n103, 294-296, 302, 304n12, 304n13, 309, 314-18, 324, 336n2, 336n2, 338n31, 338n34, 390n5, 411-12, 414; *See also* whole
Kauffman, Stuart, 49, 57n74, 69, 81, 85, 88n26, 119, 159-61, 296, 298, 304n12
Keller, Evelyn Fox, 27-28, 51n26, 362n45, 396n73
Kellogg, Vernon, 68
Kepler, Johannes, 409
Kimura, Motoo, 69
kinetics, 103; alteration, 102; chemical, 323; energy, 171; parameters, 101; properties, 108
Kirschner, Marc, 151
Kuhn, Thomas, 212, 422, 425
Kull, Kalevli, 8, 160

Lakatos, Imre, 67
Laland, Kevin, 79

Lamarck, Jean-Baptiste, 186, 188, 350, 375, 396n61
Lamarckism, 186-88, 251, 259, 261, 271, 349, 369, 373, 375, 377-79, 394n41, 396n73, 413; neo-, 26, 188-90, 252, 271, 373, 375, 381, 395n61
Lamb, Marion, 266, 377-82, 390n2, 395n56, 396n65-66
Laplace, Simon Pierre, 2, 3
Laudan, Larry, 67
Laughlin, Robert, 78
Leibniz, Gottfried, Wilhelm, 311, 315, 432, 435-36, 439
Lenton, Timothy, 242
Lewontin, Richard, 126, 155, 194, 252
Linnaean species archetype, 196
Lloyd Morgan, Conway, 37, 42, 251, 259
Longo, Giuseppe, 8
Lorenz, Konrad, 29, 54n41, 391n5
Lotze, R. H., 439
Lovelock, James, 210
Luhmann, Niklas, 110
Luisi, Pier Luigi, 78, 296
lung fish, 12-13
lungs, 12, 240, 246n33
Lyotard, Jean François, 138

machine, 4, 11, 17, 25, 28, 33-36, 38-39, 45, 51n24-25, 52n26-27, 149-50, 163, 193, 221, 236-37, 243, 293-94, 335, 359n11, 378, 380-82, 391n5, 397n81, 409-10, 414, 417, 419-20, 431-33, 439, 441-42; analogy/metaphor, 11, 25, 34-35, 46, 370, 381, 411, 416-17, 420; Bernard, 193; biological, 163, 236; epigenetic molecular, 397n81; growth maximizing, 236; homeostatic, 356; in-

formation processing, 335; living, 382; molecular, 4; part, 237; pinball, 397n81; replication, 56n64; self-moving, 34; self-regulating, 419; survival, 45, 380, 414; -thinkers, 149; Turing, 20, 22
machinery, 28, 380; cellular, 150, 155, 159; of epigenetic modifications, 378;
Mahner, Martin, 315
Malebranche, 434-35
Malthus, Thomas, 276n44, 412
marine iguana, 262-263
materialism, 34, 43, 46, 96, 126-28, 133, 187, 189-91, 238, 242, 334; scientific, 126, 187, 238
matter, 33-34, 43, 47, 52n26, 69, 75, 77-78, 86, 96, 117, 120, 126-28, 135, 137, 140, 161, 172, 185, 192, 208-09, 213, 218, 222-25, 241, 243, 246n23, 246n26, 306n31, 309, 331, 338n34, 391n5, 391n7; -energy, 69, 208-09, 222, 440; -form dualism, 158; living, 123, 125, 209, 211, 222-23; organic, 150, 441
Maturana, Humberto, 52n27, 176, 298, 301
Mayr, Ernst, 26, 42, 68, 176, 221, 228n53, 252, 261, 264, 289-90, 314, 373-77, 396n66
Mead, George Herbert, 172
mechanical agencies, 116
mechanical dynamics, 121
mechanical forces, 297
mechanical nature, 34, 53n28, 54n38
mechanical operations, 94, 96, 110
mechanical parts, 51n25
mechanical principles, 81, 116, 220, 289, 317

mechanical rule, 33
mechanical substance. *See* substance
mechanical switch. *See* switches
mechanical world, 124-25
mechanics: celestial, 2, 21 409, 416; classical, 438-39; Newtonian, 316; quantum, 3, 15-16, 81, 87n6, 96, 208; statistical, 5
mechanic of the body, 35
mechanism, 1, 4, 6-9, 11, 19, 25-34, 36-38, 40-47, 50n4, 50n 7, 52n26, 54n38-39, 57n65, 59n100, 75, 84, 94-95, 97, 101, 103, 115-16, 123-24, 236, 238, 245n9, 251-52, 261, 269, 275n44, 276n44, 287, 289, 291, 293-94, 314, 316-17, 335, 346, 349, 352-58, 362n40, 370-71, 374-75, 378, 381, 387, 391n5, 409, 411-17, 419-25, 431-40; of adaptation, 188; algorithmic, 54; blind, 419; causal, 19, 289, 381, 387; of change, 346, 415; chemical, 84, 208; control, 378, 381; of cultural tradition, 276n44; developmental, 362n40; of embryonic development, 358; of epigenetic in- heritance, 377; of epigenetics, 353, 354, 357, 370-71, 378, 381, 387, 397n82; of evolution, 252, 261, 269, 275n44, 349, 358, 374, 412-13; extrinsic, 291; of gene activity, 352; of gene expression, 352; of gene regulation, 346; of information processing and decision making, 245n9; of inheritance, 349; of Lamarckian inheritance, 188, 251-52; of life, 183, 208; molecular, 115, 371; of natural selection 412, 414, 419; of nature, 53, 317; of

negative feedback, 313; of organic causality, 335; of the origin of novel traits, 376; of phenotypic plasticity, 355; releasing, 54n41; selective, 275; survival, 423; of transmission of heritable information, 362n45
mechanistic account, 287
mechanistically oriented research, 27, 39-40, 43, 57n65
mechanistic analogy/metaphor, 28, 33, 236-38, 397n82
mechanistic biology, 97, 115, 415, 439
mechanistic Cartesian dualism, 380
mechanistic dogma, 37
mechanistic explanation, 39, 46, 50n4, 51n24, 53n28, 85, 94-95, 98, 173, 416, 439, 440
mechanistic interpretation, 37, 94, 171, 272n12
mechanistic justification, 95
mechanistic language, 45, 378, 382
mechanistic law, 7, 35, 67-68, 238, 265, 338n34
mechanistic magisteria, 184
mechanistic materialism, 242
mechanistic metaphysical framework, 29, 31-33, 38-40, 42-47, 369, 369, 379-82, 386, 388
mechanistic metaphysical lens, 25, 27, 38, 41-42
mechanistic method, 28, 42
mechanistic model, 43, 101, 236-37, 421
mechanistic narrative, 115
mechanistic ontology, 243
mechanistic philosophy, 32, 44, 432
mechanistic physicalism. *See* physicalism
mechanistic physics. *See* physics

mechanistic position, 411
mechanistic process, 316, 441
mechanistic purposiveness, 290
mechanistic reductionism, 36, 39, 42-43, 45-47
mechanistic science, 46, 58n81
mechanistic thinking, 44, 54n41, 59n100, 318
mechanistic umbrella, 193
mechanistic understanding, 40-41, 75, 236
mechanistic view of genes, 381n7
mechanistic worldview, 2, 9, 27, 93-94, 96, 409, 422, 431-33
medicine, 35, 163, 185, 215, 346, 354, 357, 362n41, 384-85; bio-, 51n25, 370, 381, 389
meiosis, 371
melanism, 214
memory, 33, 83, 101-03, 107, 184, 193-94, 237, 432; computer, 288; group, 245n12; ecological, 196; epigenetic, 372; external, 241; hereditary, 193-94; heritable, 184, 241; organismic, 111
Mendel, Gregor, 3, 26, 68, 95, 350, 357, 374, 376, 413-14
mentalism, 317-18, 333, 336
mentality, 27, 29, 34, 79, 85, 96, 128, 133, 139-41, 153, 163, 169, 171-73, 175-77, 222, 243, 260, 265, 275n44, 316-19, 324-25, 328, 333, 336, 338n34, 386, 390n2, 435; proto-, 27, 265, 311-12, 315
metabolism, 27, 78, 98-99, 107-10, 189-90, 205-11, 215-16, 220, 222, 225, 263, 294, 297, 330, 335, 353, 387, 398n105
metabolomics, 75
metaphysics, 25-26, 29-32, 34, 41, 43-44, 46-48, 53n28, 53n31,

53n34, 55n46, 58n90, 70, 80, 93, 125, 133-35, 137, 139-41, 173, 186, 189, 220, 236, 238, 241, 243, 269, 309-13, 315, 334, 336, 369, 383, 391n7, 394, 433, 437, 439; Aristotelian, 158; ecological, 125; mechanistic, 25, 31-33, 39, 40, 43-44, 369, 379-81, 386, 388; of natural complexes (Buchler), 41, 134; naturalistic, 125, 139; process-relational (Whitehead), 42, 58n90, 370, 382, 392n8; *See also* philosophy
methylation. *See* deoxyribonucleic acid (DNA); histone
Miller, Wolfgang, 214
mind, 6, 33-35, 54n38, 56n64, 69, 76, 78, 84-85, 96, 135, 141, 163, 169-70, 172, 176-77, 216, 221, 223, 237, 245n20, 246n35, 251, 290, 303, 334, 380, 417, 432, 439
mitosis, 297-98, 371
Modern Synthesis, 25-27, 32, 48-49, 67-70, 95, 147-48, 150-51, 210, 240, 251-52, 261, 289, 369, 373-81, 394n35, 395n61, 396n65, 414, 416; *See also* German Synthesis
molecules, 7-8, 13, 16-19, 38, 42, 80, 81, 83, 125, 140, 149, 151, 155, 158-60, 169, 171, 175, 206, 209, 224, 292, 295-97, 299-301, 323, 328-29, 331-32, 334-35, 356
molecular biology. *See* biology
monad, 435
Monod, Jacques, 4-5, 8, 11, 163, 314
Montevil, Maël, 8

morality, 176, 268-71, 381, 384, 422-24; proto-279n70; slave, 270, 279n74
moral relativism, 271
Moreno, Alvaro, 79, 306n31
Morgan, Thomas Hunt, 376, 391n7, 395, 413
Morito, Bruce, 276n49
Morowitz, Harold, 79, 81, 128
morphic resonance, 82
morphology, 25, 37, 46, 60n115, 80-81, 110, 147-48, 162, 212, 221, 234, 259-60, 263-64, 266, 274n29, 348, 370, 416, 419
Muller, Johannes Peter, 436
multiple realizability, 11, 335
mutation. *See* genetic mutation

natural complexes. *See* complexes
naturalism, 54n39, 125, 134-36, 139-40, 159, 205, 216, 221-23, 268-70, 279n75, 309-10, 333-36, 411, 435; neo-, 334, 336
natural selection. *See* selection
naturalistic fallacy, 6, 268
Nazis, 424
needs, 30, 97, 207, 267, 275n44, 374
neo-Darwinism, 3-4, 14, 18, 25-26, 42, 45-46, 67, 93-96, 110-11, 115, 147, 185, 188, 193, 195, 207, 212, 215-16, 218-19, 222-24, 236, 242-43, 262, 278n68, 279n75-76, 304n3, 314, 328, 335, 369-70, 373, 376, 378-81, 391n7, 394n35, 395n61, 414, 436
neo-Lamarckism. *See* Lamarckism
neo-naturalism. *See* naturalism
neo-teleologism, 311-12, 314-16, 319, 328, 331, 333-35
network, 78-79, 89n45, 98, 108, 194, 225, 239-40, 297, 323-24, 356; adaptive representational,

99; Boolean, 3; chemical reaction, 3, 7, 19, 297; of efficient causal mechanisms, 41; genetic regulatory, 3, 356; information processing, 99
New Frontiers of biology. *See* biology
Newton, Isaac, 1-4, 6-11, 14, 16, 19-21, 34, 44, 67-68, 93, 117, 118, 123-25, 151, 156-57, 165n29, 185, 252, 265, 294, 316-17, 409, 431-33, 443; of biology, 6, 16, 318; of a blade of grass, 68, 147, 317-18, 411, 414
nexus, 109, 113n37, 218, 238-39
niche construction, 14, 27, 150, 194-96, 252
niche creation, 15
Nietzsche, Friedrich, 269-71, 279n74, 336n2
Noble, Denis, 212
non-ergodic universe, 4-7, 11-12, 69
non-overlapping magisteria, 184
non-targeted effects, 40-41, 43, 266, 370, 385, 386
nothing but, 35, 42, 93, 148, 171, 173, 265, 330, 424, 44

occasionalism, 434-35
Odum, Howard T., 127-28
ontogeny, 115, 147, 345, 350, 357
ontological parity, 134, 136-37
ontology, 3, 15, 36, 44, 52n27, 76, 85, 96, 100, 101, 115, 125-26, 134-36, 138, 140, 210, 233, 238, 304n13, 318, 322, 328, 334; atomist, 67; of collective individuals, 233, 237; entitative, 170; mechanistic, 243; object-oriented, 126; substance, 44, 238; tropical rainforest, 134, 140
ordinalism, 134-35, 140
organicism, 42, 58n91, 185, 409, 416-17, 420-22, 424-25, 433-34, 436-41
organicism-mechanism debate, 422, 425, 433, 437-41
Organic Selection (Baldwin's theory of), 27, 37, 48, 259-62, 264-66, 268, 270-71, 273n12, 274n32
Origin of Species, 86, 151, 188, 276n45, 287, 306n32, 395n48, 412, 434
orthogenetic theory, 26, 375
orthoplasy, 260, 265, 270
Osborn, Henry Fairfield, 251, 259, 261, 272n12
Överkalix study, 383
Oyama, Susan, 252, 336n1

Paleontology, 12, 69, 147, 188, 259
Paley, William, 35, 56n64, 68, 186, 412
pan-experientialism, 312
pangenesis, 349-50
Panning, Thomas, 323
pan-psychism, 161, 172, 439
pan-selectionism, 266-71, 278n64-65
parasite, 162, 176, 214, 330
Partridge, Brian, 233
part-whole composition relations, 137
Pasteur, Louis, 186, 189-90
Pattee, Howard, 158, 160, 169
Pauli Exclusion Principle, 119
Peirce, Charles Sanders, 84, 116, 134, 152-53, 156-58, 164n18, 170-72, 237, 311, 336n2
Pembrey, Marcus, 383-84
peptides, 6-8, 19

pharmaepigenomics, 385
phenomenology, 32, 51n25, 54n39, 54n43, 169, 173, 183
phenotype, 11-12, 17, 29, 45, 58n90, 60n115, 70, 95, 97, 111, 147, 150-51, 173, 194, 252, 260-61, 272n12, 274n29, 314, 347, 350-55, 369-71, 375, 377-81, 386, 390n3, 392n10, 396n73; extended, 236, 240, 242; *See also* plasticity; variation
phenotypic change, 60n115
phenotypic novelty, 376
philosophy, Analytic, 41; Continental, 41; of organism (Whitehead), 41, 44, 100, 133, 241, 243, 265, 391n7
phosphorylation, 99, 106, 372, 398n103
photosynthesis, 12, 213, 243
phylogeny, 147, 345, 350, 357
physicalism, 133, 237-38, 241, 243, 310, 312, 318, 326-28, 336; mechanistic, 237
physics, 2, 4, 6, 8, 10, 14, 19, 21, 26, 33, 37-38, 56n64, 71, 78-82, 96, 117-18, 126, 140-41, 170, 173, 209-10, 222, 236, 243, 287, 290-91, 294, 317-18, 322, 324, 326-27, 334-36, 391n7, 411-12, 414, 416, 433; classical, 2, 15-16, 78, 93, 102, 116-17; Aristotle's *Physics*, 311-12, 315-16; eighteenth-century, 316, 318; laws of, 75-76, 120, 174; mechanistic, 317, 439; modern, 1, 3, 318, 324, 391n7, 433, 439, 441; nineteenth-century, 93; quantum, 15-16, 76, 78; twentieth-century, 3, 96
Pigliucci, Massimo, 70, 164n5
Pikaia gracilens, 37

Pittendrigh, Colin, 176, 290
Planck, Max, 5, 140, 320
plasticity, 261, 373, 376, 396n73; developmental, 252; evolutionary, 355; phenotypic, 70, 346, 355, 358, 378, 396n61
Plato, 311, 315, 409, 416, 418, 424
Platonic forms, 31, 70, 135, 196, 208-09, 216
pleroma, 116
pluralism, 126, 133-36, 138, 140-41, 170; causal, 68-70; ordinal, 133
Poincaré, Henri, 2-3
Polanyi, Michael, 293-94
pollution, 42, 267, 373, 384-85
Polya process, 121, 125
Polya's Urn, 121
Pope Pius XII, 225
Popper, Karl, 69-70, 157
postmodernism, 138
power, 5, 21, 27, 33, 35, 39, 44, 53, 84, 86, 141, 151, 158, 187, 196, 246n33, 287-88, 303, 380, 413, 417, 421, 440; causal, 28, 252, 381, 440; explanatory, 33, 138, 195, 336; formative, 25, 52n26, 294, 391n5; of group selection, 421; of machine metaphor, 27, 33, 416-17, 421; motive, 52n26, 294, 391n5; of natural selection, 413; of niche construction metaphor, 196; reasoning, 33; of reductionist methodology, 35; selective, 266; self-regulatory, 374; teleological, 266; *See also* will to power
predictability, 2, 3, 76, 81, 116, 137, 192, 266, 270, 374, 440
prehension (Whitehead), 59n94, 98, 100, 108, 172, 265, 392n8; positive, 100; negative, 100, 109

Price, Derek, 432
Prigogine, Ilya, 287, 304n12
Princehouse, Patricia, 69-70
Principia Mathematica: Newton, 433; Whitehead and Russell, 355
probability distribution, 3
process philosophy, 41-42, 44, 48, 58n90, 96, 98, 100, 246n23, 253, 355, 382-83, 439
prokaryote, 37, 103, 108
proteomics, 75, 357, 441
psychiatry, 141, 356
psychology, 33, 38, 79-80, 141, 157, 161, 259, 275n44, 279n76, 293, 315, 334; *See also* evolutionary psychology
pulse, 102-105, 107, 246n23
purpose, 4, 26, 31, 53n28, 54n38, 56n64, 96, 152, 156, 175-77, 183, 191, 193, 196-97, 205, 207, 209-10, 218, 220-23, 225, 239, 246n22, 252, 263, 279n76, 290, 295, 310, 311-18, 333, 338n24, 383, 389, 409-12, 419, 424, 434, 437

quantum field theory, 136 quantum mutation event, 15 quantum physics, *See* physics, mechanics
quarks, 135-36
Quine, Willard Van, 134

radiation, *See also* adaptive radiation; nuclear, 384; shortwave, 374; the sun's, 40; ultraviolet, 384
RAFs, 16-18
Rahman, Qazi, 378-79 random mutation, 211, 214 random variation, *See* variation ratchet effect, 302
reciprocal causation, *See* causality

reductionism, 2, 11, 26-27, 33, 35-43, 45-48, 57n65, 57n74, 57n78, 67, 75-76, 78-80, 86, 94, 133, 136-37, 147, 169, 170, 173-74, 177, 262, 266-71, 278n65, 279n76, 289, 322, 328, 330, 357-58, 370, 380, 414, 418-21; Cartesian, 35-38; componential, 136, 173; of epigenetic processes to mechanisms, 370; explanatory, 38, 76, 174, 289; genes only, 357-58; greedy (Dennett), 57n78; as heuristic strategy, 57n65; of morality to a function of community selection, 270, 279n76; physicalist, 328; scientific, 76, 78; *See also* mechanistic reductionism
reductionistic analysis, 47, 84
reductionistic language, 84
reductionistic method, 36, 48
reductionistic program, 57n74
reductionistic theory, 76
reductionistic thinking, 36
re-enchantment, 1-2, 6, 13-14, 21-22
Reichenbach, Carl, 436
relations: external, 43-45, 52n26, 59n100, 96; internal, 43, 45, 59n94, 59n100, 97, 99, 238-39, 377, 397n80
releasing event. *See* event
Renaissance, 439
replication. *See* self-replication
rhythm of learning and research (Whitehead), 46, 47, 382
ribosome, 150, 171
RNA, 6-7, 18, 125, 149-150, 155, 160, 169, 171, 194, 208, 346, 351, 356, 372, 376, 387; messenger, 95, 171, 353
Romantics, 417, 421
Royal Society, 58n87, 437

Royce, Josiah, 439
Russell, Bertrand, 117, 123, 355

saltation, 26, 212, 413
Salthe, Stanley, 70, 134, 137, 139
Satisfaction, 28, 97, 276n44
Schneider and Kay, 321
Schrödinger, Erwin, 3, 10, 14, 288, 291
Seeley, Thomas, 234, 236, 244n3, 245n12
selection, 45-46, 55n44, 69, 70, 95, 108, 123, 128, 149, 155, 158, 175-76, 193-96, 251, 253, 265, 268, 274n29, 274n32, 276n44, 278n64, 278n65, 314, 396n73, 413, 415, 419, 421; artificial, 26, 264, 266, 276n44; behavioral, 251, 259, 261, 264; community, 26, 45, 264, 269; conservation, 264; eugenic, 276n44, 278n65; gene, 149, 193-96, 251, 274n29; group, 26, 45, 242, 264, 421, 423, 438; habitat, 264; individual, 421, 423; intellectual, 31-32, 54n44; kin, 45, 264, 423; levels of, 69, 253, 264, 267; natural, 1, 3-4, 8-10, 12-17, 22, 26, 46, 56n64, 67-70, 84-85, 93-95, 108, 117, 128, 147-51, 157, 170, 175-76, 184, 187, 191, 193-96, 205-06, 209-10, 214, 220-21, 240-42, 251-52, 259-65, 268-69, 274n32, 276n44, 287, 291, 310-11, 314-15, 332-33, 335, 349, 352, 355, 358, 375-76, 378, 394n35, 395n48, 396n73, 412-16, 419-21, 423; nest, 245n12; objects of, 265, 274n32; *See* Organic Selection; pressure, 84-85, 123-24, 151, 251, 263-64; *See also* pan-; physiological, 274n29; process, 158, 175, 251, 264; psychosocial, 31-32, 54n44, 275n44; replicator, 289, 291; sexual, 26, 264, 415; *See also* subjects of; valuative-selective activity
selective activity. *See* valuative-selective activity
self-affirmation, 22, 270
self-assembly, 299-300, 302, 328-31, 356
self-creation, 27, 31, 52n26, 59n94, 96, 100, 297
self-formation, 27
self-generation, 27
selfhood, 170, 175-76, 287-88, 291, 293, 295, 298, 300, 302-03, 383
self-movement, 34
self-organization, 27-28, 48, 51n26, 57n74, 70, 77-78, 83-84, 89n40, 95-96, 98, 112n12, 117, 159-60, 235-36, 244n5, 253, 287, 291, 293-96, 299-303, 304n12-13, 305n24, 306n31, 310, 314-24, 326-330, 356, 381, 416, 438
self-propagation, 25, 52n26, 160, 294, 302, 391n5
self-reference, 84, 98, 100-02, 107-08, 110, 121, 159, 172
self-renewal, 27
self-repair, 27, 206
self-replication, 160, 170, 175, 294, 300, 328, 333
self-sustenance, 100
semethic interaction, 170
semiosis, 6, 153, 155-59, 161, 169-73, 177, 303; proto- 171-72, 177; *See also* biosemiosis
semiosphere, 172
semiotic activity, 172
semiotic code, 8

semiotic mutualism, 162
semiotic process, 70-71, 170-71
semiotics, 84, 147, 152-56, 159, 161-63, 169-72, 333, 336; *See also* biosemiotics; eco-semiotic interaction structure; causality
Shalizi, Cosma, 193
Shannon, Claude, 11, 288-89, 313, 334-35
Sheets-Johnstone, Maxine, 152
Sheldrake, Rupert, 82
sign, 4, 133, 150, 152-55, 161-62, 164n18-19, 169, 170-73
signal, 99, 155, 162, 244n3, 313, 326, 352, 357
signaling, 8, 104-05, 151, 172, 385; pathways, 385-86, 398n103; transduction, 99, 151, 159
sign construction, 141
sign interpretation, 152
sign process, 153
sign system, 152
Simon, Herbert, 137
Simpson, George Gaylord, 26, 68, 252, 261, 272n12, 289-90
simulation, 107, 291, 310, 322-23, 327, 331
skyhook (Dennett), 29, 58n78, 262
Smith, John Maynard, 252, 422
social order, 235-37, 239
soft inheritance, 369-70, 373, 375-80, 386, 391n7, 395n56
soma, 39, 45, 95, 141, 149, 170, 188, 235, 244n8, 349-50, 374, 394n35-36
somatic cell, 39, 251, 381
Somatic Cell Nuclear Transfer. *See* cloning
somatic inheritance, 188
somatoplasm, 373, 377, 379-80
spacetime, 140
Spencer, Herbert, 268, 279n76
Spinoza, Benedict de, 218, 224, 439

stability, 78, 81, 125, 138, 208, 238, 297, 302, 325-26, 332, 386, 399n111
Stebbins, Ledyard, 68
Steel, Michael, 16
Stent, Gunther, 116, 120
striving, 30, 123, 151, 197, 237, 279n75, 311-12, 316
struggle for existence, 30, 45, 260, 268-70, 279n74, 391n5, 412-13, 419-20, 425
subject, 44, 96, 109, 380
subjects of selection, 46, 264-65, 274n32
subjective aim (Whitehead), 108-09
subjectivity, 94, 96-97, 100-01, 108, 170, 175, 177, 312, 378
substance, 33-34, 41, 59n100, 59n103, 86, 93, 96, 111, 238, 326, 332-33, 417; chemical, 82, 323; efficient causal, 380; extended/material (*res extensa*), 33-34, 43-44, 48, 93; mechanical, 48; personal, 86; self-sufficient, 391n7; thinking / mental (*res cogitans*), 34, 93
suffering, 27, 39, 149, 163
superorganism, 170, 186, 233-236, 241, 244n3, 244n8, 245n13, 245n20, 246n22, 246n33, 246n35, 419
supervenient properties. *See* emergent properties
survival of the fittest, 9
swarm, 170, 234, 236, 244n9, 245n12
switchboard of the genome, 381-82
switches: mechanical, 28, 33, 39, 155; genetic, 387; *See also* epigenetic switches
Sykes, Bryan, 219
symbiogenesis, 48, 205-07, 211-15, 217, 225

symbiogenetics, 206-07, 211-14, 218, 225
symbionts, 213, 215-16
symbol, 4, 11, 18, 85, 110, 152-53, 158-59, 170, 218, 233, 252, 269, 377
systems biology, 3, 27, 32, 35-36, 38, 46, 48, 75, 78, 84, 86, 304n12, 322, 327, 331, 356, 358, 385, 438, 441
systems theory, 84, 110, 134, 141, 287, 314-15, 319, 323-24, 327-28, 331, 333, 335-36, 352, 355, 356
Szyf, Moshe, 385-86, 398n99

task closure, 8, 10, 13, 15, 17
taxonomy, 45, 139, 216-17
technology, 25, 28, 38-41, 43, 46, 48, 58n81, 126, 210-11, 219, 223, 225, 266, 280n79, 290, 357, 387, 389, 432; reproductive, 388; *See also* biotechnology
teleodynamics, 302-03, 328, 330-31, 333-34, 336
teleology, 4, 26, 29-32, 47-48, 54n38-39, 59n100, 97, 155, 175-76, 205, 207, 210, 216, 219-24, 278n68, 287, 289-90, 294-96, 301, 303, 304n12-13, 309-18, 327-28, 334-35, 338n34; concepts, 155, 315; cybernetic, 314; explanation, 96; language, 148, 315; mental, 312, 315, 336; naturalistic, 216, 207, 221-24; pan-, 311, 315; theistic, 223; *See also* causality; neo- teleologism
teleonomy, 4-5, 175-77, 221, 290-91, 293
telos, 157, 160, 165n32, 191, 221, 246n22, 311-12, 313, 315

Tennyson, Alfred Lord, 436
thermodynamics, 83, 107, 120, 128, 160, 191, 205-07, 220, 226n1, 236, 242, 288, 302, 321-22; first law of, 322; second law of, 120, 157, 192, 206-07, 209, 291-92, 322
thermoregulation, 210, 224, 263
Thom, René, 252, 362n48
Thompson, D'Arcy, 416
Thompson, Evan, 51n25, 54n39, 54n43, 296-98, 301-02, 305n22
Tiezzi, Enzo, 125
Toepfer, Georg, 315, 337n21
Tononi, Giulio, 77
Toxoplasma, 216-19
transgenerational effects, 377, 382-84
transhumanism, 271
Turing, Alan, 9, 20, 22
Turner, J. Scott, 49, 236, 239-43, 246n22, 246n33
Tyndall, John, 189

Uexküll, Jacob von, 169, 336
Ulanowicz, Robert, 49, 67-71, 118, 122-24, 252
Umerez, Jon, 79
United Nations, 20
unprestatability, 1, 10-11, 20-22
untargeted effects. *See also* non-targeted effects
Utilitarianism, 269, 279n76
Utricularia, 122

valuation, 27, 52n26, 59n94, 101, 176-77, 264-66, 270, 276n49, 313
valuative-selective activity, 264-65, 270, 276n49
value, 20, 176-77, 267, 270, 277n49, 279n74, 312-15, 335, 383, 422, 424-23, 443n17; aes-

thetic, 382; heuristic, 324; instrumental, 382; intrinsic, 312, 383; market, 385; survival 30, 251, 269-71, 279n76, 280n79
Varela, Francisco, 52n27, 176, 298, 302
variation, 69, 94, 119, 148, 151, 264-65, 276n44, 276n45, 326-27, 332, 335, 349, 358, 376-377, 394n35-36, 413, 415, 420; epigenetic, 356, 378, 382, 396n73; genetic, 148, 377, 420; heritable, 3, 9, 17, 148, 160, 188, 251, 373-74; phenotypic, 94, 151, 260, 375; random, 207, 211; structural, 373
velocity of light, 438
Vernadsky, Vladimir, 209
vinclozolin, 384
vitalism, 26, 29, 31, 53n28, 96, 138, 165n32, 172, 184-87, 189-91, 289, 334, 415-16, 434, 436, 440-41; crypto-, 209-10.
vorinostat, 386-87, 398n105, 399n113

Waddington, Conrad Hal, 3, 29, 41-42, 58n90, 68, 267, 345-46, 351-56, 358, 360n25-26, 364n48, 370-71, 382, 392n8, 392n10, 396n73
Wallace, Alfred Russel, 278n64, 421
Weinberg, Stephen, 2, 21
Weismann, August, 53n28, 54n38, 95, 187-90, 194, 251, 278n64, 349-50, 350n18, 373-74, 376, 378, 394n35-36
Weismannian barrier, 383
Werren, John, 214
West-Eberhard, Mary Jane, 60n115, 252, 376

Wheeler, William Morton, 234, 244n8, 419-20
Whewell, William, 68
Whitehead, Alfred North, 30, 32, 41-47, 55n46, 58n90, 59n94, 96, 98, 100, 102, 108-09, 117, 133, 172, 205-06, 237-39, 241-43, 246n23, 246n26, 246n35, 265, 276n49, 277n51, 278n63-64, 311-12, 315, 328, 336n2, 355, 370-71, 382-83, 391n7, 392n8, 443n10
whole, 5-8, 15, 32, 36-37, 39, 42-43, 48, 57n74, 70, 84, 97-99, 102, 108-110, 124, 134-41, 151, 159-61, 174, 185-86, 208, 213, 219, 234, 236, 239-40, 242-43, 265, 289, 295-96, 298, 300-01, 303, 311, 314, 316-17, 322, 324, 332, 381, 410, 412, 417-19, 421, 436; Kantian, 5-13, 15-17
Wicken, Jeffrey, 69
Wiener, Norbert, 312-13, 334-35
Wilkinson, David, 242
will to power (Nietzsche), 269-70, 279n74
Wilson, E. B., 351, 352, 360n22
Wilson, E. O., 233, 235-37, 244n3, 244n8, 245n9, 245n13, 245n20, 246n22, 419-20, 423
Wolff, Caspar Friedrich, 3, 346-48, 350, 359n6, 436
Wright, Sewall, 3, 26, 68, 147, 195, 413
Wundt, William, 439

X-bacteria. *See* bacteria
x-rays, 374

Zhang, Xiaoyu, 354, 356
Zia, Asim, 21

About the Contributors

Lawrence Cahoone, *College of the Holy Cross*
Dr. Cahoone, Professor of Philosophy at the College of the Holy Cross, received his doctorate in philosophy from Stony Brook University. He is author of: *The Dilemma of Modernity: Philosophy, Culture and Anti-Culture; The Ends of Philosophy: Pragmatism, Foundationalism, and Postmodernism; Civil Society: The Conservative Meaning of Liberal Politics; Cultural Revolutions: Reason versus Culture in Philosophy, Politics, and Jihad;* and editor of *From Modernism to Postmodernism: An Anthology*. His book *The Orders of Nature* will be published in 2013 by SUNY Press.

Tyrone Cashman
Dr. Cashman earned his Ph.D. from Columbia University. His concentration has been in issues in the philosophy of biology together with philosophy of technology and history of religion. He served in the Office of Appropriate Technology in California State government and has written and lectured widely on issues of mind in nature and on ecologically derived food and energy systems.

Philip Clayton, *Claremont School of Theology; Claremont Lincoln University*
Dr. Clayton is Provost of Claremont Lincoln University, an interreligious consortium of schools, and Dean of Claremont School of Theology. He received his Ph.D. from Yale University and has taught at Williams College and in the California State University system, as well as holding guest professorships at the University of Munich, the University of Cambridge, and Harvard Divinity School. Clayton specializes in constructive theology, the religion-science debate, and comparative religious studies. The demands of this task have led to his research and writing in a wide variety of fields: religion and theology, the philosophy of science and neuroscience, philosophy of mind, the history of philosophy, philosophy of religion, and constructive metaphysics. He has authored or edited twenty books, most recently *Adventures in the Spirit* (Fortress) and *The Predicament of Belief: Science, Philosophy, Faith* (Oxford, with Steven Knapp).

Terrence Deacon, *University of California—Berkeley*
Dr. Deacon is Chair of the department of Anthropology at the University of California, Berkeley, and member of the U C Berkeley Helen Wills Neuroscience Institute. He has held faculty positions at Harvard University, Harvard Medical School, Boston University, and the University of California. His Ph.D. from Harvard University (1984) traced the neural connections of the Broca's and Wernicke's area homologues (of human language cortex) in the macaque brain, and subsequent work further traced connectivity of efferent systems controlling tongue muscles and other related systems. He has also contributed to developmental and comparative neuroanatomy using histological, stereological, quantitative, and cross-species neural transplantation studies. His 1997 book *The Symbolic Species* synthesized much of this neuroanatomical research with a then novel evodevo and niche construction perspective on human brain and language evolution. His most recent book, *Incomplete Nature* explores the role of self-organizational processes in the evolution of teleological and intentional phenomena, including the origins of life and the emergence of consciousness.

Gernot Falkner, *Institute of Physical Biochemistry, University of Munich*
Dr. Falkner studied chemistry, physics and philosophy at the University of Vienna. After having received his Ph.D. in 1969 (with a dissertation on plant hormones, he worked for two years as research assistant with M. Klingenberg at the Institute of Physical Biochemistry of the University of Munich and for one year as post-doctoral fellow with R.H. Burris at the Department of Biochemistry of the University of Madison. Since 1973 he investigated at the Institutes of Molecular Biology and Limnology of the Austrian Academy of Sciences physiological adaptation of algae and bacteria, interrupted in 1977 by a nine-month stay at the Institute of Botany of the University of Würzburg, within the framework of cooperation with W. Simonis. After his retirement from the Institute of Limnology in 2007 he joined the Neurosignaling Unit of the University of Salzburg, where he is also involved in teaching obligations.

Renate Falkner, *Institute of Molecular Biology and Limnology of the Austrian Academy of Sciences*
Dr. Falkner studied biology and chemistry at the University of Würzburg. After having written a thesis and completed postdoctoral work on biological effects of polychlorinated biphenyls at the Institute of Botany of the University of Würzburg and the Institute of Limnology of the Austrian Academy of Sciences, she worked in collaboration with G. Falkner since 1980 on physiological adaptation of algae and bacteria at the Institute of Molecular Biology (Salzburg) and Limnology (Mondsee) of the Austrian Academy of Sciences.

Brian K. Hall, *Dalhousie University*
Dr. Hall is University Research Professor Emeritus at Dalhousie University in Halifax, Nova Scotia, Canada, and Visiting Distinguished Professor at Arizona

State University, Tempe. He received his Ph.D. in zoology from the University of New England Armidale, New South Wales, Australia, in 1969 and his Doctorate of Biological Sciences in 1978. He was Chairman of the Department of Biology at Dalhousie University from 1978 to 1985 and was the Faculty of Science Killam Professor of Biology from 1996 to 2001. In 2002, Dr. Hall was awarded the NSERC Award of Excellence in Research, and in 2005, he received the Killam Prize from the Canada Council for the Arts. Dr. Hall has authored or co-authored numerous volumes in biology, including *Evolution: Principles and Processes* (Jones & Bartlett, 2010), *The Neural Crest and Neural Crest Cells in Vertebrate Development and Evolution* (Springer, 2009), and *Strickberger's Evolution: The Integration of Genes, Organisms and Populations, 4th Edition* (Jones & Bartlett, 2009). However, most recently, he is the co-author, with Benedikt Hallgrimsson, of *Epigenetics: Linking Genotype and Phenotype in Development and Evolution* (University of California Press, 2011).

Brian G. Henning, *Gonzaga University*
Dr. Henning earned his doctorate in philosophy from Fordham University and is Associate Professor of Philosophy at Gonzaga University in Spokane, WA. He is author of nearly twenty articles or anthology chapters, two books, and three co-edited volumes, including *Beyond Metaphysics? Explorations in Alfred North Whitehead's Late Thought*, co-edited with Roland Faber and Clinton Combs (Rodopi 2010) and *Being in America: Sixty Years of the Metaphysical Society,* co-edited with David Kovacs (forthcoming, Rodopi). His 2005 book, *The Ethics of Creativity* (University of Pittsburgh), won the Findlay Book Prize from the Metaphysical Society of America.

Jesper Hoffmeyer, *University of Copenhagen*
Dr. Hoffmeyer has a Master's degree in Biochemistry from University of Copenhagen and a doctorate in philosophy from Aarhus University. He has been an associate professor at the biological institute of University of Copenhagen since 1972. He was named a Thomas Sebeok Fellow by the Semiotic Society of America on the occasion of its twenty-fifth annual meeting in 2000, and from 2007 he has served as president of the International Society for Biosemiotic Studies. He is now, since 2009, a professor emeritus at the University of Copenhagen. Author of *Biosemiotics. An Examination into the Signs of Life and the Life of Signs* and editor of *A Legacy for Living Systems: Gregory Bateson as Precursor to Biosemiotics.*

Stuart A. Kauffman, *University of Vermont*
Dr. Kauffman received his M.D. from the University of California, San Francisco in 1968, after which he transitioned from emergency medicine to academia and scientific research, pioneering some of the earliest studies for what has become today's systems biology, and advancing the theory of the spontaneous emergence of collectively autocatalytic sets. Since then, he has held appoint-

ments at the University of Chicago, the University of Pennsylvania, the Santa Fe Institute, Tampere University of Technology, Harvard Divinity School, and the University of Calgary, in diverse areas such as Biochemistry, Theoretical Biology, Biophysics, Signal Processing, and Philosophy. Dr. Kauffman has founded or co-founded four companies (Darwin, Cistern, Genesis, and BiosGroup) and has membership as a fellow of the *Royal Society of Canada* (2008 and 2009). He is currently doing research at the University of Vermont's *Complex Systems Center*. Dr. Kauffman is the author or co-author of over two hundred scholarly and/or scientific articles. His books include: *Origins of Order: Self-Organization and Selection in Evolution* (Oxford Univeristy Press, 1993), *At Home in the Universe: The Search for Laws of Self-Organization and Complexity* (Oxford University Press, 1995), *Investigations* (Oxford University Press, 2000), and *Reinventing the Sacred: A New View of Science, Reason, and Religion* (Basic Books, 2008).

Spyridon Koutroufinis, *University of California, Berkeley; Technical University of Berlin*
Dr. Koutroufinis is Visiting Associate Professor at the department of Anthropology at the University of California, Berkeley and Associate Professor of Philosophy at the Technical University of Berlin. He earned his Ph.D. at the Humboldt University of Berlin and his habilitation at the Technical University of Berlin. His main areas of specialization and teaching are the philosophy of biology, process philosophy, and the theory of complexity. He is the author of the books *Selbstorganisation ohne Selbst* (1996) and *Organismus als Prozess* (forthcoming), editor of *Prozesse des Lebendigen* (2007), and *Life and Process* (forthcoming), and author of several articles and book chapters.

Lynn Margulis (1938-2011), *University of Massachusetts, Amherst*
Lynn Margulis was Distinguished University Professor in the Department of Geosciences at the University of Massachusetts-Amherst, was one of NASA's first women Principal Investigators in Exobiology (Astrobiology), a contributor to James E. Lovelock's Gaia theory, and is perhaps best known for her theory of symbiogenesis that challenges neo-Darwinism's central tenet that evolutionary novelty (including new species) arise by "gradual accumulation of random mutations." Margulis was elected to the US National Academy of Sciences (1983) and received the Presidential Medal of Science from President Clinton (1999).

Michael Ruse, *Florida State University*
Dr. Ruse is Lucyle T. Werkmeister Professor of Philosophy and Director of the Program in the History and Philosophy of Science at Florida State University. He is the author of many articles and books on the history and philosophy of evolutionary theory as well as on the relationship between science and religion. Recent books include *Darwinism and Its Discontents, Science and Spirituality:*

Making Room for Faith in the Age of Science, and *The Philosophy of Human Evolution*.

Dorion Sagan, *Independent Author*
Sagan is Co-Director of Chelsea Green Publishing Company's Sciencewriters imprint, co-founder of Sciencewriters, and author or coauthor of twenty-four books including *What Is Life?* and *Death and Sex*. A Fellow of the Lindisfarne Association, and a columnist for *Wild River Review,* Sagan's work includes philosophical essays, most recently the introduction to a new translation of Jakob von Uexküll's "A Foray into the Worlds of Animals and Humans."

Adam C. Scarfe, *University of Winnipeg*
Dr. Scarfe is Assistant Professor of Philosophy at the University of Winnipeg. His areas of research are applied ethics, philosophy of education, Continental philosophy, and philosophy of biology. Scarfe is the Executive Director of the *International Process Network*, an organization dedicated to advancing process philosophy globally. He has published over twenty-five articles and book chapters, and is the editor and a co-author of *The Adventure of Education: Process Philosophers on Learning, Teaching, and Research* (Rodopi Press, 2009).

J. Scott Turner, *State University of New York College of Environmental Science and Forestry*
Dr. Turner is Professor of Biology at the State University of New York College of Environmental Science and Forestry. He earned his Bachelor of Arts degree from the University of California at Santa Cruz, and his Master of Science and the Doctor of Philosophy degrees from Colorado State University. He is a physiological ecologist by training with a principal research interest in the biology of social insects. His work has led him to an exploration of the fundamentals of evolution and adaptation, which he has explored in two books: *The Extended Organism. The Physiology of Animal-Built Structures* and *The Tinkerer's Accomplice: How Design Emerges from Life Itself,* both published by Harvard University Press.

Robert Ulanowicz, *University of Maryland*
Dr. Ulanowicz is a native of Baltimore, MD, where he graduated from the Baltimore Polytechnic Institute and received his Ph.D. in Chemical Engineering from the Johns Hopkins University. He was Assistant Professor at the Catholic University of America from 1969 to 1970 and then spent thirty-eight years at the Chesapeake Biological Laboratory, University of Maryland, pursuing research on networks of ecological transfers. He is currently Professor Emeritus from the University of Maryland and Courtesy Professor at the University of Florida. Ulanowicz is the author of three books, the latest, *A Third Window: Natural Life beyond Newton and Darwin* provides a process-based, nonmechanical view of causality in living systems.

Bruce H. Weber, *California State University—Fullerton*
Dr. Weber is Professor of Biochemistry Emeritus at California State University Fullerton as well as the Robert H. Woodworth Chair of Science and Natural Philosophy Emeritus at Bennington College. He obtained his Ph.D. degree in Chemistry from the University of California at San Diego in protein crystallography. He was an American Cancer Society post-doctoral fellow in the bioenergetics laboratory of Prof. Paul Boyer (Nobel-laureate 1997) at UCLA. Weber is author of over seventy scientific publications, co-editor of three collections of essays on evolution (*Evolution at a Crossroads; Entropy, Information, and Evolution;* and *Evolution and Learning*) and co-author (with Prof. David Depew of The University of Iowa) of *Darwinism Evolving: Systems Dynamics and the Genealogy of Natural Selection*, and co-author (with Dr. John Prebble of the University of London) of *Wandering in the Gardens of the Mind: Peter Mitchell and the Making of Glynn*. His research interests are on the application of complex systems dynamical theory to problems of the origin of life and emergent evolutionary phenomena more generally, as well as historical and philosophical issues about evolutionary theory; he also publishes on the conceptual development of bioenergetics.